Lecture Notes in Computer Science 14404

The series Lecture Notes in Computer Science (LNCS), including its subseries Lecture Notes in Artificial Intelligence (LNAI) and Lecture Notes in Bioinformatics (LNBI), has established itself as a medium for the publication of new developments in computer science and information technology research, teaching, and education.

LNCS enjoys close cooperation with the computer science R & D community, the series counts many renowned academics among its volume editors and paper authors, and collaborates with prestigious societies. Its mission is to serve this international community by providing an invaluable service, mainly focused on the publication of conference and workshop proceedings and postproceedings. LNCS commenced publication in 1973.

Paulo Quaresma · David Camacho · Hujun Yin ·
Teresa Gonçalves · Vicente Julian ·
Antonio J. Tallón-Ballesteros
Editors

Intelligent Data Engineering and Automated Learning – IDEAL 2023

24th International Conference
Évora, Portugal, November 22–24, 2023
Proceedings

 Springer

Editors
Paulo Quaresma 🆔
University of Évora
Évora, Portugal

David Camacho 🆔
Technical University of Madrid
Madrid, Spain

Hujun Yin 🆔
University of Manchester
Manchester, UK

Teresa Gonçalves 🆔
University of Évora
Évora, Portugal

Vicente Julian 🆔
Polytechnic University of Valencia
Valencia, Spain

Antonio J. Tallón-Ballesteros 🆔
University of Huelva
Huelva, Spain

ISSN 0302-9743   ISSN 1611-3349 (electronic)
Lecture Notes in Computer Science
ISBN 978-3-031-48231-1   ISBN 978-3-031-48232-8 (eBook)
https://doi.org/10.1007/978-3-031-48232-8

This Springer imprint is published by the registered company Springer Nature Switzerland AG
The registered company address is: Gewerbestrasse 11, 6330 Cham, Switzerland

Paper in this product is recyclable.

Preface

The International Conference on Intelligent Data Engineering and Automated Learning (IDEAL) is an annual international conference dedicated to emerging and challenging topics in intelligent data analytics and associated machine learning paradigms and systems.

The 24th edition, IDEAL 2023, was held in Évora, Portugal, on November 22–24 and it was a 100% in-person event, following IDEAL 2022 in hybrid mode in Manchester. This represented an excellent opportunity for the scientific community to meet in person and to enhance the exchange of ideas and sharing of information. Moreover, Évora, being a UNESCO World Heritage City and the 2027 European Capital of Culture, provided all the conditions to hold a welcoming and successful event.

The IDEAL conference aims to bring together researchers from all over the world, presenting the results of their work, and sharing ideas and experiences in these challenging scientific domains. The core themes of IDEAL 2023 included Big Data challenges, Machine Learning, Deep Learning, Data Mining, Information Retrieval and Management, Bio-/Neuro-Informatics, Bio-Inspired Models, Agents and Hybrid Intelligent Systems, and Real-World Applications of Intelligence Techniques and AI.

In total, 49 papers were accepted and presented at IDEAL 2023, from 77 submissions. The Programme Committee provided fundamental feedback and peer reviews to all the submissions. In addition to the IDEAL 2023 main track, there were three very interesting and relevant special sessions: Federated Learning and (Pre) Aggregation in Machine Learning; Intelligent Techniques for Real-World applications of Renewable Energy and Green Transport; and Data Selection in Machine Learning. We would like to thank the Special Sessions and Workshop Chairs Vítor Nogueira, Fernando Nuñez Hernandez, and Susana Nascimento. A special thank you also goes to our distinguished keynote speakers – Carlos Cotta, Alípio Joge, and Paolo Missier – for their outstanding lectures.

We would like to thank our partners and co-sponsors – University of Évora, IEEE CIS UK and Ireland Chapter, and Springer LNCS, and all the people involved in the organization of the conference, in particular the members of the Programme Committee, the organisers of the Special Sessions, and all the authors who contributed to the scientific quality of the event.

A final word to thank the local organising team at the University of Évora, who, with their work, made this event possible, and the team at the University of Manchester, esp.

Halil Sahin and Yating Huang, for compiling and checking the proceedings files and forms.

October 2023

Paulo Quaresma
David Camacho
Hujun Yin
Teresa Gonçalves
Vicente Julian
Antonio Tallón-Ballesteros

Organization

General Chairs

Paulo Quaresma — University of Évora, Portugal
David Camacho — Technical University of Madrid, Spain
Hujun Yin — University of Manchester, UK

Program Chairs

Teresa Gonçalves — University of Évora, Portugal
Vicente Julian — Universitat Politècnica de València, Spain
Antonio J. Tallón-Ballesteros — University of Huelva, Spain

Special Session Chairs

Vítor Beires Nogueira — University of Évora, Portugal
Fernando Nuñez Hernandez — University of Seville, Spain
Susana Nascimento — NOVA University of Lisbon, Portugal

Steering Committee

Hujun Yin — University of Manchester, UK
Guilherme Barreto — Federal University of Ceará, Brazil
Jimmy Lee — Chinese University of Hong Kong, China
John Keane — University of Manchester, UK
Jose A. Costa — Federal University of Rio Grande do Norte, Brazil
Juan Manuel Corchado — University of Salamanca, Spain
Laiwan Chan — Chinese University of Hong Kong, China
Malik Magdon-Ismail — Rensselaer Polytechnic Institute, USA
Marc van Hulle — KU Leuven, Belgium
Ning Zhong — Maebashi Institute of Technology, Japan
Paulo Novais — University of Minho, Portugal
Peter Tino — University of Birmingham, UK
Samuel Kaski — Aalto University, Finland
Vic Rayward-Smith — University of East Anglia, UK

| Yiu-ming Cheung | Hong Kong Baptist University, China |
| Zheng Rong Yang | University of Exeter, UK |

Publicity and Liaison Chairs

Bin Li	University of Science and Technology of China, China
Guilherme Barreto	Federal University of Ceará, Brazil
Jose A. Costa	Federal University of Rio Grande do Norte, Brazil
Yimin Wen Guilin	University of Electronic Technology, China

Programme Committee

Hector Alaiz Moreton	University of León, Spain
Richardo Aler	Universidad Carlos III, Spain
Romis Attux	University of Campinas, Brazil
Carmelo Bastos Filho	University of Pernambuco, Brazil
Riza T. Batista-Navarro	University of Manchester, UK
Lordes Borrajo	University of Vigo, Spain
Federico Bueno de Mata	Universidad de Salamanca, Spain
Robert Burduk	Wrocław University of Science and Technology, Poland
Anne Canuto	Federal University of Rio Grande do Norte, Brazil
Roberto Carballedo	University of Deusto, Spain
Mercedes Carnero	Universidad Nacional de Rio Cuarto, Argentina
Pedro Castillo	University of Granada, Spain
Luís Cavique	University of Aberta, Portugal
Songcan Chen	Nanjing University of Aeronautics & Astronautics, China
Xiaohong Chen	Nanjing University of Aeronautics & Astronautics, China
Stelvio Cimato	Università degli studi di Milano, Italy
Carlos Coello Coello	CINVESTAV-IPN, Mexico
Roberto Confalonieri	Free University of Bozen-Bolzano, Italy
Paulo Cortez	University of Minho, Portugal
Jose Alfredo F. Costa	Federal University of Rio Grande do Norte, Brazil
Carlos Cotta	Universidad de Málaga, Spain
Raúl Cruz-Barbosa	Universidad Tecnológica de la Mixteca, Mexico
Andre de Carvalho	University of São Paulo, Brazil
Dalila A. Durães	Universidade do Minho, Portugal
Bruno Fernandes	University of Minho, Portugal

João Ferreira — ISCTE, Portugal
Joaquim Filipe — EST-Setubal/IPS, Portugal
Felipe M. G. França — COPPE-UFRJ, Brazil
Pedro Freitas — Universidade Católica Portuguesa, Portugal
Dariusz Frejlichowski — West Pomeranian University of Technology, Poland
Hamido Fujita — Iwate Prefectural University, Japan
Marcus Gallagher — University of Queensland, Australia
Isaias Garcia — University of León, Spain
María José Ginzo Villamayor — Universidad de Santiago de Compostela, Spain
Teresa Goncalves — University of Évora, Portugal
Anna Gorawska — Silesian University of Technology, Poland
Marcin Gorawski — Silesian University of Technology, Poland
Manuel Graña — University of the Basque Country, Spain
Maciej Grzenda — Warsaw University of Technology, Poland
Pedro Antonio Gutierrez — Universidad de Cordoba, Spain
Barbara Hammer — Bielefeld University, Germany
J. Michael Herrmann — University of Edinburgh, UK
Wei-Chiang Hong — Jiangsu Normal University, China
Jean-Michel Ilie — University Pierre et Marie Curie, France
Dariusz Jankowski — Wrocław University of Science and Technology, Poland
Vicente Julian — Universitat Politècnica de València, Spain
Rushed Kanawati — Université Paris 13, France
Bin Li — University of Science and Technology of China, China
Victor Lobo — Universidade Nova de Lisboa, Portugal
Wenjian Luo — Harbin Institute of Technology Shenzhen, China
Jesús López — Tecnalia Research & Innovation, Spain
José Machado — University of Minho, Portugal
Rui Neves Madeira — Instituto Politécnico de Setúbal, Portugal
José F. Martínez-Trinidad — INAOE, Mexico
Cristian Mihaescu — University of Craiova, Romania
José M. Molina — Universidad Carlos III de Madrid, Spain
Paulo Moura Oliveira — UTAD University, Portugal
Tatsuo Nakajima — Waseda University, Japan
Susana Nascimento — Universidade Nova de Lisboa, Portugal
Grzegorz J. Nalepa — AGH University of Science and Technology, Poland
Antonio Neme — Universidad Nacional Autónoma de México, Mexico
Vitor Nogueira — Universidade de Évora

Fernando Nuñez	University of Seville, Spain
Eva Onaindia	Universitat Politècnica de València, Spain
Eneko Osaba	Tecnalia Research & Innovation, Spain
Jose Palma	University of Murcia, Spain
Carlos Pereira	Instituto Superior de Engenharia de Coimbra, Portugal
Radu-Emil Precup	Politehnica University of Timisoara, Romania
Héctor Quintián	University of A Coruña, Spain
Kashyap Raiyani	UNINOVA, Portugal
Izabela Rejer	University of Szczecin, Poland
Matilde Santos	Universidad Complutense de Madrid, Spain
Richardo Santos	Polytechnic of Porto, Portugal
Jose Santos	University of A Coruña, Spain
Ivan Silva	University of São Paulo, Brazil
Dragan Simic	University of Novi Sad, Serbia
Marcin Szpyrka	AGH University of Science and Technology, Poland
Murat Caner Testik	Hacettepe University, Turkey
Qing Tian	Nanjing University of Information Science and Technology, China
Stefania Tomasiello	University of Salerno, Italy
Alexandros Tzanetos	University of the Aegean, Greece
Eiji Uchino	Yamaguchi University, Japan
José Valente de Oliveira	Universidade do Algarve, Portugal
Gianni Vercelli	University of Genoa, Italy
Tzai-Der Wang	Cheng Shiu University, Taiwan
Michal Wozniak	Wrocław University of Technology, Poland
Hua Yang	Zhongyuan University of Technology, China
Xin-She Yang	Middlesex University London, UK

Special Session on Federated Learning and (Pre) Aggregation in Machine Learning

Organisers

Jaime Andrés Rincon	Universitat Politècnica de València, Spain
Cedric Marco-Detchart	Universitat Politècnica de València, Spain
Vicente Julian	Universitat Politècnica de València, Spain
Carlos Carrascosa	Universitat Politècnica de València, Spain
Giancarlo Lucca	Universidade Federal do Rio Grande, Brazil
Graçaliz P. Dimuro	Universidade Federal do Rio Grande, Brazil

Special Session on Intelligent Techniques for Real-World Applications of Renewable Energy and Green Transport

Organisers

J. Enrique Sierra García	University of Burgos, Spain
Matilde Santos Peñas	Complutense University of Madrid, Spain
Payam Aboutalebi	University of the Basque Country, Spain
Bowen Zhou	Northeastern University, China

Special Session on Data Selection in Machine Learning

Organisers

Antonio J. Tallón-Ballesteros	University of Huelva, Spain
Ireneusz Czarnowski	Gdynia Maritime University, Poland

Contents

**Special Session on Federated Learning and (Pre) Aggregation in
Machine Learning**

**Special Session on Intelligent Techniques for Real-World Applications
of Renewable Energy and Green Transport**

Special Session on Data Selection in Machine Learning

Main Track

Optimization of Image Acquisition for Earth Observation Satellites via Quantum Computing

Antón Makarov[1(✉)], Márcio M. Taddei[2], Eneko Osaba[3],
Giacomo Franceschetto[2,4], Esther Villar-Rodríguez[3], and Izaskun Oregi[3]

[1] GMV, 28760 Madrid, Spain
amakarov@gmv.com

[2] ICFO - Institut de Ciencies Fotoniques, The Barcelona Institute of Science
and Technology, 08860 Castelldefels, Barcelona, Spain

[3] TECNALIA, Basque Research and Technology Alliance (BRTA), 48160 Derio,
Spain

[4] Dipartimento di Fisica e Astronomia "G. Galilei", Universitá degli Studi di Padova,
35131 Padua, Italy

Abstract. Satellite image acquisition scheduling is a problem that is omnipresent in the earth observation field; its goal is to find the optimal subset of images to be taken during a given orbit pass under a set of constraints. This problem, which can be modeled via combinatorial optimization, has been dealt with many times by the artificial intelligence and operations research communities. However, despite its inherent interest, it has been scarcely studied through the quantum computing paradigm. Taking this situation as motivation, we present in this paper two QUBO formulations for the problem, using different approaches to handle the non-trivial constraints. We compare the formulations experimentally over 20 problem instances using three quantum annealers currently available from D-Wave, as well as one of its hybrid solvers. Fourteen of the tested instances have been obtained from the well-known SPOT5 benchmark, while the remaining six have been generated ad-hoc for this study. Our results show that the formulation and the ancilla handling technique is crucial to solve the problem successfully. Finally, we also provide practical guidelines on the size limits of problem instances that can be realistically solved on current quantum computers.

Keywords: Quantum Computing · Satellite Image Acquisition · Earth Observation · Quantum Annealer · D-Wave

1 Introduction

The Satellite Image Acquisition Scheduling Problem (SIASP) [2,22] is of great importance in the space industry; every day, satellite operators must face the challenge of selecting the optimal subset of images to be taken during the next

P. Quaresma et al. (Eds.): IDEAL 2023, LNCS 14404, pp. 3–14, 2023.
https://doi.org/10.1007/978-3-031-48232-8_1

orbit pass of a satellite with respect to some notion of value from a set of requests made by their clients. For any given set of requests, there are constraints such as geographical proximity incompatibilities, on-board disk space availability or special image configuration requirements, which make it impossible to collect all the images. The problem becomes even harder to solve when we consider that, since requests arrive continuously over time, the planning must be updated several times a day, which requires the planning algorithm to be re-run many times over short periods of time. This fact makes execution speed crucial. Moreover, the industry is currently facing a myriad of challenges such as extending the problem to multiple satellites [3,12] or including climatic restrictions [21], making it even more computationally expensive.

The SIASP and its extensions are extremely combinatorial in nature, which make them very challenging to solve with classical computing methods even for moderately sized instances. Traditionally, the problem has been tackled by exact integer linear programming algorithms based on branch-and-bound methods [2, 16,20] or (meta-)heuristic and hybrid algorithms, which are usually preferred [1,10] since they are the only ones able to handle large scale problems and comply with execution-time limitations.

On another note, Quantum Computing (QC) is emerging as a very promising alternative for dealing with optimization problems, where Quantum Annealing based computers such as the ones developed by D-Wave are especially suitable and might offer significant speedups in the future. The SIASP has been barely studied from the QC perspective, which makes this problem a great candidate to explore the possible near- to mid-term benefits that QC can bring to the table in the space industry.

Taking as motivation the scarcity of works conducted in this regard, our objective in this paper is to study SIASP from the QC perspective. To this end, we propose two distinct formulations to encode the variables of the problem.

Overall, 20 different problem instances have been used to test the adequacy of each formulation. On one hand, 14 of them have been directly taken from the well-known SPOT5 dataset [2], while the other six have been generated based on SPOT5 using an instance reductor implemented ad-hoc for this study. As solving alternatives, we assess the performance of four solvers available to the public from D-Wave, three purely quantum-annealing based and one hybrid. This allows us to assess the quality of our formulations as well as to test the limits of today's available hardware.

The rest of the paper is structured as follows. The following section offers some background on the problem dealt with in this paper. Next, Sect. 3 introduces the mathematical formulations employed for dealing with the problem by the quantum annealers considered. The description of the experimental setup is given in Sect. 4, along with the tests conducted and the discussion. Finally, Sect. 5 ends the paper with several conclusions and future research following up from the reported findings.

2 Background

QC is gaining significant notoriety in the scientific community, mainly due to its promising computational characteristics. It introduces new mechanisms, based on quantum mechanics, for solving a wide variety of problems more efficiently. This paradigm is expected to achieve a speed and/or precision advantage in modelling systems and in solving complex optimization problems, besides potential energetic benefits. More specifically, quantum computers have been employed in recent times for facing problems coming from diverse fields such as logistics [15], economics [13] or industry [11].

Although the problem dealt with in this research is of great industrial interest, one of the main obstacles faced by any researcher when trying to tackle it through the quantum perspective is the shortage of previous work focused on this task. This fact implies a challenge to be faced and a research opportunity, which has motivated the realization of the present paper.

There is nevertheless a specific work, published in 2020 by Stollenwerk et al. [17,18], which deals for the first time with the SIASP using the quantum paradigm. They investigate the potential and maturity of then-current quantum computers to solve real-world problems by carrying out an experimental study on a reduced number of small instances of the SIASP. To this end, they introduce a Quadratic Unconstrained Binary Optimization (QUBO) formulation, which served as inspiration for our research. In any case, since that formulation does not completely cover our needs, we have extended it in order to contemplate ternary constraints and multi-camera satellites.

The main contributions that our research provides with respect to that pioneering work can be summarized as follows:

- The dataset employed in our research has been taken from realistic simulations of a satellite mission.
- The satellite considered has three cameras (instead of the single camera used in [17]), two of which can be used in conjunction to take stereo images. This issue greatly increases the complexity of the problem, and gives rise to several possible formulations, of which two distinct ones are studied.
- In our paper, we consider ternary constraints, which require ancillary qubits to model them. We study two ways to encode those constraints and show the compressing power of one with respect to the other.
- We perform an experimental study on the performance of several quantum annealing based solvers.

Finally, it is interesting to mention the research recently published in [23] and [24]. These papers are focused on solving the SIASP using a quantum genetic algorithm. Despite the scientific contribution behind these investigations, they are outside the scope of the present paper since the methods developed fall within the area known as *quantum-inspired evolutionary computation*. Thus, these techniques are really classical algorithms with the particularity of using concepts from quantum physics for their design.

3   Mathematical Formulations of the Problem

In this section, we focus on the mathematical formulation of the SIASP treated in this paper. First of all, we go in detail on the classical formulation of the problem. After that, the section gravitates around the quantum-oriented formulations chosen for the experimentation.

3.1   Classical Formulation of the SIASP

Our classical formulation for the SPOT5 instances (we refer the reader to [2] for a complete description of the structure of the instances) is largely inspired by [19], and it can be stated in the language of mathematical programming as follows. Let $x_{i,j}$ be the binary decision variable, defined as:

$$x_{i,j} = \begin{cases} 1 & \text{if image } i \text{ is taken with camera } j, \\ 0 & \text{otherwise,} \end{cases}$$

where $i \in \{1, 2, \ldots, N\}$ is the index representing the image requests, N being the total amount of requests and $j \in \{1, 2, 3, 4\}$ the identifier of the camera. There are three physical cameras which can take mono images and we define camera 4 to represent the combined use of cameras 1 and 3 to take stereo images. Thus, the objective function to be optimized is:

$$\min \quad -\sum_i \sum_j w_i x_{i,j},$$

where w_i is the weight or value of taking image i. Note that although our task is to maximize the value, we can express it as the minimization of the negative of the objective. This optimization is subject to the following constraints:

$$\sum_j x_{i,j} \leq 1 \quad \forall i \tag{1a}$$

$$x_{p,j_p} + x_{q,j_q} \leq 1 \quad \forall \left((p, j_p), (q, j_q)\right) \in C_2 \tag{1b}$$

$$x_{p,j_p} + x_{q,j_q} + x_{r,j_r} \leq 2 \quad \forall \left((p, j_p), (q, j_q), (r, j_r)\right) \in C_3 \tag{1c}$$

$$x_{i,4} = 0 \quad \forall i \in M \tag{1d}$$

$$x_{i,j} = 0 \quad \forall i \in S, \forall j \in \{1, 2, 3\} \tag{1e}$$

$$x_{i,j} \in \{0, 1\} \tag{1f}$$

where Constraint (1a) forces the images to be taken only once. Constraint (1b) represents the incompatibilities of taking certain pairs of images (set C_2) which arise from two images being too close geographically to each other to take both. Constraint (1c) represents the ternary constraints (set C_3) related to the data flow restrictions of the satellite that do not allow to take more than two images at once. Constraints (1d) and (1e) forbid mono images (set M) to be taken as stereo and stereo (set S) images to be taken as mono, respectively.

3.2   Formulations of the SIASP for its Quantum Solving

In order to treat the problem with existing quantum-annealing hardware, we need a different formulation. Quantum annealers' QPUs are able to produce a final Hamiltonian of the form:

$$H_F = \sum_i h_i \, Z_i + \sum_{i \neq j} J_{i,j} \, Z_i Z_j \, , \qquad (2)$$

where Z_i is the z Pauli operator of the i-th qubit, h_i is its on-site energy and $J_{i,j}$ the coupling between qubits i, j. This allows us to minimize the corresponding function on classical binary variables x_i (obtained transforming $Z_i \to 1 - 2x_i$). This is by analogy called *energy*, and can be written as:

$$E(\boldsymbol{x}) = \boldsymbol{x}^T Q \, \boldsymbol{x} \, , \qquad (3)$$

where Q is a matrix that can be assumed symmetric without loss of generality. A QUBO consists in the minimization of such functions, and currently available quantum annealers such as D-Wave target this kind of problems specifically. The limitation to polynomials of order two comes from the fact that the Hamiltonian of Eq. (2) only couples qubits two by two, and is intrinsic to the hardware.

Additionally, the hardware is not able to couple every pair of qubits, hence an important feature of the QPU is its topology, i.e. the configuration of which qubits are in fact coupled, hence which off-diagonal terms of Q are nonzero and can be tuned. For this study, three different QPUs have been used, which are DW_2000Q_6 (2000Q), Advantage_system6.1 (Advantage) and Advantage2_prototype1.1 (Advantage2), have topologies called Chimera, Pegasus and Zephyr, respectively, which are increasingly interconnected [6–8]. If the topology does not present all couplings needed for the problem, a typical solution is to embed a given qubit in more than one physical qubit. This is a source of overhead in the amount of qubits needed for a given instance. We make use of D-Wave's native algorithm based on minor embedding [5].

The transformation of our original problem into a QUBO formulation presents several challenges. Firstly, QUBO is unconstrained, i.e., we cannot explicitly impose Constraints (1). Secondly, currently available quantum hardware possesses few qubits, so reducing the number of variables is of utmost importance. Thirdly, it can only tackle problems written in quadratic form.

The fact that QUBO is unconstrained is circumvented by the addition of penalty terms [9,17], which are extra terms that raise the value of $E(\boldsymbol{x})$ whenever $\boldsymbol{x}$ is not a feasible solution — i.e. when $\boldsymbol{x}$ does not obey all constraints. Importantly, imperfections in the penalization and minimization may lead to infeasible solutions.

Because of the reduced number of available qubits, we have devised a denser encoding of our problem into binary variables, mainly related to the representation of mono and stereo photos. In Sect. 3.1 we have presented the classical formulation using our first encoding, which we refer to as 4cam. It is a straightforward and fixed-length representation of the binary variables, whose allocation depends only on the total number of photos requested. In the denser encoding, which we call 3cam, if requested photo i is mono, there are three variables $x_{i,1}$, $x_{i,2}$, $x_{i,3}$, whereas if it is stereo there exists a single variable x_i. The advantages of the 3cam formulation are the reduction in the number of qubits necessary and also in the number of constraints, since Eqs. (1d, 1e) no longer need to be imposed.

Finally, the polynomial to be minimized must be of quadratic form, a limitation particularly relevant for the penalty terms relative to ternary constraints (1c). These require the introduction of additional binary variables ("slack" variables), which is another source of overhead in the amount of qubits needed. Let us momentarily simplify the notation of inequality (1c) to $x_p + x_q + x_r \leq 2$ for cleanness. For 4cam we write the corresponding penalty term directly in quadratic form, introducing two slack variables s_1, s_2 for each ternary constraint [9, Sec 5]:

$$P(x_p + x_q + x_r - 2 + s_1 + 2s_2)^2, \tag{4}$$

where $P > 0$ is a parameter. For 3cam, in the spirit of optimizing this formulation as much as possible, we take a more involved approach: initially we write a cubic penalty term $P\, x_p x_q x_r$ and then reduce it to quadratic following Boros et al. [4, Sec 4.4]: a single slack variable s_1 replaces the pair $x_q x_r$, and an extra term is added (parentheses below),

$$P\, x_p s_1 + P(x_q x_r - 2x_q s_1 - 2x_r s_1 + 3s_1). \tag{5}$$

Advantages of the latter method are fewer terms in Eq. (5) than in Eq. (4) after expansion, and avoiding the introduction of a second slack variable per constraint. Additionally, if the same pair $x_q x_r$ appears in many constraints, the same slack variable replaces it in all terms, with the same extra term added.

Figure 1 graphically shows the compression achieved with the formulation and encoding for instance 15. Notice that the representations include all variables, original and slack, and all necessary penalty terms. See Table 1 for a full breakdown of all instances.

 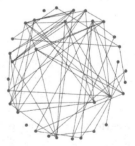

Fig. 1. Graph representation of problem instance 15 (see the details in Table 1) obtained from using the QUBO Q matrices as adjacency matrices. The one on the left corresponds to the 4cam formulation with binary expansion encoding (Eq. 4) and the one on the right to the 3cam formulation with Boros encoding (Eq. 5). For clarity, the self-loops (one for each node, corresponding to the diagonal of the matrix) have been omitted.

4 Experimentation

This section is devoted to describe the experimentation carried out. First, we detail the benchmark of instances employed for testing the formulations detailed in the previous section. After that, we delve on the experimental setup and the analysis of the results obtained.

4.1 Benchmark

To conduct our study, the well-known SPOT5 benchmark dataset introduced in [2] have been used, which is composed of 21 simulated instances of image-acquisition planning problems from the SPOT5 earth observation satellite of the French space agency. Among the 21 instances, 8 were discarded as they have capacity constraints, which consideration is out of the scope of this research. This dataset has certain characteristics that make it suitable for consideration in our study: it is open, heterogeneous, real-world oriented and widely used by the scientific community.

However, the large size of many of the instances is a limiting factor to asses the performance of the QPUs. To mitigate this, we have implemented a Python script for the automatic generation of reduced instances. This script, coined `InstanceReductor`, takes as input an existing instance of the SPOT5 dataset and the desired size for the newly generated instance. Then, the `InstanceReductor` generates a new file by randomly reducing the data in order to contemplate the number of requests introduced as input.

Overall, 20 different instances have been employed in the experimentation designed in this study. 14 of them are part of SPOT5, and their sizes range from 8 to 348 requests. The 6 remaining cases have been generated by `InstanceReductor`, and they consider a number of requests between 15 and 40. We summarize in Table 1 the characteristics of each instance considered in this paper. Lastly, and aiming to enhance the reproducibility of this study, both generated cases and `InstanceReductor` are openly available in [14].

4.2 Setup and Results

To conduct the experiments on quantum hardware, we have used our two QUBO formulations (4cam and 3cam) on the 20 instances detailed in Table 1. Additionally, for each instance and encoding, we tested four different D-Wave solvers (2000Q, `Advantage`, `Advantage2` and `HybridBQM`, where the latter stands for `hybrid_binary_quadratic_model_version2`). To account for the probabilistic nature of the solvers, we run each combination 5 times. For the three completely quantum solvers we have left all parameters as default except the number of runs, which we have set at 2000 following the advice of D-Wave. For the hybrid solver, we adopt all the default parameters. Lastly, the value of all penalty coefficients P of each instance has been set to one plus the maximum possible value that the objective function can reach. In this regard, refining the choice of P could be further investigated as it can severely affect the feasibility of the solutions obtained.

Table 1. Main characteristics of the 20 used instances, ordered by increasing number of requests. For each instance we depict the number of total and stereo requests, the amount of total and ternary constraints as well as the number of linear (L4cam, L3cam) and quadratic (Q4cam, Q3cam) terms for the two QUBO formulations. Shaded instances are generated by `InstanceReductor`.

ID	Requests	Stereo	Constraints	Ternary	L4cam	Q4cam	L3cam	Q3cam
8	8	4	7	0	32	65	16	29
15	15	6	14	3	66	147	33	62
20	20	12	23	0	80	171	36	75
25	25	5	34	0	100	213	39	84
30	30	10	69	0	120	320	52	173
35	35	23	64	4	148	331	52	119
40	40	27	139	0	160	431	56	215
54	67	35	204	23	314	997	140	544
29	82	29	380	0	328	1102	120	667
404	100	63	610	18	436	1672	176	1078
503	143	78	492	86	744	2294	310	1118
42	190	19	1204	64	888	3438	315	2026
408	200	120	2032	389	1578	6870	489	3624
28	230	35	4996	590	2100	12139	512	7501
505	240	119	2002	526	2012	8280	694	4226
412	300	160	4048	389	1978	11495	705	7823
5	309	46	5312	367	1970	18787	1052	15828
507	311	163	5421	2293	9246	24762	1053	9641
509	348	178	8276	3927	9246	39418	1431	14619
414	364	182	9744	4719	10894	46827	1643	17429

Additionally, in order to use it as a reference for the quantum experiments, we have first implemented the classical `4cam` formulation described in Sect. 3.1 and solved all the considered instances with `Google OR-Tools`, checking that our results coincide with the known optima.

Table 2 depicts the average results and standard deviations reached by `2000Q`, `Advantage`, `Advantage2` and `HybridBQM`, as well as by the classical solver. In Fig. 2 we show the detailed results of our experiments split by instance and formulation. Together with Table 2, we can see that the better-performing formulation across almost all instances and solvers is `3cam`. Also, with `3cam` we obtain solutions for certain combinations of instance and solver that are untreatable with `4cam`. This is so because it is much more efficient in terms of variables and connections to be used for modelling the problem. Additionally, the best-performing solver is the `HybridBQM`.

Table 2. Results for the considered instances by encoding and solver. Each instance was run 5 times and the values reported are the mean ± standard deviation of the objective function value. Marked in bold are the best-performing results for each problem instance ignoring the hybrid solver. Results marked with * are those for which an embedding was not found in at least one of the runs while the ones with no numerical values are the ones for which no embedding was found in any of the 5 attempts. Instances 404, 42 and 408 with the 4cam formulation and Advantage QPU had some unfeasible solutions, which were removed when computing the results shown in this Table.

ID	4cam				3cam			
	2000Q	Advantage	Advantage2	HybridBQM	2000Q	Advantage	Advantage2	HybridBQM
8	1.0 ± 0.0	1.0 ± 0.0	1.0 ± 0.0	1.0 ± 0.0	1.0 ± 0.0	1.0 ± 0.0	1.0 ± 0.0	1.0 ± 0.0
15	0.99 ± 0.02	0.90 ± 0.13	0.95 ± 0.05	1.0 ± 0.0	1.0 ± 0.0	1.0 ± 0.0	1.0 ± 0.0	1.0 ± 0.0
20	1.0 ± 0.0	1.0 ± 0.0	1.0 ± 0.0	1.0 ± 0.0	1.0 ± 0.0	1.0 ± 0.0	1.0 ± 0.0	1.0 ± 0.0
25	1.0 ± 0.0	1.0 ± 0.0	1.0 ± 0.0	1.0 ± 0.0	1.0 ± 0.0	1.0 ± 0.0	1.0 ± 0.0	1.0 ± 0.0
30	1.0 ± 0.0	1.0 ± 0.0	1.0 ± 0.0	1.0 ± 0.0	1.0 ± 0.0	1.0 ± 0.0	1.0 ± 0.0	1.0 ± 0.0
35	0.68 ± 0.07	0.62 ± 0.12	0.69 ± 0.10	1.0 ± 0.0	**0.96 ± 0.01**	0.90 ± 0.02	0.93 ± 0.04	1.0 ± 0.0
40	0.88 ± 0.05	0.88 ± 0.00	0.88 ± 0.04	1.0 ± 0.0	0.95 ± 0.03	0.91 ± 0.03	**0.96 ± 0.02**	1.0 ± 0.0
54	0.49 ± 0.03*	0.59 ± 0.04	–	1.0 ± 0.0	0.78 ± 0.04	0.73 ± 0.01	**0.82 ± 0.05**	1.0 ± 0.0
29	–	0.78 ± 0.10	–	1.0 ± 0.0	**0.95 ± 0.04**	0.93 ± 0.03	0.90 ± 0.03	1.0 ± 0.0
404	–	0.70 ± 0.08	–	1.0 ± 0.0	0.73 ± 0.04	0.74 ± 0.03	**0.82 ± 0.03**	1.0 ± 0.0
503	–	**0.76 ± 0.18**	–	1.0 ± 0.0	0.59 ± 0.05*	0.73 ± 0.05	–	1.0 ± 0.0
42	–	**0.80 ± 0.11**	–	0.98 ± 0.01	–	0.59 ± 0.02	–	1.0 ± 0.0
408	–	–	–	0.93 ± 0.13	–	**0.66 ± 0.29**	–	1.0 ± 0.0
28	–	–	–	0.78 ± 0.08	–	**0.59 ± 0.02***	–	1.0 ± 0.0
505	–	–	–	0.63 ± 0.10	–	**0.43 ± 0.08**	–	1.0 ± 0.0
412	–	–	–	0.82 ± 0.05	–	**0.34 ± 0.13***	–	1.0 ± 0.0
5	–	–	–	0.91 ± 0.02	–	–	–	0.99 ± 0.00
507	–	–	–	0.43 ± 0.14	–	–	–	1.0 ± 0.0
509	–	–	–	0.54 ± 0.15	–	–	–	1.0 ± 0.0
414	–	–	–	0.34 ± 0.07	–	–	–	1.0 ± 0.0

If we turn our attention to the purely quantum solvers, it is interesting that Advantage2, although being still a prototype at the time of writing this article, has the edge for some of the smaller instances where an embedding can be found. For larger instances, the Advantage QPU is the most reliable solver, which manages to obtain results (albeit not necessarily outstanding ones) up until instance 412, with 300 requests.

Furthermore, 2000Q cannot handle instances larger than 503 (143 requests), where the limit of Advantage2 is at instance 404 (100 requests). The evolution of the solvers seems to be clear, and we can expect that when the final version of the Advantage2 QPU is launched, we will be able to solve even larger problems with greater precision. Finally, an important note is that in instances 404, 42 and 408 there are some solutions above the optimum value, which means that some constraints were broken in the solving process. This was likely due to insufficiently large penalty values, which highlights the importance of choosing them correctly.

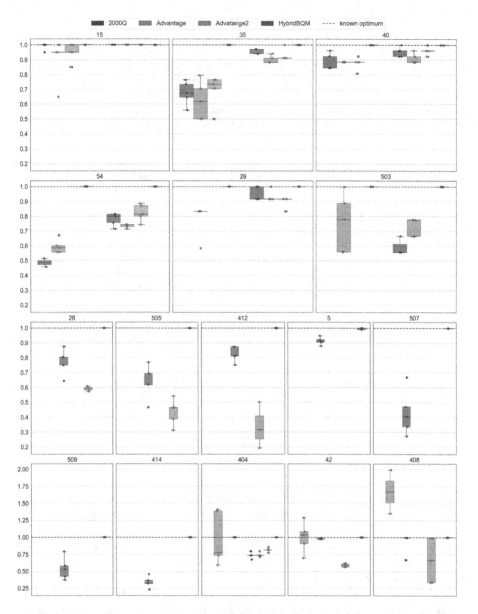

Fig. 2. Box and jitter plots of the experimental results. Each subplot shows the execution data for one instance normalized to its optimum value. The dashed green line represents the optimal solution and the colors encode the solvers. The x axis is split between the two formulations, 4cam on the left and 3cam on the right. The first three rows of the plot share the same y axis, while the last row does not due to the broken constraints in instances 404, 42 and 408. Instances 8, 20, 25 and 30, with perfect performance up to 2 decimals in all cases, have been omitted. Elsewhere, when no data is shown for a given solver and instance, no embedding was found for it in any of the 5 runs (Color figure online).

5 Conclusions and Further Work

In this paper we have experimentally assessed the performance of D-Wave's three pure quantum annealers and one hybrid solver for the SIASP using two distinct formulation approaches. We have resorted to a realistic benchmark dataset and established how the quality of the solutions degrades with problem size, imposing practical limits on instances that can currently be solved effectively.

Our results show that an efficient formulation allows to solve larger problem instances with better accuracy. This fact is key for success, which makes it a promising avenue for further research. The parameterization also influences the quality of the solutions, leading us to believe that a more exhaustive tuning of the problem penalty values as well as the solver parameters such as number of reads, chain strength, annealing time, etc. could bring us better performances overall.

On another note, future research could be focused on extending the problem to consider capacity constraints or multiple satellites, which would make it more appealing for industrial applications. Finally a study of the problem from the perspective of gate-based quantum computers, for example by means of variational quantum algorithms such as the quantum approximate optimization algorithm, would also be of significant interest.

Acknowledgments. This work was supported by the Government of Spain (Misiones CUCO Grant MIG-20211005, FIS2020-TRANQI and Severo Ochoa CEX2019-000910-S), Fundació Cellex, Fundació Mir-Puig, Generalitat de Catalunya (CERCA program), and by the European Research Council ERC AdG CERQUTE.

References

1. Barkaoui, M., Berger, J.: A new hybrid genetic algorithm for the collection scheduling problem for a satellite constellation. J. Oper. Res. Soc. **71**(9), 1390–1410 (2020)
2. Bensana, E., Lemaitre, M., Verfaillie, G.: Earth observation satellite management. Constraints **4**(3), 293–299 (1999)
3. Bianchessi, N., Cordeau, J.F., Desrosiers, J., Laporte, G., Raymond, V.: A heuristic for the multi-satellite, multi-orbit and multi-user management of earth observation satellites. Eur. J. Oper. Res. **177**(2), 750–762 (2007)
4. Boros, E., Hammer, P.L.: Pseudo-boolean optimization. Discret. Appl. Math. **123**(1), 155–225 (2002). https://doi.org/10.1016/S0166-218X(01)00341-9
5. Choi, V.: Minor-embedding in adiabatic quantum computation: II. Minor-universal graph design. Quant. Inf. Process. **10**(3), 343–353 (2011). https://doi.org/10.1007/s11128-010-0200-3
6. D-Wave Systems, Burnaby, Canada: QPU-Specific Physical Properties: Advantage2_prototype1.1 (User Manual) (2022). https://docs.dwavesys.com/docs/latest/_downloads/08c75269a89583c35e421c45c35437eb/09-1275A-B_QPU_Properties_Advantage2_prototype1_1.pdf
7. D-Wave Systems, Burnaby, Canada: QPU-Specific Physical Properties: Advantage_system6.1 (User Manual) (2022). https://docs.dwavesys.com/docs/latest/_downloads/1bfa16c9915114bdf8a37b14713c8953/09-1272A-A_QPU_Properties_Advantage_system6_1.pdf

8. D-Wave Systems, Burnaby, Canada: QPU-Specific Physical Properties: DW_2000Q_6 (User Manual) (2022). https://docs.dwavesys.com/docs/latest/_downloads/50b8fa700f78e5d5c4c3208e0a8377c9/09-1215A-D_QPU_Properties_DW_2000Q_6.pdf
9. Glover, F., Kochenberger, G., Du, Y.: Quantum bridge analytics I: a tutorial on formulating and using QUBO models. 4OR **17**(4), 335–371 (2019). https://doi.org/10.1007/s10288-019-00424-y
10. Lin, W.C., Liao, D.Y.: A tabu search algorithm for satellite imaging scheduling. In: 2004 IEEE International Conference on Systems, Man and Cybernetics (IEEE Cat. No.04CH37583), vol. 2, pp. 1601–1606 (2004). https://doi.org/10.1109/ICSMC.2004.1399860
11. Luckow, A., Klepsch, J., Pichlmeier, J.: Quantum computing: towards industry reference problems. Digitale Welt **5**(2), 38–45 (2021)
12. Malladi, K.T., Minic, S.M., Karapetyan, D., Punnen, A.P.: Satellite constellation image acquisition problem: a case study. In: Fasano, G., Pintér, J.D. (eds.) Space Engineering. SOIA, vol. 114, pp. 177–197. Springer, Cham (2016). https://doi.org/10.1007/978-3-319-41508-6_7
13. Orús, R., Mugel, S., Lizaso, E.: Quantum computing for finance: overview and prospects. Rev. Phys. **4**, 100028 (2019)
14. Osaba, E., Makarov, A., Taddei, M.M.: Benchmark dataset and instance generator for the satellite-image-acquisition scheduling problem (2023). https://doi.org/10.17632/dzwvt4bz4j.1. Online at Mendeley Data
15. Osaba, E., Villar-Rodriguez, E., Oregi, I.: A systematic literature review of quantum computing for routing problems. IEEE Access **10**, 55805–55817 (2022)
16. Ribeiro, G.M., Constantino, M.F., Lorena, L.A.N.: Strong formulation for the spot 5 daily photograph scheduling problem. J. Comb. Optim. **20**, 385–398 (2010)
17. Stollenwerk, T., Michaud, V., Lobe, E., Picard, M., Basermann, A., Botter, T.: Image acquisition planning for earth observation satellites with a quantum annealer. arXiv preprint arXiv:2006.09724 (2020)
18. Stollenwerk, T., Michaud, V., Lobe, E., Picard, M., Basermann, A., Botter, T.: Agile earth observation satellite scheduling with a quantum annealer. IEEE Trans. Aerosp. Electron. Syst. **57**(5), 3520–3528 (2021)
19. Vasquez, M., Hao, J.K.: A "logic-constrained" knapsack formulation and a tabu algorithm for the daily photograph scheduling of an earth observation satellite. Comput. Optim. Appl. **20**(2), 137–157 (2001)
20. Vasquez, M., Hao, J.K.: Upper bounds for the spot 5 daily photograph scheduling problem. J. Comb. Optim. **7**, 87–103 (2003)
21. Wang, J., Demeulemeester, E., Hu, X., Qiu, D., Liu, J.: Exact and heuristic scheduling algorithms for multiple earth observation satellites under uncertainties of clouds. IEEE Syst. J. **13**(3), 3556–3567 (2018)
22. Wang, X., Wu, G., Xing, L., Pedrycz, W.: Agile earth observation satellite scheduling over 20 years: formulations, methods, and future directions. IEEE Syst. J. **15**(3), 3881–3892 (2020)
23. Zhang, Y., Hu, X., Zhu, W., Jin, P.: Solving the observing and downloading integrated scheduling problem of earth observation satellite with a quantum genetic algorithm. J. Syst. Sci. Inf. **6**(5), 399–420 (2018)
24. Zhi, H., Liang, W., Han, P., Guo, Y., Li, C.: Variable observation duration scheduling problem for agile earth observation satellite based on quantum genetic algorithm. In: 2021 40th Chinese Control Conference (CCC), pp. 1715–1720. IEEE (2021)

Complexity-Driven Sampling for Bagging

Carmen Lancho[1][(✉)], Marcilio C. P. de Souto[2], Ana C. Lorena[3],
and Isaac Martín de Diego[1]

[1] Data Science Laboratory, Rey Juan Carlos University,
C/ Tulipán, s/n, 28933 Móstoles, Spain
{carmen.lancho,isaac.martin}@urjc.es
[2] Fundamental Computer Science Laboratory, University of Orléans,
Léonard de Vinci, B.P. 6759 F-45067 Orleans Cedex 2, France
marcilio.desouto@univ-orleans.fr
[3] Aeronautics Institute of Technology, Praça Marechal Eduardo Gomes,
50, São José dos Campos, São Paulo 12228-900, Brazil
aclorena@ita.br
https://www.datasciencelab.es, https://www.univ-orleans.fr/lifo/,
https://www.engrena.ita.br/en/ccm-ita-en/

Abstract. Ensemble learning consists of combining the prediction of
different learners to obtain a final output. One key step for their success is the diversity among the learners. In this paper, we propose to
reach the diversity in terms of the classification complexity by guiding
the sampling of instances in the Bagging algorithm with complexity measures. The proposed Complexity-driven Bagging algorithm complements
the classic Bagging algorithm by considering training samples of different
complexity to cover the complexity space. Besides, the algorithm admits
any complexity measure to guide the sampling. The proposal is tested in
28 real datasets and for a total of 9 complexity measures, providing satisfactory and promising results and revealing that training with samples
of different complexity, ranging from easy to hard samples, is the best
strategy when sampling based on complexity.

Keywords: Complexity measures · Ensemble methods · Sampling

1 Introduction

Ensemble learning can be seen as the *Machine Learning (ML)* interpretation of
the wisdom of the crowd [11]. Combining different sources of information leads
to solutions that are often better than those offered by a single one, considering
also the difficulty of finding the optimal model [2]. A necessary and sufficient
condition for an ensemble model to outperform any of its individual component
models is that they are accurate and diverse [5]. This diversity can be achieved
with different techniques [11]. For example, by manipulating the input data of
every learner, which is the case of the Bagging [1] or Boosting [4] strategies.

The aim of this paper is to achieve that diversity through complexity analysis, which is devoted to estimate the difficulty in solving a classification problem

P. Quaresma et al. (Eds.): IDEAL 2023, LNCS 14404, pp. 15–21, 2023.
https://doi.org/10.1007/978-3-031-48232-8_2

given properties of the available training data [6,9]. That is, by training the classifiers with samples of different difficulty levels, the final ensemble is enabled to cover the complexity variability space of each dataset. To this end, the sampling of the instances to obtain the training samples will be driven by complexity measures, as they characterize the underlying difficulty of the data, capturing factors (e.g., overlap or noise) that negatively impact the classification performance [6].

This work is motivated by the recent literature, since in last years distinct versions of the Bagging algorithm have been proposed by modifying the sampling probability of the instances according to their complexity level [7,12,14]. Instead of obtaining training samples by sampling from a uniform distribution, the probability of picking each instance is influenced by its complexity. As ensemble methods are susceptible to noise and outliers, Walmsley et al. [14] proposed Bagging-IH, a version of Bagging where the probability of selecting an instance is inversely proportional to its complexity, computed using *k-Disagreeing Neighbors (kDN)* [13], which is a measure that regards on the level of disagreement on the labels of an instance and those of its nearest neighbors. For multi-class imbalanced problems, Sleeman et al. [12] presented UnderBagging+ where the sampling probabilities of the instances of each class are directly proportional to their complexity (also in terms of *kDN*). Kabir et al. [7] presented a "mixed bagging" framework where the training samples can have different complexity. This is achieved by varying the weights applied to easy or hard instances. The complexity of the instances is calculated as *instance hardness* [13], that is, it is estimated as the proportion of classifiers in a pool that misclassify the instances. In [10], a method for generating classifier pools based on diversity in the classifier decision and in the data complexity spaces is introduced. To achieve this diversity, they chose samples maximizing their differences in complexity, an approach tested using different ensemble strategies, including Bagging.

The previous works are good examples of the use of complexity in ensemble learning. However, among the existing approaches for guiding the sampling in Bagging, there is no clear general agreement. Indeed, some methods emphasize easy instances, others the hard ones and others present a combination of both options. Thus, we aim to analyze different strategies in order to provide guidance on how to conduct the sampling and, also, to study how a variety of complexity measures perform in this task.

The paper is structured as follows. The proposed methodology is presented in Sect. 2. Experiments are depicted in Sect. 3. Finally, Sect. 4 concludes.

2   Methodology

This section presents the Complexity-driven Bagging algorithm, a version of the Bagging algorithm following a complexity-based sampling scheme to guarantee that the training samples have different degrees of complexity. In the Bagging case, every learner is trained with a sample obtained by sampling uniformly and with replacement the training set. This uniform sampling disregards the complexity of the data samples. The proposal complements the classic Bagging

---

Algorithm 1. Complexity-driven Bagging

Input: Dataset D, learner L, number of learners M, splits $s \geq 1, s \in \mathbb{Z}$, α for extreme weights
Output: Classifier H
1: $\mathbf{c} = Complexity(D)$ ▷ Complexity vector of the instances
2: $\mathbf{r}_e = rank(\mathbf{c}, ascending = T)$, $\mathbf{r}_h = rank(\mathbf{c}, ascending = F)$ ▷ Ranking of complexity
3: **if** $\alpha \geq 2$ **then**   ▷ If extreme weights are applied
4:     $\forall j = e, h$: $\mathbf{r}_j[\mathbf{r}_j \geq q_{75}] = \mathbf{r}_j[\mathbf{r}_j \geq q_{75}] * \alpha$ ▷ q_x is percentile x of r_j
5:         $\mathbf{r}_j[q_{50} \leq \mathbf{r}_j < q_{75}] = \mathbf{r}_j[q_{50} \leq \mathbf{r}_j < q_{75}] * \alpha/2$
6: $\mathbf{w}_e = \mathbf{r}_e/sum(\mathbf{r}_e)$, $\mathbf{w}_h = \mathbf{r}_h/sum(\mathbf{r}_h)$ ▷ Weights to emphasize easy or hard cases
7: $\mathbf{w}_u = (1/n, \ldots, 1/n)$ ▷ Uniform weights
8: $\mathbf{w}_1 = (\mathbf{w}_u - \mathbf{w}_e)/s$, $\mathbf{w}_2 = (\mathbf{w}_h - \mathbf{w}_u)/s$ ▷ Values for evolving weights
9: $W = (\mathbf{w}_e, \mathbf{w}_e + \mathbf{w}_1, \ldots, \underbrace{\mathbf{w}_e + (s-1) * \mathbf{w}_1, \mathbf{w}_e + s * \mathbf{w}_1}_{\mathbf{w}_u}, \underbrace{\mathbf{w}_u + \mathbf{w}_2, \ldots, \mathbf{w}_u + s * \mathbf{w}_2}_{\mathbf{w}_h})$
10: $i = 1$
11: **for** $m = 1, \ldots, M$ **do**
12:     **if** $i \leq len(W)$ **then**
13:         $D_w = sample(D, W[i])$; $h_m = L(D_w)$ ▷ D_w training sample, h_m trained learner
14:     **else**
15:         $i = 1$
16:         $D_w = sample(D, W[i])$; $h_m = L(D_w)$
17:         $i = i + 1$
18: $H(\mathbf{x}) = \arg\max_y \sum_{m=1}^{M} I(h_m(\mathbf{x}) = y)$ ▷ Ties are broken arbitrarily

---

by proposing to obtain the training samples not only based on the uniform distribution, but also based on the complexity of the instances. With this, the learners are trained with samples of different complexity, thus enabling the final classifier to learn the different characteristics of the complexity spectrum.

In particular, to achieve training samples with different complexity, the complexity of the original training set is first calculated, using a given measure able to estimate the hardness level of the individual observations of the dataset. Then, the instances are ranked in ascending or descending order according to their estimated complexity level to give more weight to the easy or the hard instances, respectively. To obtain a proper probability distribution for the sampling, each ranking vector is divided by the sum of all its values. With this, two vectors of weights for sampling are available: $\mathbf{w}_e$ and $\mathbf{w}_h$, giving more emphasis to easy or hard instances, respectively. To also have training samples representing the original complexity of the dataset, uniform weights $\mathbf{w}_u$ are considered too. The basic case of the proposal is to obtain training samples with these three types of weights: $\mathbf{w}_e$ for easy samples, $\mathbf{w}_u$ for regular samples and $\mathbf{w}_h$ for hard samples. In addition, two parameters are considered to better cover the complexity space. The parameter s serves to obtain intermediate weights between the already defined ones, indicating the number of splits between them. The higher the s value, the smoother the changes in complexity obtained by sampling with each type of weight. The α parameter multiplies the weight of the easiest and the hardest instances to obtain training samples with more extreme values of complexity. Algorithm 1 presents the pseudocode for a dataset $D = \{(\mathbf{x}_i, y_i) \, | \, i = 1, \ldots, n\}$ where $\{\mathbf{x}_i\}_{i=1}^n$ are the instances and $\{y_i\}_{i=1}^n$ are their corresponding labels.

Contrary to the state of the art, several complexity measures are considered in this study to analyze the differences in their performance. The complexity measures here contemplated are a representative subset of the state of the art of complexity measures defined at the instance level [8]: (1) $N1$ is the number of connections that each point has with points from other classes based on a Minimum Spanning Tree built from the dataset; (2) $N2$ confronts, for each point, the distance between the point and its nearest neighbor from its class with the distance between the point and its nearest neighbor from other class; (3) *Hostility measure* estimates the probability of misclassifying an instance based on the distribution of the classes in neighborhood of sizes adapted to the density of the data; (4) *kDN* of a point is the proportion of its nearest neighbors that belong to a different class; (5) *Local-Set average Cardinality (LSC)* of an point is the set of instances that are closer to that instance than the closest point from other class; (6) *Class Likelihood Difference (CLD)* offers the difference between the class likelihood of an instance and its maximum likelihood for the rest of classes; (7) *Tree Depth Unpruned (TDU)* is the depth of the leaf node in an unpruned *Decision Tree (DT)* where a point is classified; (8) *Disjunct Class Percentage (DCP)* of an instance is the proportion of instances from its class in its belonging disjunct (obtained with a pruned DT); and (9) $F1$ counts the number of features for which an instance is in an overlap region of the classes.

3   Experiments

In this section the performance of the proposed algorithm (with different parameter options) and the standard Bagging procedure are compared in 28 real datasets from [3]. The accuracy of both methods is obtained through a 10-fold cross-validation scheme. The code, the real datasets and the supplementary results of the several strategies analyzed to guide the sampling including, among others, sampling schemes always emphasizing the easiest or the hardest instances can be found at https://github.com/URJCDSLab/ComplexityDrivenBagging.

Table 1 shows the average proportion of times the proposal gets higher accuracy than standard Bagging with $10, 11, \ldots, 200$ learners (using unpruned DTs as base classifiers) to build the final classifier for every dataset. The darker the cell color in the table, the greater the proportion of wins registered. The datasets are divided in easy, intermediate and hard according to their difficulty in terms of the *hostility measure* (thresholds: $[0, 0.15]$, $(0.15, 0.3]$, $[0.3, 1]$) which serves as an estimation of the expected classification error [8]. Figure 1a shows, for the case of $s = 2$ and $\alpha = 4$, the difference in accuracy with standard Bagging when guiding the sampling with the 9 complexity measures. In general, the complexity measures show better performance for intermediate and hard datasets and similar to standard Bagging performance for the easy datasets. This reveals that when a dataset has a certain level of complexity, it generally benefits more from training with samples of varying complexity, but when it is easy, both methods tend to offer similar performance. Among complexity measures, the *hostility measure*, *kDN* and *CLD* outshine for their satisfactory performance normally reaching higher accuracy than standard Bagging in around 70% of cases for the intermediate and hard datasets. On the contrary, $F1$ and *TDU* have the weakest results, presenting similar results to standard Bagging.

Figure 1b depicts the complexity (using the *hostility measure*) of 200 training samples for the *teaching_assistant_LM* dataset. It reveals how the complexity of

Table 1. Proportion of times (avg.) the proposal wins standard Bagging in accuracy. The darker the color, the greater the proportion of wins.

Parameters	Data	Host.	N1	N2	kDN	LSC	CLD	TDU	DCP	F1
	hard	0.722	0.673	0.621	0.733	0.582	0.737	0.737	0.568	0.565
$s=1$	int.	0.665	0.629	0.689	0.606	0.592	0.699	0.703	0.578	0.548
	easy	0.253	0.335	0.571	0.321	0.377	0.441	0.707	0.410	0.683
	hard	0.743	0.558	0.561	0.630	0.644	0.648	0.450	0.602	0.700
$s=1,\alpha=4$	int.	0.540	0.574	0.675	0.612	0.552	0.666	0.513	0.621	0.597
	easy	0.305	0.178	0.514	0.300	0.377	0.402	0.459	0.426	0.622
	hard	0.675	0.632	0.620	0.783	0.695	0.701	0.653	0.749	0.542
$s=2$	int.	0.655	0.608	0.598	0.669	0.543	0.774	0.621	0.671	0.635
	easy	0.447	0.281	0.584	0.478	0.518	0.628	0.661	0.469	0.725
	hard	0.811	0.760	0.593	0.739	0.627	0.810	0.644	0.610	0.713
$s=2,\alpha=4$	int.	0.792	0.599	0.707	0.586	0.679	0.597	0.525	0.695	0.586
	easy	0.468	0.282	0.614	0.338	0.681	0.687	0.432	0.605	0.591
	hard	0.645	0.598	0.650	0.671	0.630	0.756	0.698	0.685	0.661
$s=4$	int.	0.775	0.582	0.602	0.707	0.596	0.696	0.667	0.704	0.640
	easy	0.389	0.415	0.555	0.506	0.353	0.503	0.683	0.383	0.726
	hard	0.742	0.715	0.678	0.601	0.656	0.727	0.560	0.684	0.548
$s=4,\alpha=4$	int.	0.686	0.549	0.637	0.633	0.507	0.718	0.643	0.682	0.617
	easy	0.498	0.366	0.506	0.450	0.748	0.823	0.483	0.558	0.638

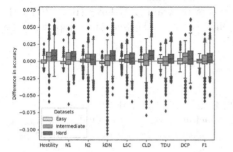
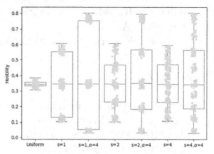

(a) Difference in accuracy with $s=2$, $\alpha=4$ (b) Hostility of 200 training samples

Fig. 1. Difference in accuracy with Bagging and complexity of training samples

the samples used for standard Bagging is totally concentrated, while the complexities of the proposal cover a wider range. The distribution of these complexities are influenced by the parameters of the algorithm. Regarding the alternatives analyzed, always giving more weight to easy or hard instances is discarded as the classifier does not learn the real data structure and there is no complexity variability. The combination of samples with different complexity levels yields better results, being the best case the here presented where the complexity of the original dataset is also considered by using uniform weights.

4   Conclusions and Future Work

Based on the state of the art, in this work, different strategies to guide the sampling in the Bagging algorithm have been studied and the best one has been presented and compared with standard Bagging. This strategy consists on complementing the uniform weighting scheme of Bagging by also considering weighting schemes giving more probability to easy or hard instances. By doing this, the complexity range covered by the training samples is wider and, as shown by the results, this is translated into better performance. Only training with samples of the same complexity does not exploit the different complexity ranges of the data and results in poorer performance. In addition, different complexity measures have been analyzed and, although all of them have shown promising results, the *hostility measure, kDN* and *CLD* stand out.

Future work will mainly focus in testing more parameters of the proposal and in comparing it with the state of the art alternatives. Besides, the idea of complexity-driven sampling will be also analyzed in other ensemble methods like Boosting, aiming to establish a general framework about sampling with complexity measures in ensemble learning. Finally, the different performance of the complexity measures deserves deeper analysis to also provide some guidance.

Acknowledgements. This research is supported by grants from Rey Juan Carlos University (Ref: C1PREDOC2020) and the Spanish Ministry of Science and Innovation, under the Knowledge Generation Projects program: XMIDAS (Ref: PID2021-122640OB-100). A. C. Lorena would also like to thank the financial support of the FAPESP research agency (grant 2021/06870-3).

References

1. Breiman, L.: Bagging predictors. Mach. Learn. **24**, 123–140 (1996)
2. Dietterich, T.G.: Ensemble methods in machine learning. In: Kittler, J., Roli, F. (eds.) MCS 2000. LNCS, vol. 1857, pp. 1–15. Springer, Heidelberg (2000). https://doi.org/10.1007/3-540-45014-9_1
3. Dua, D., Graff, C.: UCI machine learning repository (2017). https://archive.ics.uci.edu/ml
4. Freund, Y., Schapire, R.E., et al.: Experiments with a new boosting algorithm. In: ICML, vol. 96, pp. 148–156 (1996)
5. Hansen, L.K., Salamon, P.: Neural network ensembles. IEEE Trans. Pattern Anal. Mach. Intell. **12**(10), 993–1001 (1990)
6. Ho, T.K., Basu, M.: Complexity measures of supervised classification problems. IEEE Trans. Pattern Anal. Mach. Intell. **24**(3), 289–300 (2002)
7. Kabir, A., Ruiz, C., Alvarez, S.A.: Mixed bagging: a novel ensemble learning framework for supervised classification based on instance hardness. In: 2018 IEEE International Conference on Data Mining (ICDM), pp. 1073–1078. IEEE (2018)
8. Lancho, C., Martín De Diego, I., Cuesta, M., Acena, V., Moguerza, J.M.: Hostility measure for multi-level study of data complexity. Appl. Intell. **53**, 1–24 (2022)
9. Lorena, A.C., Garcia, L.P., Lehmann, J., Souto, M.C., Ho, T.K.: How complex is your classification problem? a survey on measuring classification complexity. ACM Comput. Surv. (CSUR) **52**(5), 1–34 (2019)
10. Monteiro, M., Jr., Britto, A.S., Jr., Barddal, J.P., Oliveira, L.S., Sabourin, R.: Exploring diversity in data complexity and classifier decision spaces for pool generation. Inf. Fusion **89**, 567–587 (2023)
11. Sagi, O., Rokach, L.: Ensemble learning: a survey. Wiley Interdisc. Rev. Data Min. Knowl. Discovery **8**(4), e1249 (2018)
12. Sleeman IV, W.C., Krawczyk, B.: Bagging using instance-level difficulty for multiclass imbalanced big data classification on spark. In: 2019 IEEE International Conference on Big Data, pp. 2484–2493. IEEE (2019)

13. Smith, M.R., Martinez, T., Giraud-Carrier, C.: An instance level analysis of data complexity. Mach. Learn. **95**(2), 225–256 (2014)
14. Walmsley, F.N., Cavalcanti, G.D., Oliveira, D.V., Cruz, R.M., Sabourin, R.: An ensemble generation method based on instance hardness. In: 2018 International Joint Conference on Neural Networks (IJCNN), pp. 1–8. IEEE (2018)

A Pseudo-Label Guided Hybrid Approach for Unsupervised Domain Adaptation

Eva Blanco-Mallo[1]([✉]), Verónica Bolón-Canedo[1], and Beatriz Remeseiro[2]

[1] Department of Computer Science and Information Technologies,
Universidade da Coruña, CITIC, Campus de Elviña s/n, 15071 A Coruña, Spain
{eva.blanco,vbolon}@udc.es

[2] Department of Computer Science, Universidad de Oviedo, Campus de Gijón s/n,
33203 Gijón, Spain
bremeseiro@uniovi.es

Abstract. Unsupervised domain adaptation focuses on reusing a model trained on a source domain in an unlabeled target domain. Two main approaches stand out in the literature: adversarial training for generating invariant features and minimizing the discrepancy between feature distributions. This paper presents a hybrid approach that combines these two methods with pseudo-labeling. The proposed loss function enhances the invariance between the features across domains and further defines the inter-class differences of the target set. Using a well-known dataset, Office31, we demonstrate that our proposal improves the overall performance, especially when the gap between domains is more significant.

Keywords: unsupervised domain adaptation · certainty-aware pseudo-labeling · maximum mean discrepancy

1 Introduction

Computer vision has become a constant in our lives. A multitude of applications that we use every day rely on it, from searching for products on a website to unlocking mobile devices. Computer vision heavily relies on machine learning algorithms to analyze, interpret, and extract meaningful information from images or video data. However, machine learning models are task-specific and limited to the tasks they are trained for, decreasing considerably in performance if the data distribution changes. For example, a model that identifies different objects on an artificial white background, would greatly decrease its accuracy if it had to recognize the same objects in a real environment. Therefore, there is a need for a procedure to adapt these models and reuse them to solve the same task under different circumstances. The ability to transfer knowledge implies a considerable

This work was supported by the Ministry of Science and Innovation of Spain (Grant PID2019-109238GB, subprojects C21 and C22; and Grant FPI PRE2020-092608) and by Xunta de Galicia (Grants ED431G 2019/01 and ED431C 2022/44).

saving in time and resources and can improve the final result through the use of related content. This process is known as domain adaptation, and formally, it involves reusing a model learned from a source data distribution in a different, but related, target data distribution.

This work is focused on unsupervised domain adaptation (UDA), i.e., the situation when the labels of the target set are not available. Among the different UDA approaches that can be found in the field, the adversarial-based and discrepancy-based are particularly noteworthy. The former generally consists of a network composed of a feature extractor, a label classifier, and a domain discriminator. The extractor parameters are learned by maximizing the loss of the domain discriminator and minimizing the loss of the label classifier. DANN [1] was the first network to present this architecture, and since then, several adversarial learning approaches have emerged as modifications of it [8]. The discrepancy-based approach is focused on minimizing the discrepancy between domains by optimizing a distance measure [6]. Maximum mean discrepancy (MMD) [2], proposed by Gretton et al., is gaining popularity in the context of transfer learning [9,12]. Similarly, it has been shown in the literature that pseudo-labeling improves the performance of UDA models [3,6] and is found in most of the current state-of-the-art models [10,13]. Note that pseudo-labeling involves incorporating the labels from the unlabeled set, generated by the classifier, to refine the model during training.

We propose a hybrid method that combines the two approaches mentioned above, adversarial- and discrepancy-based, with pseudo-labeling. The goal is to improve performance by reinforcing the main components that constitute the network structure. First, we incorporate pseudo-labeling in the target set to compute the classifier loss. In this way, a better fit is achieved in the classifier and the extractor, since the domain of interest is also considered when adjusting the network weights. Second, the distance between the features of both sets is minimized by using MMD. By doing so, the task of the domain discriminator is enhanced, with the aim of optimizing the similarity between the feature representations of the source and target domains.

2   Methodology

Let us consider a classification problem, with an input space X and a set of L possible labels $Y = \{0, 1, ..., L-1\}$. An unsupervised domain adaptation problem arises when there are two distinct distributions over the joint space $X \times Y$: the source domain D_s and the target domain D_t. Specifically, we have a collection of n_s labeled samples, $S = \{(x_i, y_i)\}_{i=1}^{n_s}$, which are i.i.d. drawn from D_s; and a set of n_t unlabeled samples, $T = \{(x_i)\}_{i=1}^{n_t}$, which are i.i.d. drawn from D_t.

The main goal is to learn a classifier $\eta : X \rightarrow Y$ that performs well on the target domain D_t, as measured by a certain classification loss function $\mathcal{L}(\cdot)$. Specifically, we aim to minimize the expected target risk given by:

$$PR_{(x,y) \sim D_t}[\mathcal{L}(\eta(x), y)] \tag{1}$$

2.1   Model Architecture

DANN [1] is the base architecture used in our hybrid approach. It is composed of a feature extractor $G_f(\cdot; \theta_f)$, a classifier $G_y(\cdot; \theta_y)$, and a domain discriminator $G_d(\cdot; \theta_d)$, with parameters θ_f, θ_y, θ_d, respectively (see Fig. 1a). First, the feature extractor maps the input sample x into a D-dimensional representation, which feeds both the classifier and the domain discriminator. The classifier consists of a fully connected (FC) layer of 256 units, followed by batch normalization (BN) and ReLU activation layers. Finally, an FC layer of size L follows and the softmax activation function, which provides a probability distribution over the set of possible labels Y. The discriminator is composed of two FC layers of 1024 units, followed by BN and ReLU activation layers. The output layer consists of a 1-unit FC followed by the sigmoid activation function. It is connected to the feature extractor with a gradient reversal layer $\mathcal{R}$, which acts as an identity transformation during the forward pass and reverses the gradient during the backpropagation by multiplying it by -1. By adding this layer, the domain discriminator is competing against the classifier in an adversarial manner.

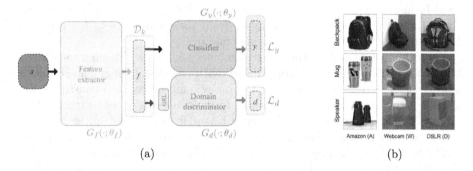

(a)   (b)

Fig. 1. (a) The model architecture for unsupervised domain adaptation. (b) A representative example of the Office31 dataset, with images of three classes (backpack, mug, speaker) in three different domains (Amazon, Webcam, DSLR).

2.2   Training Process

The proposed hybrid method combines a discrepancy-based approach with an adversarial architecture. Thus, the aim of the training procedure is minimizing the label prediction loss of the classifier $G_y(\cdot; \theta_y)$, maximizing the classification loss of the domain discriminator $G_d(\cdot; \theta_d)$, and thus reducing the feature discrepancy D_k between the two sets (see Fig. 1a). In each iteration of the training process, the model is fed with one batch from the source set and another from the target set. The classifier uses the categorical cross-entropy function ($\mathcal{L}_y$) with the predictions made from the source set, while for the target set, a certainty-aware pseudo-labeling approach is adopted. That is, the predicted labels of the target samples are added to the loss calculation if their confidence exceeds a certain threshold, considering confidence as the probability of the prediction [4]. For its part, the domain discriminator uses the binary cross-entropy ($\mathcal{L}_d$) calculated on all the samples.

Multiple kernel maximum mean discrepancy (MK-MMD) [7] is used to calculate the discrepancy between the features extracted from both domains. Using the features generated by the extractor, the function is computed as:

$$D_k(f_s, f_t) = \frac{1}{n_s^2} \sum_{i=1}^{n_s} \sum_{j=1}^{n_s} k(f_i^s, f_j^s) + \frac{1}{n_t^2} \sum_{i=1}^{n_t} \sum_{j=1}^{n_t} k(f_i^t, f_j^t) - \frac{1}{n_s n_t} \sum_{i=1}^{n_s} \sum_{j=1}^{n_t} k(f_i^s, f_j^t)$$

(2)

where k is a kernel function. Therefore, the training consists in optimizing the following function:

$$E(\theta_f, \theta_y, \theta_d) = \mathcal{L}_y + D_k - \mathcal{L}_d$$

(3)

Note that due to the gradient reversal layer $\mathcal{R}$, the gradients of the domain discriminator are subtracted rather than added, thus maximizing its loss. In addition, by incorporating pseudo-labeling during the training stage, the features of the target domain gain more weight when fine-tuning both the classifier and the feature extractor, which is particularly beneficial when the source set is significantly larger than the target set. Furthermore, the purpose of the incorporation of MK-MMD minimization is to reinforce the discriminator's task of adjusting the extractor to generate invariant features across domains.

3   Experimental Results

The effectiveness of our hybrid method was evaluated using Office31 [11], one of the most popular datasets in UDA. It is composed of 4110 images categorized into 31 classes belonging to three different domains: Amazon (A), Webcam (W), and DSLR (D), with 2817, 795, and 498 images, respectively. Figure 1b shows some representative images of objects belonging to three different classes in each domain. Certain aspects must be taken into account when dealing with this dataset. First of all, the W and D domains are remarkably similar to each other, the difference is the camera with which the images were taken (see Fig. 1b). Regarding the A domain, the images differ greatly from the previous ones, since their background is completely different. In addition, there is a large imbalance present in the dataset, being A the domain with the largest number of samples.

The implementation of DANN and MMD used in this work is available in the Transfer Learning Library [5], with the default configuration. Building upon it, we incorporate the pseudo-labeling procedure. Due to the domain shift inherent to this problem, blindly accepting all the pseudo-labels generated by the classifier may introduce noise and negatively affect the model's performance [6]. For this reason, a certainty threshold of 0.8 was selected after conducting some preliminary experiments with values $\{0.7, 0.8, 0.9, 0.99\}$. Note that selecting higher certainty threshold values yields a conservative, smaller yet potentially more accurate pseudo-labeled set, while lower threshold values generate a larger set with the possibility of introducing more noise. All experiments were performed using an NVIDIA RTX 3080.

Table 1 presents the accuracy obtained with the original model (DANN) and with our proposal (DANN+P+M), as well as the effect of the different components separately. As can be seen, the proposed model achieves superior performance compared to the original model in all cases. Mainly in those where the source and target domains are more different from each other. Note that the cases involving both the W and D domains obtain by far the highest results, reaching even 100% in the case of $W \rightarrow D$. The cases in which the transference is made between A and the other two domains entail the most difficult challenges, mainly the cases of $D \rightarrow A$ and $W \rightarrow A$. Not only are the inter-domain features more disparate, but the source set, the labeled one, is significantly smaller than the target set. Here, the proposed method achieves an increase of about 10% in accuracy. This also applies to scenarios where the source set is much larger than the target set ($A \rightarrow W$ and $A \rightarrow D$).

Table 1. Accuracy results obtained on the Office31 dataset with the original model (DANN) and with the different components of our proposal: pseudo-labeling (P) and MK-MMD (M). Best results are in bold.

	A → W	D → W	W → D	A → D	D → A	W → A	Avg
DANN	82.30	97.50	**100**	83.70	64.70	63.60	81.97
DANN+P	91.80	98.20	99.80	89.80	**76.30**	73.10	88.17
DANN+M	88.40	98.00	99.80	82.70	72.60	**74.30**	85.97
DANN+P+M	**93.20**	**98.40**	**100**	**90.60**	74.80	73.70	**88.45**

Regarding the use of pseudo-labeling, there is also a great improvement in the transfer between the most disparate domains ($A \rightarrow W$, $A \rightarrow D$, $D \rightarrow A$, and $W \rightarrow A$). The incorporation of MMD produces the same effect, except for $A \rightarrow D$. However, when it is combined with pseudo-labeling (our proposal), the highest accuracy is obtained. In $D \rightarrow W$ it is necessary to incorporate both techniques to maintain the 100% accuracy achieved by the original model.

4   Conclusions

Domain adaptation has become a critical problem in the field of machine learning due to the increasing need to generalize models across different domains. In this study, we proposed a hybrid unsupervised method that combines the two main approaches, adversarial- and distance-based, along with pseudo-labeling. Minimizing the discrepancy between both domains improves the fine-tuning process of the feature extractor used to generate cross-domain invariant features. In addition, incorporating pseudo-labeling contributes to better defining intra-class differences in the target set. We show how our proposal improves the performance of DANN, the adversarial-based method used as the baseline, especially when the differences between the source and target domains are more significant. Our results indicate that pseudo-labeling has the greatest impact on improvement. However, the best results are obtained when combining this technique with MMD minimization. As future research, it would be interesting to explore the effectiveness of the proposed approach in other scenarios and to evaluate the performance of different distance measures.

References

1. Ganin, Y., et al.: Domain-adversarial training of neural networks. J. Mach. Learn. Res. **17**(1), 2096–2130 (2016)
2. Gretton, A., et al.: A kernel statistical test of independence. In: Advances in Neural Information Processing Systems, vol. 20, pp. 585–592 (2007)
3. Gu, X., Sun, J., Xu, Z.: Spherical space domain adaptation with robust pseudo-label loss. In: IEEE/CVF Conference on Computer Vision and Pattern Recognition, pp. 9101–9110 (2020)
4. Hubáček, O., Šourek, G., Železný, F.: Exploiting sports-betting market using machine learning. Int. J. Forecast. **35**(2), 783–796 (2019)
5. Jiang, J., Chen, B., Fu, B., Long, M.: Transfer-learning-library (2020). https://github.com/thuml/Transfer-Learning-Library
6. Kang, G., Jiang, L., Yang, Y., Hauptmann, A.G.: Contrastive adaptation network for unsupervised domain adaptation. In: IEEE/CVF Conference on Computer Vision and Pattern Recognition, pp. 4893–4902 (2019)
7. Long, M., Cao, Y., Wang, J., Jordan, M.: Learning transferable features with deep adaptation networks. In: International Conference on Machine Learning, pp. 97–105 (2015)
8. Long, M., Cao, Z., Wang, J., Jordan, M.I.: Conditional adversarial domain adaptation. In: Advances in Neural Information Processing Systems, vol. 31 (2018)
9. Lu, N., Xiao, H., Sun, Y., Han, M., Wang, Y.: A new method for intelligent fault diagnosis of machines based on unsupervised domain adaptation. Neurocomputing **427**, 96–109 (2021)
10. Na, J., Jung, H., Chang, H.J., Hwang, W.: FixBi: bridging domain spaces for unsupervised domain adaptation. In: IEEE/CVF Conference on Computer Vision and Pattern Recognition, pp. 1094–1103 (2021)
11. Saenko, K., Kulis, B., Fritz, M., Darrell, T.: Adapting visual category models to new domains. In: Daniilidis, K., Maragos, P., Paragios, N. (eds.) ECCV 2010. LNCS, vol. 6314, pp. 213–226. Springer, Heidelberg (2010). https://doi.org/10.1007/978-3-642-15561-1_16
12. Wang, W., et al.: Rethinking maximum mean discrepancy for visual domain adaptation. IEEE Trans. Neural Netw. Learn. Syst. **34**, 264–277 (2021)
13. Xu, T., Chen, W., Pichao, W., Wang, F., Li, H., Jin, R.: CDTrans: cross-domain transformer for unsupervised domain adaptation. In: International Conference on Learning Representations, pp. 1–14 (2021)

Combining of Markov Random Field and Convolutional Neural Networks for Hyper/Multispectral Image Classification

Halil Mertkan Sahin$^{(\boxtimes)}$ ⓘ, Bruce Grieve ⓘ, and Hujun Yin ⓘ

Department of Electrical and Electronic Engineering, The University of Manchester, Manchester M13 9PL, UK
halil.sahin@machester.ac.uk, {bruce.grieve, hujun.yin}@manchester.ac.uk

Abstract. In the last decades, computer vision tasks such as face recognition, pattern classification, and object detection, have been extensively explored due to advances in computing technologies, learning algorithms, and availability of large amounts of labelled data. Image texture analysis is helpful and important to extracting valuable features of images in order to perform the aforementioned tasks. Markov Random Field (MRF) theory defines probabilistic relationships between the pixel value and its neighbouring pixels, which can help to produce features for classification purposes with deep learning. Convolutional Neural Network (CNN) is a popular feed-forward deep neural network, especially for handling visual tasks. Therefore, this study focuses on combining MRF and CNN in order to achieve hyper/multispectral image classification tasks. MRF images were generated to produce prefixed MRF filters for the first and/or second attention-like layers of CNNs to better extract features. Then CNNs were trained with the prefixed filters for experimentation on the UoM, BONN, and Cassava datasets. Experimental results have demonstrated that such combination of MRF and CNN can enhance classification accuracy while being more time-efficient for hyper/multispectral image classification compared to CNN structures without prefixed MRF filters.

Keywords: Markov Random Field · Convolutional Neural Networks · Hyper-/Multi-spectral Imaging · Image Classification

1 Introduction

Image texture analysis has received a great deal of attention in computer vision due to its various potential applications. Image texture models or image texture features can provide a solution for face detection, pattern recognition, object recognition, image classification, medical image analysis tasks, and more [1, 2].

There are many methods for the extraction of image texture features, which can be categorised into statistical, structural, model-based, and transform-based methods. In this work, a model-based method is employed for feature extraction in order to enhance image classification results with a deep learning algorithm.

P. Quaresma et al. (Eds.): IDEAL 2023, LNCS 14404, pp. 28–38, 2023.
https://doi.org/10.1007/978-3-031-48232-8_4

MRF is the model-based method for texture feature extraction, which has gained more attraction after the theorem provided the equivalence between MRFs and Gibbs distributions. The theorem was established by Hammersley and Clifford [3], and then it was further developed by Besag [4]. The MRF-Gibbs equivalence theorem enables to show that MRFs joint probability can be expressed as a Gibbs distribution, and for any Gibbs distribution an MRF can be modelled [5].

Over the last years, deep learning methods have proved that they can outperform previous state-of-the-art machine learning techniques in several fields. As an example of the deep learning method, CNN is one of the powerful deep learning algorithms, which most commonly takes images as inputs, in computer vision. The importance and effectiveness of CNN in computer vision were understood after the advancements in, especially, powerful computing machines, algorithms, and labelled data [6]. CNNs, as its name states, have convolutional layers that can extract learnable weights and biases to produce feature maps. Also, while 2D-CNNs extract spatial features, 3D-CNNs can perform extracting the deep spectral–spatial-combined features effectively [7].

The MRF-CNN combination has been studied from many different perspectives [8–11]. For instance, in [12], the classification result of the CNN was optimized by using MRF. However, in this work, MRF images were generated in order to produce prefixed filters for the first and/or second layers of a CNN; also, several datasets were used to test the effect of the combination of the MRF-CNN on image classification.

The 5th-order neighbourhoods have been used to generate MRF patterns, and then MRF filters have been cropped and applied as prefixed filters for the first and/or second layers of the CNN. The same CNN structures have been employed in experiments on the UoM [8, 9, 13–15], BONN [8, 9, 13–15], and Cassava datasets [16, 17]. It is found that MRF is a powerful tool for modelling and extracting image features using dependencies between neighbourhood pixels, our results have validated the effectiveness of this combined model. In addition, it has been seen that prefixed MRF filters can help reduce the training times of CNNs compared to the CNN structure without MRF filters.

Section 2 gives a brief summary of MRFs and CNNs. Information about the produced MRF filters, the structure of the CNN, and the used datasets are respectively revealed in the materials and methods section (Sect. 3). Experiments and results are given in Sect. 4, followed by conclusions in the final section.

2 Background

2.1 Markov Random Field and Gibbs Distribution

MRF theory explains the relationship between the pixel value at a point and its neighbouring pixels [18]. MRF modelling has gained popularity after a theorem established by Hammersley and Clifford [3], which demonstrates the equivalence between MRFs and Gibbs distributions. Additionally, Besag [4] later expanded on this theorem, showing that the joint distribution of an MRF can be described as a Gibbs distribution, paving the way for statistical image analysis [5]. Using this model, both spectral and spatial information can be extracted simultaneously. Furthermore, a three-dimensional (3D) extension of the MRF model can provide temporal information in addition to spectral and spatial information [1].

An MRF consists of a set of random variables denoted $F = \{F_1, F_2, \ldots, F_m\}$ indexed by the set S in which each random variable F_i that takes a value f_i from the label set $\mathcal{L}$.

Markovian properties of an MRF are as follows:

1. Positivity: Any configuration is positive.

$$P(f) > 0, \forall f \in \mathbb{F} \tag{1}$$

where $f = \{f_1, f_2, \ldots, f_m\}$ is one possible configuration and $\mathbb{F}$ is the set of all possible configurations of F. For technical reasons, this property is required [5] to obtain the joint probability $P(f)$ of any random field is uniquely determined by its local conditional probability distributions [4].

2. Markovianity: It depicts that only neighbouring labels have direct interactions with one another.

$$P\left(f_i | f_{S-\{i\}}\right) = P\left(f_i | f_j, j \in \mathcal{N}_i\right) = P(f_i | \mathcal{N}_i) \tag{2}$$

where $S - \{i\}$ denotes the set of differences (all sites excluding site i), $f_{S-\{i\}}$ is the set of labels at the sites in $S - \{i\}$ and $\mathcal{N}$ is the neighbourhood system. This property describes the local characteristics of F [5].

As mentioned, the Hammersley − Clifford theorem provides a way to estimate MRFs using Gibbs distribution. The theorem states that any MRF's joint probability may be expressed as a Gibbs distribution, and there is an MRF model for any Gibbs distribution.

A Gibbs distribution takes the form

$$P(f) = \frac{1}{Z} e^{-\frac{1}{T} U(f)} \tag{3}$$

where $P(f)$ denotes a Gibbs distribution on the set S and

$$Z = \sum_{f \in \mathbb{F}} e^{-\frac{1}{T} U(f)} \tag{4}$$

Z is a normalizing constant defined by the equation above and it is also called the partition function. T stands for the temperature and $U(f)$ is the energy function formulated in the next equation. The energy

$$U(f) = \sum_{c \in C} V_c(f) \tag{5}$$

where $V_c(f)$ is the sum of clique potentials over all possible cliques C and it is positive for all possible configurations [1, 5].

2.2 Convolutional Neural Networks (CNNs)

CNN is one of the most powerful and popular feed-forward deep neural networks. Although CNNs have been used for visual tasks since the late 1990s, rapid progresses have been shown since early 2010s thanks to the developments in computing power, improved algorithms and increasing accessibility of large amounts of labelled data [6, 19]. CNNs have a wide range of application areas such as image classification, textural feature representation, and natural language processing (NLP). There have been different versions of CNNs in the literature; however, in this paper, image-based CNN was applied for image classification with texture features being automatically extracted from images [1, 19].

A CNN architecture may consist of an input, convolution, pooling, batch normalization, and dropout, followed by fully connected layers, which are the final stages where classification results are generated. CNN is a type of Deep Neural Network and as the name suggests, layers can be added to the architecture for producing more complex cascade CNNs in order to accomplish high-demand tasks. Multiple pairs of convolution layers and pooling layers may have more learnable parameters, and it may result in higher-level feature recognition [1, 20]. For instance, while partial lines or curves of the digits or characters may be low-level features, entire characters or digits can be higher-level features [19]. The convolution operation has a key feature which is weight sharing. This feature allows observing the local feature patterns, learning spatial hierarchies of feature patterns by downsampling, and improving the model efficiency by minimising the number of parameters to learn [21].

In [10], the MRF textures were produced to generate prefixed filters, also called filter banks, which were used to train the first layer of the CNN. A 5th-order neighbourhood system was applied to create 100×100 MRF patterns using Metropolis sampler with a simulated annealing strategy. MNIST [22], EMNIST [23], and CIFAR-10 [24] datasets were tested by a combination of MRF-CNN. The MRF-CNN method has resulted in a low error rate on the datasets used in the experiments. In this paper, we applied pre-generated MRF filters of varying sizes in attention-like layers to extract features.

In [11], a convolutional translation-invariant self-organising map (SOM_{TI}) has been proposed to generate image features from produced MRF textures using Metropolis sampler with a simulated annealing strategy. The proposed method MRF_{Rot5}-SOM_{TI}-CNN, where Rot5 stands for five different rotations between $0°$ and $45°$, was tested on the MNIST [22], rotated MNIST, CIFAR-10 [24], and CIFAR-100 [24] datasets with outperforming results. Different parameters of CNN's, SOM's, and MRF's were examined to show the effects on the results. In terms of the filter size of CNN's architectures, 7×7 filters and 3×3 filters resulted in lower error rates on MNIST datasets and CIFAR datasets, respectively.

3 Materials and Methods

3.1 MRF Filter Bank

MRF patterns were generated to produce filter banks for the first and/or second layers of the CNN as mentioned before. The CNN layers were trained except for the first and/or second layers. By their nature, MRF models provide an easy way to extract features using the interaction between spatially related random variables in their neighbourhood systems. Multilevel logistic MRF models (label set $\mathcal{L} = \{0, 1\}$) have been employed with β parameters on a 5th-order neighbourhoods system as shown below (Eq. 6). Also, the simulated annealing method with Metropolis sampler is used to simulate MRF models.

$$\beta = (\beta_{11}, \beta_{12}, \beta_{21}, \beta_{22}, \beta_{31}, \beta_{32}, \beta_{41}, \beta_{42}, \beta_{51}, \beta_{52})$$

$$= \begin{bmatrix} \beta_{51} & \beta_{41} & \beta_{32} & \beta_{42} & \beta_{52} \\ \beta_{41} & \beta_{21} & \beta_{12} & \beta_{22} & \beta_{42} \\ \beta_{31} & \beta_{11} & 0 & \beta_{11} & \beta_{31} \\ \beta_{42} & \beta_{22} & \beta_{12} & \beta_{21} & \beta_{41} \\ \beta_{52} & \beta_{42} & \beta_{32} & \beta_{41} & \beta_{51} \end{bmatrix} \tag{6}$$

The procedures for generating filter banks for the first and second layers of a CNN are explained in Table 1.

Table 1. Procedures for generating filter banks

Filter bank for the first layer ($n \times n \times 1 \times d$):
1. Generate a set of 32 MRF patterns with a size of 100×100 from a uniform distribution
2. Crop $n \times n$ patches 5000 times to produce a filter from each generated MRF pattern, and repeat for each MRF pattern; where n is the filter size
3. Average 5000 $n \times n$ patches to get an MRF filter
4. Go to the first step and repeat steps 1–3 until d filters are produced where d is the number of the total filters
Filter bank for the second layer ($n \times n \times d \times d$):
1. Generate 125 sets of 32 MRF randomly with a size of 100×100 from a uniform distribution
2. Crop $n \times n$ patches 5000 times to produce a filter from each generated MRF pattern, and repeat for each MRF pattern; where n is the filter size
3. Average 5000 $n \times n$ patches to get an MRF filter
4. Go to the first step and repeat steps 1–3 until $d \times 125$ filters are produced where d is the number of the filters
5. Deploy K-means clustering algorithm to produce ($n \times n \times d \times d$) filter bank

K-means clustering algorithm is applied to reduce the total number of produced filters to the defined numbers of the filters.

A total of 32 MRF patterns were modelled by changing the value of β. While half of the MRF patterns were generated with specified parameter values in order to draw horizontal, vertical and diagonal line textures, the other half was produced using multi-level logistic distributions with random parameters.

3.2 The CNN Structure

This study examined two different CNN structures in the experiments. The first structure employed prefixed filters in both the first and second layers, while the second structure had MRF filters with different sizes in parallel in the first layer. Additionally, a 1×1 convolutional layer was integrated between layers for both structures, facilitating the learning of a non-linear combination of these fixed filters. The architectures used in the experiments are presented in Figs. 1 and 2.

For both the first and second structures, dropout layers were inserted after each batch normalization layer to prevent overfitting. Additionally, a flatten layer was used before the first dense layer, and the activation function (Softmax or Sigmoid) was applied to the last dense layer, depending on the dataset, to observe the predictions. The number of epochs and the learning rate were initialised as 100 and 0.01, respectively. Also, the learning rate was annealed by a factor drop rate per epoch. Adaptive Moment Estimation (ADAM) was deployed as an optimiser of the CNN. Over the experiments, the same architectures and hyperparameters have been applied.

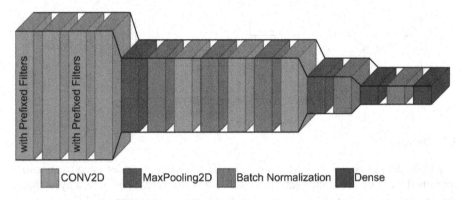

Fig. 1. Architecture of the first CNN structure

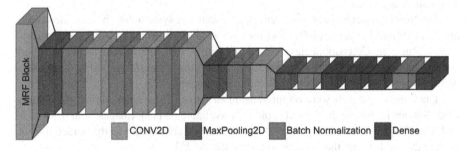

Fig. 2. Architecture of the second CNN structure

A visual representation of this MRF block is presented in Fig. 3. The input image was convolved with multiple prefixed MRF filters of sizes 3×3, 5×5, and 7×7, producing intermediate feature maps. These feature maps were then transposed and element-wise multiplied with other feature maps produced by 1×1 filters. The resulting feature maps were upsampled and combined using element-wise addition. The concatenated feature maps were then passed through a softmax activation function to obtain normalized probabilities for each pixel's class. Finally, a 1×1 convolutional layer was applied to the original input, and its output was element-wise multiplied with the softmax output to obtain the final output.

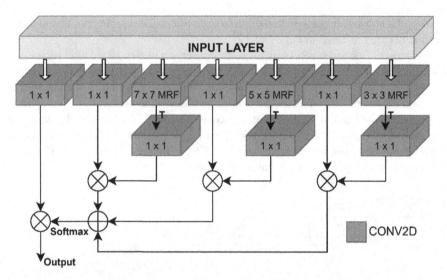

Fig. 3. Details of the MRF block

3.3 Datasets

The UoM dataset contains scans of Arabidopsis leaf samples under three treatments: controlled, cold-stressed, and heat-stressed. These scans were captured by a hyperspectral imaging device that utilises 33 spectral bands that cover a range of 400 to 720 nm [8, 9, 13–15]. For the experiments, five 40×40 patches were cropped per leaf using the application described in [25].

The Bonn dataset was acquired using a line scanning system that spanned the visible and near-infrared regions (400–1000 nm) across 211 wavebands. It consists of controlled, Rust, and Cercospora classes, captured from leaf samples taken under controlled conditions [8, 9, 13–15]. For the experiments with this dataset, nine 40×40 patches were extracted per leaf using the application explained in [25].

The Cassava datasets were acquired using an active MSI system with 15 wavebands (395–880 nm) [16] and the dataset is publicly available on [17]. Images from trials 1, 2, and 3 at 28 dpi were used, and nine 40×40 patches were automatically extracted from each leaf sample using the application explained in [25].

Data augmentation was implemented on all the datasets to increase the number of samples for CNN models and the augmentation pipeline consisted of a combination of transformations, including rotation, width and height shifts, shearing, zooming, and horizontal flipping. Also, before data augmentation, ReliefF was applied to obtain the most ranked 20 channels for both UoM and BONN datasets in order to reduce the number of features to decrease the time requirement.

Table 2. Classification results (%) (w and wo stands for with and without, respectively.)

Experiments		32 Filters	64 Filters	Comparison Results
Dataset	Filter Size	Mean + Std	Mean + Std	Mean + Std
Cassava Trial 3§	3 × 3	87.99 ± 1.47	**93.37 ± 0.90**	93.60 ± 14.80 [16]
	5 × 5	84.41 ± 2.10	92.50 ± 1.38	
	7 × 7	79.92 ± 1.80	92.43 ± 1.38	
Cassava Trial 3*	w MRF	**97.44 ± 1.70**	93.47 ± 11.08	
	wo MRF	85.75 ± 19.26	84.19 ± 20.16	
Cassava Trial 1*	w MRF	**99.87 ± 0.07**	99.87 ± 0.08	87.30 ± 17.40 [16]
	wo MRF	77.05 ± 20.05	78.74 ± 20.61	
Cassava Trial 2*	w MRF	**98.44 ± 1.36**	96.52 ± 6.76	98.50 ± 6.90 [16]
	wo MRF	87.91 ± 17.82	86.90 ± 18.56	
BONN* Cercospora vs. Control	w MRF	**97.59 ± 1.59**	96.88 ± 2.53	Cercospora & Rust vs. Control 93.90 ± 0.01 [15]
	wo MRF	92.54 ± 9.61	92.21 ± 10.22	
BONN* Rust vs. Control	w MRF	**93.97 ± 10.09**	92.61 ± 9.27	
	wo MRF	88.54 ± 16.09	90.84 ± 13.93	
UoM* Cold vs. Control	w MRF	**96.94 ± 10.36**	95.15 ± 14.00	Cold & Heat vs. Control 93.90 ± 0.01 [9]
	wo MRF	88.07 ± 20.58	91.18 ± 18.37	
UoM* Heat vs. Control	w MRF	93.87 ± 14.53	93.98 ± 14.21	
	wo MRF	92.71 ± 16.50	**94.02 ± 14.34**	

§ and * represent Structure 1 and Structure 2, respectively.

4   Experiments and Results

Classification results for the two structures with different sizes and numbers of kernels are presented in Table 2. The mean and standard deviation of these classification results were calculated based on 100 runs. In each run, 20% of the dataset was randomly selected as a test set, while the remaining 80% was used for training purposes. Consistency was rigorously maintained across these runs, ensuring that results remained directly comparable between experiments conducted with and without MRF filters. In both instances, identical test and training sets were utilized to guarantee the comparability of the results obtained.

In the experiments conducted on the Cassava datasets, two different CNN structures were tested using the 3rd trial of the Cassava dataset as explained in the Materials and Methods section. The first structure achieved its best classification result using 64 MRF filters of size 3 × 3; however, the second CNN architecture produced better results, leading to the selection of the second architecture for all remaining experiments. For the second structure, experiments without the implementation of MRF filters resulted in lower classification accuracy and higher standard deviation, while also consuming more time. For instance, in Cassava Trial 3, the accuracy rose from 85.75% to 97.44% and the standard deviation decreased from 19.26 to 1.70 when MRF was employed. This demonstrates the effectiveness of employing MRF filters in this context. As seen in Table 2, the best results for all the Cassava datasets were achieved using the second structure with 32 prefixed MRF filters.

Similarly, for the Bonn dataset, the best classification results were observed when the CNN with 32 prefixed MRF filters were applied. The MRF-based approach improved classification in both Cercospora vs. Control and Rust vs. Control cases. With 32 prefixed MRF filters, the mean accuracy reached 97.59% ($\pm$1.59) for the Cercospora vs. Control and 93.97% ($\pm$10.09) for the Rust vs. Control experiments. In contrast, without prefixed MRF filters, the mean accuracy dropped to 92.54% ($\pm$9.61) for Cercospora vs. Control and 88.54% ($\pm$16.09) for Rust vs. Control.

The best results on the UoM dataset were obtained with the 32 MRF filter applied in the CNN structure as in the other experiments except for the Hot vs. Control experiment with 64 filters which is the only case in the proposed method that produced a lower result. In the Cold vs. Control experiments, employing 32 MRF filters led to an improved mean accuracy of 96.94% ($\pm$10.36) compared to 88.07% ($\pm$20.58) without MRF. Similarly, for Heat vs. Control experiments, a higher mean accuracy of 93.87% ($\pm$14.53%) was achieved with MRF filters in contrast to 92.71% ($\pm$16.50%) without MRF.

The results demonstrate that the proposed MRF-applied CNN consistently led to improved classification performance across various experiments, with higher mean accuracy and relatively lower standard deviations compared to scenarios where MRF was not employed. The other advantage of using MRF filters is a reduction in training time. The time reduction due to prefixed filters may have a big impact on applications that are time-intensive. For instance, the average runtime for a single run of the cassava trials was around 200 s with 32 MRF filters and around 235 s without MRF filters. A similar reduction in training time was observed in the UoM and Bonn experiments. MRF filters decreased the training time for each run by approximately 20 s.

5   Conclusions

In this study, CNN and MRF combination was explored for hyper/multispectral image classification tasks. The results of extensive experiments on various hyper/multispectral datasets, including UoM, BONN, and Cassava, consistently demonstrated the effectiveness of this combination. The enhancement in accuracy was particularly pronounced in Cassava Trial 3, where classification accuracy increased significantly from 85.75% (without MRF filters) to 97.44% (with the use of MRF filters). In terms of the number of kernels, the second CNN structure with 32 MRF filters gives better classification results with lower time costs for all the experiments. It has been shown that MRF is a powerful tool that can be used as a means of modelling and extracting image features using dependencies between neighbourhood pixels. Furthermore, the combination of MRF and CNN demonstrated its potential to reduce training times, making it a time-efficient solution for image classification tasks. This reduction in training time can have practical implications for applications that demand quick results, as evidenced by the decreased runtime observed across multiple trials.

Acknowledgements. Halil Mertkan Sahin would also like to acknowledge the Scholarship provided by the Ministry of National Education of the Republic of Türkiye.

References

1. Hung, C.-C., Song, E., Lan, Y.: Image Texture Analysis. Springer International Publishing, Cham (2019). https://doi.org/10.1007/978-3-030-13773-1
2. Armi, L., Fekri-Ershad, S.: Texture image analysis and texture classification methods – A review. Int. Online J. Image Process. Pattern Recognit. **2**, 1–29 (2019). https://doi.org/10. 48550/arXiv.1904.06554
3. Hammersley, J.M., Clifford, P.: Markov Fields on Finite Graphs and Lattices. Unpubl. Manuscr. 1–26 (1971)
4. Besag, J.: Spatial interaction and the statistical analysis of lattice systems. J. R. Stat. Soc. Ser. B **36**, 192–225 (1974). https://doi.org/10.1111/j.2517-6161.1974.tb00999.x
5. Li, S.Z.: Markov Random Field Modeling in Image Analysis. Springer, London, London (2009). https://doi.org/10.1007/978-1-84800-279-1
6. Sahin, H.M., Miftahushudur, T., Grieve, B., Yin, H.: Segmentation of weeds and crops using multispectral imaging and CRF-enhanced U-Net. Comput. Electron. Agric. **211**, 107956 (2023). https://doi.org/10.1016/j.compag.2023.107956
7. Li, Y., Zhang, H., Shen, Q.: Spectral-spatial classification of hyperspectral imagery with 3D convolutional neural network. Remote Sens. **9**, 67 (2017). https://doi.org/10.3390/rs9010067
8. AlSuwaidi, A., Grieve, B., Yin, H.: Towards spectral-texture approach to hyperspectral image analysis for plant classification. In: Yin, H., et al. (eds.) Intelligent Data Engineering and Automated Learning – IDEAL 2017, pp. 251–260. Springer International Publishing, Cham (2017). https://doi.org/10.1007/978-3-319-68935-7_28
9. Alsuwaidi, A., Grieve, B., Yin, H.: Combining spectral and texture features in hyperspectral image analysis for plant monitoring. Meas. Sci. Technol. **29**, 104001 (2018). https://doi.org/ 10.1088/1361-6501/aad642
10. Peng, Y., Yin, H.: Markov random field based convolutional neural networks for image classification. In: Yin, H., et al. (eds.) IDEAL 2017. LNCS, vol. 10585, pp. 387–396. Springer, Cham (2017). https://doi.org/10.1007/978-3-319-68935-7_42
11. Peng, Y., Hankins, R., Yin, H.: Data-independent feature learning with Markov random fields in convolutional neural networks. Neurocomputing **378**, 24–35 (2020). https://doi.org/10. 1016/j.neucom.2019.03.107
12. Geng, L., Sun, J., Xiao, Z., Zhang, F., Wu, J.: Combining CNN and MRF for road detection. Comput. Electr. Eng. **70**, 895–903 (2018). https://doi.org/10.1016/j.compeleceng.2017. 11.026
13. Alsuwaidi, A., Veys, C., Hussey, M., Grieve, B., Yin, H.: Hyperspectral selection based algorithm for plant classification. In: IST 2016 – 2016 IEEE International Conference Imaging System Techniques Proceedings, pp. 395–400 (2016). https://doi.org/10.1109/IST.2016.773 8258
14. Alsuwaidi, A., Veys, C., Hussey, M., Grieve, B., Yin, H.: Hyperspectral feature selection ensemble for plant classification. In: Hyperspectral Imaging Applications (HSI 2016) (2016)
15. AlSuwaidi, A., Grieve, B., Yin, H.: Feature-ensemble-based novelty detection for analyzing plant hyperspectral datasets. IEEE J. Sel. Top. Appl. Earth Obs. Remote Sens. **11**(4), 1041–1055 (2018). https://doi.org/10.1109/JSTARS.2017.2788426
16. Peng, Y., et al.: Early detection of plant virus infection using multispectral imaging and spatial–spectral machine learning. Sci. Rep. **12**, 3113 (2022). https://doi.org/10.1038/s41598-022-06372-8

17. Peng, Y., Dallas, M.M., Ascencio-Ibáñez, J.T., Hoyer, J.S., Legg, J., Hanley-Bowdoin, L., Grieve, B., Yin, H.: Cassava-TME204-UCBSV multispectral imaging dataset for early detection of virus infection with spatial-spectral machine learning. Zenodo (2021). https://doi.org/10.5281/zenodo.4636968

18. Cross, G.R., Jain, A.K.: Markov random field texture models. IEEE Trans. Pattern Anal. Mach. Intell. **PAMI-5**(1), 25–39 (1983). https://doi.org/10.1109/TPAMI.1983.4767341

19. Rawat, W., Wang, Z.: Deep convolutional neural networks for image classification: a comprehensive review. Neural Comput. **29**, 2352–2449 (2017). https://doi.org/10.1162/neco_a_00990

20. Szirányi, T., Kriston, A., Majdik, A., Tizedes, L.: Fusion markov random field image segmentation for a time series of remote sensed images. In: Faragó, I., Izsák, F., Simon, P.L. (eds.) Progress in Industrial Mathematics at ECMI 2018, pp. 621–629. Springer International Publishing, Cham (2019). https://doi.org/10.1007/978-3-030-27550-1_79

21. Patil, A., Rane, M.: Convolutional neural networks: an overview and its applications in pattern recognition. Smart Innov. Syst. Technol. **195**, 21–30 (2021). https://doi.org/10.1007/978-981-15-7078-0_3

22. LeCun, Y., Bottou, L., Bengio, Y., Haffner, P.: Gradient-based learning applied to document recognition. Proc. IEEE **86**, 2278–2323 (1998). https://doi.org/10.1109/5.726791

23. Cohen, G., Afshar, S., Tapson, J., Van Schaik, A.: EMNIST: extending MNIST to handwritten letters. In: Proceedings of the International Joint Conference on Neural Networks, pp. 2921–2926 (2017). https://doi.org/10.1109/IJCNN.2017.7966217

24. Krizhevsky, A.: Learning Multiple Layers of Features from Tiny Images (2009)

25. Sahin, H.M., Grieve, B., Yin, H.: Automatic multispectral image classification of plant virus from leaf samples. In: Analide, C., Novais, P., Camacho, D., Yin, H. (eds.) Intelligent Data Engineering and Automated Learning – IDEAL 2020: 21st International Conference, Guimaraes, Portugal, November 4–6, 2020, Proceedings, Part I, pp. 374–384. Springer International Publishing, Cham (2020). https://doi.org/10.1007/978-3-030-62362-3_33

Plant Disease Detection and Classification Using a Deep Learning-Based Framework

Mridul Ghosh[1(✉)], Asifuzzaman Lasker[2], Poushali Banerjee[3], Anindita Manna[3], Sk Md Obaidullah[2], Teresa Gonçalves[4], and Kaushik Roy[5]

[1] Department of Computer Science, ShyampurSiddheswariMahavidyalaya, Howrah, India
mridulxyz@gmail.com
[2] Department of Computer Science and Engineering, Aliah University, Kolkata, India
[3] Department of Computer Science, Surendranath College, Kolkata, India
[4] Departamento de Informática, Universidade de Évora, Evora, Portugal
[5] Department of Computer Science, West Bengal State University, Kolkata, India

Abstract. Plant diseases pose a significant threat to agriculture, causing substantial yield losses and economic damages worldwide. Traditional methods for detecting plant diseases are often time-consuming and require expert knowledge. In recent years, deep learning-based approaches have demonstrated great potential in the detection and classification of plant diseases. In this paper, we propose a Convolutional Neural Network (CNN) based framework for identifying 15 categories of plant leaf diseases, focusing on Tomato, Potato, and Bell pepper as the target plants. For our experiments, we utilized the publicly available PlantVillage dataset. The choice of a CNN for this task is justified by its recognition as one of the most popular and effective deep learning methods, especially for processing spatial data like images of plant leaves. We evaluated the performance of our model using various performance metrics, including accuracy, precision, recall, and F1-score. Our findings indicate that our approach outperforms state-of-the-art techniques, yielding encouraging results in terms of disease identification accuracy and classification precision.

Keywords: Plant disease · Deep learning · CNN · Leaf disease classification

1 Introduction

Plant diseases pose significant threats to agricultural productivity, food security, and the global economy [1]. Detecting and diagnosing plant diseases at an early stage is crucial for effective disease management and prevention. Over the years, advances in technology and the application of various scientific methods have greatly improved the process of plant disease detection. The detection involves identifying and determining the presence of pathogens, such as bacteria, fungi, viruses, and other harmful microorganisms, that can cause diseases in plants. Timely and accurate detection enables farmers, plant pathologists, and researchers to take appropriate measures to mitigate the spread and impact of diseases, thus minimizing crop losses and ensuring sustainable agricultural practices.

© The Author(s), under exclusive license to Springer Nature Switzerland AG 2023
P. Quaresma et al. (Eds.): IDEAL 2023, LNCS 14404, pp. 39–50, 2023.
https://doi.org/10.1007/978-3-031-48232-8_5

Traditional methods of plant disease detection primarily relied on visual observations of symptoms exhibited by the infected plants. These symptoms could include wilting, discoloration, leaf spots, abnormal growth patterns, and various other physical changes. While visual inspection remains a valuable tool, it is often limited by the subjectivity of human observation and the difficulty in differentiating between similar symptoms caused by different pathogens. In recent years, technological advancements have revolutionized plant disease detection by providing more precise, rapid, and reliable methods. Here are some of the key techniques and tools used in modern plant disease detection:

1. Molecular Techniques: Polymerase Chain Reaction (PCR), DNA sequencing, and other molecular biology methods [2] are widely employed for the identification and characterization of plant pathogens. These techniques enable the detection of specific DNA or RNA sequences unique to a particular pathogen, allowing for highly accurate and targeted diagnosis.
2. Immunoassays: Enzyme-linked immunosorbent Assays (ELISA) and other immuno-logical techniques [3] are utilized to detect the presence of plant pathogens based on the specific immune response generated by the host plants. These tests rely on the recognition and binding of pathogen-specific antigens by antibodies, providing a sensitive and specific detection method.
3. Machine learning (ML) and Artificial Intelligence (AI): By leveraging ML and AI-based models [5–8], large datasets of plant images, genomic sequences, and environmental parameters can be analysed to develop predictive models for disease detection. These models can identify patterns, correlations, and anomalies that may not be apparent to human observers, improving the accuracy and efficiency of detection.

By using machine learning-based work, the performance of classifying the diseases obtained is high and the time of classification is much less in comparison with other techniques. For this, in this work, a deep learning-based [9–11] framework is designed to classify the Tomato and Potato leaf diseases. Here, CNN [12, 13] based model is developed for the classification of Tomato, Potato, and Bell pepper plant diseases.

Kaur et al. [14] employed a CNN-based framework to detect diseases in tomato, potato, and grape leaves. They utilized a dataset containing more than 4,000 images of diseased and healthy leaves and trained several CNN models to classify the images as either diseased or healthy. The best-performing model achieved an accuracy of 98.5%.

In [15], the authors proposed a technique that involves acquiring images of healthy and diseased tomato plants, pre-processing the images to remove noise and unwanted regions, extracting features from the pre-processed images, selecting relevant features, and training an SVM classifier to categorize the tomato images as healthy or diseased. The proposed technique was evaluated on a dataset of tomato images with four diseases: Bacterial Spot, Early Blight, Late Blight, and Septoria Leaf Spot. The results demonstrated an accuracy of 94.7% in detecting tomato diseases using the proposed technique. Tiwari et al. [16] developed an automated system to diagnose and classify diseases like early blight, late blight, and healthy conditions in potato leaves, offering a novel solution. The results demonstrated an accuracy of 97.8% over the test dataset, with improvements of 5.8% and 2.8%. Srinivasan et al. [17] proposed an image categorization technique to identify healthy and unhealthy leaves from a multilevel image dataset. Additionally, it identifies the specific type of disease affecting the unhealthy leaves. Using the CNN technique, they extracted 39 types of diseases in 13 crop species from the PlantVillage image dataset, achieving an accuracy of 98.75% at epoch 25.

2   Material and Method

Deep learning models based on CNN frameworks [9, 10] are specifically designed for processing and analyzing visual data, such as images. These models utilize layers that play a vital role in extracting significant features and patterns from raw input images. Here are some key CNN layers:

i. Convolutional Layer: The convolutional layer applies a set of filters or kernels to the input image, performing a mathematical operation known as convolution. This operation detects local patterns by sliding the filters across the image and calculating dot products between filter weights and pixel values. Convolutional layers are capable of capturing features like edges, textures, and shapes.

ii. Pooling Layer: Pooling layers reduce the spatial dimensions of the feature maps obtained from convolutional layers. The most common pooling operation is max pooling, which downsamples the input by selecting the maximum value within a predefined neighborhood. Pooling helps to achieve spatial invariance and reduces the computational burden by summarizing the most salient features.

iii. Fully Connected/dense Layer: The fully connected layer is typically placed after the convolutional and pooling layers. It connects all neurons from the previous layer to every neuron in the current layer. It captures high-level abstractions by combining the features learned from previous layers. The output of this layer is often fed into a Soft-Max activation function for classification.

2.1   Dataset

In this study, we utilized the publicly available PlantVillage dataset, curated by Hughes & Salathe [18] (https://www.kaggle.com/datasets/emmarex/plantdisease last visited on 31.10.2023), for the purpose of detecting plant leaf diseases. The PlantVillage dataset comprises a total of 20,639 images, thoughtfully organized into two categories: diseased and healthy leaves of various plant species. These images have been expertly classified by specialists in the field of plant pathology. The dataset primarily focuses on three plant species: Bell Pepper, Potato, and Tomato. The photographs were taken with leaves positioned against a paper sheet, which provided a consistent gray or black background. Within the PlantVillage dataset, there are 15 distinct classes of diseases, as listed in Table 1. For visual reference, a selection of sample images from the dataset can be observed in Fig. 1.

2.2   Data Pre-Processing

The pixel values of the images are rescaled to be in the range of 0 to 1 by applying the rescale argument with a value of 1/255. This step ensures that the pixel values are normalized, which can help improve the training process. The dataset is split into training and validation sets, this split allows for evaluating the model's performance on unseen

Table 1. The details specification of PlantVillage dataset [18].

Plant names	Disease names	No. of images
Bell pepper	Healthy	1478
	Diseased: Bacterial spot	997
Potato	Healthy	152
	Diseased: Early Blight	1000
	Diseased: Late Blight	1000
Tomato	Healthy	1591
	Diseased: Mosaic Virus	373
	Diseased: Bacterial spot	2127
	Diseased: Early Blight	1000
	Diseased: Late Blight	1909
	Diseased: Leaf Mold	952
	Diseased: Septoria Leaf Spot	1771
	Diseased: Two Spotted Spider mites	1676
	Diseased: Target Spot	1404
	Diseased: Yellow Leaf Curl Virus	3209

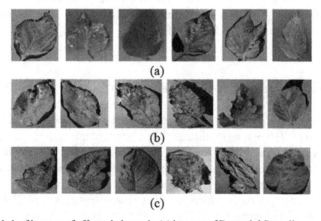

Fig. 1. Sample leaf images of affected plants: in (a) images of Bacterial Spot disease, in Bellpepper (b) Early Blight, in Potato and (c) Late Blight Tomato.

data. This resizing ensures that all input images have consistent dimensions. It helps in standardizing the input size for the model. Here, the images are resized to a target size of (224, 224). These pre-processing steps ensure that the input images are appropriately rescaled and resized before being fed into the model for training and evaluation.

2.3 Proposed Model

Plants are vulnerable to a range of disease-related disorders and infections. These issues can arise from various causes, including disturbances caused by environmental conditions such as temperature, humidity, inadequate or excessive nutrients, and light exposure. Additionally, plants commonly face bacterial, viral, and fungal diseases.

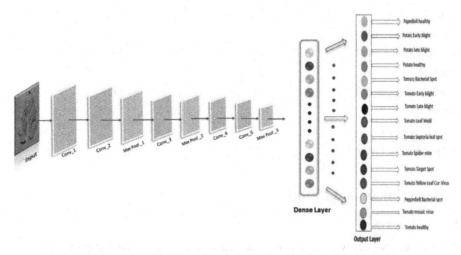

Fig. 2. The proposed C5P3D2 architecture for classification of the plant diseases

To classify the fifteen categories of potato and tomato plant disease, we propose a plant disease classification architecture called C5P3D2 which is shown in Fig. 2. This architecture consists of 5 Convolutional layers, 3 max-pooling layers, and 2 dense layers. The input image is passed through the first Convolutional layer with 128 filters of size 5 × 5. The resulting feature map is then fed into another Convolutional layer with a filter size of 64 of size 3 × 3. The output of this Convolutional layer is further processed by a max-pooling layer of filter size 2 × 2. The output of this pooling layer is further processed by another Convolutional layer with 32 filters of size 3 × 3. After this the resulting feature map is then fed into another max-pooling layer of filter size 2 × 2. Next, the output from the previous layer is inputted into a pair of Convolutional layers with filter sizes of 3 × 3 and 16, and 3 × 3 and 16, respectively. Each Convolutional layer captures different aspects of the input features. After each Convolutional layer, a max-pooling layer with a filter size of 3 × 3 is applied. Two batch normalization layers are utilized before the dense 1 and dense2 to enhance the stability and convergence of the network. Subsequently, dropout regularization is employed with dropout rates of 0.5 after the dense layers. Two dense layers are included in the architecture with sizes of 128 and 15, respectively. These layers perform weighted sums of the input features. Finally, the output layer consists of 15 neurons, representing the classification categories. The generated parameters in our model are tabulated in Table 2.

Table 2. The generated parameters in C5P3D2 model

Layers	Output dimension	#Parameter
Convolution_1	(220,220,128)	9,728
Convolution_2	(218,218,64)	73,792
Convolution_3	(107,107,52)	18,464
Convolution_4	(51,51,16)	4,624
Convolution_5	(49,49,16)	2,320
Batchnormalization_1	9216	36,864
Dense_1	128	11,79,776
Batchnormalization_2	128	512
Dense_2	15	1,935
Total Parameters	1,328,015	
Total Trainable Parameters	1,309,327	

3 Experiment and Results

3.1 System Setup

Our proposed system was tested and assessed on a GPU machine equipped with dual Intel(R) Core(TM) i5-10400H CPUs, operating at a clock speed of 2.60 GHz, accompanied by 32 GB of RAM, and powered by two NVIDIA Quadro RTX 5000 GPUs. The machine ran on Windows 10 Pro OS version 20H2 and utilized TensorFlow 2.0.0 for training and inference of deep learning models.

3.2 Training Regime

The dataset is split into train and test sets using different ratios. The dataset is split into train-test sets in CNN training to evaluate model performance on unseen data, prevent overfitting, and optimize hyperparameters. Training set used for learning, test set for unbiased evaluation. Specifically, the ratios used are 8:2, 7:3, and 6:4. These ratios determine the proportion of images allocated to the training and testing sets. For the 8:2 ratios, out of the total image pool of 20,639, 16,512 images are assigned to the training set, while 4,127 images are allocated to the testing set. For the 7:3 ratios, 14,444 images are used for training, and 6,192 images are used for testing. For the 6:4 ratios, 12,383 images are used for training, and 8,256 images are used for testing. The training and testing sets are used independently of each other to avoid data leakage. This means that the models are trained only on the training set and evaluated only on the testing set.

3.3 Evaluation Protocols

There are four different evaluation protocols are considered in this work. Apart from accuracy precision, recall, and F1Score are measured to evaluate the model's performance. The Equations of the corresponding evaluation parameters are presented below:

$$\text{Accuracy} = (TruePositivies + TrueNegatives)/(TruePositivies + \\ TrueNegatives + FalsePostivies + FalseNegatives) \tag{1}$$

$$\text{Precision} = TruePositivies/(TruePositivies + FalsePositivies) \tag{2}$$

$$\text{Recall} = TruePositives/(TruePositives + FalseNegatives) \tag{3}$$

$$\text{F1 score} = 2 * (precision * recall/precision + recal) \tag{4}$$

In our C5P3D2 architecture, Categorical Cross-Entropy Loss function was used which can be described by Eq. 5.

$$L = -N_1 \sum_{i=1}^{N} \sum_{j=1}^{C} y_{ij}\log(p_{ij}) \tag{5}$$

Here, N is the number of samples in the batch, is the number of classes in the classification problem, y_{ij} is an indicator function that is 1 if the sample i belongs to class j and 0 otherwise, and p_{ij} is the predicted probability that sample i belongs to class j.

3.4   Result and Discussions

3.4.1   Ablation Study

We performed a rigorous experiment to establish a suitable framework for our problem. We have used different combinations of convolutional, pooling, and dense layers along with the different dimensional filters. In Table 3 the outcome of these experiments is tabulated considering 8:2 train-test set, batch size of 32, and 100 epoch trial runs. Finally, C5P3D2 architecture was considered for our work.

Table 3. Ablation study to build a CNN-based framework to get higher performance

Architecture	Convolutional	Pooling	Dense	Accuracy (%)
CPD	5×5	3×3	256	81.54
CCPPDD	$5 \times 5, 3 \times 3$	$3 \times 3, 2 \times 2$	256,128	84.02
CPPDD	3×3	$2 \times 2, 2 \times 2$	256, 128	85.05
CCPDD	$5 \times 5, 3 \times 3$	3×3	256, 128	79.52
CCPPD	$3 \times 3, 5 \times 5$	$3 \times 3, 2 \times 2$	256	86.35
C5P3D2	$5 \times 5, 3 \times 3, 3 \times 3, 3, 3 \times 3$	$2 \times 2, 3 \times 3, 2 \times 2$	128, 15	94.95

3.4.2 Our Results

In Table 4 the performance of our C5P3D2 architecture was measured by changing the epoch values of 20, 50, and 100 for different train-test ratios and batch size of 32 at the beginning. The maximum accuracy of 94.95% has been obtained by running 100 epochs in 80:20 ratios. Considering 100 epochs, the rest of the experiments were performed. Figure 3 represents the training performance of C5P3D2 architecture with training and validation curves.

Table 4. The accuracies for different train-test ratios for 20, 50 and 100 epochs run

Train-test Ratio	Epoch	Accuracy (%)
80:20	20	77.58
	50	94.00
	100	94.95
70:20	20	93.76
	50	94.45
	100	94.62
60:40	20	91.85
	50	92.43
	100	94.13

From Table 5 it is observed that in 100 epochs the highest accuracy and precision are obtained 94.95% and 94.94%, respectively. So, we keep this epoch constant for the rest of our experiment. Considering the epoch value as 100 and batch size 32, different train-test ratios were considered. We further investigated by changing the batch size of 32, 64, and 128 considering the 100 epochs and 80:20 train-test ratio which is tabulated in Table 6. It is seen that there are improvements in accuracy, precision, recall, and f-score, respectively. The train-test accuracy and loss curves for 100 epochs 128 batch size and 80:20 train test ratio are shown in Figs. 3 (a) and (b), respectively.

Table 5. For the 8:2 train-test ratio the performance measurement by changing the epochs

Epochs	Accuracy	Precision	Recall	F1-Score
20	77.85%	82.35%	77.58%	75.07%
50	94%	94%	94%	93%
100	94.95%	94.94%	94.95%	94.91%

Table 6. For 100 epochs, 8:2 train-test ratio for different batch size

Batch size	Accuracy	Precision	Recall	F1-score
32	94.95%	94.94%	94.95%	94.91%
64	95.41%	95.43%	95.41%	94.46%
128	**95.51%**	**95.51%**	**95.24%**	**95.47%**

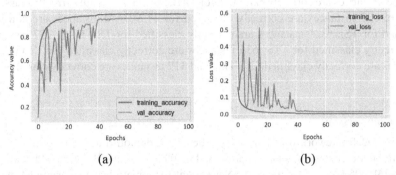

(a)   (b)

Fig. 3. Training and validation accuracy (a) and loss curves (b) for 100 epochs 128 batch size and 80:20 train test ratio.

Table 7. The confusion matrix of the highest result i.e., corresponding to the accuracy of 95.51% for 128 batch size in 100 epochs of our proposed model is presented

	PbBs	Pbh	PEb	PLb	Ph	TBs	TEb	TLb	TLM	TSls	TSmTssm	TTS	TTYCV	TTmv	Th
PbBs	185	5	0	1	0	1	0	0	0	5	0	1	0	0	0
Pbh	1	292	0	0	0	0	0	1	0	1	0	0	0	0	0
PEb	0	0	195	1	0	0	0	4	0	0	0	0	0	0	0
PLb	0	0	1	193	0	0	0	6	0	0	0	0	0	0	0
Ph	0	0	0	5	25	0	0	0	0	0	0	0	0	0	0
TBs	0	0	0	1	0	417	1	1	0	1	0	0	4	0	0
TEb	1	0	1	1	0	6	156	16	1	2	0	12	4	0	0
TLb	1	2	1	3	0	0	7	362	3	0	0	1	1	0	0
TLM	2	0	0	0	0	0	0	3	177	7	1	0	0	0	0
TSls	1	1	2	1	0	2	5	3	7	330	0	1	0	1	0
TSmTssm	0	0	0	1	0	0	1	0	0	0	321	11	0	1	0
TTs	0	0	0	1	1	1	1	0	2	5	10	257	0	1	1
TTYCV	0	0	0	0	0	2	0	2	0	0	2	0	635	0	0
TTmv	0	0	0	0	0	0	0	0	0	0	0	0	0	74	0
Th	0	0	0	0	0	0	0	0	0	0	0	1	0	0	317

Here full names corresponding to the acronyms are:
PbBs = Peeper_bell_Bacterial_spot, Pbh = Peeper_bell_healthy,
PEb = Potato_Early_blight, b = Poato_Late_blight, Ph = Potato_healthy,
Tbs = Tomato_Bacterial_spot, Tb = Tomato_Early_blight,
TLb = Tomato_Late_blight,TLM = Tomato_Leaf_Mold,

Tls = Tomato_Septoria_leaf_spot, TSmTssm = Tomato_Spider_mites_Two _spotted_spider_mites, TTS = Tomato_Target_Spot, TTYCV = Tomato_Tomato_YellowLeaf_Curl_Virus, TTmv = Tomato_Tomato_mosaic _virus, Th = Tomato_healthy.

It is observed from the confusion matrix shown in (Table 7). That for the class PbBs, 185 instances are correctly classified, 292 instances are correctly classified for Pbh,, 195 instances are correctly classified for Potato_Eb, 193 instances are correctly classified for PLb, 25 instances are correctly classified for Ph, 417 instances are correctly classified for TBs, 156 instances are correctly classified for TEB, 362 instances are correctly classified for TLb, 177 instances are correctly classified for TLM, 330 instances are correctly classified for TSls, 321 instances are correctly classified for TSmTssm, 257 instances are correctly classified for TTS, 635 instances are correctly classified for TTYCV, 74 instances are correctly classified for TTmv and 317 instances are correctly classified for Th.

3.4.3   Comparison

Our C5P3D2 framework for plant disease detection demonstrates competitive results compared to other studies, which is presented in Table 8. While Prajwala TM et al. [19] achieved 94–95% accuracy with a LeNet variant for tomato leaves, our model, which extends to multiple plants, shows a gain in accuracy of more than 1%. Salih et al.'s [20] CNN model yielded a gain in accuracy of less than 1%, despite using several levels of deep learning approaches. Basavaiah and Anthony's [21] multifeature fusion method did not achieve a significant improvement in accuracy compared to our approach. It is seen that our method produced 18.5% higher accuracy compared to the MobileNetV4.

Table 8. Comparison of our method with standard CNN on the PlantVillage dataset

Methods	Accuracy	Recall	Precision	F1-Score
VGG-16	52.50%	51.50%	50.02%	52.49%
MobileNetV3	71.40%	56.10%	70.41%	82.43%
MobileNetV4	77.01%	66.05%	65.02%	66.50%
Proposed	95.51%	95.51%	94.24%	95.47%

4   Conclusion and Future Scope

We have designed a C5P3D2 framework for the detection of plant diseases in pepper bell, tomato, and potato leaves. The system successfully takes image inputs from the user and provides output indicating the detected disease. This enables farmers to take appropriate preventive measures and use the correct pesticides. This proposed framework can be extended to other crops that suffer from diseases, given the availability of a sufficiently large dataset for that specific crop. In the future, we will consider improving the performance of the method by utilizing more advanced deep learning-based techniques. We plan to deploy the system as a GUI-based platform. The web interface may also include a forum for farmers to discuss the current trends they face in different diseases. The future of plant disease detection using CNNs involves expanding and di-versifying training datasets, optimizing CNN architectures for enhanced performance, integrating real-time monitoring technologies, exploring multimodal analysis techniques, improving the interpretability of CNN models, and addressing practical challenges for widespread deployment in agricultural settings.

References

1. Anand, G., Rajeshkumar, K.C.: Challenges and threats posed by plant pathogenic fungi on agricultural productivity and economy. In: Fungal Diversity, Ecology and Control Management, pp. 483–493. Springer Nature, Singapore (2022)
2. Joshi, M., Deshpande, J.D.: Polymerase chain reaction: methods, principles and application. Int. J. Biomed. Res. 2(1), 81–97 (2010)
3. Tijssen, P., Adam, A.: Enzyme-linked immunosorbent assays and developments in techniques using latex beads. Curr. Opin. Immunol. 3(2), 233–237 (1991)
4. Flores, A.M., Demsas, F., Leeper, N.J., Ross, E.G.: Leveraging machine learning and artificial intelligence to improve peripheral artery disease detection, treatment, and outcomes. Circ. Res. 128(12), 1833–1850 (2021)
5. Ghosh, M., Obaidullah, S.M., Gherardini, F., Zdimalova, M.: Classification of geometric forms in mosaics using deep neural network. J. Imaging 7(8), 149 (2021)
6. Ghosh, M., Mukherjee, H., Obaidullah, S.M., Roy, K.: STDNet: a CNN-based approach to single-/mixed-script detection. Innovations Syst. Softw. Eng. 17(3), 277–288 (2021)
7. Ghosh, M., Roy, S.S., Mukherjee, H., Obaidullah, S.M., Santosh, K.C., Roy, K.: Understanding movie poster: transfer-deep learning approach for graphic-rich text recognition. The Visual Comput. 38(5), 1645–1664 (2021). https://doi.org/10.1007/s00371-021-02094-6
8. Ghosh, M., Mukherjee, H., Obaidullah, S.M., Santosh, K.C., Das, N., Roy, K.: LWSINet: a deep learning-based approach towards video script identification. Multimed. Tools Appl. 80(19), 29095–29128 (2021)
9. Ghosh, M., Roy, S.S., Mukherjee, H., Obaidullah, S.M., Gao, X.Z., Roy, K.: Movie title extraction and script separation using shallow convolution neural network. IEEE Access 9, 125184–125201 (2021)
10. Lasker, A., Ghosh, M., Obaidullah, S.M., Chakraborty, C., Roy, K.: LWSNet-a novel deep-learning architecture to segregate Covid-19 and pneumonia from x-ray imagery. Multimed. Tools Appl. 82(14), 21801–21823 (2023)
11. Lasker, A., Ghosh, M., Obaidullah, S.M., Chakraborty, C., Goncalves, T., Roy, K.: Ensemble stack architecture for lungs segmentation from X-ray images. In: Yin, H., Camacho, D., Tino, P. (eds.) IDEAL 2022. LNCS, vol. 13756, pp. 3–11. Springer, Cham (2022). https://doi.org/10.1007/978-3-031-21753-1_1
12. Lasker, A., Ghosh, M., Obaidullah, S.M., Chakraborty, C., Roy, K.: A deep learning-based framework for COVID-19 identification using chest X-Ray images. In: Advancement of Deep Learning and its Applications in Object Detection and Recognition, pp. 23–46. River Publishers (2023)
13. Ghosh, M., Roy, S.S., Mukherjee, H., Obaidullah, S.M., Santosh, K.C., Roy, K.: Automatic text localization in scene images: a transfer learning based approach. In: Babu, R.V., Prasanna, M., Namboodiri, V.P. (eds.) NCVPRIPG 2019. CCIS, vol. 1249, pp. 470–479. Springer, Singapore (2020). https://doi.org/10.1007/978-981-15-8697-2_44

14. Kaur, S., Pandey, S., Goel, S.: Plants disease identification and classification through leaf images: a survey. Arch. Comput. Methods Eng. **26**, 507–530 (2019)

15. Vetal, S., Khule, R.S.: Tomato plant disease detection using image processing. Int. J. Adv. Res. Comput. Commun. Eng. **6**(6), 293–297 (2017)

16. Tiwari, D., Ashish, M., Gangwar, N., Sharma, A., Patel, S., Bhardwaj, S.: Potato leaf diseases detection using deep learning. In 2020 4th International Conference on Intelligent Computing and Control Systems (ICICCS), pp. 461–466. IEEE (2020)

17. Srinivasan, R., Santhanakrishnan, C., Iniyan, S., Subash, R., Sudhakaran, P.: CNN-based plant disease identification in crops from multilabel images using contextual regularization. J. Surv. Fish. Sci. **10**(2S), 522–531 (2023)

18. Hughes, D., Salathé, M.: An open access repository of images on plant health to enable the development of mobile disease diagnostics. arXiv preprint arXiv:1511.08060 (2015)

19. Tm, P., Pranathi, A., SaiAshritha, K., Chittaragi, N.B., Koolagudi, S.G.: Tomato leaf disease detection using convolutional neural networks. In: 2018 Eleventh International Conference on Contemporary Computing (IC3), pp. 1–5. IEEE (2018)

20. Salih, T.A.: Deep learning convolution neural network to detect and classify tomato plant leaf diseases. Open Access Libr. J. **7**(05), 1 (2020)

21. Basavaiah, J., Arlene Anthony, A.: Tomato leaf disease classification using multiple feature extraction techniques. Wirel. Pers. Commun. **115**(1), 633–651 (2020)

Evaluating Text Classification in the Legal Domain Using BERT Embeddings

José Alfredo F. Costa(✉)(iD), Nielsen Castelo D. Dantas(iD),
and Esdras Daniel S. A. Silva(iD)

Universidade Federal do Rio Grande do Norte, Natal, RN 59078-900, Brazil
alfredo.costa@ufrn.br

Abstract. Brazil's justice system faces grave case backlogs stemming from surging caseloads. This research explores automated text classification to expedite legal document analysis. Supervised machine learning approaches leveraging BERT embeddings fine-tuned on Brazilian legal text were evaluated using a 30,000 document dataset encompassing ten motion types from the Rio Grande do Norte Court of Justice. Documents were encoded into semantic representations via a BERT model adapted to local jurisprudence. The resulting optimized embeddings were used to train and benchmark models including KNN, Naive Bayes, SVM, neural networks, CNNs, and deep learning architectures. Despite BERT's state-of-the-art capabilities, TF-IDF outperformed neural techniques across considered metrics. High similarity between certain classes was hypothesized to hinder BERT's contextual embedding.

Keywords: Legal text classification · Bert embeddings · Natural language processing · Machine learning

1 Introduction

The digital revolution has led to an explosion of data across all sectors, including legal systems. This deluge of information poses major challenges for countries like Brazil, where the judicial system is struggling to effectively store and process the surging volume of cases. According to the National Justice Council's annual Justice in Numbers report [9], the number of cases pending in the Brazilian Judiciary in 2021 was 77.3 million. Among them, 15.3 million (19.8% of the total) are suspended, stayed or temporarily archived cases, awaiting some future legal situation. The number of new cases in 2021 was 27.7 million, which represents a 10.4% increase compared to 2020.

With the Brazilian judiciary's widespread adoption of digital platforms like PJe for fully electronic case management, the majority of this legal data is now digitized. This combination of a staggering case volume and comprehensive digital records creates ideal conditions for applying machine learning techniques like supervised text classification that can enhance efficiency. Some recent Brazilian initiatives have been described in a study that presents an in-depth analysis of

P. Quaresma et al. (Eds.): IDEAL 2023, LNCS 14404, pp. 51–63, 2023.
https://doi.org/10.1007/978-3-031-48232-8_6

artificial intelligence (AI) systems in Brazil's judicial system, focusing on projects at the National Council of Justice (CNJ) and key federal courts [14].

Classification methods that automatically categorize cases based on their textual content could help streamline document routing, analysis and decision-making. By training algorithms to assign case documents to relevant judicial departments and procedural classes, the justice system can alleviate pressures from overloaded dockets. Automated text classification has the potential to accelerate processing, reduce duplicated work, and increase access to timely rulings.

Text classification has emerged as a promising technique to categorize legal documents according to relevant topics, with recent research exploring advanced NLP methods like BERT embeddings for legal text classification [7]. However, few studies have evaluated BERT for legal text classification in Portuguese. This paper employs NLP and BERT embedding techniques to categorize 30,000 Brazilian legal documents from Tribunal de Justiça do Rio Grande do Norte (TJRN) [3]. Documents were encoded into semantic representations using a BERT model fine-tuned on local jurisprudence. The resulting embeddings were used to train and compare KNN, Naive Bayes, SVM, Neural Networks, Convolutional Networks, Deep Learning, and TF-IDF models, evaluated on accuracy, F1, confusion matrix analysis, and other metrics. Legal texts present linguistic challenges due to length, terminology, and argument complexity, requiring advanced NLP methods to uncover semantic relationships within large legal corpora [7,34].

The rest of the paper is structured as follows. The next section provides related works on text classification and BERT derived models. Section 3 describes materials and methods. Results and discussions are presented in Sect. 4 and Sect. 5 presents conclusion, limitations and possible extensions of this work.

2   Related Works

2.1   Natural Language Processing for Legal Document Analysis and Classification

The application of natural language processing for analyzing and classifying legal texts has been an active area of research since the 1990s [23]. Early work focused on using knowledge engineering and rules-based systems to classify legal documents, extract key information, and generate summaries [13,14]. These systems were limited in their flexibility and dependence on manual encoding of rules. With the rise of statistical and machine learning approaches to NLP, researchers began exploring techniques like naive Bayes, support vector machines (SVMs), and neural networks for legal text analysis tasks [5,19]. Key challenges in legal text processing include the complex ontology, domain-specific terminology, and nuanced semantics [4]. A variety of features and representations have been proposed to address these challenges, from bag-of-words and TF-IDF to more advanced embeddings learned from the legal corpora [13,21]. Topic modeling approaches like latent Dirichlet allocation (LDA) have also been effective for capturing thematic patterns in legal texts. More recent work has leveraged deep learning methods like recurrent neural networks (RNNs), convolutional neural networks (CNNs) and attention mechanisms to achieve state-of-the-art results in legal document classification [20].

Viegas et al. [30] proposed an innovative approach for pre-training BERT models with Brazilian legal texts. The JurisBERT model generated by the authors is used in the present paper to extract the textual embeddings presented. The pre-training of JurisBERT with legislation, jurisprudence and legal doctrine allowed obtaining textual representations more adequate to the legal domain, resulting in significant improvements in the performance of the semantic textual similarity task between acórdão summaries. The use of pre-trained in this work aims to take advantage of its enhanced representational capabilities for legal text in Portuguese, generating richer and more informative embeddings to feed the textual clustering algorithms. Therefore, the model proposed by Viegas et al. [30] is employed here as a fundamental resource to enable semantic similarity analysis in the jurisprudential domain.

2.2   Embedding Models

BERT leverages the Transformer architecture and its self-attention mechanism to efficiently capture long-range dependencies in text. A key innovation is BERT's bidirectional pretraining, unlike left-to-right or right-to-left training of prior models. By conditioning on both preceding and succeeding context, BERT derives richer contextualized word embeddings. After pretraining on large corpora, BERT is fine-tuned on downstream tasks via labeled data to adapt the pretrained representations to target domain specifics in a transfer learning approach. This enables better performance with less task-specific data. BERT's bidirectional pretraining and fine-tuning overcome limitations of sequential architectures like RNNs and CNNs for NLP. This drives BERT's state-of-the-art capabilities on tasks like text classification, named entity recognition, and question answering. All models can be downloaded at https://huggingface.co/alfaneo.

MBert. The multilingual BERT base model [11], which contains 110 million parameters and a vocabulary of 120,000 tokens. It was pre-trained on Wikipedia in 104 languages, totaling over 2.5 billion training tokens. This multilingual model allows inference in more than 100 different languages.

Bertimbau. The BERT-Base model bert-base-portuguese-cased ([28]) was trained exclusively on a 1 billion token Portuguese corpus from Wikipedia and news websites, better capturing linguistic nuances compared to multilingual models. It outperformed the multilingual BERT on Portuguese NLP tasks like sentiment analysis, cloze test, and anaphora resolution, demonstrating the benefits of pre-training BERT models in the specific application language.

Jurisbert. This Jurisbert model [30] has around 400 Mb and a size of 768. The particular model was trained on a large corpus of legal texts in Portuguese, with the goal of capturing specific knowledge and linguistic patterns found in this domain. It utilizes a pre-training approach through contextual representation learning, where the model learns to predict words or chunks of text based on the context in which they appear. With "jurisbert", it is possible to leverage its prior knowledge about legal terms, sentence structure, and nuances of legal language. This can be useful for tasks such as document classification, extraction of legal information, among others.

MBertSTS. Another pre-trained model used was bert-base-multilingual-sts [30], focused on textual semantic similarity (STS) tasks. It was trained on multilingual corpora to learn semantic representations that capture similarity between sentences. The training data included public STS datasets like STS Benchmark and translated sentences from SNLI. Compared to the original multilingual BERT base, this model achieved better results on tasks like semantic similarity determination, coreference resolution, and multilingual text alignment.

BertimbauSTS. This textual semantic similarity specialized model is called bertimbau-base-portuguese-sts [30]. This model was pre-trained specifically on STS tasks for the Portuguese language. Compared to the multilingual model, bertimbau-base-portuguese-sts achieved better performance on monolingual Portuguese STS tasks, by being pre-trained on the specific language. It serves as a complementary specialized model in our pipeline.

JurisbertSTS. This pre-trained model we use is jurisbert-base-portuguese-sts [30], specialized in textual semantic similarity for the legal domain in Portuguese. This JurisBERT model has 110 million parameters and a vocabulary of 50,000 tokens. It was pre-trained on annotated legal corpora for textual semantic similarity, including legislation, jurisprudence, and legal doctrine. Compared to generic BERT models, JurisBERT achieved better performance on STS tasks with legal texts, by being trained on the specific domain. Its legal specialization makes it suitable to identify semantic similarity between legal articles and paragraphs.

TF-IDF. Term frequency-inverse document frequency (TF-IDF) [1] is a technique for generating semantic vector representations of text documents. TF-IDF assigns weights to terms based on term frequency (TF) in the document and inverse document frequency (IDF) in the corpus. TF quantifies term relevance to a document, while IDF amplifies rare, distinctive words. This assigns greater weights to terms characterizing document semantics. TF-IDF transforms text into semantic feature vectors for analysis.

2.3   Classification of Legal Texts Using BERT Embeddings

Contextualized word embeddings from pretrained language models like BERT have become ubiquitous in NLP, leading to their adoption in legal text classification tasks as well [7, 22]. Fine-tuning BERT has been shown to outperform classic machine learning algorithms by learning both generic and domain-specific linguistic patterns from large unlabeled legal corpora [7]. For instance, Chalkidis et al. [7] demonstrated that fine-tuned BERT achieved over 10% higher micro-F1 than SVM and kNN for multi-label classification of legal briefs. The semantic representations learned by BERT have also proven beneficial for related legal NLP problems such as information retrieval, question answering, and prediction [2, 16]. BERT embeddings capture syntactic and semantic characteristics useful for identifying relevant legal documents, answering natural language queries, and anticipating court judgments. Fine-tuning multilingual BERT models further extends the approach to legal texts in different languages [16].

3   Materials and Methods

Text embeddings represent documents in a dense vector space where proximity indicates semantic similarity [26]. This allows identifying groups of related cases to support tasks like information retrieval and precedent analysis [6, 17]. However, embedding quality strongly impacts classification accuracy [31]. A supervised classifier is trained over these BERT embeddings to categorize cases based on relevant topics and issues. We evaluate and compare classification performance across neural architectures including CNNs, RNNs, and Transformer encoders [8, 33].

3.1   Preprocessing

The documents went through several preprocessing steps in Python for standardization and cleaning before applying the classification methods. Procedural numbers, monetary values, and other numbers were replaced with standardized terms; dates, numbers spelled out, accents, and stopwords were removed; isolated words and footers with dates were removed. In addition, techniques were applied to treat compound words, ensuring that different ways of expressing them were considered equivalent. The goal was to ensure text consistency and quality, removing terms irrelevant to the semantic analysis while maintaining the essential meaning of the documents.

3.2   Generation of Embeddings

Advanced data transformation techniques using 6 embedding models and TF-IDF were applied to generate robust, semantically rich vector representations (embeddings) for legal texts. Approaches were based on SentenceTransformers in Keras using pre-trained BERT models like SBERT and BERTimbau that encapsulate semantic/syntactic knowledge of Portuguese. Additional contextualized embeddings were generated from other BERT models using Transformer encoding and Pooling aggregation. Models followed a standardized interface for consistent application across various BERT versions. Key hyperparameters manipulated include BERT model size (Base or Large) and maximum input sequence length (typically 768 tokens). The pre-trained model JurisBERT Base Portuguese STS was trained on a 1 billion token jurisprudential corpus, implementing a 12 layer BERT Base architecture with 110 million parameters. Pre-training on semantic textual similarity aims to generate embeddings that capture meaning in legal domains, with superior performance on similarity tasks compared to generic BERTs.

3.3 Classification Models

Classification models are pivotal tools in machine learning and AI. In recent years, neural network architectures have achieved immense success on complex categorization tasks. However, conventional techniques remain vital benchmarks and practical alternatives.

This research employed established classifiers from Python's scikit-learn library [24] for legal text categorization. The k-nearest neighbor algorithm predicts based on proximity to training examples. In this application, the knn algorithm was trained with 'K' = 3, meaning each test sample was classified based on the 3 nearest training examples. Naive Bayes uses Bayes' theorem to model class probabilities. Support vector machines [10] maximize margins between classes, the training parameters used were the defaults of the LinearSVC class from the sklearn library. Decision trees [27] partition data to infer rules, the algorithm was trained with the parameters 'maximum depth' = 10 and the criterion used to determine the quality of a split was 'entropy'. Random forests [18] ensemble decision trees to reduce overfitting. Despite simplicity, these fundamental techniques can effectively categorize complex data. Conventional classifiers provide useful baselines to evaluate more complex neural methods. Next, some neural network-based models used in this research are briefly introduced.

CB. CatBoost [25] is an open-source machine learning algorithm developed by Yandex for classification and regression tasks. It is part of the family of boosting algorithms, along with XGBoost, LightGBM and others. Unlike other boosters, CatBoost can natively handle categorical data by ordering the categorical features and using that ordering to influence the construction of the decision trees. This removes the need to create dummy variables and allows complex relationships between categories to be captured. CatBoost was trained using the CatBoostClassifier class and the following parameters: iterations = 100, depth = 6 (Maximum depth of each decision tree), learning rate = 0.1, loss function = 'MultiClass'. These settings allow CatBoost to capture complex relationships in the categorical data to make accurate predictions while maintaining good computational performance.

Neural 1. The neural network 1 was built using Keras' Sequential API, which allows the architecture definition in a linear sequence of layers [15]. The first layer is a Dense (fully connected) layer with 100 units and ReLU activation. This layer extracts 100 features from the input data X_{train}. The input to this layer has dimension equal to the number of features in the X_{train} set. After the first Dense layer, a Dropout layer with rate 0.1 is added for regularization. Dropout randomly disables 10% of the units during training to prevent overfitting. The second Dense layer has 50 units and also uses ReLU activation to introduce non-linearity. Another Dropout layer with rate 0.1 is added. The final layer is a Dense layer with 10 units and softmax activation, as this is a multi-class classification problem with 10 classes. The softmax activation ensures the outputs sum to 1 and can be interpreted as probabilities.

Neural 2. The Neural Network 2 was built using a sequential architecture composed of LSTM (Long Short-Term Memory) [29], GRU (Gated Recurrent Unit), Dense and Dropout layers. The first layer is an LSTM with 256 units, ReLU activation and return sequences enabled. This layer processes the input time sequence and extracts sequential patterns. After the LSTM, a Dropout layer with rate 0.2 is used for regularization. Then, a GRU layer with 128 units and ReLU activation is added to learn temporal representations. Another Dropout layer with rate 0.2. After that, there is a sequence of fully connected Dense layers interleaved with Dropout to extract features and perform classification. The layers have 64, 32 and 10 units, with ReLU activation, except the last one which uses softmax.

Neural 3. This recurrent neural network architecture is suitable for univariate time series classification. It combines LSTM and GRU layers to model sequential relationships in the data, with dense layers for feature extraction and final classification.

The first layer is an LSTM (Long Short-Term Memory) with 256 units and ReLU activation. It processes the input time series and learns temporal dependencies. Return sequences is enabled. After the LSTM, a Dropout layer regularizes the model with a rate of 0.2. Then, a GRU (Gated Recurrent Unit) layer with 128 units and ReLU activation extracts further sequential patterns. Another Dropout layer follows. Reshape layer converts the 3D representation of the RNN to 2D. GlobalAveragePooling1D aggregates the data across time for each step of the series. Dense blocks of 64, 32 and 10 units with Dropout in between classify the aggregated data. The last Dense layer uses softmax activation for multi-class classification. The model is compiled with categorical crossentropy loss, Adam optimizer and accuracy metric. This combination of RNNs for sequence modeling and dense networks for classification is effective and found in various works [12].

Neural 4. This recurrent neural network architecture [32] with attention layer is suitable for time series classification, combining sequential modeling capabilities and extraction of discriminative features. The first layer is an LSTM (Long Short-Term Memory) with 256 units and ReLU activation. It processes the input time series and learns sequential representations, capturing temporal dependencies in the data. Return sequences is enabled. After the LSTM, a Dropout layer with rate 0.2 is used for regularization, improving generalization. The next layer is a GRU (Gated Recurrent Unit) with 128 units and ReLU activation. It extracts further complementary sequential patterns from the time series. Return sequences is also enabled here. Another Dropout layer with rate 0.2 is added after the GRU for more regularization.

The output of the GRU is passed to a Scaled Dot-Product Attention layer. This layer generates importance weights for each time step of the series, highlighting informative parts.

The representation enhanced by the Attention layer is reshaped from 3D to 2D by a Reshape layer. GlobalAveragePooling1D aggregates the sequential data over time. Dense blocks with 64, 32 and 10 units with Dropout 0.2 in between for feature extraction and classification. The last Dense layer has softmax activation for multi-class classification. The model is compiled with categorical crossentropy loss, Adam optimizer and accuracy metrics. This hybrid architecture of RNNs and attention extracts discriminative features from time series for accurate classification. It captures temporal dependencies and highlights informative parts through the attention mechanism.

3.4   Application of Classifiers

The classification methods were applied to the generated textual embeddings to evaluate and compare performance across models and vectorization techniques.

5-fold stratified cross-validation was employed to robustly estimate classifier accuracy. The dataset was split into 5 subsets, with one held out as the test set and the remainder used for training in each iteration. Performance was averaged across the 5 folds to balance computational efficiency and stability. Cross-validation efficiently leverages all available data, while stratification maintains class balance across splits, minimizing bias and variance.

3.5   Evaluation of Results

To evaluate the performance of the different classification models, standardized metrics were used. Accuracy measures the fraction of correct predictions out of the total. The F1-score consists of the weighted harmonic mean between precision and recall, measuring the balance between these indicators. These metrics were calculated using a holdout approach, in which part of the data is reserved exclusively for testing. In addition, k-fold cross validation, described above, was also used for more robust and less biased metrics.

The balanced combination of these metrics allows evaluating the models under different perspectives, identifying possible trade-offs and limitations. This provides insights into both the generalization capability and specific error types of the models.

3.6   Databases

This study utilizes a dataset of 30,000 legal documents evenly divided into 10 classes representing different procedural motions [3]. The documents were extracted from the PJe electronic case management system from TJRN. Motions include ending trial execution, granting declaratory relief, denying declaratory relief, trial merit decisions, liminar injunctions, case abandonment, and lack of standing.

4   Results and Discussion

In this study, a diversity of classification algorithms were evaluated with different pre-trained embeddings. Tables 1 and 2 present the results of classifiers in terms of the averages of the F1-score and accuracy metrics, respectively. The best performances in terms of F1-Score were obtained by the TF-IDF model (0.9467).

Table 1. Classifiers' F1-Score with different embeddings

Algorithm	Mbert	Bertimbau	Jurisbert	MBertSTS	BertimbauSTS	JurisbertSTS	TF-IDF
KNN	0.8136	0.8684	0.8650	0.4873	0.7691	0.7972	0.8272
NB	0.4973	0.6676	0.6561	0.2866	0.5991	0.6117	0.8322
DT	0.5798	0.6813	0.7267	0.3296	0.5735	0.6137	0.8710
RF	0.7936	0.8533	0.8640	0.4946	0.7641	0.7852	0.9467
SVM	0.8568	0.8789	0.8894	0.6143	0.8187	0.8357	0.9397
Neural1	0.8323	0.8673	0.8771	0.5623	0.8151	0.8372	0.9376
Neural2	0.8557	0.8833	0.8806	0.5828	0.8137	0.8530	0.9406
Neural3	0.8354	0.8791	0.8991	0.5833	0.8230	0.8489	0.9342
Neural4	0.8487	0.8880	0.8827	0.5882	0.8113	0.8458	0.9342
CB	0.7540	0.8190	0.8282	0.4589	0.7280	0.7570	0.9113

It is observed that, in general, the models based on the textual embeddings of the BERT models (MBERT, BERTimbau, JurisBERT) obtained superior performance in the classification task compared to the other evaluated embeddings. In particular, the JurisBERT model presented the best indices in most metrics, indicating its effectiveness in capturing relevant semantic and legal knowledge in procedural texts. Its F1-score values exceeded 0.86 in most of the tested configurations.

Among the traditional classifiers, RF, KNN and SVM exhibited good performance, particularly when RF was paired with TF-IDF document representation. Neural models performed well mainly when using with Bertimbau, Jurisbert and TF-IDF embeddings. The embeddings based exclusively on textual semantic similarity (MBERT-STS, BERTimbau-STS), on the other hand, exhibited the worst results among the fine tunned BERT models. This suggests that, although useful for STS tasks, they do not adequately encode the thematic and legal aspects necessary for the proposed classification.

Table 2. Classifiers' Accuracy with different embeddings

Algorithm	Mbert	Bertimbau	Jurisbert	MBertSTS	BertimbauSTS	JurisbertSTS	TF-IDF
KNN	0.8142	0.8678	0.8658	0.4823	0.7711	0.7998	0.8119
NB	0.5073	0.6786	0.6692	0.3035	0.6031	0.6189	0.8336
DT	0.5813	0.6801	0.7259	0.3346	0.5754	0.6134	0.8680
RF	0.7942	0.8527	0.8640	0.4946	0.7643	0.7864	0.9467
SVM	0.8570	0.8795	0.8896	0.6172	0.8188	0.8370	0.9399
Neural1	0.8338	0.8674	0.8767	0.5598	0.8170	0.8374	0.9377
Neural2	0.8558	0.8840	0.8812	0.5884	0.8156	0.8522	0.9409
Neural3	0.8359	0.8798	0.9000	0.5877	0.8212	0.8488	0.9344
Neural4	0.8482	0.8876	0.8836	0.5907	0.8149	0.8464	0.9344
CB	0.7553	0.8208	0.8283	0.4657	0.7282	0.7574	0.9113

The TF-IDF model achieved good performance, with F1-score and accuracy above 0.90 in most classifiers. TF-IDF's effectiveness highlights the power of heuristic term frequency analysis on lengthy, information-rich documents. By emphasizing rare, topically-relevant words through inverse document frequency weighting, TF-IDF distills key semantic content from noisy texts. With fewer parameters versus complex neural networks, TF-IDF also avoids overfitting on limited training data. This parsimony and generalizability underpin its competitive accuracy despite simplicity. Among BERT models, monolingual Portuguese BERTimbau outperformed multilingual BERT, underscoring the benefits of language-specific conditioning. Additionally, legal-domain fine-tuning enhanced JurisBERT over generic BERT, evidencing gains from domain adaptation. However, no BERT configuration exceeded TF-IDF, likely due to high inter-class similarity impeding contextual modeling. With greater data diversity, deep semantic representations may eventually surpass count-based approaches. The performance of the BERT-based models is likely due to overlap between the used text data, which are relatively short motion documents, generally one or two pages long. The motion types 'Preliminary' and 'Interlocutory remedy' have significant term overlap as well as the classes 'Abandonment of the claim' and 'Discontinuance' motions.

5   Conclusions

This study delved into the application of automated text classification to expedite legal document analysis within the Brazilian justice system, which faces challenges related to case backlogs. We conducted a thorough evaluation of supervised machine learning approaches, employing BERT embeddings fine-tuned specifically for the legal domain on a diverse ten-class dataset consisting of 30,000 Brazilian judicial moves from TJRN Court.

Our research findings demonstrated that despite the high potential of BERT embeddings, the TF-IDF method outperformed BERT-derived models in the considered metrics. One of the primary reasons for this observation lies in the dataset's characteristic of having several very similar classes, making it challenging for BERT-derived models to capture the subtle semantic differences between these classes effectively.

Notably, the TF-IDF method showcased its suitability for this specific task by effectively addressing the challenge of high class similarity. Although the results revealed TF-IDF's superiority in this context, it is crucial to recognize that BERT embeddings remain powerful tools for natural language processing tasks and have shown exceptional performance in various applications. It is planned for future work to evaluate the models on datasets containing longer text documents covering a wider variety of topics.

Acknowledgement. Research in part supported by CAPES - Brazilian Agency of Superior Education.

References

1. Aizawa, A.: An information-theoretic perspective of tf-idf measures. Inf. Process. Manag. **39**(1), 45–65 (2003)
2. Aletras, N., Tsarapatsanis, D., Preotiuc-Pietro, D., Lampos, V.: Predicting judicial decisions of the European court of human rights: a natural language processing perspective. PeerJ Comput. Sci. **2**, e93 (2016)
3. Araújo, D.C., Lima, A., Lima, J.P., Costa, J.A.: A comparison of classification methods applied to legal text data. In: Marreiros, G., Melo, F.S., Lau, N., Lopes Cardoso, H., Reis, L.P. (eds.) EPIA 2021. LNCS (LNAI), vol. 12981, pp. 68–80. Springer, Cham (2021). https://doi.org/10.1007/978-3-030-86230-5_6
4. Boer, A., Engers, T.M.v., Winkels, R.: Using ontology's for an intelligent legal information system. In: Proceedings of the Workshop on Legal Ontologies (1997)
5. Bourcier, D., Mazzega, P.: Toward measures of legal predictability. In: Proceedings of the 11th International Conferenec on Artificial Intelligence and Law, pp. 207–215 (2007)
6. Chalkidis, I., Androutsopoulos, I.: A deep learning approach to contract element extraction. arXiv preprint arXiv:2005.07033 (2020)
7. Chalkidis, N., Fergadiotis, M., Malakasiotis, P., Aletras, N., Androutsopoulos, I.: Extreme multi-label legal text classification: a case study in EU legislation. In: Proceedings of the Natural Legal Language Processing Workshop 2019, pp. 78–87 (2019)
8. Conneau, A., Kiela, D., Schwenk, H., et al.: Supervised learning of universal sentence representations from natural language inference data. In: EMNLP (2017)
9. Conselho Nacional de Justiça: Justiça em números 2022: ano-base 2021 (2022). Accessed 16 Feb 2023. https://www.cnj.jus.br/wp-content/uploads/2022/09/sumario-executivo-jn-v3-2022-2022-09-15.pdf
10. Cortes, C., Vapnik, V.: Support-vector networks. Mach. Learn. **20**(3), 273–297 (1995)
11. Devlin, J., Chang, M.W., Lee, K., Toutanova, K.: Bert: pre-training of deep bidirectional transformers for language understanding. In: Proceedings of the 2019 Conf. of the North American Chapter of the Association for Computational Linguistics: Human Language Technologies (NAACL-HLT), vol. 1, pp. 4171–4186 (2019)

12. Ismail Fawaz, H., Forestier, G., Weber, J., Idoumghar, L., Muller, P.-A.: Deep learning for time series classification: a review. Data Min. Knowl. Disc. **33**(4), 917–963 (2019). https://doi.org/10.1007/s10618-019-00619-1

13. Fawei, D., Wanjun, Z., Linjun, L., Yangjun, L., Xiaobing, L., Xiaolong, L.: Legal document clustering based on latent semantic analysis. In: Proceedings of the 2008 IEEE International Conference on Information Reuse and Integration, pp. 475–480. IEEE (2008)

14. Ferraz, L.S.F., Salomão, T.S., Salomão, C.T.S., Nunes, D.J.C., da Silva, F.B., et al.: Relatório de inteligência artificial e o poder judiciário. Accessed 17 Feb 2023. https://ciapj.fgv.br/sites/ciapj.fgv.br/files/relatorio_ia_3a_edicao_0.pdf

15. Kim, Y.: Convolutional neural networks for sentence classification. arXiv preprint arXiv:1408.5882 (2014)

16. Le, H., Vial, L., Frej, J., Segonne, V., Coavoux, M., Lecouteux, B., et al.: Flaubert: unsupervised language model pre-training for french. arXiv preprint arXiv:1912.05372 (2019)

17. Le, H., Vial, L., Frej, J., Segonne, V., Coavoux, M., et al.: Flaubert: unsupervised language model pre-training for french. arXiv preprint arXiv:1912.05372 (2020)

18. Liaw, A., Wiener, M., et al.: Classification and regression by randomforest. R News **2**(3), 18–22 (2002)

19. Lodder, A., Oskamp, A.: DSS for law; opportunities and dangers. In: Proceedings of the 5th International Conference on Artificial Intelligence and Law, pp. 154–163 (1995)

20. Long, W., Lu, Q., Zhang, X.: A hierarchical attention model for legal judgment prediction. IEEE Access **7**, 47275–47283 (2019)

21. Moens, M.F., Uyttendaele, C., Dumortier, J.: Abstracting of legal cases: the potential of clustering based on the selection of representative objects. J. Am. Soc. Inf. Sci. **50**(2), 151–161 (1999)

22. Otake, Y., Suzuki, J., Kajiwara, H., Kuribayashi, T.: Legal article classification using only a small dataset based on BERT. In: Proceedings of the Natural Legal Language Processing Workshop 2020, pp. 46–51 (2020)

23. Palmirani, M., Benyovskyi, L., Ceci, A., Mazzei, A., Biagioli, C.: NLP tools for legal text analysis. Artif. Intell. Law **25**(2), 161–183 (2017)

24. Pedregosa, F., Varoquaux, G., Gramfort, A., Michel, V., Thirion, B., Grisel, O., et al.: Scikit-learn: machine learning in python. J. Mach. Learn. Res. **12**, 2825–2830 (2011)

25. Prokhorenkova, L., Gusev, G., Vorobev, A., et al.: CatBoost: unbiased boosting with categorical features. In: Advances in Neural Information Processing Systems, pp. 6638–6648 (2018)

26. Reimers, N., Gurevych, I.: Sentence-bert: Sentence embeddings using siamese bert-networks. In: Proceedings of the 2019 Conference on Empirical Methods in Natural Language Processing. Association for Computational Linguistics (2019)

27. Safavian, S.R., Landgrebe, D.: A survey of decision tree classifier methodology. IEEE Trans. Syst. Man Cybern. **21**(3), 660–674 (1991)

28. Souza, F., Nogueira, R., Lotufo, R.: BERTimbau: pretrained BERT models for Brazilian Portuguese. In: Cerri, R., Prati, R.C. (eds.) BRACIS 2020. LNCS (LNAI), vol. 12319, pp. 403–417. Springer, Cham (2020). https://doi.org/10.1007/978-3-030-61377-8_28

29. Sundermeyer, M., Schlüter, R., Ney, H.: LSTM neural networks for language modeling. In: Proceedings of Interspeech 2012, pp. 194–197 (2012).https://doi.org/10.21437/Interspeech.2012-65

30. Viegas, C.F.O., Costa, Bruno C., Ishii, R.P., et al.: Jurisbert: a new approach that converts a classification corpus into an STS one. In: Gervasi, O., et al. (eds.) Computational Science and Its Applications-ICCSA 2023. ICCSA 2023. LNCS, vol. 13956, pp. 349–365. Springer, Cham (2023). https://doi.org/10.1007/978-3-031-36805-9_24
31. Wang, Y., Subramanian, N., Wang, E., Lou, C., et al.: LEGAL-BERT: the muppets straight out of law school. In: FAcct (2019)
32. Yang, Z., Yang, D., Dyer, C., et al.: Hierarchical attention networks for document classification. In: Proceedings of NAACL-HLT 2016, pp. 1480–1489. Association for Computational Linguistics (2016)
33. Zhang, X., Li, J.J., Huang, S.J., et al.: SGM: sequence generation model for multi-label classification. In: Proceedings of the 27th (COLING), pp. 3915–3926 (2018)
34. Zhong, X., Guo, J., Tu, K., et al.: Legal judgment prediction via topological learning. In: EMNLP, pp. 3540–3549. Association for Computational Linguistics (2018)

Rapid and Low-Cost Evaluation of Multi-fidelity Scheduling Algorithms for Hyperparameter Optimization

Aakash Mishra[✉][iD], Jeremy Hsu[iD], Frank D'Agostino[iD], Utku Sirin, and Stratos Idreos

Harvard University, Cambridge, MA 02138, USA
{amishra,jeremyhsu,fdagostino}@college.harvard.edu,
utkusirin@g.harvard.edu, stratos@seas.harvard.edu

Abstract. Hyperparameter optimization (HPO), the process of searching for optimal hyperparameter configurations for a model, is becoming increasingly important with the rise of automated machine learning. Using HPO can be incredibly expensive as models increase in size. Multi-fidelity HPO schedulers attempt to more efficiently utilize resources given to each model configuration by determining the order they are evaluated in and the number of epochs they are run for. Pre-tabulated benchmarks are often used to reduce the compute power required to evaluate state-of-the-art schedulers. However, over-reliance on evaluation using these benchmarks can lead to overfitting. To solve this problem, we introduce a Platform for Hyperparameter Optimization Search Simulation (PHOSS), that enables rapid HPO scheduler evaluation by dynamically generating surrogate benchmarks. PHOSS uses a computationally efficient and expressive exponential decay parametric modeling approach to accurately generate surrogate benchmarks from real-world dataset samples without having to train an expensive surrogate model. We demonstrate that PHOSS can simulate several state-of-the-art schedulers on real-world benchmarks $4.5\times$ faster while also dramatically reducing multi-GPU compute requirements by enabling full testing of HPO schedulers on a single commodity CPU.

Keywords: Multi-fidelity Schedulers · Hyperparameter Optimization · Neural Architecture Search · Surrogate Models

1 Introduction

1.1 Rising Prevalence of Hyperparameter Optimization

The fast-pace of machine learning (ML) model design has created a need for quickly fine-tuning models. HPO plays a major role in this process [1,7]. As the workloads and constraints change, ML models need to be re-designed, re-tuned, and re-trained, which requires further searching for the ML model with an optimal set of hyperparameters. Hence, fast and efficient HPO is key to finding and using successful ML models [1,3,11,13].

ⓒ The Author(s), under exclusive license to Springer Nature Switzerland AG 2023
P. Quaresma et al. (Eds.): IDEAL 2023, LNCS 14404, pp. 64–69, 2023.
https://doi.org/10.1007/978-3-031-48232-8_7

1.2 Multi-fidelity Hyperparameter Optimization Schedulers

Distributed hyperparameter optimization (HPO) evaluation strategies aim to develop resource-efficient and robust solutions for sorting out desirable hyperparameter configurations for a specific dataset, agnostic of the hyperparameter space. State-of-the-art optimization strategies aim to use compute-power more efficiently through early-stopping, or the termination of poor performing configurations during training. These methods are known as multi-fidelity gradient-free optimizers [6]. Finding the ideal optimizer for hyperparameter optimization is important as some optimizers are better than others in finding the best model [1].

1.3 Pre-tabulated Benchmarks are Not a Generic Solution for Evaluating HPO Schedulers

Pre-tabulated benchmarks, such as Neural Architecture Search (NAS) Benchmarks [5], aim to reduce the resource requirements for NAS researchers by pre-tuning, training, and storing model configurations and evaluations (i.e., their training, validation, and test loss and accuracy values per epoch). These pre-tabulations help with quickly evaluating the efficiency of HPO optimizers (multi-fidelity schedulers). The inherent problem with pre-computed benchmarks is their dependence on a fixed dataset. Since each dataset has its own intrinsic characteristics, HPO developers who aim to create schedulers that perform well on these pre-tabulated benchmarks risk over-fitting their HPO development strategy and losing generalizability to new problem spaces. As an example, the best multi-fidelity optimizer that can find the best model for classification on CIFAR-10 will not necessarily be the best optimizer for the ImageNet10k dataset.

1.4 Problem: Scheduler Evaluation Time is Slow

It can take weeks in GPU time to run HPO schedulers on real-life datasets with large models (500 MB >) and that restricts smaller scale research teams and firms from participating in HPO research. Moroever, there is no general consensus on the most optimal scheduler and we have yet to see a scheduler that performs optimally for every dataset. As the data-size keeps growing, the required amount of optimization time easily exceeds the limits of many small-to-mid-sized organizations. Therefore, choosing the best scheduler to optimize a HPO problem becomes crucial. The time cost of evaluation raises the barrier of entry for testing novel schedulers. Currently, evaluating a new algorithm for multi-fidelity optimization is intractable for a researcher with limited compute resources. This culminates into two main problems: choosing the right scheduler for a HPO problem (including the scheduler's hyperparameters) and resource constrained development of novel scheduling algorithms. We aim to solve both of these problems with one solution.

2   A Platform for Hyperparameter Optimization Search Simulation (PHOSS)

2.1   Surrogate Modelling Approach

Our novel method is powerful because we recognize that evaluating the performance of the scheduler does not require training each model configuration on the dataset to get the learning curves. Rather we just need to capture the likeness of the learning curve topology with a synthetic model. We interpolate learning curves by employing simple parametric models that are fit on a small sample of models from the search space that were partially-trained. This allows us to fit the surrogate model on the sample of real learning curves in order to then utilize the surrogate to generate synthetic learning curves from its latent distribution to evaluate our hyperparameter schedulers as shown in Fig. 1. We reduced the complexity of our learning curve (surrogate) model as compared to past surrogate model approaches such as in NAS-301 in order to improve user-interpretability and increase generalizability. While other surrogate models exist for providing learning curve interpolation, they are not generalizable and are often solely designed to match a specific search space's models trained on supervised datasets. Inspired by the pow_4 parametric model used in [4], we propose a novel parametric loss-curve model such that each fitted-curve follows the following formula: $\ell_k(\tau|\boldsymbol{\theta}) = \ell_k(\tau|\boldsymbol{\beta}, \alpha) = \beta_0 + \frac{\beta_1}{(\beta_2+\tau)^\alpha+\beta_3}$ where k is a hyperparameter configuration, τ is the epoch number, $\boldsymbol{\beta}$ is the vector of β values, and α is the exponential parameter. We generate values of $\theta \in \{\beta_0, \beta_1, \beta_2, \beta_3, \alpha\}$ such that $\beta_i \sim \mathcal{N}(\mu_{\beta_i}, \sigma_{\beta_i}^2)$ and $\alpha \sim \mathcal{N}(\mu_\alpha, \sigma_\alpha^2)$, where $\mathcal{N}$ is the normal distribution, μ and σ are the mean and standard deviations of the normal distribution. The user sets the σ values. To find the μ, we solve a nonlinear least-squares problem where we fit to the median of a selected sample of learning curves using the Trust Region Reflective algorithm [2]. We choose the median as opposed to the mean since the median is robust to outliers and can be more indicative of the typical behavior of the learning curves. As such, we have that

$$\mu_\theta = \left[\underset{\hat{\boldsymbol{\beta}},\hat{\alpha}}{\arg\min} \left(\text{median}_i(\boldsymbol{y}) - \ell_k(\tau|\hat{\boldsymbol{\beta}}, \hat{\alpha}) \right) \right]_\theta$$

for our estimate, where $\hat{\boldsymbol{\beta}}, \hat{\alpha}$ are the estimated parameters from the least-squares fitting and $\boldsymbol{y}$ is the observed learning curve data across all i samples. Moreover, to make more expressive learning curves, we allow for the addition of multiple parametric curve configurations such that a configuration C is given by: $C(\beta_1, \beta_2, \beta_3, \alpha) = \frac{\beta_1}{(\beta_2+\tau)^\alpha+\beta_3}$. Stacking parametric configurations allows for the same flexibility as a polynomial function. In order to make our simulation more realistic, we also added the ability for users to specify the "noise schedule" for the decay/growth of the noise as the number of epochs increases. This is motivated by a proposition by [8] when looking at the heteroskedasticity of their model parameters. Therefore, the user can specify a mean and standard deviation for both the starting noise loss and ending noise loss, such that they input the means and standard deviations $(\mu_s, \sigma_s, \mu_e, \sigma_e)$ for the noise.

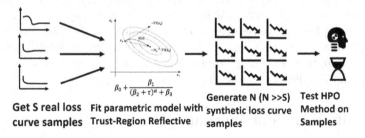

Get S real loss curve samples   **Fit parametric model with Trust-Region Reflective**   **Generate N (N >>S) synthetic loss curve samples**   **Test HPO Method on Samples**

Fig. 1. Simulating learning curves with the parametric model reduces the required number of resource-intensive real-life samples to test a HPO scheduler.

2.2   Implementation Details

PHOSS is built upon Ray Tune [10], a library that executes scalable hyperparameter tuning and experiments. It provides integrations with PyTorch and Tensorflow, and implements distributed versions of most state-of-the-art schedulers such as ASHA, Hyperband, and PBT. We generate a learning curve for each hyperparameter configuration in our search space, and then have our experiment runner execute the scheduler on these synthetic learning curves. However, note that the scheduler's role is to find the optimal hyperparameter configuration using the most efficient path. Crucially, it does not care about what that configuration is. Thus, in PHOSS, each hyperparameter configuration is represented as an index $i = 0, \ldots, \texttt{num\_samples} - 1$, where $\texttt{num\_samples}$ is passed in by the user. At each epoch, the scheduled worker pulls a value from the pre-generated learning curve for the index it is assigned to evaluate. Overall, PHOSS exposes a simple API for users to quickly test and compare schedulers on a variety of simulated learning curves and resource configurations. Furthermore, the code is designed to offer a level of extensibility such that one could very easily add support for existing state-of-the-art schedulers implemented in Ray, define their own custom schedulers, specify new metrics to monitor during experiments, configure their workloads to mimic configurations on different hardware, and more. We provide the open-source code for the PyPI package here https://github.com/aakamishra/phoss.

3   Related Work

Complex models of learning curves have high compute overhead. In previous work, researchers characterized parametric models to describe the behavior of learning curves. Domhan et al. [4] describes a set of 11 model types, expressed as parametric functions $\phi_i(\theta_i, t)$, to extrapolate learning curves as ensemble weighted model for time series prediction. They opt for a Markov Chain Monte Carlo (MCMC) inference to solve the optimization problem. MCMC, however, is not efficient for fast learning-curve modeling, which makes scaling the learning curves difficult. [12]. We use a more simplistic approach in learning curve modeling than [4] to reduce the computational burden and improve user-interpretability. We further apply our novel learning curve modeling to HPO scheduler evaluation. By doing so, we demonstrate that we can achieve up to 4.5x speed up in scheduler evaluation.

4   Results and Contributions

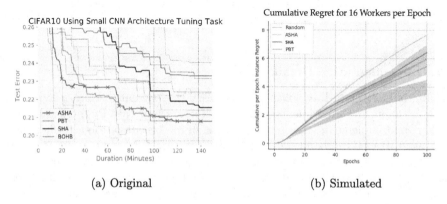

(a) Original   (b) Simulated

Fig. 2. PHOSS successfully replicates results for tuning a ConvNet architecture on CIFAR-10 HPO Benchmark in significantly less time: (a) Original results by [9], (b) Our simulated results.

Our experiment reveals our method achieves a 14x speedup over conventional scheduler evaluation. We simulate experiments constructed as benchmarks for four state-of-the-art schedulers discussed in [9]: Successive Halving Algorithm (SHA), Asynchronous SHA (ASHA), Population-Based Training (PBT), and Random. For demonstration, we display the time series plot of validation accuracy error (Test Error) from the original ASHA paper (Fig. 2a) and a plot of cumulative regret from our much faster synthetic experiments (Fig. 2b) to compare.

PHOSS Successfully Replicates the CIFAR-10 ConvNet Architecture Tuning Task. First, we construct a VGG3 model similar to the ConvNet architecture used by [9] on the CIFAR-10 dataset. We sampled 20 configurations which were run to completion for $R = 20$ epochs and fit our parametric learning curve model on the data points. We generate a synthetic test set of 1000 learning curves for $R = 100$ epochs and calculate the performance of each scheduler with multiple random seeds. In our simulated environment, we observe that ASHA's cumulative regret is the lowest, followed by PBT, SHA, and Random, as seen in Fig. 2b. This result matches with the real-world published results on Fig. 2a, where the final test error of ASHA is the lowest, followed by PBT and SHA. Our results show that, by using our platform, we are able to provide guidance on which scheduler algorithm we should choose for hyperparameter optimization. This way, we can save hundreds of GPU-hours of training and reduce the required compute power. To illustrate, our real-world experiment took 140 min of training and scheduling per scheduler on 25 V100 GPU threads, whereas our simulated experiment took about 10 min per scheduler on 16 CPU threads. We note the one-time training cost for getting the 20 additional samples on a single V100 GPU took 81 min. In total, this is a 4.5× speedup of the real-world execution with significantly less compute resources. For organizations and companies with a small number of compute resources, our platform increases accessibility for the evaluation of HPO schedulers.

References

1. Bischl, B., et al.: Hyperparameter optimization: foundations, algorithms, best practices and open challenges. CoRR abs/2107.05847 (2021)
2. Branch, M.A., Coleman, T.F., Li, Y.: A subspace, interior, and conjugate gradient method for large-scale bound-constrained minimization problems. SIAM J. Sci. Comput. **21**(1), 1–23 (1999)
3. Cubuk, E.D., Zoph, B., Mane, D., Vasudevan, V., Le, Q.V.: AutoAugment: learning augmentation policies from data. CoRR abs/1805.09501 (2018)
4. Domhan, T., Springenberg, J.T., Hutter, F.: Speeding up automatic hyperparameter optimization of deep neural networks by extrapolation of learning curves. In: IJCAI (2015)
5. Dong, X., Yang, Y.: Nas-bench-201: extending the scope of reproducible neural architecture search. CoRR abs/2001.00326 (2020)
6. Eggensperger, K., et al.: HPOBench: a collection of reproducible multi-fidelity benchmark problems for HPO. CoRR abs/2109.06716 (2021)
7. Elsken, T., Metzen, J.H., Hutter, F.: Neural architecture search: a survey. J. Mach. Learn. Res. **20**(55), 1–21 (2019)
8. Klein, A., Falkner, S., Springenberg, J.T., Hutter, F.: Learning curve prediction with Bayesian neural networks. In: ICLR (2017)
9. Li, L., et al.: A system for massively parallel hyperparameter tuning. In: MLSys (2020)
10. Liaw, R., Liang, E., Nishihara, R., Moritz, P., Gonzalez, J.E., Stoica, I.: Tune: a research platform for distributed model selection and training. In: ICML AutoML Workshop (2018)
11. Liu, H., Simonyan, K., Yang, Y.: DARTS: differentiable architecture search. In: International Conference on Learning Representations (ICLR) (2019)
12. Salvatier, J., Wiecki, T.V., Fonnesbeck, C.: Probabilistic programming in Python using PyMC3. PeerJ Comput. Sci. **2**, e55 (2016)
13. Zeiler, M.D.: ADADELTA: an adaptive learning rate method. CoRR abs/1212.5701 (2012)

The Applicability of Federated Learning
to Official Statistics

Joshua Stock[1(✉)], Oliver Hauke[2], Julius Weißmann[2], and Hannes Federrath[1]

[1] Universität Hamburg, Hamburg, Germany
joshua.stock@uni-hamburg.de
[2] Federal Statistical Office (Destatis), Wiesbaden, Germany

Abstract. This work investigates the potential of Federated Learning (FL) for official statistics and shows how well the performance of FL models can keep up with centralized learning methods. FL is particularly interesting for official statistics because its utilization can safeguard the privacy of data holders, thus facilitating access to a broader range of data. By simulating three different use cases, important insights on the applicability of the technology are gained. The use cases are based on a medical insurance data set, a fine dust pollution data set and a mobile radio coverage data set–all of which are from domains close to official statistics. We provide a detailed analysis of the results, including a comparison of centralized and FL algorithm performances for each simulation. In all three use cases, we were able to train models via FL which reach a performance very close to the centralized model benchmarks. Our key observations and their implications for transferring the simulations into practice are summarized. We arrive at the conclusion that FL has the potential to emerge as a pivotal technology in future use cases of official statistics.

1 Introduction

The aim of national statistical offices (NSOs) is to develop, produce and disseminate high-quality official statistics that can be considered a reliable portrayal of reality [14]. In order to effectively capture our rapidly changing world, NSOs are currently undergoing a process of modernization, leveraging new data sources, methodologies and technologies.

NSOs have effectively extracted information from new data sources, such as scanner data[1] or Mobile Network Operator (MNO) data[2]. However, the potential of numerous other data sources, including privately held data or data from certain official entities, remains largely untapped. Legal frameworks, which are fundamental to official statistics, only adapt slowly to changing data needs.

---

[1] Scanner data in consumer price statistics https://www.destatis.de/EN/Service/EXSTAT/Datensaetze/scanner-data.html, accessed on September 29, 2023.

[2] Use of MNO data https://cros-legacy.ec.europa.eu/content/12-use-mno-data_en, accessed on September 29, 2023.

© The Author(s), under exclusive license to Springer Nature Switzerland AG 2023
P. Quaresma et al. (Eds.): IDEAL 2023, LNCS 14404, pp. 70–81, 2023.
https://doi.org/10.1007/978-3-031-48232-8_8

Cooperation with potential data donors faces restrictions due to concerns about privacy, confidentiality, or disclosing individual business interests.

In the meantime, the methodology employed by NSOs is evolving, with machine learning (ML) gaining substantial popularity and undergoing a process of establishment. ML has been applied in various areas of official statistics (e.g. [1,3]), and new frameworks such as [14] address the need to measure its quality.

Federated learning (FL) is an emerging approach within ML that provides unexplored potential for official statistics. It addresses the challenge of extracting and exchanging valuable global information from new data sources without compromising the privacy of individual data owners. Introduced in [7], FL enables collaborative model training across distributed data sources while preserving data privacy by keeping the data localized. In scenarios where external partners are unwilling to share information due to regulatory or strategic considerations, but still aim to disseminate global insights in their field of application, NSOs can offer trustworthy solutions by utilizing FL. In return, FL empowers contributing NSOs to integrate new data sources into statistical production.

To the best of our knowledge, besides our own work only one currently presented study investigates the applicability of FL in official statistics. In a proof of concept by the United Nations, FL is applied to estimate human activity based on data collected from wearable devices [2,11]. Their work emphasizes operative aspects of FL coordinating multiple NSOs and benefits of additional privacy enhancing technologies. In contrast, we focus on measuring the numerical predictive performance and reproducibility by openly sharing our code.

Our main contribution lies in presenting three applications of FL that address current data needs representative for official statistics (see Sect. 3). The first two simulations focus on assessing the performance achieved by FL in comparison to models trained in a centralized way. The third simulation presents valuable insights and lessons learned from the implementation of FL, involving the active participation of an external partner. We draw conclusions on the applicability of FL in NSOs in Sect. 4, which are summarized in Sect. 5.

2   Background

Before presenting the simulated use cases in Sect. 3, this section provides an overview of FL and privacy challenges with ML.

2.1   Federated Learning

In FL, a centralized server (or aggregator, in our case a NSO) coordinates the process of training a ML model by initializing a global model and forwarding it to the data owners (clients). In each training round, clients train the model with their private data and send the resulting model back to the server. The server uses a FL method to aggregate the updates of the clients into the next iteration of the global model and starts the next round by distributing the updated model to the clients. This process is repeated to improve the performance of the model.

NSOs primarily strive to generate global models that accurately represent the available data, which, in our setting, is distributed among multiple clients. Thus, we compare the performance of FL to models with access to the combined data of all clients. Alternatively, if upcoming applications seek to supply each client with an optimized individual model by leveraging information from the other clients, *personalized* FL can be used. This approach is not covered in this paper but can be found in [4,5].

2.2 Privacy Challenges with Machine Learning

When training data for a ML model is distributed among multiple parties, the data traditionally needs to be combined on a central server prior to training an ML model. FL has become a popular alternative to this approach, as it allows to train a model in a distributed way from the start, without the need to aggregate training data first. Thus, using FL has the privacy advantage that there is no need to exchange private training data.

But although FL makes sharing private training data obsolete, there are other privacy challenges inherent to ML which have also been observed for FL. As an example, *property inference* attacks [8,12] allow to deduce statistical properties of a trained model's training data, in settings where an attacker has only access to the model itself, but not its training data.

The applicability of such attacks depends on the concrete use case, the type of model and other factors. Analyzing the individual privacy leakage of the simulated use cases in this paper are out of scope. Nonetheless, raising awareness to these issues, e.g., by communicating potential risks to clients in an FL scenario, should not be neglected. Beyond this, strategies under the umbrella term *privay-preserving machine learning* (PPML) can help to mitigate these risks [13].

2.3 Frameworks

In our simulations, we use the frameworks TensorFlow for neural networks and TensorFlow Federated[3] for FL. We use PyCaret[4] for automizing benchmark experiments in the centralized settings and scikit-learn[5] for data processing.

The code we have written for this work is openly available on GitHub[6].

---

[3] TensorFlow Federated https://tensorflow.org/federated, accessed on September 29, 2023.

[4] PyCaret https://pycaret.org/, accessed on September 29, 2023.

[5] scikit-learn https://scikit-learn.org/, accessed on September 29, 2023.

[6] Code repository for this paper: https://www.github.com/joshua-stock/fl-official-statistics, accessed on September 29, 2023. Note that for the mobile radio coverage simulation, the code has only been executed locally on the private data set, hence it is not included in the repository.

3   Simulations

Most relevant for NSOs is *cross-silo* FL, where a few reliable clients train a model, e.g. official authorities. In contrast, *cross-device* FL uses numerous clients, e.g. smartphones, to train a model. For each of our three use cases, we first compute benchmarks by evaluating centralized ML models, i.e., models which are trained on the whole data set. Afterwards, we split the data set and assign the parts to (simulated) FL clients for the FL simulation. This way, we have a basis for interpreting the performance of the model resulting from the FL training simulation. The performance metrics of the trained ML models (including coefficient of determination R^2 or accuracy) are computed on test sets of each data set.

3.1   Medical Insurance Data

The demand for timely and reliable information on public health is steadily increasing. The COVID-19 pandemic has significantly accelerated this trend, raising questions about the financial feasibility of our healthcare system and the availability of medical supplies.

Thus, our first experiment focuses on modeling a regression problem related to healthcare by considering the following question: Given an individual's health status characteristics, what is the magnitude of their insurance *charges*? We aim to address two primary questions. Firstly, we explore the suitability of neural networks in comparison to other models for the regression task. Secondly, we assess the feasibility of utilizing a simulated decentralized data set in an FL setting to tackle the problem.

Data Set. The given data set links medical insurance premium *charges* to related individual attributes[7]. Considered are the six features *age*, *sex*, *bmi* (body mass index), *children* (count), *smoker* (yes/no) and four *regions*. The feature *region* is excluded during FL training and solely utilized for partitioning the data within the FL setting. In total, the data set consists of 1338 complete records, i.e. there are no missing or undefined values. Also, the data set is highly balanced: The values in *age* are evenly dispersed, just as the distribution of male and female records is about 50/50 (attribute *gender*) and each *region* is represented nearly equally often. The origin of the data is unknown, however its homogeneity and integrity suggest that it has been created artificially.

Data Preprocessing. We encode the attributes *sex* and *smoker* into a numeric form (0 or 1). The attributes *age*, *bmi* and *children* are scaled to a range from 0 to 1. In the centralized benchmarks, the attribute *region* is one-hot-encoded.

---

[7] US health insurance dataset https://www.kaggle.com/datasets/teertha/ushealthinsurancedataset, accessed on September 29, 2023.

Setup. We aim to investigate the suitability of neural networks for estimating insurance *charges* and explore the extent to which this problem can be addressed using a FL approach. To achieve this, we compare different models and evaluate their performance.

A basic fully connected neural network architecture with five input features is used. The network consists of three hidden layers with 40, 40, and 20 units in each respective layer. Following each layer, a Rectified Linear Unit (ReLU) activation function is applied. The final output layer comprises a single neuron. To optimize the network, the Adam optimizer with a learning rate of 0.05 is employed. In the federated setting, we utilize the same initial model but integrate FedAdam for server updates. This decision is based on previous research [9], which emphasizes the benefits of adaptive server optimization techniques for achieving improved convergence.

In the centralized approach, we allocate a training budget of 100 epochs. In contrast, the federated approach incorporates 50 rounds of communication between the client and server during training. Each round involves clients individually training the model for 50 epochs. To track the running training, 10% of the data is used as evaluation data in the FL setting, respectively 20% in the centralized scenario. The remaining shallow learning models undergo hyperparameter optimization (HPO) using a random search approach with a budget of 100 iterations. We evaluate all models using 5-fold cross validation.

Table 1. Performance comparison of different prediction models for the medical insurance use case. The performance is quantified using R^2 in %, along with the corresponding standard deviation (std). Additionally, the relative loss to the best centralized model (rel. loss) is reported.

Model	$R^2(\pm$ std)	Rel. loss (%)
neural network	81.5(4.01)	3.5
neural network (federated)	78.4(3.13)	7.2
random forest	84.5(4.73)	0.0
XGBoost	84.3 (3.96)	0.2
decision tree	84.1 (4.23)	0.5
k-nearest neighbors	74.4 (5.53)	12.0
linear regression	72.8 (6.07)	13.8

Results. We conduct a performance comparison of the models based on their 5-fold cross-validation R^2 scores and consider their standard deviation (see Table 1). The random forest model achieves the highest performance with an R^2 of 84.5%, closely followed by XGBoost and Decision Tree, which scores 0.2 and 0.5% points lower, respectively.

The neural network model achieves an R^2 of 81.5%, indicating a performance 3.5% worse than the best model. However, it still provides a reasonable result compared to K-Nearest Neighbors (KNN) and Linear Regression, which obtain significantly lower R^2 scores of 12% and 13.8%, respectively.

The Federated neural network demonstrates an R^2 of 78.4%, slightly lower than the centralized neural network but 7.2% worse than the random forest model. Notably, the Federated neural network exhibits a lower standard deviation of 3.99 compared to the centralized neural network (4.92) and also outperforms the random forest model (4.73) in this regard.

Discussion. Based on the research questions, we can draw clear conclusions from the findings presented in Table 1. Initially, we compare the performance of different models, including a simple neural network. Although the random forest model outperforms others, its performance is only 3.5% higher, distinguishing it significantly from models such as KNN and linear regression, which performs 12% and 13.8% worse than the random forest, respectively.

The observed performance decrease from 81.5% to 78.4% in the FL approach can be attributed to the training process and falls within a reasonable range. Considering the privacy advantages of FL, the 7.2% accuracy loss compared to the best model is acceptable, particularly when taking into account the reduction in standard deviation from 4.92 to 3.99.

Although this example is hypothetical, it highlights the potential benefits and importance of FL in official statistics. It showcases how FL provides access to crucial data sets for ML while maintaining nearly negligible loss in accuracy compared to a centralized data set.

3.2   Fine Dust Pollution

Reducing air pollution is a significant part of the Sustainable Development Goals (SDGs) established by the United Nations[8]. To measure progress toward achieving SDGs, NGOs and other data producing organizations developed a set of 231 internationally comparable indicators, including *annual mean levels of fine particulate matter* (e.g. $PM_{2.5}$ and PM_{10}). [4] showed that personalized FL can be used to extract timely high frequent information on air pollution more accurately than models using centralized data.

In our second use case, we again provide a comparison between centralized and FL models (without personalization). We utilize a slightly different data set and methodology compared to [4], which we explain at the end of this section. We model a classification task in which the current fine dust pollution is inferred based on meteorological input data. More precisely, 48 consecutive hourly measurements are used to make a prediction for the current $PM_{2.5}$ pollution (the total weight of particles smaller than $2.5\,\mu m$ in one m^3). The predictor classifies the given inputs in one of the classes *low*, *medium* or *high*. The thresholds for these classes are chosen such that the samples of the whole data set are distributed evenly among the three classes.

---

[8] Air quality and health https://www.who.int/teams/environment-climate-change-and-health/air-quality-and-health/policy-progress/sustainable-development-goals-air-pollution, accessed on September 29, 2023.

Data Set. The data set is a multi-feature air quality and weather data set[9]. It consists of hourly measurements of 12 meteorological stations in Beijing, recorded between 2013 and 2017. It consists of more than 420 000 data records.

Data Preprocessing. To compensate missing data records, we use linear interpolation. The wind direction attribute is one-hot encoded, all other features are scaled to a range from 0 to 1. The attributes PM_{10}, SO_2, NO_2, CO and O_3 are highly correlated with the target attribute, thus we exclude them from training. 80% of the data are used as training data, the rest is used as test data.

Setup. As in the first use case, we implement a centralized learning benchmark and compare it with a FL approach. We model one FL client per meteorological station and split the data accordingly, while the benchmark model is trained with data from all 12 stations. In both settings, we use neural networks with long-short term memory (LSTM) layers and apply 5-fold cross validation. The architecture of the networks is similar across both settings and has been manually tuned to reach a good performance: The input layer is followed by a 10-neuron LSTM layer, a dropout layer with a dropout rate of 25%, a 5-neuron LSTM layer, another dropout layer with a rate of 35% and a 3-neuron dense output layer. For the same reasons as in the first use case, we use the Adam optimizer and apply a learning rate of 0.05 on the server and 0.005 on the client. The client learning rate is decreased every 64 epochs by a factor of 10 to facilitate fine-tuning in later stages of the training. The total training budget is 10 epochs for centralized learning and 200 epochs for FL (with a single round of local training per epoch).

Results. A summary of our results for the fine dust pollution use case is provided in Table 2. Depicted are the means of our 5-fold cross validation experiments.

The centralized learning benchmark reaches a mean accuracy of 72.4%, with similarly high numbers for precision and recall (72.8%, respectively 72.3%). In comparison, the FL classifier reaches a performance of both an accuracy and a recall of 68.0% and a precision of 67.9%. The relative standard deviation is higher in the FL scenario for all three metrics, reaching from +2.67% points (accuracy) to +2.9% points (both precision and recall).

Table 2. Performance in the fine dust pollution simulation. The span of the relative loss refers to all three metrics.

Model	Acc. ($\pm$ std)	Prec. ($\pm$ std)	Rec. ($\pm$ std)	Rel. loss (%)
neural net	72.4% (4.92)	72.8% (8.66)	72.3% (8.10)	0.0
neural net. (fed.)	68.0% (7.59)	67.9% (10.05)	68.0% (9.59)	5.9–6.7

---

[9] Beijing multi-site air-quality data set https://www.kaggle.com/datasets/sid321axn/beijing-multisite-airquality-data-set, accessed on September 29, 2023.

Discussion. Compared to the first use case, the training database is significantly larger. With 12 clients, there are also four times as many participants in the FL scenario as in the first use case. Still, the performance decrease is small, with an accuracy of 68.0% (FL) compared to 72.4% in the centralized training scenario.

Apart from preprocessing the data set, another time-consuming part of the engineering was tuning the hyperparameters of the FL training. Tools for an automated HPO for FL were out of scope for this work, thus it was necessary to manually trigger different trial runs with varying hyperparameters.

Comparison with Literature. The authors of [4] compare the results of their personalized FL strategy "Region-Learning" to a centralized learning baseline and standard FL. Although their personalized FL approach outperforms the other two approaches (averaged over the regions by 5% points compared to standard FL), we want to stress that Region-Learning has another goal than standard FL–namely multiple specialized models, and not one global model as in standard FL and most use cases for official statistics (also see Subsect. 2.1).

Furthermore, Hu et al. have not provided sufficient information to retrace their experiments. Especially the number of classes for $PM_{2.5}$ classification and information on the training features are missing, which complicates a meaningful comparison. Also, we have no information on whether cross validation was applied in the work of Hu et al. Hints in the paper [4] further suggest that they have used a slightly different data set than we have: The data set they describe includes "more than 100 000" data records from 13 meteorological stations in Beijing, while our data set contains more than 420 000 records from 12 stations.

One consistency across both works is the accuracy drop from centralized learning to FL, with 4, respectively 4.4% points in [4] and our work.

3.3   Mobile Radio (LTE)

Mobile Network Operator (MNO) data is a valuable source for obtaining high-frequency and spacial insights in various fields, including population structure, mobility and the socio-economic impact of policy interventions. However, a lack of legal frameworks permitting access to data of all providers constrain the quality of analysis [10]. Accessing only data of selected providers introduces biases, making FL an attractive solution to enhance the representativeness by enabling the aggregation of insights from multiple MNOs.

Thus, our third use case is based on private MNO data owned by the company umlaut SE[10]. Different from the first two use cases, we had no direct access to the data, just as the aggregator in realistic FL settings. Hence, the focus of this use case is more on practical engineering issues of FL and less on optimal results.

The data set contains mobile communication network coverage data, each linked to the mobile LTE devices of individual users, a timestamp and GPS coordinates, such that a daily "radius of action" can be computed for each user. This radius describes how far a user has moved from their home base within

---

[10] umlaut website https://www.umlaut.com/, accessed on September 29, 2023.

one day. We define a home base as the place where most data records have been recorded. The ML task we model in this use case is to estimate the daily radius of action for a user, given different LTE metrics of one particular day (see below).

Data Set. The whole data set originally contains 286 329 137 data records. The following features of the data set have been aggregated for each day and user: *radius of action* in meters, *share of data records with Wi-Fi connection* and the variance and mean values for each of the following LTE metrics: *RSRQ, RSRP, RSSNR* and *RSSI*. The date has been encoded into three numeric features (*calendar week, day of the week* and *month*) and the boolean feature *weekend*.

Data Preprocessing. We set a time frame of six months and a geofence to filter the data. Additionally, each user in the database needs to have data for at least 20 different days and 10 records on each of these days–otherwise, all records of a user are discarded. After filtering, there are 1 508 102 records in the data set. All features are scaled to a range from 0 to 1. 60% of the data are used as training data, 20% are used as validation data and the remaining 20% as test data. For FL, we have divided the data set according to the three major mobile network operators (MNOs) of the users.

Setup. We use two centralized learning benchmarks: a random forest regressor and a neural network, which have both been subject to a hyperparameter search before training. The network architecture for both the centralized benchmark neural network and the FL training process is the same: The first layer consists of 28 dense-neurons and the second layer consists of 14 dense-neurons, which lead to the single-neuron output layer. All dense layers except for the output layer use the ReLU activation function. For FL, we use the SGD optimizer with a server learning rate of 3.0, a client learning rate of 0.8 and a batch size of 2.

Results. The benchmarks of the centralized learning regressors are R^2 values of 0.158 (random forest), 0.13 (neural network) and 0.13 (linear regression). For the neural network trained in the FL scenario, we achieve a slightly lower R^2 value of 0.114 (see Table 3).

Table 3. Performance in the mobile radio simulation.

Model	R^2	Rel. loss (%)
neural network	0.130	17.7
neural network (federated)	0.114	27.8
random forest	0.158	0.0

Discussion. The reasons behind the weak performance of the benchmark models (R^2 of 0.158 and 0.13) are not clear. The hyperparameters might not be optimal, since not many resources could be spent on HPO due to time constraints of the data owner. Another reason might be the that the modeled task (estimating the radius of action based on LTE connection data) is inherently hard to learn. With an R^2 of 0.114, we were able to reproduce this performance in the FL setting.

Since the private data set in this use case has not left company premises, there are important lessons to be learned from a practical perspective:

1. Even if the data set is not directly available during the model engineering process, it is crucial to get basic insights on the features and statistical properties before starting the training. Essential decisions, such as the type of model to be trained, can be made based on this.
2. A thorough HPO is needed to obtain useful results, which might take a lot of time and computational resources.
3. Technical difficulties while creating the necessary APIs and setting up the chosen ML framework at the clients can further slow down the process. Without access to the data, it might be hard to reproduce technical errors.

While all points mentioned above were encountered in the third simulation, there was only *one* party who held all data. In real FL scenarios with multiple data holders, the process might get much more complicated.

3.4 Key Observations

Our simulations lead to the following key observations:

Models trained via FL can reach a performance very close to models trained with centralized ML approaches, as we have shown in all three use cases. While the performance gap itself is not surprising (since the FL model has been exposed to the complete data set only indirectly), we want to stress that without FL, many ML scenarios might not be possible due to privacy concerns, trade secrets, or similar reasons. This is especially true for health care data (see Subsect. 3.1).

While the random forest regressor has demonstrated superior performance compared to other centralized learning benchmarks in all three simulations, exploring the potential of tree-based models within a FL context [6] could be a promising avenue for further investigation. The improved interpretability over many other model types is another advantage of tree-based models.

On the other hand, random forest regressors are not suitable for complicated tasks. Also, their architecture, i.e., many decision trees which may be individually overfitted to parts of the training data, can facilitate the extraction of sensitive information of the training data and thus pose an additional privacy risk.

All FL simulations were performed on the machine which also had access to the complete data set. In a real-world application with clients on distinct machines, other frameworks might be more practical than TensorFlow Federated.

Last but not least, we want to emphasize that FL, despite its privacy-enhancing character, may still be vulnerable to some ML privacy issues (see Subsect. 2.2). Hence, analyzing and communicating these risks is an important step before an application is rolled out in practice.

4 Implications for Official Statistics

Official statistics are characterized by high accuracy while underlying strict standards in confidentiality and privacy. In this setting, our findings indicate that FL bears significant potential to support statistical production. Specifically, FL can help NSOs to address pressing data needs, empowering them to generate reliable models that accurately capture global relations while ensuring quality standards.

If upcoming applications require to optimize an individual model for each participating party, personalized FL can be used to generate potentially improved models tailored to individual clients [5]. This increases the interest to cooperate for each participating party, as it offers to enhance the analytic potential for each client and the server. However, it is important to note that this customization may come at the cost of global predictive performance.

In cases where legislative changes prove impractical, FL provides a crucial pathway to assess and prepare for regulations' modernization. By showcasing the advantages and implications of accessing new data sources before legal frameworks permit, FL not only significantly accelerates and relieves statistics production but also occasionally enables it.

Our observations indicate that FL generally requires a greater number of training epochs compared to achieve similar performance levels to centralized learning. Thus, we see a need for the development of infrastructure for seamless sending and receiving ML model updates. Additionally, it is crucial to provide partners with the necessary tools to solve issues like harmonizing client data without directly accessing it and finding a suitable model architecture.

The efficiency of practical FL frameworks is expected to be further optimized in the future. Similarly, we expect the development of more usable PPML algorithms including Secure Multi-Party Computation (SMPC) and Homomorphic Encryption (HE) - allowing for provably secure collaborative ML [13]. Although such PPML methods have been proposed and frameworks exist, their performance today is often unacceptable for many practical applications [11,13].

In summary, FL should indeed be recognized as an important technology that can facilitate the modernization of legal frameworks for official statistics. It enables NSOs to safely use publicly relevant information, ultimately enhancing the quality and relevance of official statistics. However, further development is still required to fully realize the potential of FL in this context.

5 Conclusion

In scenarios where external partners are unwilling to share individual-level information but still aim to provide global insights in their data, FL can help to make an exchange possible. We have shown across a range of three simulated use cases that FL can reach a very similar performance to centralized learning algorithms. Hence, our results indicate that if classic (centralized) ML techniques work sufficiently well, FL can possibly produce models with a similar performance.

One of the next steps to transfer FL into the practice of official statistics could be to conduct pilot studies. These could further showcase both the applicability and challenges of FL beyond a simulated context. Another focus of future work in this area could be the analysis of privacy risks in FL scenarios of official statistics and potential mitigation strategies. This would be an important stepping stone in ensuring the privacy protection of involved parties, on top of the privacy enhancement by using FL. Just as in countless other domains, we expect FL to become a relevant technology for official statistics in the near future.

References

1. Beck, M., Dumpert, F., Feuerhake, J.: Machine learning in official statistics. In: arXiv preprint arXiv:1812.10422 (2018)
2. Buckley, D.: 15. United Nations Economic Commission for Europe: trialling approaches to privacy-preserving federated machine learning (2023)
3. United Nations Economic Commission for Europe. Machine Learning for Official Statistics. In: UNECE Machine Learning Group (2022)
4. Hu, B., et al.: Federated region-learning: an edge computing based framework for urban environment sensing. In: IEEE GLOBECOM (2018)
5. Kulkarni, V., Kulkarni, M., Pant, A.: Survey of personalization techniques for federated learning. In: WorldS4. IEEE (2020)
6. Li, Q., Wen, Z., He, B.: Practical federated gradient boosting decision trees. In: AAAI (2020)
7. McMahan, B., et al.: Communication-efficient learning of deep networks from decentralized data. In: Artificial Intelligence and Statistics, pp. 1273–1282. PMLR (2017)
8. Melis, L., et al.: Exploiting unintended feature leakage in collaborative learning. In: SP, pp. 691–706. IEEE (2019)
9. Reddi, S., et al.: Adaptive federated optimization. In: arXiv preprint arXiv:2003.00295 (2020)
10. Saidani, Y., Bohnensteffen, S., Hadam, S.: Qualität von Mobilfunkdaten - Projekterfahrungen und Anwendungsfälle aus der amtlichen Statistik. Wirtschaft und Statistik 5, 55–67 (2022)
11. Santos, B., et al.: Insights into privacy-preserving federated machine learning from the perspective of a national statistical office. In: Conference of European Statistics (2023)
12. Stock, J., et al.: Lessons learned: defending against property inference attacks. In: SECRYPT (2023). https://doi.org/10.5220/0012049200003555
13. Yin, X., Zhu, Y., Hu, J.: A comprehensive survey of privacy-preserving federated learning: a taxonomy, review, and future directions. ACM Comput. Surv. (CSUR) 54(6), 1–36 (2021)
14. Yung, W., et al.: A quality framework for statistical algorithms. Stat. J. IAOS 38(1), 291–308 (2022)

Generating Wildfire Heat Maps with Twitter and BERT

João Cabral Pinto, Hugo Gonçalo Oliveira, Alberto Cardoso, and Catarina Silva[✉]

Department of Informatics Engineering, CISUC, University of Coimbra, Coimbra, Portugal
joaopinto@student.dei.uc.pt, {catarina,alberto,hroliv}@dei.uc.pt

Abstract. Recent developments in Natural Language Processing (NLP) have opened up a plethora of opportunities to extract pertinent information in settings where data overload is a critical issue. In this work, we address the highly relevant scenario of fire detection and deployment of firefighting resources. Social media posts commonly contain textual information during a fire event, but their vast volume and the necessity for swift actionable information often precludes their effective utilization. This paper proposes an information extraction pipeline capable of generating a wildfire heat map of Portugal from Twitter posts written in Portuguese. It uses a fine-tuned version of a BERT language model to extract fire reports from large batches of recent fire-related tweets as well as the spaCy NLP library to query the location of each recently reported fire. Wildfire locations are plotted to a colored map indicating the most probable fire locations, which could prove useful in the process of allocating firefighting resources for those regions. The system is easily adaptable to work with any other country or language, provided compatible BERT and spaCy models exist.

Keywords: Natural Language Processing · Volunteered Geographic Information · Fire Detection · BERT · Wildfire Heat Maps

1 Introduction

In recent years, wildfires have become increasingly frequent, large, and severe[1], resulting in great loss of human lives, as well as the destruction of natural ecosystems and man-made infrastructure. It is therefore of great importance to create early detection mechanisms to reduce their spread and negative effects.

Traditionally, wildfire detection methods rely on physical devices capable of sensing heat or other signs indicative of a fire burst. In many cases, though, this approach may prove ineffective, as it requires significant investment to cover all susceptible areas.

Fire detection using intelligent methods can be extremely helpful in supporting fire management decisions [6], namely in resource deployment for firefighting. As fire can be considered one of the leading hazards endangering human life, the economy, and the environment [8], identifying a forest fire using such methods in its early stages is

---

[1] https://www.epa.gov/climate-indicators/climate-change-indicators-wildfires (accessed 2023-07-23).

P. Quaresma et al. (Eds.): IDEAL 2023, LNCS 14404, pp. 82–94, 2023.
https://doi.org/10.1007/978-3-031-48232-8_9

essential for a faster and more effective response from fire and civil protection authorities. An effective initial response can significantly decrease fire damage and, in certain cases, can also prevent the loss of lives and property, e.g., houses, and other belongings. Recently, climate change awareness has amplified the interest in fire detection and different techniques are being pursued.

Citizen science [15] can be described as the participation of individuals in scientific research with the goal of offering advancements by improving awareness, participation, and knowledge of science by the public. Usually, it is applied in solving problems that technology still struggles. For firefighting, crowdsourcing, or citizen science, can be extremely important in early detection, when processing and communication systems are in place for such reporting [9]. However, such contributions are sometimes overwhelming in volume and the need for swift actionable information often preempts their effective usage.

A typical example of this readily available and cheap information is social media that offers a large amount of accessible constantly updating information provided by users all across the globe. For this reason, it has become increasingly used in the field of disaster detection, and, more specifically, fire detection.

Language models like Bidirectional Encoder Representations from Transformers (BERT) [4] are being increasingly utilized in a wide range of scenarios, from natural language understanding and generation to recommendation systems and machine translation. In the context of fire detection a possible application is the analysis of social media posts to extract relevant information, thus enabling faster response times and more efficient deployment of firefighting resources.

In this paper we propose an information extraction pipeline capable of generating a wildfire heat map of Portugal from Twitter posts written in Portuguese. In the architecture described in the rest of this paper, Natural Language Processing (NLP) techniques are applied to large batches of recently posted tweets to generate a wildfire heat map of a country, which could serve as an additional tool for fire departments to allocate resources to areas in need in a timely manner.

The remainder of the paper is organized as follows. Section 2 starts off by providing a quick rundown of some of the work published in the area of fire burst detection. In Sect. 3, each step of the proposed pipeline is detailed. Section 4 follows with an empiric evaluation of the pipeline. The article is concluded in Sect. 5, where the main takeaways of this work are discussed, as well as some of the possible developments it could benefit from.

2 Related Work

Automatic wildfire detection has already been subjected to a significant amount of research over the years. This section summarizes some of the most relevant publications in the field, primarily focusing on those that use social media and natural language processing to detect wildfire occurrences. It is important to keep in mind that while these architectures have been designed for the English language and English-speaking regions, they should still be applicable to other linguistic contexts, such as Portugal. Also discussed is the recent history of language models that led to the development of BERT and its later application in disaster detection.

2.1 Fire Burst Detection

Traditionally, fire burst detection techniques rely on the presence of sensing equipment present near the emergency. In particular, video-based fire detection (VFD) from surveillance cameras is reported to be beneficial compared to using heat sensors as it can be used to monitor much larger areas with less equipment and does not suffer from the transport delay that heat sensors do. In [12], B. Töreyin et al. use a set of clues to identify fire pixels in surveillance footage. To determine if a region of an image is moving between frames, the authors employ a hybrid background estimation method. They also verify that pixels match a range of fire colors derived from a sample image containing a fire region. Additionally, flame flicker is detected in the video through both spatial and temporal wavelet domain analysis. More recent approaches resort to CNN as in [1,5], where a CNN-based architecture based on GoogleNet for detecting fires in surveillance footage is presented. The authors aim to improve the efficiency of previous feature extraction methods in order to facilitate use in real-time. Experimental results on benchmark fire datasets show that the proposed framework is effective in detecting fires in video compared with previously available methods and validate its suitability for fire detection in surveillance systems.

C. Boulton et al. [2] demonstrate that social media activity about wildfires is both temporally and spatially correlated with wildfire events in the contiguous United States. The authors collect data from both Twitter and Instagram by querying their public Search APIs for posts containing fire-related keywords or hashtags, respectively. Fire related terms include "wildfire", "forest fire", "bush fire", among others. This data is then compared to two official sources of wildfire occurrence data in the United States: MODIS and FPA. The authors start by analyzing the daily frequencies of fire-related posts and fire occurrences over time and find a significant positive correlation between the two over the whole of the US, although it is not consistent from state to state. To test for spatial correlation, the authors restrict their analysis to the California-Nevada region and measure the distance between any given geotagged Twitter post and its closest fire on the same day. They find that this distance is generally quite small, implying that tweets about wildfire events are most often posted by users close to them. It is important to note that only 0.51% of Twitter posts were geotagged, thus wasting a significant amount of data. This problem calls for an alternative solution, such as geoparsing, to identify the location of fires reported in Twitter posts.

During the Oklahoma Grassfires in April 2009, Vieweg et al. [14] analyzed Twitter posts to extract information that would assist in providing common situational awareness. They found that 40% of all on-topic tweets contained some sort of geolocation information. This data corroborates the notion that information systems supporting emergency response teams may benefit from the information exchanged within the local Twitter community.

K. Thanos et al. [11] propose a pipeline similar to the one described in this document. The authors employ an LSTM-CNN hybrid classifier to determine whether a fire-related tweet is authentic or not. A set of NLP procedures is applied to each tweet that is deemed valid, in order to transform it into a tagged sentence form. These procedures include tokenization, part-of-speech tagging, entity recognition, and relation recognition. Fire locations are extracted from the resulting tagged sentences using a set of regular expressions defined at the training stage. The authors aggregate the extracted locations with DBSCAN and filter out invalid reports using a mathematical reliability model to eliminate those that do not correspond to real fires.

2.2 Language Models

The field of natural language processing underwent a paradigm shift in 2017, when A. Vaswani et al. [13] introduced the original transformer architecture for machine translation, outperforming the then-prevailing Recurrent Neural Network (RNN) encoder-decoder models, both in speed and performance. One of the key characteristics of the transformer model is that it is designed to process sequential input data, such as natural language, concurrently rather than sequentially. The transformer is comprised of an encoder and a decoder stack. The encoder stack applies a multi-headed self-attention mechanism to enrich each word embedding with contextual information derived from the entire input. The resulting contextual embeddings are fed into the decoder stack, responsible with generating text in an iterative manner. For additional details, readers are strongly encouraged to consult the source material, which is beyond the scope of this article.

In 2018, A. Radford et al. [7] solidified the practice of generative pre-training (GPT) and discriminative fine-tuning of a language model. In an initial stage, the authors pre-train a transformer decoder model on a large unlabeled corpus of tokens $\mathcal{U} = \{u_1, \ldots, u_n\}$ to maximize the objective function $L_1(\mathcal{U}) = \sum_i \log P(u_i | u_{i-k}, \ldots, u_{i-1})$, where k is the size of the context window. It is then possible to fine-tune the model to perform a classification task on a labeled dataset $\mathcal{C}$, where each dataset instance consists of a sequence of tokens $\{x_1, \ldots, x_n\}$ and a label y, by maximizing $L_2(\mathcal{C}) = \sum_{(x,y)} \log P(y | x_1, \ldots, x_n)$. In practice, the model is fine-tuned by maximizing $L_3(\mathcal{C}) = L_2(\mathcal{C}) + \lambda L_1(\mathcal{C})$, as this method promotes better generalization and faster convergence. The authors achieve improved results over the contemporary state of the art on 9 out of 12 natural language processing tasks.

A significant advantage of the pre-training and fine-tuning approach over other deep learning methods is that it requires only a relatively small amount of manually labeled data, making it suitable for applications across a wide range of domains where annotated resources are scarce. Moreover, since less data is required to fine-tune the model, training time after the pre-training stage is significantly reduced compared to other approaches.

The following year, J. Devlin et al. [4] release a paper in which they argue that unidirectional approaches such as the one proposed by A. Radford et al., where each token only attends to the tokens on its left, are sub-optimal in tasks where context from both directions is essential. Instead, the authors introduce Bidirectional Encoder Representations from Transformers (BERT), which overcomes this constraint through the use of a masked language model (MLM) as a pre-training objective. In particular, they mask random tokens in a large unlabeled corpus and train a transformer encoder model to predict the masked tokens based on their surrounding context. The model is additionally pre-trained towards the task of next sentence prediction, which is demonstrated to improve performance in both question answering (QA) and natural language inference (NLI). To allow BERT to be fine-tuned for classification, every sentence fed into the

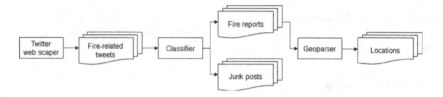

Fig. 1. Proposed architecture: (1) tweets containing fire-related keywords are obtained via web scraping; (2) tweets are classified as fire reports or junk tweets; (3) fire reports are geoparsed to obtain fire locations..

transformer is prepended with a special classification token ([CLS]) whose output can be fed into a classification layer. This architecture is shown to advance the state of the art on a large suite of sentence-level and token-level tasks.

S. Deb and A. Chanda [3] compare the contextual embeddings generated with BERT with previous context-free embedding techniques on predicting disaster in Twitter data. The authors experiment with feeding the embeddings into either a softmax or a Bi-LSTM layer. The authors found that contextual models outperformed context-free models in this domain.

3   Proposed Approach

This section details the proposed approach, comprised of three main stages. Firstly, the Twitter social network is queried to obtain posts containing fire-related words. The obtained collection is further filtered with a fine-tuned BERT classifier to solely contain fire reports. Lastly, each fire report is geoparsed to obtain the location of each fire and mark it on the map for the user to inspect. This pipeline is summarized in Fig. 1 and the corresponding Python code was made available[2].

3.1   Data Fetching

Twitter is a microblogging and social networking platform with 229 million monetizable daily active users, sending 8,817 tweets every second[3], making it an ideal platform to perform social sensing. It offers public Application Programming Interfaces (API) through which to query its database, but it can be inadequate for large data collection tasks, because it limits the number of extracted tweets to 450 every 15 min. Instead, the unofficial API SNScrape[4] was preferred for this work, as it is able to extract a virtually unlimited amount of tweets for a given query by web scraping Twitter's search page. In order to obtain data containing wildfire reports, tweets were queried to be written in Portuguese and contain one or more fire related keywords, such as "fogo" and "incêndio", which are approximate Portuguese equivalents to "fire" and "wildfire".

---

[2] https://github.com/cabralpinto/wildfire-heat-map-generation (accessed 2023-07-23).

[3] https://www.bestproxyreviews.com/twitter-statistics (accessed 2023-07-23).

[4] https://github.com/JustAnotherArchivist/snscrape (accessed 2023-07-23).

3.2 Classification

Fetching tweets containing fire-related words alone is not enough to restrict the collected data to fire reports, since no keyword is strictly used in posts reporting fires. This may be due to the keyword having more than one connotation in the language – such as "fogo", which is also used as an interjection in Portuguese – or because the tweet it is in mentions a fire but is not an actual fire report – as in "Houve um grande incêndio nos anos 90" (Translation: "There was a great fire in the 90s").

This makes it necessary to employ some sort of model capable of distinguishing between real fire reports from these particular cases. To this end, after some preliminary testing, the Portuguese version of the language model BERT, dubbed BERTimbau [10], emerged as the obvious choice. BERT was fine-tuned for the purpose by adding a feed-forward layer downstream of the pre-trained model and training it to perform the classification task. Compared to training an entire model from scratch, fine-tuning has the significant advantage of massively reducing training times without any significant loss in performance.

3.3 Geoparsing

In this stage, the fire location is extracted from each fire report to then include it in the heat map. Assuming the fire report was tweeted near the location of the fire, a simple technique to achieve this would be to use the geographical metadata in the tweet in question, as is done by Boulton et al. [2]. This method, however, poses two issues: i) a significant portion of tweets containing fire reports are from news agencies, whose locations are fixed and do not coincide with the locations of the fires they report; ii) only a residual portion of tweets are in fact geotagged (0.51%) [2], meaning that a large majority of the queried tweets, possibly containing valuable information, would not contribute to the final result. The solution is to once again resort to a language model, this time to extract the location of reported fires from the text of the corresponding tweets.

To simplify this task, the assumption is made that any location name mentioned in a fire report corresponds to the location of the fire, which seems to be quite often the case. Taking this premise as true, fire locations can be extracted through geoparsing, which is the process of parsing text to detect terms associated with geographic places. Generally, the first step of geoparsing is Named Entity Recognition (NER), which is the process of locating and classifying named entities mentioned in unstructured text into pre-defined categories, including locations. NER is applied to the text of a tweet using the spaCy Python library and results in a set of location names which are concatenated to form a preliminary geocode. The geocode is sent as a search request to the Nominatim API[5] to obtain the geometry of the corresponding region in the world map.

The final step of the pipeline is to detect intersections between the regions extracted from each fire report and the regions in a predefined area of interest such as Portugal, in the case of this work. Regions with a higher intersection count appear with a stronger orange tone in the resulting map, meaning that there is a high volume of recent tweets

---

[5] https://nominatim.org/release-docs/latest/api/Overview (accessed 2023-07-23).

reporting nearby fires. Figure 2a depicts an example of a heat map generated for June 18, 2017, in Portugal, showing the catastrophic fire that occurred in Pedrógão Grande (central region of Portugal). Figures 2b, 2c, 2d, 2e and 2f present other examples of heat maps generated for critical wildfire days in 2018, 2019, 2020, 2021 and 2022, respectively. Some districts in the maps are depicted in brighter orange tones, which coincides with a higher number of fires being reported in those areas.

4   Experimentation and Results

This section details the experimental setup used, which evaluates the proposed architecture by separately assessing the performance of the classifier and geoparser components. All training and testing was conducted on a modified version of a dataset previously created in the context of the FireLoc[6] project containing news headlines from various Portuguese media outlets. The original 2,263 dataset entries were manually annotated by volunteers according to three labels. The first label indicates whether the corresponding headline pertains to fire, while the second label specifies whether it is a fire report. Evidently, the latter implies the former, that is, a fire report always relates to the subject of fire. The third label links fire reports to specific known fire occurrences, whenever applicable. As part of this work, this label has been replaced by a set of coordinates indicating where in the map the fire occurred. A representative excerpt of the dataset is shown in Table 1, in which the original headlines have been truncated and translated to English for the reader's convenience.

Table 1. Dataset excerpt

Text	Related	Report	Latitude	Longitude
Building on fire on Rua da Moeda	1	1	40.21	−8.43
Firefighting forces to be reinforced with 75 rescue ambulances by the end of the year	1	0	–	–
New wildfire covers Leiria in smoke	1	1	39.87	−8.9
Man accused of killing ex-wife has sentence postponed	0	0	–	–
Nineteen found dead after the fire that broke out this afternoon	1	1	–	–

4.1   Classifier

The first experiment aimed to assess the utility of BERTimbau in identifying fire report in text. To accomplish this, the dataset was randomly split into training and testing sets, in a nine to one proportion. Stratified sampling was used to ensure that the resulting sets

---

[6] https://fireloc.org/?lang=en (accessed 2023-07-23).

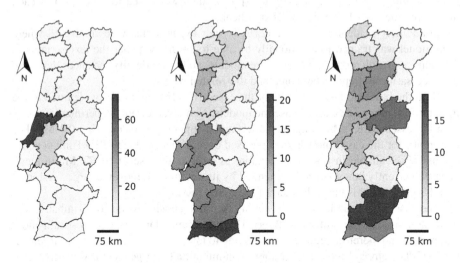

(a) Wildfire heat map of Portugal (Pedrógão Grande fire on June 18, 2017.

(b) Wildfire heat map of Portugal (Large wildfire on June 13, 2018, in Monchique).

(c) Wildfire heat map of Portugal (Large wildfire in Alentejo July 3, 2019).

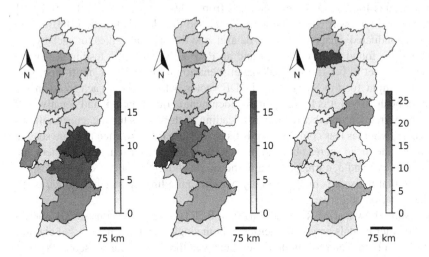

(d) Wildfire heat map of Portugal (August 22, 2020).

(e) Wildfire heat map of Portugal (July 31, 2021).

(f) Wildfire heat map of Portugal (August 7, 2022).

Fig. 2. Wildfire heat maps.

accurately reflected the label distribution in the original dataset[7]. This approach should prepare the model to be applied in real-life scenarios. The dataset was split thirty times to obtain the same number of train-test splits, which were used to train and test each model, in order to achieve statistical significance.

In addition to identifying fire reports, further experiments were conducted where each model was trained to additionally be able to predict whether the text pertains to fire. This approach aimed to enhance the models' ability to identify instances that were not necessarily fire reports but contained fire-related words.

Both the base and large versions of the cased BERTimbau model were fitted to the training sets, alongside a set of baseline models. Namely, the same experiment was run with the cased BERT Multilingual models and a support vector machine (SVM). The preference for the cased models over uncased ones is merely due to BERTimbau being exclusively available in the cased version, as well as the use of BERT Multilingual uncased currently being not recommended by its authors. Moreover, a large version of BERT Multilingual is unavailable as of the time of writing.

The base models underwent fine-tuning for three epochs, while BERTimbau Large was fine-tuned for an additional three. This is because larger models contain more parameters and therefore require more training to minimize the loss function. Moreover, due to GPU memory limitations, gradient accumulation for large models was performed over two training steps instead of one, and the batch size was set to 4 rather than the standard 8. A preliminary grid search was run to obtain optimal hyperparameters for the SVM classifier. The search space included its three main parameters: C, γ, and kernel type. The values for C and γ were sampled from $\{0.001, 0.01, 0.1, 1, 10, 100, 1000\}$ while the kernel type was chosen between linear and radial basis function (RBF). The search was conducted once on the entire dataset, and the optimal values for each hyperparameter were found to be $C = 100$ and an RBF kernel for both testing scenarios, while γ is best set to 0.1 in the single-label scenario and to 0.01 for multi-label prediction. The input for the SVM is a TF-IDF vector, which scores each word in the vocabulary positively if it is frequent in the document in question in it and negatively if it is generally frequent in other documents, leading to stopwords having a near null relevancy score, whereas less common words like "fire" are favored when they are seen in a document. The mean scores obtained in each experiment, after removing outliers, are shown in Tables 2 and 3, corresponding to the single and multi-label scenarios, respectively. Note that the scores in Table 3 only pertain to the fire report label, which is the main classification objective.

The results show that BERTimbau Base and BERTimbau Large consistently outperformed the other models across all metrics in both single-label and multi-label classification tasks. In both scenarios, both models achieved the same accuracy score (96.4%). However, while BERTimbau Large achieved the highest precision score (83.6% and 85.1%), it was outperformed by the Base version in terms of recall (89.5% and 84.0%). Somewhat surprisingly, the lowest performing model was BERT Multilingual, which was outperformed by the SVM across most metrics. The reason for the superior performance of the SVM may be due to it taking a TF-IDF vector as input and the dataset

---

[7] For further details, see https://scikit-learn.org/stable/modules/generated/sklearn.model_selection.StratifiedShuffleSplit.html.

Table 2. Single-label classification results

Model	Accuracy	Precision	Recall	F1-score
SVM	94.8 ± 0.2	79.7 ± 1.1	73.4 ± 1.2	76.3 ± 0.8
BERT Multilingual Base	93.6 ± 0.4	74.8 ± 2.3	70.2 ± 4.0	70.7 ± 2.7
BERTimbau Base	**96.4 ± 0.2**	80.8 ± 0.9	**89.5 ± 0.9**	84.9 ± 0.7
BERTimbau Large	**96.4 ± 0.3**	**83.6 ± 1.8**	87.3 ± 1.3	**85.0 ± 0.9**

Table 3. Multi-label classification results

Model	Accuracy	Precision	Recall	F1-score
SVM	95.1 ± 0.2	80.5 ± 1.5	76.2 ± 1.0	78.1 ± 0.7
BERT Multilingual Base	90.7 ± 0.8	58.7 ± 2.8	80.5 ± 1.9	67.2 ± 1.9
BERTimbau Base	**96.4 ± 0.1**	84.6 ± 0.7	**84.0 ± 0.7**	**84.3 ± 0.5**
BERTimbau Large	**96.4 ± 0.2**	**85.1 ± 1.2**	83.8 ± 1.3	84.2 ± 0.9

being comprised of news headlines, some of which do not include fire-related words. Hence, the SVM may developed a strategy during training of triggering a positive outcome when a fire-related word has a high score in the input vector. Evidently, this classification strategy would not be transferable to the use case at hand, in which tweets to be classified always contain fire-related words. BERT Multilingual, on the other hand, is theoretically better equipped to handle the nuances of natural language and may be more easily applicable to the use case, leading to better overall performance. Nevertheless, further investigation is required in order to determine the reason why BERT Multilingual performed poorly on this particular dataset.

Table 4. Geoparsing results

Quantile	Latitude	Longitude
0%	0.000000	0.000000
10%	0.000000	0.000000
20%	0.000000	0.000000
30%	0.000000	0.000000
40%	0.000000	0.000000
50%	0.000354	0.000179
60%	0.013298	0.010896
70%	0.016979	0.030434
80%	0.065864	0.065530
90%	0.119159	0.134588
100%	18.040577	94.980643

Comparing the Tables 2 and 3, one can observe that the F1-score achieved is generally lower for multi-label classification. This is the case of the BERTimbau models, where a decrease in recall is observed alongside a less pronounced increase in precision. This makes sense, as training the models to differentiate between tweets that mention fire and actual fire reports should reduce the number of fire-related tweets mistakenly classified as fire reports, thereby reducing the number of false positives. However, this seems to come at the cost of recall, due to the models wrongfully classifying some fire reports as tweets that merely discuss the subject of fire. This trade-off is not observed in the case of the SVM, not penalized in terms of recall, the only model with a higher F1-score in multi-label scenario. Furthermore, BERT Multilingual sees a significant improvement in recall but a decrease in precision in the multi-label scenario, which warrants further investigation. In any case, it is worth noting that the trade-off observed in the Portuguese models is not desirable in the field of disaster detection, where false negatives could result in missed disasters and potentially risk lives.

4.2 Geoparser

To ascertain the effectiveness of the geoparsing technique, the method was applied to each entry in the dataset annotated with a set of coordinates. The coordinates for each geocode were obtained with the Nominatim API and compared against the annotations. The resulting errors for the latitude and longitude values are described in Table 4.

While there is certainly room for improvement, the employed technique allowed to successfully geoparse the majority of the fire reports in the dataset, with a median error of only 0.000354 for latitude and 0.000179 for longitude. Nevertheless, some extreme outliers could be found in the uppermost percentiles, indicating imperfections in the simple geoparsing process. To eliminate these outliers, a viable approach would be to fine-tune a BERT instance for the specific task of locating fire report sites. This approach warrants further investigation.

5 Conclusions and Future Work

Wildfires are a significant and ongoing concern in many parts of the world, including Portugal. Social media platforms like Twitter can provide a wealth of information in real-time during wildfire events, including eyewitness accounts, photos, and videos. Recent breakthroughs in Natural Language Processing (NLP) have opened up the opportunities to automatically extract pertinent information for scenarios like these.

In this work, an NLP pipeline was developed to transform batches of recent fire-related Tweets into wildfire heat maps. It could aid fire departments by quickly detecting fire bursts and allowing faster action. We propose a novel information extraction pipeline that utilizes the state-of-the-art language model BERTimbau (Portuguese BERT) to generate a wildfire heat map of Portugal from Twitter posts written in Portuguese.

The classifier component of the pipeline distinguishes tweets containing fire reports from other tweet categories. BERTimbau was fine-tuned for this task and achieves a much higher recall than an SVM baseline model trained on the same data, suggesting that it is the preferable option in NLP classification tasks, especially with imbalanced data.

Given the opportunity to further develop this work, it is a must to develop a comprehensive data set containing tweets containing fire-related words instead of news headlines. Such data set would better suit the data queried from Twitter and should therefore lead to higher classification scores. Additionally, each reported fire in the data set should be annotated with a geometric shape representing the affected area, allowing us to empirically evaluate the geoparsing stage of the pipeline.

Acknowledgments. This work is funded by the FCT – Foundation for Science and Technology, I.P./MCTES through national funds (PIDDAC), within the scope of CISUC R&D Unit - UIDB/00326/2020 or project code UIDP/00326/2020. This work is also funded by project Fire-Loc, supported by the Portuguese Foundation for Science and Technology (Fundação para a Ciência e a Tecnologia – FCT) under project grants PCIF/MPG/0128/2017.

References

1. Altowaijri, A.H., Alfaifi, M.S., Alshawi, T.A., Ibrahim, A.B., Alshebeili, S.A.: A privacy-preserving IoT-based fire detector. IEEE Access **9**, 51393–51402 (2021). https://doi.org/10.1109/ACCESS.2021.3069588
2. Boulton, C.A., Shotton, H., Williams, H.T.P.: Using social media to detect and locate wildfires. In: Proceedings of the International AAAI Conference on Web and Social Media (2016)
3. Deb, S., Chanda, A.K.: Comparative analysis of contextual and context-free embeddings in disaster prediction from Twitter data. Mach. Learn. Appl. **7**, 100253 (2022). https://doi.org/10.1016/j.mlwa.2022.100253. https://www.sciencedirect.com/science/article/pii/S2666827022000032
4. Devlin, J., Chang, M.W., Lee, K., Toutanova, K.: BERT: pre-training of deep bidirectional transformers for language understanding. In: Proceedings of the 2019 Conference of the North American Chapter of the Association for Computational Linguistics: Human Language Technologies (Volume 1: Long and Short Papers), Minneapolis, Minnesota, pp. 4171–4186. Association for Computational Linguistics (2019). https://doi.org/10.18653/v1/N19-1423. https://aclanthology.org/N19-1423
5. Muhammad, K., Ahmad, J., Mehmood, I., Rho, S., Baik, S.W.: Convolutional neural networks based fire detection in surveillance videos. IEEE Access **6**, 18174–18183 (2018). https://doi.org/10.1109/ACCESS.2018.2812835
6. Phelps, N., Woolford, D.G.: Comparing calibrated statistical and machine learning methods for wildland fire occurrence prediction: a case study of human-caused fires. Int. J. Wildland Fire **30**, 850–870 (2021). https://doi.org/10.1071/WF20139
7. Radford, A., Narasimhan, K., Salimans, T., Sutskever, I.: Improving language understanding by generative pre-training (2018)
8. Saponara, S., Elhanashi, A., Gagliardi, A.: Real-time video fire/smoke detection based on CNN in antifire surveillance systems. J. Real-Time Image Proc. **18**(3), 889–900 (2020). https://doi.org/10.1007/s11554-020-01044-0
9. Silva, C., Cardoso, A., Ribeiro, B.: Crowdsourcing holistic deep approach for fire identification. In: IX International Conference on Forest Fire Research (2022)
10. Souza, F., Nogueira, R., Lotufo, R.: BERTimbau: pretrained BERT models for Brazilian Portuguese. In: Cerri, R., Prati, R.C. (eds.) BRACIS 2020. LNCS (LNAI), vol. 12319, pp. 403–417. Springer, Cham (2020). https://doi.org/10.1007/978-3-030-61377-8_28

11. Thanos, K.G., Polydouri, A., Danelakis, A., Kyriazanos, D., Thomopoulos, S.C.: Combined deep learning and traditional NLP approaches for fire burst detection based on Twitter posts. In: Abu-Taieh, E., Mouatasim, A.E., Hadid, I.H.A. (eds.) Cyberspace, chap. 6. IntechOpen, Rijeka (2019). https://doi.org/10.5772/intechopen.85075

12. Töreyin, B.U., Dedeoğlu, Y., Güdükbay, U., çetin, A.E.: Computer vision based method for real-time fire and flame detection. Pattern Recogn. Lett. **27**(1), 49–58 (2006). https://doi.org/10.1016/j.patrec.2005.06.015. https://www.sciencedirect.com/science/article/pii/S0167865505001819

13. Vaswani, A., et al.: Attention is all you need. In: Guyon, I., et al. (eds.) Advances in Neural Information Processing Systems, vol. 30. Curran Associates, Inc. (2017). https://proceedings.neurips.cc/paper/2017/file/3f5ee243547dee91fbd053c1c4a845aa-Paper.pdf

14. Vieweg, S., Hughes, A.L., Starbird, K., Palen, L.: Microblogging during two natural hazards events: what twitter may contribute to situational awareness. In: Proceedings of the SIGCHI Conference on Human Factors in Computing Systems (2010)

15. Zanca, D., Melacci, S., Gori, M.: Toward improving the evaluation of visual attention models: a crowdsourcing approach. In: 2020 International Joint Conference on Neural Networks, IJCNN 2020, Glasgow, United Kingdom, 19–24 July 2020, pp. 1–8. IEEE (2020). https://doi.org/10.1109/IJCNN48605.2020.9207438

An Urban Simulator Integrated with a Genetic Algorithm for Efficient Traffic Light Coordination

Carlos H. Cubillas[1]([envelope]), Mariano M. Banquiero[1][iD], Juan M. Alberola[2][iD],
Victor Sánchez-Anguix[3][iD], Vicente Julián[2,4][iD], and Vicent Botti[2,4][iD]

[1] Universitat Politècnica de València, Camino de Vera, s/n, 46022 Valencia, Spain
chcub@doctor.upv.es, mbanqui@upv.es
[2] Valencian Research Institute for Artificial Intelligence (VRAIN), Universitat Politècnica de València, Camino de Vera, s/n, 46022 Valencia, Spain
{jalberola,vjulian,vbotti}@upv.es
[3] Instituto Tecnológico de Informática, Grupo de Sistemas de Optimización Aplicada, Ciudad Politécnica de la Innovación, Edificio 8g, Universitat Politècnica de València, Camino de Vera s/n, 46022 Valencia, Spain
vicsana1@upv.es
[4] ValgrAI (Valencian Graduate School and Research Network of Artificial Intelligence), Universitat Politècnica de València, Camino de Vera, s/n, 46022 Valencia, Spain

Abstract. Emissions from urban traffic pose a significant problem affecting the quality of cities. The high volume of vehicles moving through urban areas leads to a substantial amount of emissions. However, the waiting time of vehicles at traffic lights results in wasted emissions. Therefore, efficient coordination of traffic lights would help reduce vehicle waiting times and consequently, emissions. In this article, we propose a GPU-based simulator with an integrated genetic algorithm for traffic lights coordination. The key advantage of this genetic algorithm is its compatibility with an optimized urban traffic simulator designed specifically for calculating emissions of this nature. This, together with the efficiency of the simulator facilitates the processing of large amount, enabling simulation of large urban areas such as metropolitan cities.

Keywords: Urban traffic · GPU · urban traffic simulator · traffic lights coordination · genetic algorithm · large urban areas

1 Introduction

The increase in carbon dioxide (CO_2) and nitrogen dioxide (NO_2) emissions in urban environments is a growing problem that affects both air quality and the health of residents in these areas. In recent years, several scientific studies have been conducted to better understand this problem and propose effective solutions [1–4]. CO_2 and NO_2 emissions are strongly associated with urban traffic.

© The Author(s), under exclusive license to Springer Nature Switzerland AG 2023
P. Quaresma et al. (Eds.): IDEAL 2023, LNCS 14404, pp. 95–106, 2023.
https://doi.org/10.1007/978-3-031-48232-8_10

The increase in the number of vehicles and traffic congestion are key factors contributing to these emissions [5]. In this regard, the time that a vehicle spends waiting at a traffic light is idle time that does not contribute to reaching the destination, thus emissions are related to this waiting time [6]. Hence, optimizing the red/green light timing of traffic signals is important as a strategy to reduce pollutant emissions, as it allows for better traffic flow and a reduction in abrupt stops and accelerations of vehicles.

Some of the techniques widely applied for optimization of emissions are genetic algorithms. Recent papers reviewing and comparing models indicate that genetic algorithms tend to yield the best results for this type of optimization problem [7]. Numerous studies have utilized genetic algorithms to minimize carbon dioxide emissions resulting from urban traffic [8–10].

In the literature, several proposals are focused to the optimization of traffic lights, both at the junction level and the city level. At the junction level, some proposals are aimed at optimizing traffic lights at individual intersections. For example, particle swarm optimization and simulation tools like SUMO and VANET have been employed to optimize traffic light timings [11]. Other proposals introduce an active control system for traffic lights utilizing genetic algorithms and the AIMSUM simulator [12, 13]. These studies target the reduction of delays and enhancement of traffic flow efficiency at specific intersections. At the city level, there exist proposals that concentrate on optimizing traffic lights across multiple intersections to enhance overall traffic performance. For instance, some present optimization approaches for urban traffic light scheduling [14]. Similarly, harmony search and artificial bee colony algorithms have been applied to tackle the urban traffic light scheduling problem [15, 16]. These studies strive to minimize delays and ameliorate traffic conditions throughout the city. Additionally, some proposals address both the junction level and the city level within their optimization strategies. For instance, [17, 18] employ multi-agent systems and ant colony optimization to address the traffic light problem, taking into account both individual intersections and the overall traffic network. Furthermore, there are even other articles that focus on roundabouts optimization [19, 20].

Overall, the literature review reveals a combination of studies focusing on traffic light optimization methods at both the junction level and the city level, highlighting the importance of addressing traffic flow challenges at different scales. The use of urban traffic simulators can help compare different signal coordination strategies. There are several works focused on urban traffic simulators that have been applied to better understand the complex interactions between traffic, emissions, and traffic signals [21, 22]. Hence, urban traffic simulators are often designed to be highly accurate and simulate every vehicle movement in great detail. However, this level of precision comes with a high computational cost, making it challenging to simulate large-scale areas such as metropolitan cities. Thus, most of the proposals are focused on simulating small areas, such as some intersections [23–26].

In this paper, we propose a urban simulator with an integrated genetic algorithm for efficient traffic light coordination. This integration facilitates the processing of large amounts of data and adjustments. Optimal coordination of traffic lights reduces congestion and waiting times, thereby minimizing overall emissions of harmful gases. To achieve this, the GPU-based simulator is designed to run extremely quickly, even in large regions like entire cities.

The rest of the paper is organized as follows. Section 2 briefly describes the features of the urban simulator used. Section 3 provides a description of the genetic algorithm for traffic light coordination. Section 4 presents different experiments conducted to test the algorithm's performance. Finally, Sect. 5 provides concluding remarks and outlines future directions for research.

2   Urban Simulator

One of the key contributions of our proposal is the implementation of a GPU-based simulator designed for rapid execution, capable of covering extensive urban areas. The strength of our approach lies in its high level of parallelization within a computing grid. Consequently, to a certain extent, the grid size has a minimal impact on performance due to the inherently parallel nature of our algorithm. This simulator leverages the Compute Quad (CQ) GPU architecture to perform intensive calculations efficiently by harnessing the parallel processing power of the GPU. Its primary objective is to approximate and rapidly compute the total amount of CO_2 generated by traffic in a city, based on a given map and specific information about traffic lights and flow. The computation takes into account the configuration of signal timings for the traffic lights.

Unlike other simulators, this simulator is specifically designed to integrate a metaheuristic algorithm that requires computing a fitness function (Fig. 1). The main advantage of this simulator is its speed, allowing simulations of large regions to be computed within a reasonable time frame.

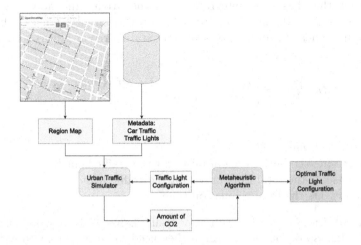

Fig. 1. Architecture of the urban simulator

The traffic simulation is discretely conducted by dividing the city map into a fixed number of cells. Each cell corresponds to a pixel on the CQ, representing an approximate area of 10×10 m in the city or map. These cells store geographic information and metadata for the corresponding portion. With a grid size of 2048×2048 and a precision of 10×10 m, a map of approximately $200 \, \text{km}^2$ can be stored, providing higher precision in the simulation.

A series of 2D textures (maps) is utilized for information storage, where the primary map contains the geographical data of the simulated region. This information can be algorithmically generated for test cases or imported from sources such as OpenStreetMap[1], in which case the data needs to be preprocessed to generate additional direction mask information. The map also stores the information of the traffic lights, representing the duration for which the light will remain red and an offset to keep track of the initial time.

When a vehicle is unable to move due to a red traffic light or if its destination is occupied by another vehicle, a certain amount of CO2 is generated, which accumulates in an output texture. Both stationary and moving vehicles generate CO2 emissions, with a larger quantity produced when they are stopped.

Once all threads in the CQ have completed their computations, the output texture contains updated information regarding the different vehicles and the quantity of CO2. However, this information is stored per texel, and to make it useful for the genetic algorithm, the total sum of all texels is required. To accomplish this, a parallel reduce operation is applied to the texture, effectively aggregating the values from all texels. The resulting sum is then transferred to the CPU, where it can be accessed and utilized by the genetic program.

3   Genetic Algorithm

In this section, we present the proposed genetic algorithm for traffic light management that has been integrated into the urban simulator. We define $L = \{l_1, \cdots, l_n\}$ as the set of traffic lights in a given city. Each traffic light l_i in a given moment t can be in two different status (red or green) according to Eq. 1.

$$status(l_i, t) = \begin{cases} red & \text{if} \quad (t + off_i)\%max < red_i \\ green & \text{otherwise} \end{cases} \qquad (1)$$

where:

- l_i, off_i, red_i, t, $max \in \mathbb{N}$
- off_i represents the initial time or offset for the traffic light l_i.
- red_i represents the total time in red for the traffic light l_i.
- max represents the total simulation time.

We also define $S_j \subset T$ as a subset of traffic lights that are complementary. Two traffic lights are complementary if they need to be synchronized with each

---

[1] https://www.openstreetmap.org/.

other, such as traffic lights at the same intersection. This means that their status must be the same at every moment t of the simulation or, in contrast, their status must be different at every moment of the simulation t:

$$\forall l_i, l_k \in S_j, (\mathit{off}_i = \mathit{off}_k) \wedge ((red_i = red_k) \vee (red_i = max - red_k)) \qquad (2)$$

$$\rightarrow (status(l_i, t) = status(l_k, t)) \vee (status(l_i, t) \neq status(l_k, t)) \forall t \in \mathbb{N}$$

As an example, Fig. 2 shows four subsets of complementary traffic lights.

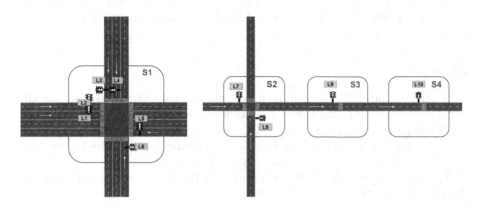

Fig. 2. Example of subsets of traffic lights

Each subset of traffic lights S_j is composed by at least one traffic light $l_i \in T$, and each traffic light l_i belongs to one and only one subset:

$$T = \bigcup_{j=1}^{n} S_j \quad \text{where} \quad S_i \cap S_j = \emptyset \quad \text{for} \quad i \neq j$$

Therefore, all traffic lights in the set T are represented in complementary subsets of traffic lights, where a traffic light cannot belong to two subsets. Thus, for each subset S_j, a representative traffic light l_r is defined which manages the status according to the Eq. 1. The remaining traffic lights $l_i \in S_j$ such that $l_i \neq r$ manage their status according to the status of l_r. If l_i is located in the same lane and direction as l_r, the status of l_i will be the same as l_r for any moment t. Otherwise, the status of l_i will be the opposite as l_r.

3.1 Representation

Each individual or chromosome represents the ordered sequence of complementary subsets of traffic lights in a city. In this context, each subset of traffic lights S_j, whose representative is l_r, will be a gene in this ordered sequence, and

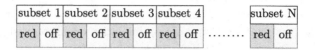

Fig. 3. Individual Encoding

the information associated with this gene will define the values of red time (red) and offset (off) of the l_r according to Eq. 2, as shown in Fig. 3.

To generate the initial population, values of offset and red time are randomly set for each subset of complementary traffic lights. The value red time is always restricted the values of $0 < red < max-1$ in order to prevent a traffic light to be in red or green status in the whole simulation, and thus, causing the causing the blockage of a street.

3.2   Fitness

As mentioned above, each individual represents the simulation's red time and offset for each subset of traffic lights. This configuration is input into the simulator, which computes the total amount of CO2 emitted during the entire simulation time for a specific number of vehicles and at a specific speed. In this regard, the fitness function represents the total amount of emitted CO2.

The selection of individuals to advance to the next generation follows an elitist model. In other words, individuals with the best fitness are selected.

3.3   Evolutionary Operations

The evolutionary operations that allow generating new individuals are as follows:

- **Simple Mutation**. It involves changing, with a certain probability, the value of off or red for a specific gene. This value is set randomly.
- **Inversion**. It involves flipping the value of red time for a specific gene: $red = max - red$.
- **One-point crossover**. It generates two offspring individuals from 2 parent chromosomes by combining genes from both parents. This operation randomly selects a crossover point on the chromosome, as illustrated in Fig. 4. The first offspring inherits the first part of the chromosome from one parent up to the crossover point, and the rest of the chromosome from the other parent. Similarly, the second offspring inherits genes in the opposite manner.
- **N-point crossover**. It generates one offspring individual from 2 parent chromosomes by combining genes from both. The N-point crossover operation randomly selects values from either parent with a 50% probability. Figure 5 shows an example of N-point crossover.

It should be noted that the selection of parents for crossover operations is randomly performed from the set of individuals whose fitness exceeds a threshold defined as a parameter.

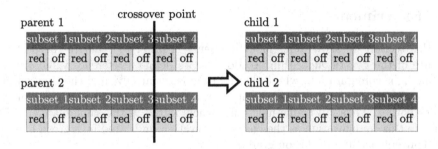

Fig. 4. One-point Crossover

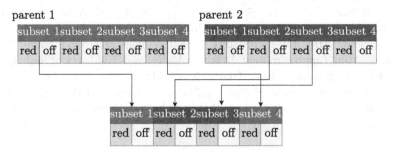

Fig. 5. N-point crossover

3.4   Genetic Algorithm Parameters

Next, we define certain parameters that allow controlling the behavior of the genetic algorithm, related to the population and genetic operations:

Probability of crossover represents the probability of performing a crossover between two selected individuals.

Type of crossover represents the probability associated with the type of crossover to perform (1-point or N-point).

Number of crossovers represents the number of crossovers to be performed in each iteration (before considering the probability of crossover).

Probability of Simple Mutation Defines the probability of mutation for each gene.

Population size configures the number of individuals in the population.

Number of duplicates defines the maximum number of individuals with the same fitness. Once this value is exceeded, a process is triggered to eliminate and replace all individuals considered identical. The goal is to maintain diversity in the population.

Fitness threshold for selection defines the threshold from which individuals will be selected for the next generation.

Maximum number of iterations indicates the maximum number of iterations to terminate the algorithm.

4   Experiments

In this section we show different experiments in order to evaluate the performance of the genetic algorithm. To do this, we define a reference value of the traffic light configuration, where the red time is set at 50% and the time offset is zero. This reference value represents a situation where any street has priority over another. Thus, all traffic lights are synchronized in the same manner. We refer to this configuration as the base configuration, indicated by a green line in the convergence and evolution graphs.

We are interested in compare the performance of the genetic algorithm with this base configuration. The objective is to analyse both the convergence and the best fitness. To do this, we define three different cities for applying the algorithm: Madrid City Center (approximately 50.000 vehicles and 1.207 traffic lights), Buenos Aires (approximately 70.000 vehicles and 5.978 traffic lights), and Manhattan (approximately 70.000 vehicles and 5.551 traffic lights). The maps of these cities were loaded into the simulator from OpenStreetMap. For this experiment we considered only 1-point crossover with a 10% of probability. The population size was defined as 1000 and the maximum number of iterations was 60.

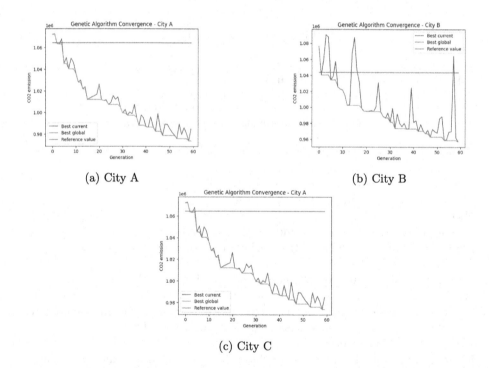

(a) City A   (b) City B

(c) City C

Fig. 6. Genetic algorithm performance

Figure 6 displays the results of this experiment. It can be observed that in all cases, an improvement over the reference value is achieved, and a convergence of the solution towards an optimum is observed. The approach to the optimum suggests that with more iterations, further slight improvements could be achieved.

4.1   Parameters Analysis

We will now present different experiments to analyze the behavior of the algorithm under different parameter changes. First, we will analyze the convergence of the algorithm based on population size. We use the population parameter (equal to the number of traffic lights divided by 2) as the study variable.

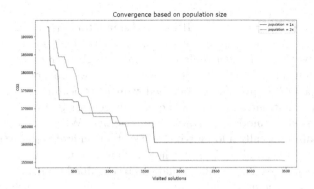

Fig. 7. Analysis of Convergence Based on Population Size

To evaluate the convergence speed, Fig. 7 shows the visited solutions, which corresponds to the number of times the fitness function is called. The fitness function decodes a chromosome to compute the total amount of emitted CO_2. The graph depicts how the fitness evolves with the visited solutions for a normal population size (1x) and for a double size (2x). In the case of a population size of 1x, the algorithm quickly converges to a local optimum but is later surpassed by 2x, which achieves a better value albeit requiring more visited solutions. This situation is quite common, as increasing the population size makes it more likely to find better solutions, but it also requires a larger number of fitness evaluations, potentially making the algorithm computationally challenging.

Next, we show the behavior of the algorithm by varying the mutation probability (Fig. 8). On one hand, we can observe that without mutation, the algorithm's performance is poor as it tends to converge to local maximum and fails to explore new search spaces. On the other hand, the results suggest that intermediate mutation probabilities enhance both the convergence speed and the best fitness achieved.

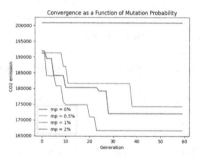

Fig. 8. Convergence based on Mutation Probability

5   Conclusions

In this paper, we have presented a GPU-based simulator with a genetic algorithm for traffic signal coordination aimed at reducing CO2 emissions in large urban areas. The advantage of this simulator lies in its high degree of parallelization within a computing grid. Unlike other simulators, we do not focus on exact calculations of CO2 emissions or detailed vehicle trajectories. Our goal is to obtain a 'score' that, although simplified, is representative of real emissions and can be optimized by modifying traffic light timings. What is more, instead of manually or automatically adding trips, our system generates these trips in real-time based on pre-established traffic flows. This is crucial as it allows us to simulate scenarios based on real data that can be obtained from publicly available traffic records, enhancing the adaptability and relevance of our proposal in real-world scenarios.

The algorithm has been validated in various environments, demonstrating its ability to reduce CO2 emissions by acting on traffic signals. These results are promising and suggest the need for further in-depth testing to validate the applicability of the model and refine its parameters. However, there are some limitations to consider as future work, such as incorporating different vehicle speeds and addressing more complex signal scenarios. In addition, we plan to address these aspects and conduct a more comprehensive evaluation of the genetic algorithm, comparing its efficiency with other metaheuristic algorithms such as GRASP.

Acknowledgements. This work was partially supported with grant DIGITAL-2022 CLOUD-AI-02 funded by the European Commission; grant PID2021-123673OB-C31 funded by MCIN/AEI/ 10.13039/501100011033 and by "ERDF A way of making Europe"; and Cátedra Telefónica Smart Inteligencia Artificial.

References

1. Lu, J., Li, B., Li, H., Al-Barakani, A.: Expansion of city scale, traffic modes, traffic congestion, and air pollution. Cities **108**, 102974 (2021)

2. Gualtieri, G., Brilli, L., Carotenuto, F., Vagnoli, C., Zaldei, A., Gioli, B.: Quantifying road traffic impact on air quality in urban areas: a Covid19-induced lockdown analysis in Italy. Environ. Pollut. **267**, 115682 (2020)
3. Popoola, O.A., et al.: Use of networks of low cost air quality sensors to quantify air quality in urban settings. Atmos. Environ. **194**, 58–70 (2018)
4. Huang, Y., et al.: A review of strategies for mitigating roadside air pollution in urban street canyons. Environ. Pollut. **280**, 116971 (2021)
5. Fujdiak, R., Masek, P., Mlynek, P., Misurec, J., Muthanna, A.: Advanced optimization method for improving the urban traffic management. In: 2016 18th Conference of Open Innovations Association and Seminar on Information Security and Protection of Information Technology (FRUCT-ISPIT), pp. 48–53. IEEE (2016)
6. Tikoudis, I., Martinez, L., Farrow, K., Bouyssou, C.G., Petrik, O., Oueslati, W.: Ridesharing services and urban transport CO2 emissions: simulation-based evidence from 247 cities. Transp. Res. Part D: Transp. Environ. **97**, 102923 (2021)
7. Abu-Shawish, I., Ghunaim, S., Azzeh, M., Nassif, A.B.: Metaheuristic techniques in optimizing traffic control lights: a systematic review. Int. J. Syst. Appl. Eng. Dev. **14**, 183–188 (2020)
8. Abdullah, A.M., Usmani, R.S.A., Pillai, T.R., Marjani, M., Hashem, I.A.T.: An optimized artificial neural network model using genetic algorithm for prediction of traffic emission concentrations. Int. J. Adv. Comput. Sci. Appl. **12**, 794–803 (2021)
9. Jan, T., Azami, P., Iranmanesh, S., Ameri Sianaki, O., Hajiebrahimi, S.: Determining the optimal restricted driving zone using genetic algorithm in a smart city. Sensors **20**(8), 2276 (2020)
10. Bagheri, M., Ghafourian, H., Kashefiolasl, M., Pour, M.T.S., Rabbani, M.: Travel management optimization based on air pollution condition using Markov decision process and genetic algorithm (case study: Shiraz city). Arch. Transp. **53** (2020)
11. Jia, H., Lin, Y., Luo, Q., Li, Y., Miao, H.: Multi-objective optimization of urban road intersection signal timing based on particle swarm optimization algorithm. Adv. Mech. Eng. **11**(4), 1687814019842498 (2019)
12. Sánchez-Medina, J.J., Galán-Moreno, M.J., Rubio-Royo, E.: Traffic signal optimization in "La Almozara" district in saragossa under congestion conditions, using genetic algorithms, traffic microsimulation, and cluster computing. IEEE Trans. Intell. Transp. Syst. **11**(1), 132–141 (2009)
13. Kesur, K.B.: Advances in genetic algorithm optimization of traffic signals. J. Transp. Eng. **135**(4), 160–173 (2009)
14. Gao, K., Zhang, Y., Sadollah, A., Lentzakis, A., Su, R.: Jaya, harmony search and water cycle algorithms for solving large-scale real-life urban traffic light scheduling problem. Swarm Evol. Comput. **37**, 58–72 (2017)
15. Gao, K., Zhang, Y., Sadollah, A., Su, R.: Optimizing urban traffic light scheduling problem using harmony search with ensemble of local search. Appl. Soft Comput. **48**, 359–372 (2016)
16. Dell'Orco, M., Baskan, O., Marinelli, M.: A harmony search algorithm approach for optimizing traffic signal timings. PROMET-Traffic Transp. **25**(4), 349–358 (2013)
17. Baskan, O., Haldenbilen, S.: Ant colony optimization approach for optimizing traffic signal timings. In: Ant Colony Optimization-Methods and Applications, pp. 205–220 (2011)
18. Sattari, M.R.J., Malakooti, H., Jalooli, A., Noor, R.M.: A dynamic vehicular traffic control using ant colony and traffic light optimization. In: Swiątek, J., Grzech, A., Swiątek, P., Tomczak, J.M. (eds.) Advances in Systems Science. AISC, vol. 240, pp. 57–66. Springer, Cham (2014). https://doi.org/10.1007/978-3-319-01857-7_6

19. Srivastava, S., Sahana, S.K.: Nested hybrid evolutionary model for traffic signal optimization. Appl. Intell. **46**(1), 113–123 (2017)
20. Putha, R., Quadrifoglio, L., Zechman, E.: Comparing ant colony optimization and genetic algorithm approaches for solving traffic signal coordination under oversaturation conditions. Comput.-Aided Civil Infrastruct. Eng. **27**(1), 14–28 (2012)
21. Gonzalez, C.L., Zapotecatl, J.L., Alberola, J.M., Julian, V., Gershenson, C.: Distributed management of traffic intersections. In: Novais, P., et al. (eds.) ISAmI2018 2018. AISC, vol. 806, pp. 56–64. Springer, Cham (2019). https://doi.org/10.1007/978-3-030-01746-0_7
22. Mohan, R., Eldhose, S., Manoharan, G.: Network-level heterogeneous traffic flow modelling in VISSIM. Transp. Developing Econ. **7**, 1–17 (2021)
23. Stanciu, E.A., Moise, I.M., Nemtoi, L.M.: Optimization of urban road traffic in intelligent transport systems. In: 2012 International Conference on Applied and Theoretical Electricity (ICATE), pp. 1–4. IEEE (2012)
24. Dezani, H., Marranghello, N., Damiani, F.: Genetic algorithm-based traffic lights timing optimization and routes definition using Petri net model of urban traffic flow. IFAC Proc. Volumes **47**(3), 11 326–11 331 (2014)
25. Castro, G.B., Hirakawa, A.R., Martini, J.S.: Adaptive traffic signal control based on bio-neural network. Procedia Comput. Sci. **109**, 1182–1187 (2017)
26. Iskandarani, M.Z.: Optimizing genetic algorithm performance for effective traffic lights control using balancing technique (GABT). Int. J. Adv. Comput. Sci. Appl. **11**(3) (2020)

GPU-Based Acceleration of the Rao Optimization Algorithms: Application to the Solution of Large Systems of Nonlinear Equations

Bruno Silva[1,2] and Luiz Guerreiro Lopes[3](✉)

[1] Doctoral Program in Informatics Engineering, University of Madeira, Funchal, Madeira Is., Portugal
bruno.silva@madeira.gov.pt
[2] Regional Secretariat for Education, Science and Technology, Regional Government of Madeira, Funchal, Portugal
[3] Faculty of Exact Sciences and Engineering, University of Madeira, 9020-105 Funchal, Madeira Is., Portugal
lopes@uma.pt

Abstract. In this paper, parallel GPU-based versions of the three Rao metaphor-less optimization algorithms are proposed and used to solve large-scale nonlinear equation systems, which are hard to solve with traditional numerical methods, particularly as the size of the systems get bigger. The parallel implementations of the Rao algorithms were performed in Julia and tested on a high-performance GeForce RTX 3090 GPU with 10 496 CUDA cores and 24 GB of GDDR6X VRAM using a set of challenging scalable systems of nonlinear equations. The computational experiments carried out demonstrated the efficiency of the proposed GPU-accelerated versions of the Rao optimization algorithms, with average speedups ranging from 57.10× to 315.12× for the set of test problems considered in this study.

Keywords: Computational intelligence · Metaheuristic optimization · Rao algorithms · Parallel GPU algorithms · Nonlinear equation systems

1 Introduction

Large systems of nonlinear equations (SNLEs) are common in areas like physics, chemistry, engineering, and industry [6]. These systems are characterized by their nonlinearity and complex interactions between variables, making their resolution a difficult task, particularly in terms of the required computational effort.

Using parallel algorithms, it is possible to leverage the power of modern parallel computing architectures, such as graphics processing units (GPUs), to execute multiple computations concurrently and significantly accelerate the problem-solving process. This reduction in overall computation time is achieved

P. Quaresma et al. (Eds.): IDEAL 2023, LNCS 14404, pp. 107–119, 2023.
https://doi.org/10.1007/978-3-031-48232-8_11

by decomposing the problem into smaller, more manageable sub-problems and distributing the computational load across multiple processor units.

This paper presents GPU-based parallel implementations of the three Rao metaphor-less optimization algorithms [7] using the Compute Unified Device Architecture (CUDA), an NVIDIA parallel computing platform for general-purpose GPU computing. Unlike metaphor-based metaheuristic algorithms, these powerful and straightforward algorithms do not require tuning of specific parameters to provide effective and desirable optimization performance. For this reason, the Rao algorithms have been successfully employed to solve numerous problems with different degrees of difficulty and complexity in a variety of fields of knowledge, and a number of variants of the original Rao algorithms were developed and studied.

In comparison to their sequential counterparts running on a central processing unit (CPU), the proposed GPU-based parallel algorithms are able to solve large SNLEs with high efficiency. These parallel implementations have advantages beyond mere speedup gains. They also provide a scalable solution for increasingly complex equation systems as their size and, consequently, the computing resources required to solve them increase.

2   Related Work

Although there has been no previous parallelization of the Rao algorithms, other metaphor-less optimization algorithms such as Jaya and the Best-Worst-Play (BWP) algorithm have been successfully parallelized [3,9,11,13], as have other parameter-less algorithms such as the Teaching-Learning-Based Optimization (TLBO) algorithm [8,12] and the Gray Wolf Optimizer (GWO) algorithm [10].

The resulting GPU-accelerated algorithms have shown that this class of metaheuristics can be effectively adapted to take advantage of the modern GPU's massively parallel processing capability and achieve significant performance gains compared to their sequential implementations.

Taking into account the efficiency of optimization and ease of configuration of this relatively recent class of population-based metaheuristic algorithms, as well as the significant performance gains observed in the aforementioned studies, the GPU-based parallel acceleration of such algorithms allows for significant expediting of the resolution of large, difficult optimization problems and the exploration of vast solution spaces that would not be possible by resorting to traditional sequential computation.

3   The Sequential Rao Algorithms

Ravipudi Venkata Rao proposed three optimization algorithms called Rao-1, Rao-2, and Rao-3 [7] in an effort to continue developing simple and effective metaphor-less optimization methods that do not require algorithm-specific control parameters, but rather only common control parameters like the population size *popSize* and the maximum number of iterations *maxIter*.

Although with variations, the three algorithms were developed to influence the population to move towards the best solution by adding to each candidate solution the product of the difference between the best and worst current solutions and a random number in each iteration.

Taking into account the population index $p \in [1, popSize]$, a design variable index $v \in [1, numVar]$, where $numVar$ is the number of decision variables, and an iteration $i \in [1, maxIter]$, let $X_{v,p,i}$ be the value of the vth variable of the pth population candidate in the ith iteration. Then, the updated value $(X_{v,p,i}^{new})$ is calculated by the Rao-1 algorithm as follows:

$$X_{v,p,i}^{new} = X_{v,p,i} + r_{1,v,i} \left(X_{v,best,i} - X_{v,worst,i} \right), \tag{1}$$

where $r_{1,v,i}$ is a uniformly distributed random number in $[0, 1]$ for the vth variable during the ith iteration, and $X_{v,best,i}$ and $X_{v,worst,i}$ are the population candidates with the best and worst fitness values, respectively.

As indicated by Eqs. (2) and (3), both the Rao-2 and Rao-3 algorithms are extended variations of the Rao-1 algorithm:

$$X_{v,p,i}^{new} = X_{v,p,i} + r_{1,v,i} \left(X_{v,best,i} - X_{v,worst,i} \right)$$
$$+ \begin{cases} r_{2,v,i} \left(|X_{v,p,i}| - |X_{v,t,i}| \right) & \text{if } f(X_{v,p,i}) \text{ better than } f(X_{v,t,i}) \\ r_{2,v,i} \left(|X_{v,t,i}| - |X_{v,p,i}| \right) & \text{otherwise,} \end{cases} \tag{2}$$

$$X_{v,p,i}^{new} = X_{v,p,i} + r_{1,v,i} \left(X_{v,best,i} - X_{v,worst,i} \right)$$
$$+ \begin{cases} r_{2,v,i} \left(|X_{v,p,i}| - X_{v,t,i} \right) & \text{if } f(X_{v,p,i}) \text{ better than } f(X_{v,t,i}) \\ r_{2,v,i} \left(|X_{v,t,i}| - X_{v,p,i} \right) & \text{otherwise,} \end{cases} \tag{3}$$

where $X_{v,t,i}$ represents a randomly selected candidate solution from the population (i.e., $t \in [1, popSize]$) that is distinct from $X_{v,p,i}$ (i.e., $t \neq p$).

The initial component of the Rao-2 and Rao-3 equations is identical to Rao-1, and the two algorithms are distinguished from Rao-1 by having different equation segments that are determined based on whether the fitness value of the current candidate solution $f(X_{v,p,i})$ is better or not than the fitness value of the randomly chosen solution $f(X_{v,t,i})$.

Algorithm 1 provides a unified description of the original sequential versions of the Rao-1, Rao-2, and Rao-3 algorithms.

4   GPU Acceleration of the Rao Algorithms

CUDA is an application programming interface that allows to use GPU hardware to accelerate certain functions (i.e., program tasks) by utilizing the parallel computational elements found in a GPU. These functions are called *kernels*, and, being a heterogeneous computing system, these kernels are launched by the host (i.e., the CPU) and executed in parallel by multiple threads on the device (i.e., the GPU).

Algorithm 1. Sequential Rao-1, Rao-2, and Rao-3

1: /* *Initialization* */
2: Initialize $numVar$, $popSize$ and $maxIter$;
3: Generate initial population X;
4: Evaluate fitness value $f(X)$;
5: $i \leftarrow 1$;
6: /* *Main loop* */
7: **while** $i \leq maxIter$ **do**
8:     Determine $X_{v,best,i}$ and $X_{v,worst,i}$;
9:     **for** $p \leftarrow 1, popSize$ **do**
10:         **if** $algorithm =$ 'Rao-2' **or** 'Rao-3' **then**
11:             $t \leftarrow$ Random candidate solution different than p;
12:         **end if**
13:         **for** $v \leftarrow 1, numVar$ **do**
14:             **if** $algorithm =$ 'Rao-1' **then**
15:                 Update population $X_{v,p,i}^{new}$ by Eq. (1);
16:             **else if** $algorithm =$ 'Rao-2' **then**
17:                 Update population $X_{v,p,i}^{new}$ by Eq. (2);
18:             **else if** $algorithm =$ 'Rao-3' **then**
19:                 Update population $X_{v,p,i}^{new}$ by Eq. (3);
20:             **end if**
21:         **end for**
22:         Calculate $f(X_{v,p,i}^{new})$;
23:         **if** $f(X_{v,p,i}^{new})$ is better than $f(X_{v,p,i})$ **then**
24:             $X_{v,p,i} \leftarrow X_{v,p,i}^{new}$;
25:             $f(X_{v,p,i}) \leftarrow f(X_{v,p,i}^{new})$;
26:         **else**
27:             Keep $X_{v,p,i}$ and $f(X_{v,p,i})$ values;
28:         **end if**
29:     **end for**
30:     $i \leftarrow i + 1$;
31: **end while**
32: Report the best solution found;

CUDA abstracts the GPU hardware, composed of multiple streaming multiprocessors with thousands of cores each, into a hierarchy of blocks and threads called a grid. The thread hierarchy refers to the arrangement of the grid (i.e., the number and size of blocks, as well as the number of threads per block). This arrangement is determined by the parallelization requirements of the kernel function (such as the data that is being processed) and is limited by the characteristics of the GPU hardware used. Blocks and threads can be arranged into one-, two-, or three-dimensional grids. Understanding the thread hierarchy is essential for writing efficient and high-performance CUDA kernels, since it determines the optimal GPU hardware occupancy.

Algorithm 2 provides a unified description of the parallel GPU-based versions of the Rao-1, Rao-2, and Rao-3 optimization algorithms, which were developed by adapting the original (sequential) algorithms to make use of the large number of parallel threads available on a GPU.

Algorithm 2. GPU-based parallel Rao-1, Rao-2, and Rao-3

1: /* Initialization */
2: Initialize $numVar$, $popSize$ and $maxIter$;
3: $X \leftarrow$ GENERATE_INITIAL_POPULATION_KERNEL();
4: EVALUATE_FITNESS_VALUES_KERNEL(X);
5: $i \leftarrow 1$;
6: /* Main loop */
7: **while** $i \leq maxIter$ **do**
8:     Determine $X_{best,i}$ and $X_{worst,i}$;   ▷ Device
9:     **if** $algorithm =$ 'Rao-2' **or** 'Rao-3' **then**
10:         $t_i \leftarrow$ SELECT_RANDOM_SOLUTIONS_KERNEL();
11:     **end if**
12:     $X_i^{new} \leftarrow$ UPDATE_POPULATION_KERNEL(X_i);
13:     EVALUATE_FITNESS_VALUES_KERNEL(X_i^{new});
14:     $X_i \leftarrow$ SELECT_BEST_BETWEEN_KERNEL(X_i, X_i^{new});
15:     $i \leftarrow i + 1$;
16: **end while**
17: Determine the best solution found and report it;   ▷ Device and host

---

When comparing the main loops of the sequential and parallel versions of each Rao algorithm, it is observable that the massively parallel implementations compute the necessary data during each iteration all at once rather than by population element. This strategy enables the GPU-based algorithms to take advantage of the GPU's large number of threads while performing efficient parallel computations.

As this is a heterogeneous computing approach, both the host and the device are utilized. The host executes the sequential portion of the algorithm, including the main loop and kernel calls, while the device manages the computationally intensive tasks in parallel. In Algorithm 2, tasks that run on the device are identified with the word *kernel* or marked with the comment *device*. All the remaining ones run on the host.

As the host and device have separate memory spaces, data cannot be accessed directly between them and must be transferred explicitly. This results in a strong negative impact on computing performance. The proposed GPU-based implementations address this issue by constantly generating and storing all computation-related data in the device's main memory.

As a result, two (small) data transfers are required. One from the host to the device at the start of the algorithm, containing the $numVar$ and $popSize$ variables for the generation of the initial population (line 3 of Algorithm 2), and another from the device to the host at the end of the run, containing the best solution found (line 17 of Algorithm 2). This final step is highlighted as being handled by the host, as it is responsible for generating the report (output) of the best solution found.

As previously stated, the Rao-2 and Rao-3 algorithms require the selection of random solutions from the population. Considering the characteristics of the parallel algorithms presented here, these random solutions must be determined for all candidate solutions simultaneously during each iteration. In Algorithm 3, this process is described in detail.

Algorithm 3. Kernel to select random solutions from the population

---

1: **function** SELECT_RANDOM_SOLUTIONS_KERNEL()
2:      /* *Device code* */
3:      Determine *index* using *blockDim.x*, *blockIdx.x*, and *threadIdx.x*;
4:      **if** *index* <= *popSize* **then**
5:          $t[index] \leftarrow rand(1, popSize)$
6:          **while** $t[index] \neq index$ **do**
7:              $t[index] \leftarrow rand(1, popSize)$
8:          **end while**
9:      **end if**
10: **end function**

This kernel is designed to run with a one-dimensional thread hierarchy (i.e., with a 1D block and thread arrangement), as the objective is to generate a vector equal in size to the population (i.e., *popSize*) containing the random solutions.

Consequently, only the x dimension of the block and thread data is utilized to determine the index used to map each individual data element to a unique GPU core (line 3 of Algorithm 3).

The function UPDATE_POPULATION_KERNEL is responsible for generating the new candidate solutions, and its operation is described in a unified way in Algorithm 4.

This section of the GPU-based implementation of the Rao algorithms is responsible for generating all candidate solutions for each iteration at once. Therefore, its output is a matrix whose size is a function of the initial population size and problem dimension (i.e., *popSize* × *numVar*). To achieve this in parallel, the kernel requires a two-dimensional (2D) thread hierarchy to index the data to the GPU threads.

Using the x and y coordinates of block dimension (*blockDim*), block index (*blockIdx*), and thread index (*threadIdx*) to identify the data matrix row and column (lines 3 and 4 of Algorithm 4), the kernel is able to assign specific computations to individual threads.

5   Numerical Experiments

The performance of both the parallel and sequential Rao algorithms was evaluated using 10 challenging scalable SNLEs. Table 1 lists the selected problems,

---

Algorithm 4. Kernel to update population

1: **function** UPDATE_POPULATION_KERNEL(X)
2:     /* Device code */
3:     Determine row using $blockDim.x$, $blockIdx.x$, and $threadIdx.x$;
4:     Determine col using $blockDim.y$, $blockIdx.y$, and $threadIdx.y$;
5:     **if** $row <= popSize$ **and** $col <= numVar$ **then**
6:         **if** $algorithm = $ 'Rao-1' **then**   ▷ Eq. (1)
7:             $X^{new}[row, col] \leftarrow X[row, col] + rand() \times (X_{best}[col] - X_{worst}[col])$;
8:         **else if** $algorithm = $ 'Rao-2' **then**   ▷ Eq. (2)
9:             **if** $f(X)[row]$ is better than $f(X)[t[row]]$ **then**
10:                 $X^{new}[row, col] \leftarrow X[row, col] + rand() \times (X_{best}[col] - X_{worst}[col]) + rand() \times (abs(X[row, col]) - abs(X[t[row], col]))$;
11:             **else**
12:                 $X^{new}[row, col] \leftarrow X[row, col] + rand() \times (X_{best}[col] - X_{worst}[col]) + rand() \times (abs(X[t[row], col]) - abs(X[row, col]))$;
13:             **end if**
14:         **else if** $algorithm = $ 'Rao-3' **then**   ▷ Eq. (3)
15:             **if** $f(X)[row]$ is better than $f(X)[t[row]]$ **then**
16:                 $X^{new}[row, col] \leftarrow X[row, col] + rand() \times (X_{best}[col] - abs(X_{worst}[col])) + rand() \times (abs(X[row, col]) - X[t[row], col])$;
17:             **else**
18:                 $X^{new}[row, col] \leftarrow X[row, col] + rand() \times (X_{best}[col] - abs(X_{worst}[col])) + rand() \times (abs(X[t[row], col]) - X[row, col])$;
19:             **end if**
20:         **end if**
21:     **end if**
22: **end function**

---

along with their domain D and additional parameters. The selected SNLEs were transformed into nonlinear minimization problems by using the sum of the absolute values of the residuals as the objective function to minimize.

To evaluate the performance of the algorithms with large-scale SNLEs, problem sizes of 500, 1000, 1500, and 2000 were chosen for the test problems. The population size was determined to be 10× the problem dimension (i.e. 5000, 10 000, 15 000, and 20 000). For each combination of algorithm, test problem, and dimension, 51 independent runs with 1000 iterations each were carried out.

The sequential and parallel Rao algorithms were implemented using Julia version 1.9.0 and double-precision floating-point arithmetic.

An Intel Core i7-5700HQ CPU with 4 cores, 8 threads, and 16 GB RAM, as well as a GeForce RTX 3090 GPU with 10 496 CUDA cores and 24 GB VRAM, were used to run the sequential and parallel implementations, respectively.

Table 1. Test problems (for each problem, $n = 500, 1000, 1500, 2000$)

No.	Problem name, domain, and parameters	Ref.
1	Broyden tridiagonal function $D = ([-1, 1], \ldots, [-1, 1])^T$	[5]
2	Discrete boundary value function $D = ([0, 5], \ldots, [0, 5])^T$ $h = \frac{1}{n+1}$, $t_i = ih$	[5]
3	Extended Powell singular function $D = ([-5, 5], \ldots, [-5, 5])^T$	[5]
4	Modified Rosenbrock function $D = ([-10, 10], \ldots, [-10, 10])^T$	[2]
5	Powell badly scaled function $D = ([0, 100], \ldots, [0, 100])^T$	[2]
6	The beam problem $D = ([-100, 100], \ldots, [-100, 100])^T$ $\alpha = 11$, $h = \frac{1}{n+1}$	[4]
7	The Bratu problem $D = ([-100, 100], \ldots, [-100, 100])^T$ $\alpha = 3.5$, $h = \frac{1}{n+1}$	[4]
8	Extended Rosenbrock function $D = ([-100, 100], \ldots, [-100, 100])^T$	[5]
9	Schubert–Broyden function $D = ([-100, 100], \ldots, [-100, 100])^T$	[1]
10	Martínez function $D = ([-100, 100], \ldots, [-100, 100])^T$	[14]

6   Results and Discussion

To ascertain the computational speed increase (i.e., speedup) of the parallel versions when compared to their sequential counterparts, the execution times of all the combinations of algorithms, test problems, problem sizes, and population sizes were measured, and the average results are presented in Tables 2 and 3.

Results demonstrated that the parallel versions of the Rao algorithms are repeatedly and consistently faster than their sequential implementations. The mean speedup improvement achieved by the GPU-based algorithms ranged from 57.10× faster (Rao-1 with problem 6 and dimension 500) to 315.12× (Rao-3 with problem 5 and dimension 2000) when compared to the same test running with the sequential version of the same algorithm.

Figure 1 depicts the mean run time for each Rao algorithm by test problem and dimension for both the sequential and parallel implementations, as well as an assessment of the overall performance in terms of the obtained speedup.

It is noticeable that the more sophisticated algorithms, namely Rao-2 and Rao-3, are able to achieve the highest speedup gains as the problem dimension increases. Because these algorithms require more processing power, they are better suited to take advantage of the GPU's parallel computing capabilities.

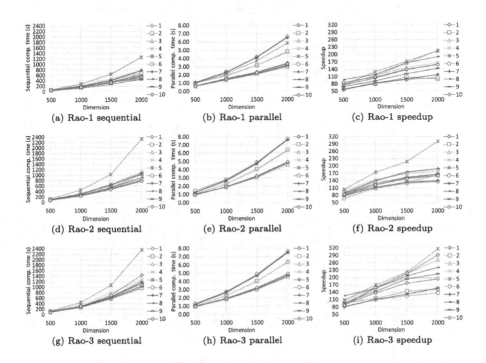

Fig. 1. Mean run time and speedup per algorithm, problem, and dimension.

Table 2. Mean computational results for the sequential and parallel versions (test problems 1 to 5)

Test prob.	Prob. dim.	Pop. size	Rao-1			Rao-2			Rao-3		
			Seq. time (s)	Par. time (s)	Mean speedup	Seq. time (s)	Par. time (s)	Mean speedup	Seq. time (s)	Par. time (s)	Mean speedup
1		5000	49.4379	0.6592	75.00	86.4834	0.9891	87.44	89.4150	0.9928	90.06
		10000	161.7995	1.5570	103.92	248.5498	1.9357	128.40	265.0238	1.9355	136.93
		15000	323.6259	2.3415	138.22	494.0552	3.2175	153.55	568.1254	3.2157	176.67
		20000	546.6453	3.4151	160.07	810.5704	4.9015	165.37	950.1452	4.9088	193.56
2		5000	60.7627	1.0138	59.93	98.9314	1.0572	93.58	101.0978	1.0585	95.51
		10000	157.1090	1.9699	79.75	250.9204	2.2951	109.33	263.1616	2.2845	115.20
		15000	338.8612	3.1608	107.21	537.5740	4.0434	132.95	578.9714	4.0381	143.38
		20000	477.3850	4.8265	98.91	1026.0120	6.3755	160.93	967.5272	6.3402	152.60
3		5000	50.1080	0.6245	80.24	87.2606	0.9892	88.21	88.9686	0.9546	93.20
		10000	153.1390	1.3646	112.22	259.6328	1.8838	137.82	292.7916	1.8404	159.09
		15000	343.7550	2.0854	164.84	523.3190	3.0415	172.06	644.8720	2.9938	215.40
		20000	635.5644	2.9544	215.13	862.6644	4.5711	188.72	1233.2630	4.5364	271.86
4		5000	75.4860	0.9799	77.03	103.9892	1.0428	99.72	110.0230	1.0452	105.27
		10000	216.6560	1.8868	114.83	325.6054	2.2640	143.82	362.6156	2.2581	160.59
		15000	460.5100	3.1350	146.89	666.3052	3.9650	168.05	785.5864	3.9714	197.81
		20000	789.2128	4.7167	167.32	1117.6504	6.2658	178.37	1271.1780	6.4119	198.25
5		5000	74.2212	1.0002	74.21	129.3172	1.2414	104.17	131.7592	1.2180	108.18
		10000	282.1038	2.1695	130.03	453.7638	2.6128	173.67	441.9993	2.6052	169.66
		15000	637.2358	3.7125	171.64	1019.6798	4.6930	217.28	1062.7754	4.7929	221.74
		20000	1252.9360	5.8406	214.52	2325.4875	7.7418	300.38	2351.7712	7.4632	315.12

Table 3. Mean computational results for the sequential and parallel versions (test problems 6 to 10)

Test prob.	Prob. dim.	Pop. size	Rao-1			Rao-2			Rao-3		
			Seq. time (s)	Par. time (s)	Mean speedup	Seq. time (s)	Par. time (s)	Mean speedup	Seq. time (s)	Par. time (s)	Mean speedup
6	500	5000	59.7980	1.0472	57.10	98.7276	1.4586	67.68	107.4524	1.3112	81.95
	1000	10000	191.9136	2.3633	81.21	306.6606	2.8138	108.98	299.6054	2.7297	109.76
	1500	15000	405.6558	4.1647	97.40	623.0466	4.8753	127.80	607.7208	4.8810	124.51
	2000	20000	694.3020	6.5943	105.29	1045.5734	7.7215	135.41	1035.4674	7.6288	135.73
7	500	5000	60.5558	1.0556	57.36	96.9984	1.2515	77.51	98.0176	1.2045	81.38
	1000	10000	194.8890	2.3273	83.74	304.9806	2.7093	112.57	302.7142	2.7335	110.74
	1500	15000	408.0658	4.1187	99.08	633.5888	4.7683	132.88	620.1944	4.7128	131.60
	2000	20000	771.2358	6.5304	118.10	1054.6742	7.6159	138.48	1176.2498	7.5530	155.73
8	500	5000	61.4826	0.6334	97.07	90.7712	1.0070	90.14	120.0028	0.9549	125.67
	1000	10000	169.8058	1.4270	118.99	268.8390	1.9190	140.09	291.7120	1.8752	155.56
	1500	15000	356.6162	2.1500	165.87	546.1792	3.1170	175.23	635.5638	3.0775	206.52
	2000	20000	585.9220	3.0747	190.56	909.1296	4.8505	187.43	1118.3004	4.6482	240.59
9	500	5000	43.5858	0.6358	68.55	80.3696	1.0090	79.65	94.0714	0.9727	96.71
	1000	10000	131.4822	1.4271	92.13	236.3964	1.9278	122.62	271.1516	1.8848	143.86
	1500	15000	267.8194	2.2047	121.48	477.1398	3.1636	150.82	602.1014	3.1135	193.39
	2000	20000	452.2118	3.1620	143.01	782.8032	4.7699	164.11	1017.8658	4.7439	214.56
10	500	5000	56.0084	0.6516	85.95	80.8860	0.9616	84.11	84.8656	0.9798	86.61
	1000	10000	161.1298	1.4975	107.60	236.1096	1.9310	122.27	304.9506	1.9369	157.44
	1500	15000	320.9534	2.2932	139.96	473.0514	3.2127	147.24	697.3074	3.2201	216.55
	2000	20000	524.2662	3.3191	157.95	778.6362	4.9005	158.89	1431.2838	4.9235	290.70

7   Conclusion

Despite the enormous computing potential that GPU acceleration offers, it is critical to address and overcome the difficulties involved in the successful use of these computational capabilities. Algorithms may be effectively accelerated on GPUs and achieve significant performance benefits by carefully taking into account several matters such as memory management, workload allocation, thread synchronization, and algorithm design.

In terms of computation time required for the same workload, the GPU-based parallelization of the Rao algorithms presented in this paper outperformed the original sequential versions by an average factor of 138.23 for the three algorithms. This clearly shows that the massively parallel versions of the Rao algorithms are unequivocally superior to their original counterparts in terms of computational efficiency.

In addition, the results obtained showed that the GPU parallel algorithms scale in a fairly predictable manner within the dimensions tested, indicating that they are better equipped to cope with the computational complexity of handling large-scale SNLEs.

The time required to compute problems with larger dimensions on the GPU varies over equal intervals at a more stable and predictable rate when compared to original (sequential) versions of the Rao algorithms running on the CPU.

Acknowledgements. The authors would like to thank Emiliano Gonçalves for providing access to the GPU hardware used in the computational experiments.

References

1. Bodon, E., Del Popolo, A., Lukšan, L., Spedicato, E.: Numerical performance of ABS codes for nonlinear systems of equations. arXiv:math/0106029 (2001)
2. Friedlander, A., Gomes-Ruggiero, M.A., Kozakevich, D.N., Martínez, J.M., Santos, S.A.: Solving nonlinear systems of equations by means of quasi-Newton methods with a nonmonotone strategy. Optim. Methods Softw. **8**(1), 25–51 (1997). https://doi.org/10.1080/10556789708805664
3. Jimeno-Morenilla, A., Sánchez-Romero, J.L., Migallón, H., Mora-Mora, H.: Jaya optimization algorithm with GPU acceleration. J. Supercomput. **75**(3), 1094–1106 (2018). https://doi.org/10.1007/s11227-018-2316-7
4. Kelley, C.T., Qi, L., Tong, X., Yin, H.: Finding a stable solution of a system of nonlinear equations. J. Ind. Manag. Optim. **7**(2), 497–521 (2011). https://doi.org/10.3934/jimo.2011.7.497
5. Moré, J.J., Garbow, B.S., Hillstrom, K.E.: Testing unconstrained optimization software. ACM Trans. Math. Softw. **7**(1), 17–41 (1981). https://doi.org/10.1145/355934.355936
6. Pérez, R., Lopes, V.: Recent applications and numerical implementation of quasi-Newton methods for solving nonlinear systems of equations. Numer. Alg. **35**(2), 261–285 (2004). https://doi.org/10.1023/B:NUMA.0000021762.83420.40
7. Rao, R.V.: Rao algorithms: three metaphor-less simple algorithms for solving optimization problems. Int. J. Ind. Eng. Comput. **11**(1), 107–130 (2020). https://doi.org/10.5267/j.ijiec.2019.6.002

8. Rico-Garcia, H., Sanchez-Romero, J.L., Jimeno-Morenilla, A., Migallon-Gomis, H., Mora-Mora, H., Rao, R.V.: Comparison of high performance parallel implementations of TLBO and Jaya optimization methods on manycore GPU. IEEE Access **7**, 133822–133831 (2019). https://doi.org/10.1109/ACCESS.2019.2941086

9. Silva, B., Lopes, L.G.: An efficient GPU parallelization of the Jaya optimization algorithm and its application for solving large systems of nonlinear equations. In: 3rd International Conference on Optimization, Learning Algorithms and Applications (OL2A), Ponta Delgada, Portugal (2023, to appear)

10. Silva, B., Lopes, L.G.: A GPU-based parallel implementation of the GWO algorithm: application to the solution of large-scale nonlinear equation systems. In: Eleventh International Symposium on Computing and Networking Workshops (CANDARW), Matsue, Japan (2023, to appear)

11. Silva, B., Lopes, L.G.: A massively parallel BWP algorithm for solving large-scale systems of nonlinear equations. In: 27th Annual IEEE High Performance Extreme Computing Virtual Conference (HPEC) (2023, to appear)

12. Silva, B., Lopes, L.G.: Massively parallel GPU implementation of the TLBO algorithm for solving high-dimensional systems of nonlinear equations. In: International Conference on Computational Intelligence, New Delhi, India (2023, to appear)

13. Wang, L., Zhang, Z., Huang, C., Tsui, K.L.: A GPU-accelerated parallel Jaya algorithm for efficiently estimating Li-ion battery model parameters. Appl. Soft Comput. **65**, 12–20 (2018). https://doi.org/10.1016/j.asoc.2017.12.041

14. Ziani, M., Guyomarc'h, F.: An autoadaptative limited memory Broyden's method to solve systems of nonlinear equations. Appl. Math. Comput. **205**(1), 202–211 (2008). https://doi.org/10.1016/j.amc.2008.06.047

Direct Determination of Operational Value-at-Risk Using Descriptive Statistics

Peter Mitic[1,2]($\boxtimes$) (ID)

[1] Department Computer Science, UCL, Gower Street, London WC1E 6BT, UK
p.mitic@ucl.ac.uk
[2] Santander UK, 2 Triton Square, Regent's Place, London NW1 3AN, UK
https://www.ucl.ac.uk/computer-science/ucl-computer-science

Abstract. Regression and machine learning methods are applied to the problem of *Value-at-Risk* determination in the context of financial Operational Risk, in order to determine an optimal technique that agrees sufficiently well with established Monte Carlo analyses. The annualised sum of operational losses is identified as the most significant statistical influence on *Value-at-Risk*, and a technique using it as a proxy for measured *Value-at-Risk* in a Test environment is formalised. The optimal standalone model is Generalized Additive, with approximately 61% success. The success rate can be enhanced to approximately 65% using a stacked model.

Keywords: Operational Risk · Pickands-Balkema-deHaan · Descriptive Statistics · Value-at-Risk · Generalized Additive Model · Loss Distribution

1 Introduction and Motivation

Intuitively, descriptive statistics of the financial data should provide a broad indication of the financial risk associated with the data. Surprisingly, little attention has been paid to using data properties directly to measure financial risk. In this paper we attempt to relate the descriptive statistics of *Operational*[1] losses ("OpRisk") to *Value-at-Risk* (VaR): a financial risk metric originating from the *J.P. Morgan/Reuters* "RiskMetrics" measure from the 1990s [11].

The most generally applicable way to determine VaR in the context of OpRisk is the Monte-Carlo-based *Loss Distribution Approach* (LDA [7]). Using the LDA has been a persistent problem because it is nearly always possible to find, for any single data set, multiple disparate 'solutions'. Then, which to choose is unclear. In this paper we explore a range of regression and machine learning (ML) techniques in order to find an optimal method to estimate VaR using descriptive statistics of the data. Any selected technique should be quick to use, interpretable, and agree tolerably well with the LDA-calculated value.

---

[1] The risk of financial loss due to flawed or failed processes, policies, systems or events that disrupt business operations.

P. Quaresma et al. (Eds.): IDEAL 2023, LNCS 14404, pp. 120–129, 2023.
https://doi.org/10.1007/978-3-031-48232-8_12

2   Literature Review

We concentrate on the few attempts to relate OpRisk VaR to properties of the underlying data. Curti and Migueis [5] and [6] test whether metrics derived from past VaR calculations can be used to predict future VaR calculations, using quantile regression. They use explanatory features such as loss frequency and mean total loss, with balance sheet co-variates such as Market Capitalisation. Some characteristics of our analysis are proposed by Chavez-Demoulin et al. [3]. They implement a bespoke *GAM*, using economic co-variates as distribution hyper-parameters. The statistical properties used are *Mean, Standard Deviation, Median, Third Quartile, Maximum, Skewness, Number of Losses Above 1€m.* Consequently, *Tail* properties[2] were missing. No methodological comparisons are given. Very recently, Mitic [10] established an upper bound for VaR as $7\frac{1}{3} \times S/Y$, where S is the loss sum and Y is the number of years spanned by those losses.

ML methods have been applied in the context of OpRisk, but not directly to predict VaR. Chen and Wen [4] used simulated data to predict OpRisk losses with operational control explanatory variables. Pakhchanyan et al. [12] used SVM and naive Bayes algorithms to assign data to specific risk classes. Aziz and Dowling [1], and Carrivick and Westphal [2], note the use of ML methods in OpRisk control (e.g. fraud prevention). Pena et al. [13] is a rare example of a ML method for calculating OpRisk capital. Diverse data sources are integrated using a convolutional neural network empowered by fuzzy cognitive maps. In common with the LDA, the inputs to their model are OpRisk frequency and severities.

3   Optimal Methodology Determination for VaR

Our approach is to apply *Linear Regression* (LR) and ML methods to a common data set that would be appropriate for a mid-to-large western European bank, optimised using a *Proxy* (described in Sect. 3.4).

3.1   Data and Pre-processing

Approximately 1100 random samples of sizes between 200 and 1000 were generated using appropriate 'fat-tailed' distributions (LogNormal, Weibull, Generalised Pareto etc.). The time span was a nominal 5 years. A single 66.7% Training set was determined by sampling and optimising *Mean Absolute Error* (MAE) using (LR). LR is thereby set as a stringent base from which to judge other results.

The Training and Test set features were standardised, separately, to mean 0 and standard deviation 1. We have found that results were impaired by either normalising data to [0, 1], or by removing significantly correlated features. For each data set, the following principal statistics were calculated.

---

[2] The largest $p\%$ of losses, with, typically, $p \in (1, 10)$.

- *Mean, SD, Skewness, Kurtosis, Maximum, Sum* (applied to all data)
- *Mean Tail SD Tail, Skewness Tail, Kurtosis Tail* (applied to the data Tail)
- Quantiles *Q10-Q90* in steps of 10, Quantiles *Q91-Q99* in steps of 1

The *Capital* (VaR at 99.9%) for each data set was calculated by fitting all distributions, and selecting the optimal distribution with respect to the *TNA* 'best fit' statistic [8], which is robust with respect to data set size.

3.2   Candidate VaR Assessment Models

A range of regression, ML, and other models were applied to the common data set (Sect. 3.1). The names below are used in the results tables in Sect. 4.

- LR, acting as a base reference method
- Other regressions: *BAYES, RIDGE, LASSO, LOESS*
- ML: Neural Network (*NN*), Random Forest (*RF*), Support Vector Machine (*SVM*), Bagging (*BAG*)
- Boost methods: *XG*, Adaptive Gradient (*ADA*), Gradient (*GRAD*)
- Others: *GLM, GAM*, k-Nearest Neighbours (*KNN*)
- *STACK*: the optimised (Sect. 3.4) mean of a subset of preceding models

3.3   Success Metrics

Established metrics (R^2, *MAE*) do not provide a sufficiently precise comparison of predicted and actual results. We use them mainly in parameter optimisations. Instead, a multi-part metric is used to measure prediction "success". The percentage of instances for which the predicted value differs from the LDA-calculated value by 10%, 25%, 50%, 75% and 90% provides a *Progressive Success* (abbreviated to *PS*) metric. In particular, the 10%, 25% figures correspond approximately to tolerable error bounds in the context of OpRisk VaR calculations.

3.4   Optimisation Using a Proxy

We have found that routine application of the methods of Sect. 3.2 does not provide sufficient success when the multi-part metric in Sect. 3.3 is applied. The primary problem is for small (<€20 m) VaR. The prediction error can the very large in comparison with the 'actual' value, and the 10% component of the multi-part success metric is particularly sensitive in that range. The *Proxy*, in which a known feature replaces an unknown feature, improves results considerably. It works by scaling predicted results in the Test environment using scale factors calculated in the Training environment, and using a *known* feature of the Test data in place of "actual" (but unknown) Test VaR values.

Proxy Correction

Recent work [9] has shown that feature *Annualised Sum* (i.e. the sum of all data divided by the number of years spanned by the data) is highly correlated with OpRisk VaR, so we define and use a *Proxy* for "actual" Test VaR values using

Annualised Sum. We call it *AnnSum* for short, and stress that *AnnSum* is always known for Test data, whereas "actual" Test VaR is not. The *Proxy* is used to modify Test predictions in a way that is beneficial in Training. The graphic in Fig. 1 shows the principal stages in *Proxy* optimisation. Processes are applied in the order indicated by the *STEP* numbers. Indexing is done for over- and under-predictions (relative to the *Proxy*). In Fig. 1, the resulting indices are denoted by I_U and I_L, and scale factors derived using them are denoted by K_U and K_L.

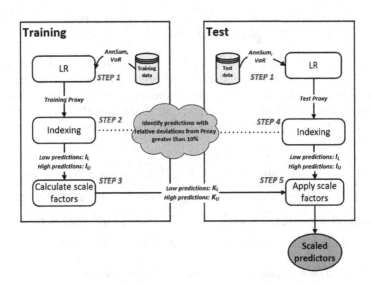

Fig. 1. The principal stages in *Proxy* optimisation. *STEP* labels refer to the detailed steps in Sect. 3.4

Proxy Correction: Details

The formal explanation below follows the same steps as in Fig. 1, and uses notation that makes a distinction between the Training and Test environments: superscript **Tr** for the Training environment, superscript **Te** for the Test environment, **P** for *Proxy*, **Y** for actual VaR, **Z** for predicted VaR, and subscript **opt** for optimised variates.

STEP 1. Determine a linear fit of the *Proxy* feature to the LDA-calculated VaR in the Training and Test environments separately. The intercepts $+1$ represent *de minimis* VaR for zero loss.

$$\mathbf{P}^{(\mathbf{Tr})} = a^{(Tr)}\mathbf{Y}^{(\mathbf{Tr})} + 1; \qquad \mathbf{P}^{(\mathbf{Te})} = a^{(Te)}\mathbf{Y}^{(\mathbf{Te})} + 1; \qquad (1)$$

STEP 2. In the main optimisation stage, the instances that have a 10% difference for the Training Proxy relative to the predicted VaR are identified, and indexed by indices $I_l^{(Tr)}$ and $I_u^{(Tr)}$.

$$I_l^{(Tr)} = \frac{-\mathbf{Z}^{(\mathbf{Tr})} + \mathbf{P}^{(\mathbf{Tr})}}{\mathbf{P}^{(\mathbf{Tr})}}; \qquad I_u^{(Tr)} = \frac{\mathbf{Z}^{(\mathbf{Tr})} - \mathbf{P}^{(\mathbf{Tr})}}{\mathbf{P}^{(\mathbf{Tr})}} \qquad (2)$$

STEP 3. Scaling constants K_l and K_u are found by optimising MAE for the indexed predicted ("Z") and Proxy ("P").

$$\mathbf{Z}_{\mathbf{opt}}^{(\mathbf{Tr})} = \min_{(0,1)} \mathbb{E}\left[\left|K_l\mathbf{Z}^{(\mathbf{Tr})} - \mathbf{P}^{(\mathbf{Tr})}\right|\mathbb{I}(I_l^{(Tr)}) + \left|K_u\mathbf{Z}^{(\mathbf{Tr})} - \mathbf{P}^{(\mathbf{Tr})}\right|\mathbb{I}(I_u^{(Tr)})\right] \quad (3)$$

STEP 4. Indices marking 10% relative deviations of Test predictions from the Test *Proxy* values are defined in a similar way to *STEP 2*.

$$I_l^{(Te)} = \frac{-\mathbf{Z}^{(\mathbf{Te})} + \mathbf{P}^{(\mathbf{Te})}}{\mathbf{P}^{(\mathbf{Te})}}; \quad I_u^{(Te)} = \frac{\mathbf{Z}^{(\mathbf{Te})} - \mathbf{P}^{(\mathbf{Te})}}{\mathbf{P}^{(\mathbf{Te})}} \quad (4)$$

STEP 5. The constants K_l and K_u feed through to the Test environment. The final Test predictor is the model-derived predictor (index m in Eq. 5), with the Proxy optimisations applied on the sets indexed u or l. No transformation is applied to Test predictions that correspond to the complement of the indices from *STEP 4* (these are the "acceptable" predictions).

$$\begin{aligned}
\text{Under predictions:} \quad & \mathbf{Z}_{\mathbf{opt}}^{(\mathbf{Te})}\mathbb{I}(I_l^{(Te)}) = 1 + K_l\mathbf{Z}^{(\mathbf{Te})}\mathbb{I}(I_l^{(Te)}) \\
\text{Over predictions:} \quad & \mathbf{Z}_{\mathbf{opt}}^{(\mathbf{Te})}\mathbb{I}(I_u^{(Te)}) = 1 + K_u\mathbf{Z}^{(\mathbf{Te})}\mathbb{I}(I_u^{(Te)}) \\
\text{Acceptable predictions index:} \quad & I_m^{(Te)} = \mathbb{I}(\sim I_u^{(Te)} \,\&\, \sim I_l^{(Te)}) \\
\text{Acceptable predictions unscaled:} \quad & \mathbf{Z}_{\mathbf{opt}}^{(\mathbf{Te})}\mathbb{I}(I_m^{(Te)}) = \mathbf{Z}^{(\mathbf{Te})}\mathbb{I}(I_m^{(Te)})
\end{aligned} \quad (5)$$

Section 4.2 shows empirical results for the Proxy/Actual relationship.

4   Results

Overall results are presented first, and are followed by results specific to application of the Proxy correction. We concentrate on results using all available explanatory features in standardised form, since that configuration usually admits superior performance over others.

4.1   Model Prediction Results

Table 1 shows ordinates of the *Progressive Success* metric per model, in the Test environment, in decreasing order of the 10% and 25% components of the metric. Their ranges (22–35% and 56–63% respectively) indicate reasonable success, given the stringency of the metric and Training set selection method. The LR model using normalised uncorrelated features out-performed the LR model using standardised correlated features. Both LR results are shown.

Table 1. Test environment *Progressive Success* metric ordinates per model, with *Proxy*: retaining correlated standardised features in all cases except for LR Normalised Uncorrelated.

Metric	0.1	0.25	0.5	0.75	0.9	MAE	R^2
STACK	0.35	0.61	0.84	0.96	0.97	36.57	0.84
LR Norm, Uncorr	0.31	0.63	0.90	0.97	0.97	29.85	0.90
GAM	0.31	0.63	0.88	0.97	0.97	33.95	0.84
BAYES	0.29	0.63	0.87	0.96	0.97	34.63	0.84
XG	0.29	0.59	0.85	0.95	0.96	36.51	0.81
LR Std, Corr	0.28	0.62	0.87	0.95	0.97	35.93	0.83
NN	0.28	0.61	0.86	0.96	0.96	35.26	0.83
RIDGE	0.28	0.62	0.88	0.96	0.97	35.14	0.84
LOESS	0.28	0.61	0.89	0.98	0.99	34.16	0.86
RF	0.26	0.62	0.86	0.95	0.97	34.85	0.85
KNN	0.26	0.60	0.84	0.96	0.96	41.67	0.79
SVM	0.26	0.61	0.88	0.98	0.98	35.06	0.85
GRAD	0.26	0.54	0.88	0.96	0.96	33.28	0.87
LASSO	0.25	0.62	0.86	0.96	0.97	38.30	0.80
GLM	0.24	0.59	0.86	0.99	0.99	40.73	0.78
ADA	0.23	0.60	0.88	0.97	0.97	34.72	0.85
BAG	0.22	0.56	0.88	0.97	0.99	34.65	0.85

A selection of scatter plots is shown in Fig. 2. Each node is an LDA-calculated/ prediction pair. The four plots illustrate optimal cases. There are two contrasting 'simplest' LR models, one with standardised features and the other without. Normalising features usually results in more accurate predictions for LDA-calculated VaR >€500m, at the expense of impaired overall success, and this is apparent in the upper right scatter. The *GAM* produced the highest individual 10% and 25% components of the *Progressive Success* metric using all features, standardised. The *Stack Model* produced the highest overall 10% and 25% components, again using all features, standardised.

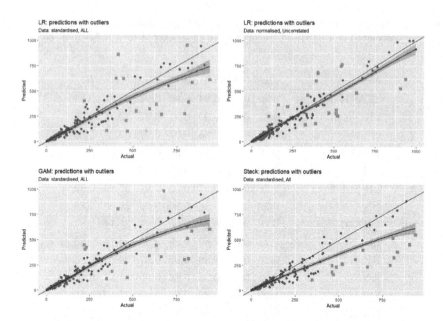

Fig. 2. Result scatters: Test environment. Upper Left: LR All features/Standardised. Upper Right: LR Uncorrelated features/Normalised. Lower Left: *GAM* all features/Standardised. Lower Right: STACK all features/Standardised. Filled circles are nodes for which VaR predictions agree with actual VaR with 95% 2-tail confidence. Filled squares are outliers.

4.2 The Effect of the Proxy

Feature importance may be assessed by the *Feature Difference* method, noting the effect on MAE of successively excluding each feature using LR. When applied to the Training set, *Sum* is the most significant feature (54.1% of the variance). Table 2 shows the effect in percentage terms of the Proxy in boosting components of the *PS* metric. The large percentage improvements are notable in nearly all cases. They come at the expense of reduced MAE and R^2 scores in most cases.

Table 2. Percentage boosting of components of the *PS* metric by applying the Proxy, expressed as the *Annual Sum* variate. Negatives indicate performance impairment.

Metric	0.1	0.25	0.5	0.75	0.9	MAE	R^2
GAM	38.57	56.8	33.66	24.79	18.75	−12.66	0.93
BAYES	100	79.82	49.73	30.13	25.83	−18.01	1.4
XG	36.92	39.69	15.72	10.86	9.49	2.67	−5.18
LR	81.25	81.13	41.67	27.71	25.21	−17.27	1.4
NN	44.25	30.16	10.67	8.71	5.68	−2.95	−3.24
RIDGE	55.23	72.44	41.3	27.07	21.48	−13.79	0.42
LOESS	34.38	33.1	27.52	18.99	14.5	−16.11	2.26
RF	47.5	43.25	32.59	23.52	17.69	−16.02	1.66
KNN	95.24	74.07	32.32	27	22.86	−19.23	−3.16
SVM	38.98	29.93	15.48	11.31	9.29	−7.69	0.02
GRAD	9.59	21.43	10	7.19	3.83	−5.99	−0.31
LASSO	182.14	128.24	68.55	44.02	36.65	−15.25	−1.71
GLM	0	0.55	0.75	0.65	0.65	−6.5	4.92
ADA	40.39	61.34	29.38	21.97	18.06	−8.19	−0.48
BAG	20.9	18.18	9.62	6.92	4.22	−6.46	−0.24

Figure 3 shows how much difference the *Proxy* can make. Nodes with low value (<€100m) VaR are shown with and without the *Proxy*. In the former case, nodes are concentrated on the diagonal ("predicted = actual") line, but some under-prediction is evident. In the latter case, there are multiple disparate predictions for individual nodes with low-valued actual VaR.

5   Discussion

The motivation for this study was to determine if OpRisk VaR can be estimated directly from the statistical properties of the underlying data. The principal model for doing that was to be multivariate linear regression. The results using that method, measured using the *PS* metric, proved to be lower than expected. That prompted alternative models to be used. It also raises several issues.

1. The values obtained using the LDA may not be optimal as far as VaR assessment is concerned, even though their fitted distributions are optimal. Considerations such as shape of the model QQ curve and consistency with results from previous years are also important.
2. The variability of LDA-derived results suggests that a simple model (such as LR) could be used for validation, or even as an LDA replacement. In the former case, LDA-derived distributions could be rejected based on an alternative assessment.

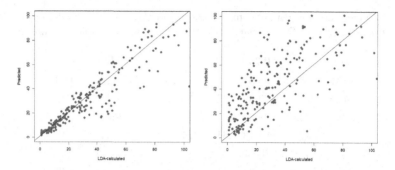

Fig. 3. Linear Model actual/predicted scatter. Left: with *Proxy*. Right: without.

3. Possibly the descriptive statistics of the data are not adequate for VaR determination. Suitable alternative are not apparent.
4. The Proxy correction can only be applied if a suitable proxy can be found.

6   Conclusion

Any of the methods discussed in this paper should be used to validate, or even replace, an LDA-derived VaR, provided that we can be confident that whichever is selected is a correct reflection of the underlying data. Given the multiple alternatives, which, if any, should be selected? There is compelling pressure to select a simple method that is explainable to practitioners in terms of data features. For this reason, LR is a clear choice.

References

1. Aziz, S., Dowling, M.: Machine learning and AI for risk management. In: Lynn, T., Mooney, J.G., Rosati, P., Cummins, M. (eds.) Disrupting Finance. PSDBET, pp. 33–50. Springer, Cham (2019). https://doi.org/10.1007/978-3-030-02330-0_3
2. Carrivick, L., Westphal, A.: Machine learning in operational risk. ORX White Paper (2019). https://managingrisktogether.orx.org/free-resources
3. Chavez-Demoulin, V., Embrechts, P., Hofert, M.: An extreme value approach for modeling operational risk losses depending on covariates. J. Risk Insur. **83**(3), 735–776 (2016). http://www.jstor.org/stable/43998282
4. Chen, Q., Wen, Y.: A BP-neural network predictor model for operational risk losses of commercial bank. In: Third International Symposium on Information Processing, Qingdao, China, pp. 291–295 (2010). https://doi.org/10.1109/ISIP.2010.43
5. Curti, F., Migueis, M.: Predicting operational loss exposure using past losses. In: Finance and Economics Discussion Series 2016–002. Washington: Board of Governors of the Federal Reserve System (2016). https://doi.org/10.17016/FEDS.2016.002

6. Curti, F., Migueis, M.: The information value of past losses in operational risk. In: Finance and Economics Discussion Series 2023–003. Washington: Board of Governors of the Federal Reserve System (2023). https://doi.org/10.17016/FEDS.2023.003

7. Frachot, A., Georges, P., Roncalli, T.: Loss distribution approach for operational risk. Working paper, Groupe de Recherche Operationnelle, Credit Lyonnais, France (2001). http://ssrn.com/abstract=1032523

8. Mitic, P.: Improved goodness-of-fit tests for operational risk. J. Oper. Risk **15**(1), 77–126 (2015). https://doi.org/10.21314/JOP.2015.159

9. Mitic, P.: Reasonableness and correctness for operational value-at-R. Econ. Anal. Lett. **2**(3), 35–44 (2023). https://doi.org/10.58567/eal02030005

10. Mitic, P.: Credible value-at-risk. J. Oper. Risk (2023, to appear)

11. Morgan, J.P.: Reuters: RiskMetrics-Technical Document, 4th edn. (1996). https://www.msci.com/documents/10199/5915b101-4206-4ba0-aee2-3449d5c7e95a

12. Pakhchanyan, S., Fieberg, C., Metko, D., Kaspereit, T.: Machine learning for categorization of operational risk events using textual description. J. Oper. Risk **17**(4), 37–65 (2022). https://doi.org/10.21314/JOP.2022.026

13. Pena, A., Patino, A., et al.: Fuzzy convolutional deep-learning model to estimate the operational risk capital using multi-source risk events. Appl. Soft Comput. **107**(107381) (2021). https://www.sciencedirect.com/science/article/pii/S1568494621003045

Using Deep Learning Models to Predict the Electrical Conductivity of the Influent in a Wastewater Treatment Plant

João Pereira[1](✉)(iD), Pedro Oliveira[1](iD), M. Salomé Duarte[2,3](iD),
Gilberto Martins[2,3](iD), and Paulo Novais[1](iD)

[1] LASI/ALGORITMI Centre, University of Minho, Braga, Portugal
joaoprp111@gmail.com, pedro.jose.oliveira@algoritmi.uminho.pt,
pjon@di.uminho.pt
[2] CEB - Centre of Biological Engineering, University of Minho, Campus de Gualtar,
4710-057 Braga, Portugal
salomeduarte@ceb.uminho.pt, gilberto.martins@deb.uminho.pt
[3] LABBELS - Associate Laboratory, Braga/Guimarães, Portugal

Abstract. Nowadays, human population face increasing water pollution problems, so treating and managing this resource is crucial. Wastewater Treatment Plants (WWTPs) provide essential services for human life since they treat wastewater and monitor its parameters to preserve water quality standards. One of these parameters is electrical conductivity, essential in quantifying water salinity levels. Therefore, this paper aims to forecast the influent conductivity in a WWTP for the next two timesteps. Hence, several experiments were conducted, considering the use of Transformers and Long Short-Term Memory (LSTMs) candidate models that were developed, tuned, and evaluated, utilising a recursive multi-step forecasting approach. The best candidate model was based on a Transformer architecture with encoding and obtained a RMSE of 155.2 μS/cm.

Keywords: Deep Learning · Influent Conductivity · Time Series · Wastewater Treatment Plants

1 Introduction

During the last decades, human activities resulted in the pollution of an exceedingly significant natural resource, specifically water. Since water is essential to human life, it is crucial to combat its pollution [1]. Thus, Wastewater Treatment Plants (WWTPs) are essential for reducing water courses contamination and monitoring its quality. WWTPs can identify and develop the necessary treatments to transform the influent wastewater into a higher quality water [2]. Therefore, one of the essential tasks in a WWTP is monitoring influent wastewater indicators to control its concentrations since the wastewater is later discharged to the environment [3]. Besides, influent flow variables information

P. Quaresma et al. (Eds.): IDEAL 2023, LNCS 14404, pp. 130–141, 2023.
https://doi.org/10.1007/978-3-031-48232-8_13

can help adjusting, in advance, the pumps required in the treatment processes [4]. Finally, the monitoring process can be supported by Deep Learning (DL), regarding forecasting problems, since this technique allows to check indicators values in advance, thus helping managing and making correct decisions during wastewater treatment [5].

One of the parameters found in wastewaters, which can play an essential role in WWTPs, is electrical conductivity. It is relevant in a WWTP because it can provide insights about the chemical processes of these infrastructures [6]. Conductivity describes the salinity level of the water, which can help detect seawater intrusions, measure the concentration of ionized chemicals in the water, and identify possible illegal water discharges sources, through seasonality [7]. Herewith, conductivity can present higher or lower values than usual, which can help detect water pollution and environmental changes. These variations in the values can be influenced by some factors, such as temperature, inorganic dissolved solids, the geology of the area where the water flows, and sea tides [8].

Therefore, the present article consists of developing, optimising and evaluating some DL candidate models to forecast the influent conductivity in a WWTP for the next two days. Predicting more than one day ahead can provide generalization in the models' overall predictions, and support decision making tasks regarding treatment adjustment and optimisation in wastewater companies. Thus, the selected models for this task were based on Long Short-Term Memory (LSTMs) and Transformers, utilising a multivariate recursive multi-step forecasting approach. The former model can be advantageous in this work because of its capacity to learn sequences and long-term dependencies [9], while the latter can capture non-sequential dependencies and process the elements of the sequence in parallel [10]. Many experiments were considered to obtain the best hyperparameter set for each base model. The succeeding sections of this document are organised as follows: the next section describes the literature review, considering the influent conductivity forecasting in WWTPs. The third section describes all the steps related to data collection, manipulation, conceived DL models, and evaluation metrics. The various experiments and the discussion of the results are presented in the fourth and fifth sections, respectively. Finally, the last section presents the conclusions and future work.

2   State of the Art

The prediction of the conductivity parameter in a WWTP is a topic barely explored. However, some studies present interesting approaches for conductivity prediction in a WWTP, Water Treatment Plant (WTP), or Water Quality Station [11–13].

In a study carried out by Maleki et al. [11], the work consisted in forecasting the influent characteristics of a WTP, using an Artificial Neural Network (ANN) based model. The dataset contains daily records (collected over 2 years) of some influent parameters, like alkalinity and electrical conductivity, of the Sanandaj WTP located in Iran. Both parameters, alongside other influent characteristics, were used as inputs, and the goal of the model was to predict each input parameter for the next day of the sequence. Linear interpolation technique was utilized to fill missing data. Besides, there is no reference to other techniques, e.g. feature engineering. The dataset was partitioned into 70% for training, 10% for validation and 20% for testing, and the implemented model is based on a Nonlinear Autoregressive (NAR) neural network. Regarding the training phase, there is no reference for cross-validation, overfitting analysis or hyperparameter optimization. Finally, the coefficient of determination (R^2) was the evaluation metric used, and through the obtained results, the electrical conductivity achieved a R^2 of 0.55.

Another work by Fu et al. [12] focused on predicting wastewater quality parameters using a wavelet de-noised Adaptive Neuro-fuzzy Inference System (ANFIS). This study has access to influent parameters timeseries, such as Total Dissolved Solids (TDS) and conductivity, measured every 2 weeks in Las Vegas. An Improved Wavelet-ANFIS (IWT-ANFIS) model was conceived to predict TDS and conductivity in a multivariate approach. With this in mind, the input data is selected considering Pearson Correlation coefficients between some chemical parameters, such as chloride and fluoride. Some important details were not mentioned like data manipulation. Additionally, it is not mentioned how the dataset was splitted. Finally, techniques like cross-validation are not mentioned. To complete the study, the final model is compared to five types of models, namely Multiple Linear Regressor (MLR), Multilayer Perceptron with ANN (MLP-ANN), ANFIS, ANN-ANFIS, and WT-ANFIS. Mean Absolute Percentage Error (MAPE) and R^2 are the evaluation metrics used. Furthermore, the final results, in terms of MAPE and R^2, reach the scores of 0.577 and 0.985, respectively.

Najah et al. [13] proposed a study to forecast water quality parameters utilising a Wavelet (WT) ANFIS-based model with hold-out validation. They used monthly measurements of water quality indicators, such as temperature, conductivity, turbidity and TDS. These registers correspond to a Water Quality Station in Malaysia. Furthermore, the main goal was to forecast these indicators using a multivariate approach, with parameters such as pH and temperature. Different architectures of the WT-ANFIS model took place, for each predicted parameter. Using the hold-out validation technique and overfitting analysis, the model outperformed the traditional ANFIS. Although the authors never mention the units of measurement for conductivity, the results show a Mean Absolute Error (MAE) of 30.6 in testing phase. Additionally, 20% of the data was allocated for testing, 8% for validating and 72% for training. Other essential techniques like hyperparameter optimisation and feature engineering are not mentioned.

Considering the mentioned studies and particularly the cross-validation techniques, one specific technique, namely *TimeSeriesSplit* (suited for time series problems), was not utilised. Besides, other relevant techniques such as feature engineering and hyperparameter optimization, were not mentioned. Additionally, data preparation tasks and underfitting/overfitting analysis are topics that were not considered in all studies. When facing a time series problem, most of these techniques are essential for conceiving the models. Finally, the DL models, such as LSTMs, not considered in any of the studies, can be an advantage in this work because of their capacity to handle temporal dependencies.

3   Materials and Methods

This section provides information about the materials used in this study, namely data collection, exploration and manipulation, the devised DL models, and the evaluation metrics. In addition, elucidation is provided regarding details of the datasets.

3.1   Data Collection

Concerning the data collection step, the dataset was made available by a portuguese multimunicipal company treating urban wastewaters. In addition, this data represents real events that occurred in a WWTP situated near the coast, and it contains data between January 2nd, 2019 and October 31th, 2022, registered every 2 days. Another dataset was utilised regarding the weather data in the exact location of the WWTP. This one was provided by *OpenWeatherMap*, in the same date range as the first dataset but with a different periodicity. The dataset contained hourly records between January 2nd, 2019 and October 31th, 2022.

3.2   Data Exploration

Given the data points, in the first dataset analysis, it is possible to verify that the WWTP parameters dataset has a total of 700 observations, while the weather dataset contains 33576. Table 1 presents the features of the WWTP parameters, and Table 2 describes some of the weather features. A short description, the data type and the units of measure are also presented. These tables show that most features contain *doubles*, except the features that represent the measurement date.

Table 1. Features of the WWTP parameters dataset.

#	Feature	Description	Data type	Unit of measure
1	Conductivity	Influent conductivity value	*double*	μS/cm
2	Flow Rate	Influent flow rate value	*double*	m^3/day
3	Date	Date of measurement	*timestamp*	Date

In the WWTP parameters dataset, each observation contains two features representing wastewater indicators: electrical conductivity and water flow rate. Another feature is the time label, which represents the observation timestamp.

Conversely, the weather dataset contains thirteen features representing weather characteristics, including air temperature, atmospheric pressure, wind speed and clouds. As the first dataset mentioned, each observation is associated with the respective timestamp, represented by the *date_time* feature.

Table 2. Some features of the weather dataset.

#	Feature	Description	Data type	Unit of measure
1	temp	Temperature	*double*	*Celsius*
2	pressure	Atmospheric pressure	*double*	hPa
3	humidity	Humidity	*double*	%
4	wind_speed	Wind speed	*double*	m/s
5	wind_deg	Wind direction	*double*	Weather degrees
6	date_time	Date of measurement	*timestamp*	Date

Considering the features distribution, we performed a statistical analysis and the Kolmogorov-Smirnov test to find if the data distribution is related to a Gaussian distribution, and consequently find the most appropriate correlation coefficient test technique. The final conclusion was that none of the WWTP parameters and weather features present a Gaussian data distribution, because p value obtained from the test was lower than 0.05.

Considering the statistical details regarding conductivity, it yields a *skewness* value equal to 5.42 and 63.7 for *kurtosis*, which means the distribution is right skewed and leptokurtic, respectively. Besides, the maximum, minimum, standard deviation, and mean values are 6320, 5, 389.57, and 1136.55, respectively.

Regarding the date range of the collected data, since this is a time series problem, it is essential to analyze if there are missing timestamps or missing values. So, after the exploration, no missing timestamps are found in WWTP parameters and weather datasets. However, there are missing values in the WWTP parameters dataset. Therefore, these values needed to be filled, and this treatment is explained in detail in the data preparation section.

Finally, a plot was developed to verify the monthly average conductivity values for each year to complete the data exploration. Afterwards, this exploration task consisted of finding patterns between the different years to search for seasonality or cyclic behaviours in the data.

As shown in Fig. 1, it is possible to get an overall look at the monthly average variations of the values. There is a similarity in the values between May and December because there is a predominant increase between May and September, followed by a fall between September and December every year. Thus, this could indicate a small amount of seasonality at this time. Furthermore, it is possible to note that the graph presents the highest average conductivity values in September 2019 and the lowest in February 2021. Hence, this could be an insight into high and low values tendencies.

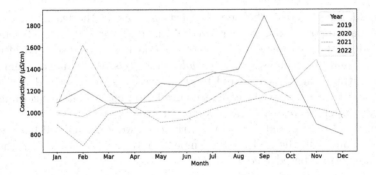

Fig. 1. Monthly average of the conductivity values for each year.

3.3 Data Preparation

Another essential step to having a clean source of knowledge for the learning process of the candidate models is data preparation. The first task consisted of changing the weather dataset date range to allow its joining with the WWTP parameters dataset. Since this data was registered hourly, the average of the 24 daily values was calculated to get the daily values.

Subsequently, the goal was to identify possible insertion errors in both datasets. To complete this task, the data was carefully explored, where the values presented significant deviations from the majority, i.e., near the maximum and minimum data points. Conductivity and flow rate columns appeared to have insertion errors in some cases, that is to say, some digits could be missing. However, there were no apparent insertion errors in each weather column. So, regarding conductivity and flow rate columns, it was decided that the possible insertion errors would be treated later as outliers.

As mentioned above, some missing values exist in the WWTP parameters dataset. There is a total of 13 for conductivity and 2 for flow rate. With this in mind, the linear interpolation technique filled the missing data of the WWTP parameters features. Afterwards, the data treatment process replaced the data points that could be outliers. Hence, to find those data points, it was necessary to analyze in detail the data distribution. Then, maximum and minimum thresholds were selected, based on the visualization of boxplots, for each WWTP parameter to define outliers' limits. Finally, the values greater or lower than the maximum and minimum thresholds, respectively, were replaced by the last four registers mean.

The following step consisted in joining both datasets utilising the inner join approach. In this way, datasets were concatenated by the timestamp, and consequently, some weather dataset timestamps were discarded. Subsequently, new features related to the timestamp, namely the day, month, week, year, season, semester and trimester, were added to the dataset. Finally, to select the best features of the dataset, we performed a correlation analysis between all of the features. As mentioned in the previous section, none of the features presented Gaussian data distribution, so we utilized Spearman's correlation coefficient test.

Regarding the weather data, most of the features were discarded after the correlation analysis, such as *clouds*, *visibility* and *feels_like*. Thus, the remaining weather features slightly correlated to conductivity were described in Table 2. Afterwards, the features with moderate correlation coefficients concerning conductivity (target) were season and temperature. Hence, all the other features were removed from the dataset. Finally, the data were normalized between the values of -1 and 1 in the LSTM-based models and between 0 and 1 in the case of the Transformer-based models.

3.4 Transformers

Transformers are a DL model architecture introduced in 2017 [14]. The vanilla architecture consists of an encoder and decoder, both composed of self-attention layers and feed-forward neural networks. Self-attention allows to capture dependencies between different positions in the sequence, while feed-forward networks process and transform the representations. Transformer has been widely used in Natural Language Processing (NLP) tasks and has achieved state-of-the-art performance in various domains, including machine translation and language generation [15]. This algorithm has also shown effectiveness in time series forecasting tasks, for example, in energy consumption prediction [16], since it can capture non-sequential long-range dependencies and process sequences in parallel. These characteristics can help achieve better results when compared to Recurrent Neural Networks (RNNs) [17]. Although, its performance can vary depending on the specific task and the dataset's quality, among other factors.

3.5 LSTMs

LSTMs are a variant of RNNs frequently applied due to their ability to capture order dependencies in sequential data problems. Additionally, they solve RNNs vanishing gradient problems handling long-term dependencies [18]. In their architecture, cells contain properties that allow to remember or forget information over time. These cells, or units, have three gates, namely input, output, and forget gates. First gate controls how much information is added to the cells. Output gate controls information that continues to the subsequent layers. Finally, forget gate manages information that is discarded. So, LSTM is prepared to manage and process information over long data sequences [19]. It can remember information over a long period because it can capture essential features from past inputs and preserve the information taken from it [20].

3.6 Evaluation Metrics

Since the problem faced in this work is a regression one, two evaluation metrics were considered. These are MAE and Root Mean Squared Error (RMSE). MAE measures the average errors magnitude between predicted and real values. Furthermore, its calculations help to know if the models' predictions match the true values. MAE is less affected by outliers compared to other metrics like RMSE. The following formula shows how MAE is calculated:

$$MAE = \frac{\sum_{i=1}^{n} |yi - yi_{pred}|}{n} \tag{1}$$

RMSE is similar to MAE, it calculates the square root of mean squared differences between predicted and true values. It considers both magnitude and direction errors. Thus, it is more sensitive to outliers. The formula is the following:

$$RMSE = \sqrt{\frac{\sum_{i=1}^{n} (yi - yi_{pred})^2}{n}} \tag{2}$$

In Eqs. 1 and 2, yi, yi_{pred} and n represent the actual values, predicted values, and the number of data points, respectively.

4 Experiments

Several experiments were developed to predict influent conductivity. The best hyperparameter combination was obtained considering the GridSearch technique. Additionally, multivariate recursive multi-step forecasting approach was also utilised, during model training and evaluation. During this process, season and temperature data were also considered for forecasting. Table 3 summarizes the list of values utilised for each hyperparameter during GridSearch.

Table 3. Hyperparameters' searching space for each model.

Parameters	Transformer value range	LSTM value range
Activation Function	[ReLU, tanh]	[ReLU, tanh]
Batch Size	[10, 20]	[10, 20]
Dropout	[0.0, 0.5]	[0.0, 0.5]
Encoding Layers	[4, 8]	–
Epochs	[50]	[25]
Layers	–	[3, 4, 5]
MLP Layers	[3, 4]	–
Neurons	[64, 128]	[32, 64, 128]
Number of heads	[4, 8]	–
Timesteps	[4, 6, 8]	[4, 6, 8]

Different hyperparameter values were tested for both base models. One important hyperparameter in this context is the number of timesteps. They are given as input to the candidate models, and the values 4, 6 and 8 were considered in both models. Considering 4 timesteps as an example, a sequence of 4 records (recorded every 2 days) is utilized to predict the following two conductivity registers (with a periodicity of 2 days). Additionally, some of the other hyperparameters are specific to the model, like *head size* and *ff_dim*, which are used only in Transformer, with the values of 128 and 512, respectively. During the experiments, all candidate models had the same seed value, in this case, 91195003. Conceived tests were supported by plotting the learning curves to prevent overfitting and underfitting situations. Afterwards, these curves were analyzed to choose the number of epochs. In addition, it was crucial to use the *TimeSeriesSplit* cross-validator with a k value equal to 3, since we are dealing with a time series problem. Concerning the technologies used, Python 3.10, was used for conceiving, tuning and evaluating the different candidate models. Also, libraries like Numpy, Pandas, Matplotlib, and scikit-learn, among others, were considered. Regarding the development of ML models, TensorFlow v2.0.0 was used. Finally, all hardware used in the development of this study was provided by Google's Collaboratory.

5   Results and Discussion

After all the experiments were performed, it was necessary to analyze their results. Table 4 summarizes the results according to the model's nature, namely Transformer and LSTM candidate models. In this table, it is possible to verify the hyperparameters space, utilized in each of the top-five candidate models and their scores in terms of RMSE, MAE, and training time.

Table 4. Top-five candidate models results, for Transformers and LSTMs.

Timesteps	Batch	Layers	Neurons	Dropout	Activation	Epochs	Encoding layers	Heads	RMSE	MAE	Time(s)
Transformer candidate models											
4	10	4	64	0.0	tanh	50	8	8	155.2	135.4	355.6
6	10	3	128	0.5	tanh	50	4	8	156.0	141.6	199.0
4	20	3	128	0.5	tanh	50	4	4	156.6	146.0	119.7
8	20	3	64	0.0	tanh	50	4	4	157.7	141.7	133.5
6	10	4	128	0.5	tanh	50	4	8	159.2	149.6	203.2
LSTM candidate models											
4	10	4	128	0.0	tanh	25	–	–	179.9	162.4	69.4
4	10	4	128	0.5	tanh	25	–	–	187.1	170.4	72.1
4	10	5	128	0.0	tanh	25	–	–	187.9	170.3	82.0
4	20	4	128	0.0	relu	25	–	–	188.2	169.9	57.8
4	20	3	128	0.5	relu	25	–	–	190.0	172.6	50.6

Regarding the Transformer-based models, some hyperparameters utilised are specific to Transformers, like the number of encoding layers and the number of heads, since Transformer uses an encoding layer, and LSTM does not. Considering these candidate models, the best one achieved a RMSE of 155.2 μS/cm and a MAE of 135.4 µS/cm. In addition, the activation function considered was the *tanh*, the dropout rate was set to 0.0, and the number of layers and neurons was 4 and 64, respectively. Considering the number of timesteps utilised as input, it was 4. Concerning the top-five Transformer-based models, there are some predominant hyperparameters such as the layers (3), encoding layers (4), and activation function (*tanh*).

When considering the LSTM-based models, the best one achieved a RMSE of 179.9 µS/cm and a MAE of 162.4 µS/cm. This candidate model needed 4 timesteps as input to achieve the best performance and utilized 0.0 as the dropout value. Additionally, *tanh* was utilized as the activation function, and the number of neurons was 128. Finally, the batch size was 10, and the number of layers of this model was 4. Considering the top-five LSTM-based models, there are homogeneous hyperparameters, namely the timesteps (4) and the number of neurons (128). Apart from this, the other parameters are not homogeneous but show some predominance in the values, like the dropout (0.0) in the majority of the candidate models and the activation function (*tanh*).

Finally, considering both LSTM-based and Transformer-based models, it is possible to observe that Transformer performed better in this time series task when considering top-five candidate models. However, these models required longer times to train (more than 100 s), and the number of epochs was superior to the ones utilized in the LSTM-based models (50 instead of 25). Finally, in the fourth Transformer-based model of the top five, the required number of input timesteps was more significant (8). Figure 2 illustrates four random predictions of the influent conductivity of the best candidate model, the Transformer-based one. Red dots represent predicted values of conductivity between July 21st, 2020 and July 27th, 2020. As can be observed in the figure, the number of timesteps used as input was 4 to forecast the next four timesteps recursively.

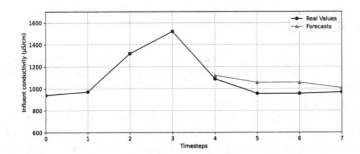

Fig. 2. Four random multi-step forecasts of influent conductivity.

6 Conclusions

Monitoring the water conductivity in a WWTP can be extremely helpful during wastewater treatment and management tasks since it can give important information about the water's salinity and help detect problems in water quality. In fact, it can be possible to detect saltwater intrusions, and consequently, adjust the wastewater treatment, since this phenomen might dilute the influent and present higher flow rates. Influent conductivity, in particular, provides earlier insights about the overall water quality entering the WWTP. Besides, WWTP operators can make adjustments to make the treatment process more efficient in the following stages of the WWTP. With this in mind, this work consisted of forecasting the next two timesteps (in a periodicity of two days) regarding the influent conductivity using LSTM and Transformer-based models.

Therefore, several experiments were considered to achieve the best hyperparameter set and performance for LSTM and Transformer architectures. Additionally, two features of the weather dataset were considered in the conception and training of the models, namely the temperature and season of the year, which presented a correlation with conductivity. The best model was based on a Transformer with encoding, achieving an RMSE of 155.2 µS/cm and an MAE of 135.4 µS/cm.

In the future, this work could be improved by utilizing more models for comparison with the models utilized in this study, such as CNNs and ANFIS. More specifically, hybrid models that might achieve better performances in influent conductivity forecasting, such as CNN-LSTM and GRU-LSTM. Additionally, new features, such as TDS, could be used to find if the influent conductivity forecasting performance can improve.

Acknowledgements. This work is financed by National Funds through the Portuguese funding agency, FCT - Fundação para a Ciência e a Tecnologia within project 2022.06822.PTDC. The work of Pedro Oliveira was supported by the doctoral Grant PRT/BD/154311/2022 financed by the Portuguese Foundation for Science and Technology (FCT), and with funds from European Union, under MIT Portugal Program.

References

1. Chaudhry, F.N., Malik, M.F.: Factors affecting water pollution: a review. J. Ecosyst. Ecography **7**(1), 225–231 (2017). https://doi.org/10.4172/2157-7625.1000225
2. Salgot, M., Folch, M.: Wastewater treatment and water reuse. Curr. Opin. Environ. Sci. Health **2**, 64–74 (2018). https://doi.org/10.1016/J.COESH.2018.03.005
3. Ahmed, U., Mumtaz, R., Anwar, H., Mumtaz, S., Qamar, A.M.: Water quality monitoring: from conventional to emerging technologies. Water Supply **20**(1), 28–45 (2020). https://doi.org/10.2166/ws.2019.144
4. Oliveira, P., Fernandes, B., Aguiar, F., Pereira, M.A., Analide, C., Novais, P.: A deep learning approach to forecast the influent flow in wastewater treatment plants. In: Analide, C., Novais, P., Camacho, D., Yin, H. (eds.) IDEAL 2020. LNCS, vol. 12489, pp. 362–373. Springer, Cham (2020). https://doi.org/10.1007/978-3-030-62362-3_32
5. Duarte, M.S., et al.: A review of computational modeling in wastewater treatment processes. ACS ES&T Water (2023). https://doi.org/10.1021/acsestwater.3c00117

6. Banna, M.H., et al.: Miniaturized water quality monitoring pH and conductivity sensors. Sens. Actuators B Chem. **193**, 434–441 (2014). https://doi.org/10.1016/j.snb.2013.12.002

7. Rusydi, A.F.: Correlation between conductivity and total dissolved solid in various type of water: a review. In: IOP Conference Series: Earth and Environmental Science, vol. 118, p. 012019. IOP Publishing (2018). https://doi.org/10.1088/1755-1315/118/1/012019

8. Ramos, P.M., Pereira, J.D., Ramos, H.M.G., Ribeiro, A.L.: A four-terminal water-quality-monitoring conductivity sensor. IEEE Trans. Instrum. Meas. **57**(3), 577–583 (2008). https://doi.org/10.1109/TIM.2007.911703

9. Elsworth, S., Güttel, S.: Time series forecasting using LSTM networks: a symbolic approach. arXiv preprint arXiv:2003.05672 (2020)

10. Yang, Y., Lu, J.: A fusion transformer for multivariable time series forecasting: the Mooney viscosity prediction case. Entropy **24**(4), 528 (2022). https://doi.org/10.3390/e24040528

11. Solaimany-Aminabad, M., Maleki, A., Hadi, M.: Application of artificial neural network (ANN) for the prediction of water treatment plant influent characteristics. J. Adv. Environ. Health Res. **1**(2), 89–100 (2013). https://doi.org/10.22102/jaehr.2013.40130

12. Fu, Z., Cheng, J., Yang, M., Batista, J.: Prediction of industrial wastewater quality parameters based on wavelet de-noised ANFIS model. In: 2018 IEEE 8th Annual Computing and Communication Workshop and Conference (CCWC), pp. 301–306. IEEE (2018). https://doi.org/10.1109/CCWC.2018.8301761

13. Najah, A.A., El-Shafie, A., Karim, O.A., Jaafar, O.: Water quality prediction model utilizing integrated wavelet-ANFIS model with cross-validation. Neural Comput. Appl. **21**, 833–841 (2012). https://doi.org/10.1007/s00521-010-0486-1

14. Vaswani, A., et al.: Attention is all you need. In: Advances in Neural Information Processing Systems, vol. 30 (2017). https://doi.org/10.48550/arXiv.1706.03762

15. Zhao, Y., Zhang, J., Zong, C.: Transformer: a general framework from machine translation to others. Mach. Intell. Res. 1–25. https://doi.org/10.1007/s11633-022-1393-5

16. Saoud, L.S., Al-Marzouqi, H., Hussein, R.: Household energy consumption prediction using the stationary wavelet transform and transformers. IEEE Access **10**, 5171–5183 (2022). https://doi.org/10.1109/ACCESS.2022.3140818

17. Zhou, H., et al.: Informer: beyond efficient transformer for long sequence time-series forecasting. In: Proceedings of the AAAI Conference on Artificial Intelligence, vol. 35, no. 12, pp. 11106–11115 (2021). https://doi.org/10.1609/aaai.v35i12.17325

18. Rehmer, A., Kroll, A.: On the vanishing and exploding gradient problem in Gated Recurrent Units. IFAC-PapersOnLine **53**(2), 1243–1248 (2020). https://doi.org/10.1016/j.ifacol.2020.12.1342

19. Siami-Namini, S., Tavakoli, N., Namin, A.S.: A comparison of ARIMA and LSTM in forecasting time series. In: 2018 17th IEEE International Conference on Machine Learning and Applications (ICMLA), pp. 1394–1401. IEEE (2018). https://doi.org/10.1109/ICMLA.2018.00227

20. Siami-Namini, S., Tavakoli, N., Namin, A.S.: The performance of LSTM and BiLSTM in forecasting time series. In: 2019 IEEE International Conference on Big Data (Big Data), pp. 3285–3292. IEEE (2019). https://doi.org/10.1109/BigData47090.2019.9005997

Unsupervised Defect Detection for Infrastructure Inspection

N. P. García-de-la-Puente[✉], Rocío del Amor, Fernando García-Torres,
Adrián Colomer, and Valery Naranjo

Instituto Universitario de Investigación en Tecnología Centrada en el Ser Humano,
Universitat Politècnica de València, Valencia, Spain
napegar@upv.es, vnaranjo@dcom.upv.es

Abstract. Artificial Intelligence (AI) provides a fundamental aid in
building operations, allowing infrastructure inspection and compliance
with safety standards. In the collaborative tasks involved, detecting
areas of interest, such as surface defects, is crucial. A drawback of
supervised AI-based approaches is that they require manual annotation,
which entails additional costs. This paper presents a novel unsupervised
anomaly detection approach for locating defects based on generative
models that learn the distribution of defect-free images. Using attention maps to validate in a subset, we propose a formulation that does
not require accessing labelled images, enabling task automation, maintenance optimisation and cost reduction.

Keywords: Visual Inspection · Infrastructure Inspection · Defects ·
Unsupervised Segmentation

1 Introduction

Collaborative Networks (CN) are alliances of entities working together to solve
complex problems that a single individual or organisation cannot efficiently
tackle. Three factors have led Artificial Intelligence (AI) to show promising
results in favour of CNs. Firstly, the increase in data availability and lower
procurement hardware prices. Secondly, the development of techniques that
allow processing large datasets to identify patterns, correlations and anomalies.
Finally, the integration and interoperability, bridging the gap between systems
and sharing knowledge. When combined, CN and AI can solve complex problems
requiring human expertise and the automation of algorithms.

In the world of AI, one field of great interest is Computer Vision (CV). It is a
huge field focused on automatically extracting information from visual content,
such as images or videos. The implementation of CV has significantly increased
because of the integration of AI and Deep learning (DL) techniques in the areas
of biomedicine, the food industry, the agricultural sector and the development of
autonomous cars, among others [1–3]. One of the most widespread applications

P. Quaresma et al. (Eds.): IDEAL 2023, LNCS 14404, pp. 142–153, 2023.
https://doi.org/10.1007/978-3-031-48232-8_14

is related to product inspection in the industry, either during the production phase of the parts or in the operation of products and structures [4].

Various algorithms have been used in CV for industry manufacturing and component performance monitoring: supervised, semi-supervised and unsupervised [5]. Supervised algorithms learn to recognise patterns in images, which allows to predict the presence of defects in unseen images. However, supervised learning requires large amounts of data and duly labelled by experts, which entails considerable time and economic costs. In contrast, unsupervised learning is used when the data is unlabelled. The goal is to find hidden patterns or structures in the data without prior knowledge. Unsupervised anomaly detection can be employed when labelled data is unavailable. This approach is based on a few assumptions, including that most samples in the dataset are expected not defective and defective instances are rare occurrences within the dataset [6]. Semi-supervised algorithms combine features of supervised and unsupervised algorithms. While unsupervised learning has shown promise in some CV applications, it is less commonly used for defect inspection than supervised learning, given that the latter has traditionally performed better.

Inspecting parts and detecting defects in the industry are generally conducted using Non-Destructive Testing (NDT). A diverse range of quality assessment and defect inspection methods are available. Their selection depends on the specific product being examined and the significance of the information obtained from each technique in a particular scenario. The most commonly used methods are visual inspection, thermography, acoustic emissions, electromagnetic testing, ultrasound or X-rays. CV has been successfully integrated into NDT, highlighting its use with thermal imaging cameras to detect heat anomalies that may indicate defects [7] or with X-ray imaging to detect internal flaws in components [8]. Combining CV technology with other NDT methods can improve defect detection capabilities, reduce the risk of product failure and increase overall product quality.

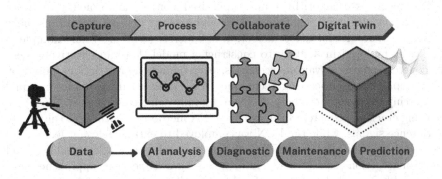

Fig. 1. Digital Twin flow as a non-destructive test for defect inspection.

Digital Twin (DT) technology is also being explored as a NDT for defect inspection. Creating a virtual replica of a product or component can simulate various operating conditions and analyse the impact of defects on product performance without physical testing. This approach can significantly reduce the time and cost associated with traditional testing methods while providing more accurate and detailed data. Leveraging DT as a NDT for defect inspection can enhance product quality, minimise the risk of failure, and increase efficiency in the manufacturing process [9]. AI can be employed to analyse the data obtained from DT (Fig. 1), facilitating diagnostics and preventive maintenance. Techniques such as explainable AI can be incorporated for data analysis and to predict potential risks [10].

The present research work is part of a more comprehensive development project intended to add value to the construction and operation of the building. Specifically, the aim is to develop a model for inspecting infrastructures to ensure compliance with safety standards. As a novelty, unsupervised techniques are to be used with an end-to-end approach to perform semantic segmentation and classification at the same time. This allows circumventing the usual lack of labelled data and provide a method that can achieve a high level of accuracy on unlabelled data, thus avoiding the costs and time associated with supervised tasks. To validate the accuracy and performance of the model proposed, a labelled subset of the data is intended to be used.

2   Related Work: Defect Detection in Civil Infrastructures

Defect detection with AI-based CV models has the potential to revolutionise the way that civil infrastructure and construction are monitored and maintained. DL is becoming increasingly popular for detecting and assessing civil infrastructure and construction defects. It has been the focus of research studies by many authors [11,12]. Several DL algorithms have been used or developed for the classification, localisation, or segmentation of defects, mainly consisting of cracks in concrete or steel structures and, to a lesser extent, corrosion [13].

The most used algorithms are supervised convolutional neural network (CNN) variations with end-to-end DL approaches. These algorithms have reached great results for defect detection in concrete structures. For example, a CNN was employed in a study to construct a model that can effectively identify concrete cracks in various scenarios, overcoming the restrictions posed by conventional image-processing techniques [14]. Yang et al. introduced a dual CNN architecture composed of a CNN and a fully connected network (FCN) for detecting cracks on concrete bridges. The CNN method was utilised to eliminate interference signals, while the FCN was employed to extract relevant features of the cracks [15]. Cha et al. used CNNs to detect and classify cracks, reaching good results in detection accuracy [16]. Nevertheless, these supervised approaches have the crucial barrier of the absence of publicly available datasets with appropriate annotations due to the necessity for time-consuming and labour-intensive work for data preparation [17].

Regarding the segmentation of cracks using CV, some authors studied the localisation and delineation of the damaged shape using deep CNN. Islam et al.

developed a transfer-learning approach to identify cracks using four pretrained models [18]. Zhang et al. used a pixel-wise CNN and an FCN to segment defects such as concrete cracks, concrete spalling, exposed reinforcement bars, steel corrosion, steel fracture and fatigue cracks, and asphalt [19].

Unsupervised machine learning methods have been used for crack detection in construction in the last years, such as Principal Component Analysis (PCA) and several clustering methods, but they need more robustness. As far as we are concerned, there need to be more approaches for training DL algorithms without annotation that have been used for defect detection in other industrial fields but to a lesser degree in construction and civil infrastructure [20]. Although some studies use DL algorithms, there are yet to be state-of-the-art studies that perform the classification and segmentation of cracks [21,22]. Thus, the application of unsupervised learning in this field is still relatively new and more research is needed to explore its full potential.

3   Methodology

An overview of our proposed method for defect detection is depicted in Fig. 2. In the following, we describe the problem formulation and each proposed component.

Problem Formulation: Under the paradigm of unsupervised anomaly detection, we denote the set of unlabeled training images as $X_T = \{x_n\}_{n=1}^N$, composed of only normal images, i.e., images without defects. We now define an encoder in charge of transforming the input data $\mathbf{X}_T$ into a latent representation (with a lower dimensionality) $\mathbf{Z}$ through a non-linear mapping function, $\mathbf{Z} = f_\phi(\mathbf{X}_T)$, where ϕ are the learnable parameters of the encoder architecture. The decoder stage produces the reconstruction of the data based on the features embedded in the latent space, $\mathbf{R} = g_\theta(\mathbf{Z})$. The reconstructed representation $\mathbf{R}$ is required to be as similar to $\mathbf{X}_T$ as possible. During the inference, we use a set of unlabeled images $X_I = \{x_m\}_{m=1}^M$, composed of images without and with defects to differentiate them and provide the map where the defect is found.

3.1   Variational Autoencoder

Variational autoencoder (VAE) is an unsupervised approach composed of an encoder-decoder architecture commonly used for anomaly detection [23]. With a VAE, the input data is coded as a multivariate normal distribution $p(z|x)$ around a point in the latent space. In this way, the encoder part is optimized to obtain a multivariate normal distribution's mean and covariance matrix. The VAE algorithm assumes no correlation exists between latent space dimensions; therefore, the covariance matrix is diagonal. In this way, the encoder only needs to assign each input sample to a mean and a variance vector. In addition, the logarithm of the variance is set, as this can take any real number in the range $(-\infty, \infty)$, matching the natural output range from a neural network, whereas that variance values are always positive. To provide continuity and completeness

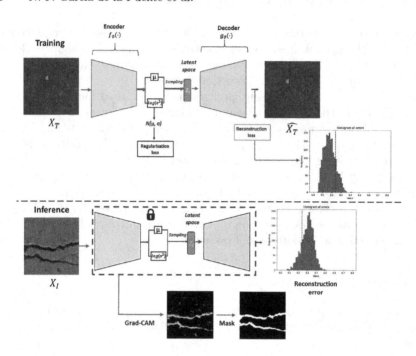

Fig. 2. Method overview. In line with the established literature on anomaly detection, the Variational Autoencoder (VAE) is optimized to maximize the evidence lower bound (ELBO). Furthermore, we incorporate an attention constraint through a size-constrained loss, which compels the network to explore the entire image. During inference, the attention map is thresholded to generate the final segmentation mask.

to the latent space, it is necessary to regularize both the logarithm of the variance and the mean of the distributions returned by the encoder. This regularisation is achieved by matching the encoder output distribution to the standard normal distribution ($z_\mu = 0$ and $z_\sigma = 1$). After obtaining and optimizing the parameters of the mean and variance of the latent distributions, it is necessary to take samples of the learned representations to reconstruct the original input data. Samples of the encoder output distribution are obtained as follows:

$$Z \approx p(z|x) = z_\mu + z_\sigma \cdot \epsilon \tag{1}$$

where ϵ is randomly sampled from a standard normal distribution and $\sigma = \exp(\frac{\log(\sigma^2)}{2})$.

The minimized loss function in a variational autoencoder comprises two terms: (1) a reconstruction term that compares the reconstructed data to the original input to get as effective encoding-decoding as possible and (2) a regularisation term in charge of regularizing the latent space organization. The regularisation term is expressed as the *Kulback-Leibler* (KL) divergence that measures the difference between the predicted latent probability distribution of the data and the standard normal distribution [24]:

$$D_{KL}[N(z_\mu, z_\sigma)||N(0,1)] = \frac{1}{2}\sum(1 + log(z_\sigma^2) - z_\mu^2 - z_\sigma^2) \tag{2}$$

The KL function is minimised to 0 if $\mu = 0$ and $log(\sigma^2) = 0$ for all dimensions. As these two terms differ from 0, the variational autoencoder loss increases. The compensation between the reconstruction error and the KL divergence is a hyper-parameter to be adjusted in this type of architecture.

Since training a VAE consists in minimizing a two-term loss function, this is equivalent to maximize the evidence lower-bound (ELBO):

$$\mathcal{L}_{VAE} = \mathcal{L}_R(x_T, \hat{x}_T) + \beta\mathcal{L}_{KL}(p(z|x)||p(z)) \tag{3}$$

where β is a weighting factor to optimize.

3.2 Anomaly Segmentation via Grad-CAMs

Several works based on unsupervised anomaly detection use attention maps to mimic the segmentation mask of the anomaly. In particular, attention maps $\mathbf{a} \in \mathbb{R}^{\Omega_i}$ are generated from the mean latent vector z_μ, by using Grad-CAM [25] via backpropagation to the encoder block output. Therefore, given an image ($\mathbf{x}$), the attention maps is calculated as follows:

$$a = \sigma\left(\sum_k^K \alpha_k f_\phi(x)_k\right) \tag{4}$$

where K is the total number of filters of the encoder layer, σ is the sigmoid function, and α_k is the generated gradient such that: $\alpha_k = \frac{1}{|a|}\sum_{t\in\Omega_T}\frac{\partial z_\mu}{\partial a_{k,t}}$, where Ω_T is the spatial feature domain.

The idea underlying the Grad-CAM is to enforce them to cover the whole free-defect image without showing high activations concentrated in some areas. In inference, the activations will be concentrated in the area with defects. Therefore, it is necessary to introduce a constraint related to the Grad-CAM in the global loss function. Following the method show in [26], we use a log-barrier extension function with a single global constraint to achieve maximum coverage of class-activation maps over the whole image. Thus, we can formally define the approximation of log-barrier as:

$$\psi_t(z) = \begin{cases} -\frac{1}{t}log(-f_c(a)) & \text{if } f_c(a) \leq -\frac{1}{t^2} \\ tf_c(a) - \frac{1}{t}log(\frac{1}{t^2}) + \frac{1}{t} & \text{otherwise,} \end{cases} \tag{5}$$

where t controls the barrier during training, and $f_c(a) = \left(1 - \frac{1}{|\Omega_T|}\sum_{l\in\Omega} a_l\right)$ is the constraint over the attention map from the jth image, which enforces the generated attention map to cover the whole image, where Ω is the spatial features domain.

$$\mathcal{L} = \mathcal{L}_{VAE} + \lambda \sum_{n=1}^{N} \psi_t (1 - \frac{1}{|\Omega_T|} \sum_{l \in \Omega_i} a_l) \tag{6}$$

In this scenario, for a given t, the optimizer will try to find a solution with a good compromise between minimizing the loss of the VAE and satisfying the constraint $f_c(a)$.

3.3 Inference

During inference, we obtain the reconstruction of the inferred images X_I using the VAE trained with defect-free images (X_T). After obtaining the reconstructed images, we calculate the error concerning the original images. According to our hypothesis, we consider that those inferred images with a high reconstruction error will be defective. Therefore, we establish a threshold by considering the mean and standard deviation of the errors in the defect-free images used for validation. Inferred images with an error exceeding the threshold will be classified as defective, while those falling below the threshold will be classified as non-defective. Additionally, we use the anomaly saliency map as a segmentation map. During the experimental stage, we found that anomalies produce larger activation on attention maps than the constrained normal samples. Then, the map is thresholded to create an anomaly mask of the image.

4 Experimental Setting

4.1 Dataset and Evaluation Metrics

The dataset utilized in this study [27] comprises concrete images which were sourced from multiple buildings within the METU Campus. The dataset contains 20,000 negative and 20,000 positive crack images, enabling image classification tasks. These images have 227×227 pixels and are represented in RGB channels. The images exhibit variations in surface finish and illumination conditions, contributing to a diverse and realistic representation of real-world scenarios. To conduct our experiments, we selected a representative part of the dataset. To validate the segmentation output, 200 mask segmentations were created manually. These masks serve as ground truth (GT) references for evaluating the segmentation algorithm.

To evaluate the performance of the proposed approach regarding the classification task, we compute accuracy, precision, F1-score, recall and confusion matrix. To assess the semantic segmentation predicted by the model, we use some metrics such as AUC ROC, AU PRC, DICE and IoU [28].

4.2 Implementation Details

We trained all our models using the dataset described during 300 epochs with the Adam optimizer, a learning rate of 0.00001 and batch size of 32. We defined

5 convolutional blocks in the VAE determining 32 as the dimension of the latent coding space. Variable t has a value of 20 in log-barrier loss. The binary cross-entropy loss was computed for segmentation and L2 in classification. The threshold value optimized and used to classify the images was 0.2366. Regarding the hardware, we used NVIDIA RTX 3090 24 GB $\times$ 1, 525.60.11 drivers & CUDA 12.0, MSI Z270 Gamming PRO Carbon (MS-7A63); 32 GB and Intel i7-7700K (4.2 GHz), whereas the software used was Pytorch for building and training the models, and Sci-kit learn for evaluation.

5   Results

The results of the experimentation will be discussed below according to the methodology proposed.

5.1   Classification

After the model is trained, we use a not seen 200-image dataset without defects to calculate the error between the images and the reconstructed images ($x - \hat{x}$). We hypothesise that the optimized threshold will have the compromise to assume that those images with reconstruction error above the threshold will be anomalies. To do so, we use the mean error plus the standard deviation error and the result is 0.2366. We plot the histograms (Fig. 3, left) to see which is the distribution and where is located the calculated error.

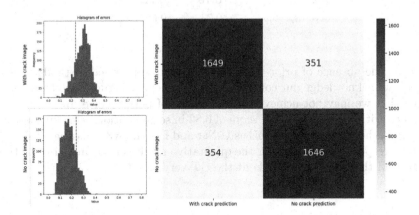

Fig. 3. Histograms of errors in crack images (up left) in images without cracks (down left) and confusion matrix (right).

The confusion matrix is included (Fig. 3, right), considering 4,000 test unseen images. Furthermore, metric results to evaluate the classification are considered in Table 1.

Table 1. Metrics for the crack-non crack classification using the optimized threshold.

Accuracy	Precision	Recall	F1-score
0.823	0.823	0.824	0.823

Since there is no previous work using unsupervised methods on the subject, we compare ourselves with supervised studies, bearing in mind that we will not be up to the task. In [29], STCNet I model accuracy over the same dataset is 99.71% and in [18] VGG16 gets an average accuracy of 99.61% using transfer learning.

5.2   Unsupervised Defect Segmentation

To assess the model performance in unsupervised segmentation, we consider an iteration as one full training with a singular seed. The quantitative results by iteration and averaged of AU_ROC, AU_PRC, DICE, IoU are shown in Table 2.

Table 2. Three Iterations Results over 200 unseen images and Average.

metric	AU_ROC	AU_PRC	DICE	IoU
Iter 0	**0.97**	0.76	0.71	**0.55**
Iter 1	0.95	0.64	0.65	0.48
Iter 2	0.96	**0.78**	**0.72**	0.54
Average	0.96	0.73	0.69	0.53

Given the absence of prior research utilizing unsupervised methods for this subject, we acknowledge our comparison with supervised studies while acknowledging that we may not achieve the same level of performance. In [30] a CNN combined with ViT shows a IoU score of 0.82 in segmentation using our dataset. They add other supervised models as U-Net and DeepLabv3+ getting a IoU 0.79 and 0.82 respectively. Regarding the qualitative results of our model (Fig. 4), it can be said that the Grad-CAMs fit the GT very closely.

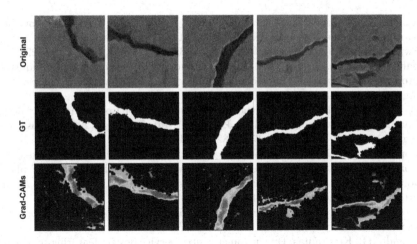

Fig. 4. Qualitative assessments of various examples: original, GT and Grad-CAMs.

6   Conclusion

A relevant body of literature on defect detection in the industry requires manual annotation to train deep learning-based models. The annotation process is an expensive task that increases the costs of the industrial process. To overcome this issue, in this work, we propose an unsupervised anomaly detection algorithm able to detect and find the crack location in concrete images. The proposed method based on VAE and Grad-CAMs allows the detection and segmentation of defects showing promising results. Our approach aims to serve as a foundational step towards achieving zero-defect manufacturing, providing a holistic solution to minimize deviations in the operation of buildings and manufacturing processes.

Funding. This work has received funding from Horizon Europe, the European Union's Framework Programme for Research and Innovation, under Grant Agreement No. 101058054 (TURBO) and No. 101057404 (ZDZW). The work of Rocío del Amor has been supported by the Spanish Ministry of Universities (FPU20/05263).

References

1. Gupta, A.: Current research opportunities of image processing and computer vision. Comput. Sci. (2019). https://doi.org/10.7494/csci.2019.20.4.3163. ISSN 2300-7036, 1508-2806
2. Tian, H., Wang, T., Liu, Y., Qiao, X., Li, Y.: Computer vision technology in agricultural automation—a review. Inf. Process. Agric. (2020). https://doi.org/10.1016/j.inpa.2019.09.006
3. Paneru, S., Jeelani, I.: Computer vision applications in construction: current state, opportunities & challenges. Autom. Constr. (2021). https://doi.org/10.1016/j.autcon.2021.103940. Accessed 12 May 2023

4. Taheri, H., Gonzalez Bocanegra, M., Taheri, M.: Artificial intelligence, machine learning and smart technologies for nondestructive evaluation. Sensors (Basel, Switzerland) **22** (2022). https://doi.org/10.3390/s22114055. ISSN 1424-8220

5. Saberironaghi, A., Ren, J., El-Gindy, M.: Defect detection methods for industrial products using deep learning techniques: a review. Algorithms (2023). https://doi.org/10.3390/a16020095. ISSN 1999-4893

6. Bhatt, P.M., Malhan, R.K., Rajendran, P., et al.: Image-based surface defect detection using deep learning: a review. J. Comput. Inf. Sci. Eng. (2021). https://doi.org/10.1115/1.4049535. ISSN 1530-9827, 1944-7078

7. Bommes, L., Pickel, T., Buerhop-Lutz, C., Hauch, J., Brabec, C., Peters, I.M.: Computer vision tool for detection, mapping, and fault classification of photovoltaics modules in aerial IR videos. Progress Photovoltaics: Res. Appl. (2021). https://doi.org/10.1002/pip.3448. ISSN 1099-159X

8. Mery, D., Arteta, C.: Automatic defect recognition in X-ray testing using computer vision (2017). https://doi.org/10.1109/WACV.2017.119

9. Fragidis, G., Konstantas, D.: Customer-centric service design: featuring service use in life practices. In: Camarinha-Matos, L.M., Ortiz, A., Boucher, X., Osário, A.L. (eds.) PRO-VE 2022. IFIPAICT, vol. 662, pp. 182–193. Springer, Cham (2022). https://doi.org/10.1007/978-3-031-14844-6_15

10. Enrique, D.V., Soares, A.L.: Cognitive digital twin enabling smart product-services systems: a literature review. In: Camarinha-Matos, L.M., Ortiz, A., Boucher, X., Osório, A.L. (eds.) PRO-VE 2022. IFIPAICT, vol. 662, pp. 77–89. Springer, Cham (2022). https://doi.org/10.1007/978-3-031-14844-6_7

11. Tang, Y., Lin, Y., Huang, X., Yao, M., Huang, Z., Zou, X.: Grand challenges of machine-vision technology in civil structural health monitoring. Artif. Intell. Evol. (2020). https://doi.org/10.37256/aie.112020250. ISSN 2717-5952

12. Fang, W., Ding, L., Love, P.E.D., et al.: Computer vision applications in construction safety assurance. Autom. Constr. (2020). https://doi.org/10.1016/j.autcon.2019.103013. ISSN 0926-5805

13. Spencer, B.F., Hoskere, V., Narazaki, Y.: Advances in computer vision-based civil infrastructure inspection and monitoring. Engineering **5** (2019). https://doi.org/10.1016/j.eng.2018.11.030. ISSN 2095-8099

14. da Silva, W.R.L., de Lucena, D.S.: Concrete cracks detection based on deep learning image classification. In: Proceedings, vol. 2 (2018). https://doi.org/10.3390/ICEM18-05387. ISSN 2504-3900

15. Yang, L., Li, B., Li, W., Liu, Z., Yang, G., Xiao, J.: Deep Concrete Inspection Using Unmanned Aerial Vehicle Towards CSSC Database (2017)

16. Cha, Y.-J., Choi, W., Büyüköztürk, O.: Deep learning-based crack damage detection using convolutional neural networks. Comput.-Aided Civil Infrastruct. Eng. (2017). https://doi.org/10.1111/mice.12263

17. Ai, D., Jiang, G., Lam, S.-K., He, P., Li, C.: Computer vision framework for crack detection of civil infrastructure—a review. Eng. Appl. Artif. Intell. (2023). https://doi.org/10.1016/j.engappai.2022.105478. ISSN 0952-1976

18. Islam, M.M., Hossain, M.B., Akhtar, M.N., Moni, M.A., Hasan, K.F.: CNN based on transfer learning models using data augmentation and transformation for detection of concrete crack. Algorithms **15**(8), 287 (2022)

19. Zhang, A., Wang, K.C.P., Li, B., et al.: Automated pixel-level pavement crack detection on 3D asphalt surfaces using a deep-learning network. Comput.-Aided Civil Infrastruct. Eng. (2017). https://doi.org/10.1111/mice.12297

20. Pei, L., Sun, Z., Xiao, L., Li, W., Sun, J., Zhang, H.: Virtual generation of pavement crack images based on improved deep convolutional generative adversarial network. Eng. Appl. Artif. Intell. (2021). https://doi.org/10.1016/j.engappai.2021.104376

21. Chow, J.K., Su, Z., Wu, J., Tan, P.S., Mao, X., Wang, Y.H.: Anomaly detection of defects on concrete structures with the convolutional autoencoder. Adv. Eng. Inform. (2020). https://doi.org/10.1016/j.aei.2020.101105

22. Rastin, Z., Ghodrati Amiri, G., Darvishan, E.: Unsupervised structural damage detection technique based on a deep convolutional autoencoder. Shock Vibration (2021). https://doi.org/10.1155/2021/6658575

23. Titus, A.J., Wilkins, O.M., Bobak, C.A., Christensen, B.C.: Unsupervised deep learning with variational autoencoders applied to breast tumor genome-wide DNA methylation data with biologic feature extraction. Bioinformatics (2018). https://doi.org/10.1101/433763

24. Foster, D.: Generative deep learning: teaching machines to paint, write, compose, and play (2019)

25. Selvaraju, R.R., Cogswell, M., Das, A., Vedantam, R., Parikh, D., Batra, D.: Grad-CAM: visual explanations from deep networks via gradient-based localization, pp. 618–626 (2017)

26. Silva-Rodrıguez, J., Naranjo, V., Dolz, J.: Constrained unsupervised anomaly segmentation. Med. Image Anal. **80**, 102 526 (2022)

27. Özgenel, Ç.F., Sorguç, A.G.: Performance comparison of pretrained convolutional neural networks on crack detection in buildings. In: ISARC. Proceedings of the International Symposium on Automation and Robotics in Construction, vol. 35, pp. 1–8. IAARC Publications (2018)

28. Mahmoudi, R., Benameur, N., Mabrouk, R., Mohammed, M.A., Garcia-Zapirain, B., Bedoui, M.H.: A deep learning-based diagnosis system for COVID-19 detection and pneumonia screening using CT imaging. Appl. Sci. (2022). https://doi.org/10.3390/app12104825

29. Ye, W., Deng, S., Ren, J., Xu, X., Zhang, K., Du, W.: Deep learning-based fast detection of apparent concrete crack in slab tracks with dilated convolution. Constr. Build. Mater. **329**, 127 157 (2022)

30. Shamsabadi, E.A., Xu, C., Rao, A.S., Nguyen, T., Ngo, T., Dias-da-Costa, D.: Vision transformer-based autonomous crack detection on asphalt and concrete surfaces. Autom. Constr. **140**, 104 316 (2022)

Generating Adversarial Examples Using LAD

Sneha Chauhan[1,2]($\boxtimes$), Loreen Mahmoud[1], Tanay Sheth[3],
Sugata Gangopadhyay[1], and Aditi Kar Gangopadhyay[4]($\boxtimes$)

[1] Department of Computer Science and Engineering, IIT Roorkee, Roorkee, India
{schauhan1,loreen_fm,sugata.gangopadhyay}@cs.iitr.ac.in
[2] Department of Computer Science and Engineering, NIT Uttarakhand,
Srinagar, India
[3] BITS Pilani Goa Campus, Sancoale, India
[4] Department of Mathematics, IIT Roorkee, Roorkee, India
aditi.gangopadhyay@ma.iitr.ac.in

Abstract. Nowadays, Machine learning models are widely used in many fields and employed to solve problems from different sectors. However, we often face issues when running these models in case the training data is insufficient. These issues happen when the dataset available is small or only part of it is available because of its sensitive content or even because the dataset is imbalanced. Thus, synthetic data generation is needed to provide data similar to actual data. We have proposed the use of the Logical Analysis of Data methodology to generate adversarial data from any given dataset. For our study, we have used an intrusion detection dataset, and the results demonstrate the potential of Logical Analysis of Data by evaluating adversarial datasets using various machine learning classifiers.

Keywords: Adversarial Examples · Logical Analysis of Data · Machine Learning

1 Introduction

Our lives have been taken over by the Internet in the present world. Everyday new devices and technologies are being added to the internet. With the rise of autonomous devices, security and integrity of information and devices from attacks and threats is also a severe challenge. Cyber criminals devise new techniques to gain control of network to access, modify data, inject viruses, malware to corrupt the system and disrupt services.

Intrusion Detection Systems (IDS) are often used in the network to monitor the traffic and detect attacks. Many Machine learning methods are used for developing intrusion detection systems. These models are trained using large datasets to learn the attack signatures so that they can distinguish an attack from normal activity. These models have excellently performed in the intrusion

P. Quaresma et al. (Eds.): IDEAL 2023, LNCS 14404, pp. 154–165, 2023.
https://doi.org/10.1007/978-3-031-48232-8_15

detection. However, training of these models depend on the dataset provided. Obtaining large dataset relevant to an attack scenario is difficult. Many times organizations are hesitant to share the actual data due to confidentiality issues and privacy concerns. Publicly available dataset may not be large enough for the model to give good results. Sometimes publicly available dataset are not a good option as they are imbalanced. They do not have characteristics of all the classes as in case of real world data [24]. Another issue with the machine learning based IDS model is that they are vulnerable to incorrect data fed for training, thus exploiting the model which results in wrong classification of new network traffic [17,21].

In this work, we have tried to use Logical Analysis of Data (LAD) to produce adversarial examples. Logical Analysis of Data is a data analysis method that uses logical patterns to know the nature of an activity i.e. the class to which that activity belongs. A detailed explanation of LAD technique is discussed in the Sect. 1.1. Our study goes through the following steps:

- The LAD process is applied on the dataset to produce negative patterns that represent normal behaviour of observations.
- The LAD classifier is constructed using these negative patterns. This classifier can classify the observations into positive and negative. This is shown in the paper [7].
- In order to produce adversarial attack observations, features that are present in the LAD patterns are used to modify their values in the original dataset.
- Different machine learning classifiers are implemented which are then trained and tested on this adversarial dataset. Two datasets: UNSW-NB15 and NSL-KDD are used in this study. The results show that the performance of these ML models improved when trained on a new dataset.

The rest of the paper describes the LAD technique used for generation of adversarial examples, validation of the adversarial data and performance metrics of the machine learning classifiers. The paper is concluded by stating the results and future scope.

1.1 Logical Analysis of Data

The data used for our study consists of observations that belong to the subspace R^d where d is the number of features. These observations may represent data related to a patient's medical report, network activity, probes in an oil field, etc. These observations belong to two classes i.e. positive or negative, for example, patient may be healthy or sick, there may be an abnormal or normal activity in the network, etc. The data analysis methodology tries to provide logical explanations that can differentiate positive observations from negative observations. The paper [11] showed that the partially defined Boolean function can achieve this for the binary data.

Consider a set $B = \{0, 1\}$. The set B^n consists of combinations of 0 and 1 with each vector having length n. A Boolean function is a function g that maps the n-dimensional vectors in B^n to the set B i.e. $g : B^n \to B$. A binary vector $v = v_1, v_2, \ldots, v_n$ is considered as a true point if $g(v) = 1$, and it is considered as a false point if $g(v) = 0$. We denote the sets of true and false points as $T(g)$ and $F(g)$, respectively. All the n-dimensional vectors of B^n will either fall into the set of true points $T(g)$ or the set of false points $F(g)$. Both of these sets are disjoint. Consider another set T and F such that $T \subset T(g)$ and $F \subset F(g)$. Such a pair (T, F) is called a partially defined Boolean function (pdBf) which does not cover all of the 2^n vectors of B^n [18]. The table 1 shows that there are 6 variables but the partially defined Boolean function is defined only on 7 vectors of the hypercube B^6.

Table 1. A partially defined Boolean function

x_1	x_2	x_3	x_4	x_5	x_6	$f(x_1, x_2, \ldots, x_6)$
1	0	1	1	0	1	1
0	0	1	1	0	1	0
1	0	0	0	0	1	1
1	0	1	1	0	0	1
1	0	1	1	0	1	0
1	0	0	1	0	1	0

There exists one or more Boolean function that may agree with a pdBf, and it is called as an extension of pdBf f. The logical analysis of data focuses on finding an extension f' of pdBf f which can distinguish positive and negative observations and is approximately equivalent to f [12].

The Logical Analysis of Data (LAD) is a data analysis technique originally developed by Peter L. Hammer in 1986 [16] for finding new rules or patterns from known observations that can classify unknown observations. LAD involves three important steps for classification which are described in detail in the following sections.

1.2 Binarization

Initially, LAD was used for the analysis of datasets having attributes that can take values 0 and 1 only. Later, LAD technique was extended to numerical datasets using a process called Binarization. In this process, each observation $a = (a_1, a_2, \ldots)$ of the real valued dataset is transformed into a binary vector $x(a) = x_1, x_2, \ldots$ where $x_i = \{0, 1\}$. This transformation is defined as follows: $x_1 = 1$ if $a_1 \geq \alpha_1$, $x_2 = 1$ if $a_2 \geq \alpha_2$, and so on. It ensures that if a represents a positive observation point and b represents a negative observation point, then their corresponding binary vectors x(a) and x(b) will be different. The binary variables x_i, where $i = 1, \ldots, n$, linked to the real attributes, are referred to as indicator variables. Additionally, the real parameters α_i, where $i = 1, ..., n$, utilized in the described procedure are known as cutpoints [2].

The binarization process can be illustrated by an example. The Table 2b shows a sample dataset having 3 attributes F1, F2 and F3. The features F1 and F2 are numerical while the feature F3 is nominal. The cutpoints are introduced for each numerical feature. Let $\alpha_1 = 3.05$ and $\alpha_2 = 5.6$ are the cutpoints for the features F1 and F2. The numerical features are transformed into binary variables using these cutpoints as, if $F1 \geq \alpha_1$ then $B1 = 1$ else $B1 = 0$ and if $F2 \geq \alpha_2$ then $B2 = 1$ else $B2 = 0$ for all the observations. The last feature contains only two values i.e. yes and no so it can be represented by a single binary variable where yes corresponds to 1 and no is denoted by 0. This transformation is shown in Table 2c.

The Table 2c can be viewed as a truth table of a partially defined Boolean function (pdBf) having three variables B1, B2 and B3. The binary vectors having class label as 1 are true points (T) and those with class label 0 are false points (F). Since here only 5 vectors are present it is a pdBf. If all the vectors are there then it is a Boolean function of three variables. The pdBf in Table 2c can be represented by a function:

$$f = B1\overline{B2} \vee \overline{B2}B3$$

This is a logical explanation of given pdBf. The vector will be a positive observation when $B1 = 1$ and $B2 = 0$ or $B2 = 0$ and $B3 = 1$. In terms of original attributes, we can say that if $F1 \geq \alpha_1$ and $F2 < \alpha_2$ or if $F2 < \alpha_2$ and $F3 = yes$ then it is a positive vector. In our example, we used only one cutpoint for each feature. There is no limitation in number of cutpoints generated for a feature. Each cutpoint is calculated by taking mean of two consecutive values of the feature provided that both values belong to different class labels. For a categorical feature, the number of binary variables is equal to the number of unique values of that feature.

Table 2. Binarization Example

(a)

Attributes	F1	F2	F3
Positive examples	1.5	4	Yes
	3.6	2.7	No
	4	5	Yes
Negative examples	2.5	6.2	Yes
	2.3	3.5	No

(b)

F1	F2	F3	Class label
1.5	4	Yes	1
3.6	2.7	No	1
4	5	Yes	1
2.5	6.2	Yes	0
2.3	3.5	No	0

(c)

B1	B2	B3	Class label
0	0	1	1
1	0	0	1
1	0	1	1
0	1	1	0
0	0	0	0

1.3 Support Set Generation

The dataset is partitioned into two classes: positive and negative. During binarization, it is made sure that no observation belongs to both classes simultaneously. Our binarization method maintains this property. Many binary variables are produced during binarization using the cutpoint method. Some of the binary variables may be redundant and irrelevant for classification. Thus, a process called as support set generation is used to select the best features that have an impact on classification. A support set is a set of binary features which is selected by removing redundant features. After obtaining the support set, the dataset should remain contradiction-free which means that there are no observations that belong to both classes simultaneously [3].

A support set is termed as minimal when the elimination of any attribute from the support set results in the intersection of positive and negative set of observations. The support set generation problem can be considered similar to the set covering problem [1,16]. Any irredundant set of attributes selected as support set is said to cover all observations of the given partially defined Boolean function as well as the vectors of the extended Boolean function. Some algorithms have been discussed to find the support set in [1]. To apply LAD in our study, we have used information gain ratio of each feature to select the best features in the support set as given in [7]. The feature with highest information gain ratio is selected for the support set. This process continues till no more samples are left for partition.

1.4   Pattern Generation

A pattern is a combination of different features or their complements that can label an observation as positive or negative. There are two types of patterns: positive and negative patterns. While generating patterns, it is important to ensure that no best pattern is left out. A combinatorial enumeration technique is followed to generate patterns [3]. It works on two approaches: top down and bottom up strategy.

In the top down strategy, we start with a characteristic term of a positive observation. This term is a pattern as it covers one positive observation. Literals are removed one by one from this term until a pattern remains. This is done to reduce the degree of pattern.

In case of bottom up approach, a reverse procedure is applied. It begins with a term having a single literal that cover few observations. If no negative observation is covered then this term becomes a positive pattern, otherwise literals are added one by one until the term becomes a pattern.

The aim of the pattern generation procedure is to generate small degree patterns and also every observation of the dataset is covered by a pattern. To achieve this, both bottom up and top down approach is followed in the given order. The paper [12] describes the algorithm for pattern generation which uses breadth first enumerative technique. The positive and negative patterns are generated of degree upto 4. The degree is kept smaller in order to minimize the computational complexity. In order to enhance the coverage of observations, an adaptation is included in the procedure. A threshold value is used to decide a pattern. If a term covers at least some number of positive observations and less than few negative observations then it is treated as a positive pattern. This is done to avoid overlapping of patterns and to cover large number of observations with few patterns. Impure patterns that cover very few negative observations also helps in reducing overfitting. By using thresholds, the theory formation step is no longer needed to select best patterns.

1.5   LAD Classifier

The patterns generated by the pattern generation procedure discussed above are used for the classification of observations. These patterns are transformed into rules to construct the classifier. Each rule or pattern is expressed as an if else statement and these statements are combined to form a classifier. A new observation is assessed by evaluating it against the conditions specified in the classifier. The observation is considered positive (negative) if all the conditions of positive (negative) patterns are met, otherwise it is labelled as negative (positive). A hybrid classifier can be created by integrating both positive and negative patterns.

2   Related Work

Alexander et al. proposed a data programming paradigm to create and model training datasets. In this approach, the labeling of data points is accomplished through the use of label functions, which essentially consist of sets of heuristic rules [22]. Another method to create training sets through programming is distant supervision [19]. This involves extracting textual features from an unlabeled dataset to train a relation classifier. The paper [25] used deep learning methods to build an anomaly based network intrusion detection system. The results showed that the model is able to classify and adapt well using incomplete data. Though machine learning and deep learning models perform with significantly high accuracy, they are also vulnerable to attacks where data provided is incorrect [21]. The authors of [15] argued that the neural networks failed classification of adversarial examples due to their linear nature. They used this explanation to suggest a fast way of generating adversarial examples and experimented on MNIST dataset leading to reduced test set error. An understanding of adversarial machine learning methods and how adversarial attacks can pass through machine learning based IDS without getting noticed is discussed in the paper [20]. The authors evaluate the robustness of existing machine learning and deep learning IDS models against the Bot-Iot dataset. Label noise adversarial attacks are created by random or target label flipping and fed to the SVM model to evaluate its performance. A Fast Gradient Sign Method (FGSM) is used to generate adversarial examples for evaluation of Artificial Neural Network.

3   Performance Evaluation

The generation of adversarial examples is carried out for intrusion detection models. Many datasets are available for implementation of intrusion detection systems. We have selected UNSW-NB15 and NSL-KDD dataset for our study. UNSW-NB15 has modern types of cyberattacks included in it. NSL-KDD dataset is a refined version of KDD99 dataset. Both the datasets have different classes of attacks. Since Logical Analysis of Data is a binary classification technique, we have combined all the attack categories together and labelled them as 'Attack' and rest of the observations are labelled as 'Normal'.

3.1   Adversarial Attack Generation

The Fig. 1 depicts the entire process of generation of adversarial data. The first step is to develop LAD classifier using the existing dataset. Since the dataset is numerical in nature, binarization process is applied to get binary dataset. The binarized dataset may contain many irrelevant and redundant attributes. A support set of features is selected from the binarized dataset that consists of irredundant features having high impact on the classification. Information gain ratio method as described in Sect. 1.3 is used to choose best binary variables. Using the support set, patterns are formed in the pattern generation step of LAD. These patterns are then written as if-else rules to construct the classifier. The testing dataset can be used to validate the LAD classifier.

A new dataset is created by selecting only normal observations of the UNSW-NB15 dataset. This dataset will now be modified to generate adversarial attack examples. The LAD classifier or the patterns that were generated in the third step of LAD are used to flip the feature values to make the normal observation into an attack observation. Each pattern or rule consists of three to four features along with their threshold values. In order to create new adversarial examples, feature value of each observation is changed according to the negative patterns produced by LAD. These negative patterns always give value true for normal observation and false for attack observation. So the feature values are complemented to convert a normal observation into an attack one. This process is repeated for every observation and then we get an adversarial dataset having all attack observations.

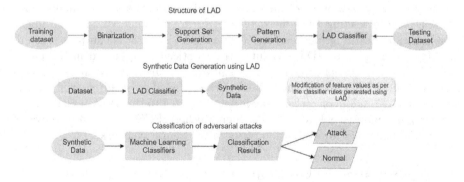

Fig. 1. Steps for Adversarial Attack Generation

3.2 Experimental Results on Adversarial Dataset

After generating the adversarial data, experiments are performed to validate our study. Classical classifiers used to analyze the performance of adversarial data are described below:

Decision Trees: Decision Trees create a hierarchical structure where each internal node represents a feature, and each leaf node represents a class label. The tree is built by recursively splitting the data based on the feature that maximally reduces impurity at each node. Classification is done by traversing the tree from the root to a leaf, following the appropriate branch at each node [6].

Random Forest: Random Forest combines multiple decision trees, each trained on a random subset of data and features, to reduce overfitting and improve prediction accuracy. It aggregates the predictions of individual trees using majority voting for classification or averaging for regression, resulting in a robust and accurate ensemble model [5].

Gradient Boosting: Gradient Boosting builds an ensemble of weak learners, typically decision trees, sequentially. Each new tree corrects the errors made by the previous ones by updating the weights of data points. This iterative process creates a powerful model that captures complex relationships in the data and minimizes the overall prediction error [14].

K Nearest Neighbors (KNN): K Nearest Neighbors classifies a data point by assigning it the majority class of its K nearest neighbors in the training set. The "closeness" is determined by a distance metric such as Euclidean distance. KNN's simplicity allows it to adapt to complex decision boundaries but may suffer from higher computational costs with large datasets [9].

XGBoost (eXtreme Gradient Boosting): XGBoost is an optimized version of gradient boosting that employs data parallelism and model parallelism techniques. It uses a regularization term in the objective function to control overfitting and handles missing values efficiently. XGBoost's superior scalability and performance have made it popular for various machine-learning tasks [8].

Logistic Regression: Logistic Regression models the probability of a binary outcome using a logistic function. It fits a linear decision boundary to the data and maps the output to a probability score between 0 and 1. Adjusting the threshold classifies data points into the appropriate binary class [10].

AdaBoost (Adaptive Boosting): AdaBoost sequentially builds an ensemble of weak classifiers (often decision trees). Each iteration focuses more on the misclassified data points, adjusting their weights to improve accuracy. The final prediction is determined by a weighted majority vote of the weak classifiers [13].

Bagging: Bagging creates an ensemble of classifiers, such as decision trees, by training them independently on random subsets of the training data with replacement. The final prediction is made by aggregating the individual predictions, reducing variance, and improving model robustness [4].

Multilayer Perceptron (MLP) Classifier: MLP Classifier is an artificial neural network consisting of multiple layers of interconnected nodes (neurons). Each neuron applies a non-linear activation function to its input, enabling the network to learn complex patterns and make predictions for classification tasks [23].

There are two versions of testing performed for the adversarial dataset. In first case, we have mixed the adversarial data with the original UNSW-NB15 dataset. 70-30 random train-test split is performed on this modified dataset. All the above mentioned classifiers are trained on this dataset and their results on the test dataset are shown in the Table 3. Here Original Dataset denotes the UNSW-NB15 dataset and Original with Adversarial Dataset is the modified combined version of the dataset. We achieved results that were mixed but skewed positively to introducing the adversarial dataset. Classifiers like decision tree, random forest, gradient boost, XGBoost, Adaboost, Bagging classifier performed well on the adversarial data as compared to the original data. This proves that the model adapted well to the synthetic data. However, accuracy of Logistic Regression, KNN and MLP classifier reduced due to inclusion of adversarial examples. The last column of Table 3 shows the increase in true positive rate of the classifier after inclusion of adversarial data in the original dataset. The increase in true positives proves that the models did adapt to the synthetic data and detected attacks correctly.

Table 3. Results of ML classifiers on adversarial dataset of UNSW-NB15

Model	Dataset	Accuracy	Precision	Recall	F1 Score	Increase in True Positives
Logistic Regression	Original Dataset	0.855	0.842	0.970	0.901	
	Original with Adversarial Dataset	0.771	0.799	0.912	0.852	14%
DecisionTree Classifier	Original Dataset	0.949	0.963	0.963	0.963	
	Original with Adversarial Dataset	0.954	0.969	0.967	0.968	22%
RandomForest Classifier	Original Dataset	0.959	0.962	0.979	0.970	
	Original with Adversarial Dataset	0.963	0.968	0.981	0.974	22%
GradientBoosting Classifier	Original Dataset	0.946	0.941	0.982	0.961	
	Original with Adversarial Dataset	0.952	0.948	0.987	0.967	22%
KNearestNeighbors Classifier	Original Dataset	0.896	0.905	0.946	0.925	
	Original with Adversarial Dataset	0.780	0.849	0.847	0.848	9%
XGB Classifier	Original Dataset	0.959	0.963	0.977	0.970	
	Original with Adversarial Dataset	0.963	0.969	0.979	0.974	22%
AdaBoost Classifier	Original Dataset	0.941	0.945	0.970	0.957	
	Original with Adversarial Dataset	0.945	0.952	0.973	0.962	22%
Bagging Classifier	Original Dataset	0.956	0.959	0.978	0.968	
	Original with Adversarial Dataset	0.960	0.965	0.981	0.973	22%
MLP Classifier	Original Dataset	0.869	0.879	0.936	0.907	
	Original with Adversarial Dataset	0.763	0.771	0.957	0.854	24%

The Table 4 shows the performance of all the machine learning classifiers on a different setup. Here, training of all the classifiers is done using a mix of adversarial and original data. But the testing is done on the original data. This is done to analyse how the classifiers respond in case of adversarial attack. Again, the Original Dataset and Original with Adversarial Dataset represents the original dataset and modified adversarial data, respectively. From the results, we can deduce that classifiers such as decision tree, random forest, XGBoost, Bagging classifier improved their accuracy which means by introduction of adversarial data in the training set resulted in more accurate classification. Logistic Regression, KNN and MLP classifiers saw a drop in their performance metrics resulting in more false positives. In this case also, we see that the true positives have

Table 4. Results of ML classifiers trained on Adversarial Dataset of UNSW-NB15

Model	Dataset	Accuracy	Precision	Recall	F1 Score	Increase in True Positives
Logistic Regression	Original Dataset	0.855	0.842	0.970	0.901	
	Original with Adversarial Dataset	0.804	0.780	0.993	0.874	2%
DecisionTree Classifier	Original Dataset	0.949	0.963	0.963	0.963	
	Original with Adversarial Dataset	0.976	0.980	0.985	0.983	2%
RandomForest Classifier	Original Dataset	0.959	0.962	0.979	0.970	
	Original with Adversarial Dataset	0.979	0.979	0.991	0.985	1%
GradientBoosting Classifier	Original Dataset	0.946	0.941	0.982	0.961	
	Original with Adversarial Dataset	0.946	0.939	0.985	0.961	0%
XGB Classifier	Original Dataset	0.959	0.963	0.977	0.970	
	Original with Adversarial Dataset	0.966	0.969	0.982	0.975	1%
AdaBoost Classifier	Original Dataset	0.941	0.945	0.970	0.957	
	Original with Adversarial Dataset	0.938	0.941	0.969	0.955	0%
Bagging Classifier	Original Dataset	0.957	0.958	0.980	0.969	
	Original with Adversarial Dataset	0.977	0.976	0.991	0.983	1%
MLP Classifier	Original Dataset	0.850	0.836	0.970	0.898	
	Original with Adversarial Dataset	0.782	0.759	0.996	0.862	3%

increased a bit which concludes that the adversarial data generated by Logical Analysis of Data helps in enhancing the model's performance.

The Table 5 presents the outcomes of applying multiple machine learning classifiers to an adversarial dataset generated using Logical Analysis of Data (LAD) on NSL-KDD Dataset. The classifiers, including Logistic Regression, Decision Tree, Random Forest, K Neighbors, XGBoost, and Gradient Boosting, were evaluated on both the original and adversarial datasets. For training the ML classifiers, we have used a mix of original and adversarial data and for testing original dataset is used. Notably, across all classifiers, the introduction of adversarial data led to a consistent improvement in performance metrics. This enhancement is particularly evident in metrics such as accuracy, precision, and recall, indicating the effectiveness of LAD in enhancing model robustness. The most substantial improvements were observed in the Decision Tree and Gradient Boosting classifiers, which exhibited increase in true positives ranging from 18% to 20%. These results underscore the potential of adversarial data augmentation as a means to fortify the performance of machine learning models in real-world scenarios.

Table 5. Results of ML classifiers on adversarial dataset generated with NSL-KDD Dataset

Model	Dataset	Accuracy	Precision	Recall	Increase in True Positives
LogisticRegression	Original Dataset	0.704	0.880	0.555	
	Original with Adversarial Dataset	0.715	0.679	0.947	39%
DecisionTreeClassifier	Original Dataset	0.786	0.967	0.646	
	Original with Adversarial Dataset	0.858	0.970	0.774	13%
RandomForestClassifier	Original Dataset	0.775	0.968	0.626	
	Original with Adversarial Dataset	0.834	0.962	0.738	11%
KNearestNeighborsClassifier	Original Dataset	0.772	0.965	0.622	
	Original with Adversarial Dataset	0.821	0.951	0.723	10%
XGBClassifier	Original Dataset	0.784	0.968	0.642	
	Original with Adversarial Dataset	0.819	0.971	0.704	6%
GradientBoostingClassifier	Original Dataset	0.789	0.969	0.651	
	Original with Adversarial Dataset	0.855	0.962	0.776	13%

4   Conclusion

The paper highlights the generation of adversarial data using Logical Analysis of Data. LAD makes generation of adversarial data easier by use of logical patterns. Adversarial examples of UNSW-NB15 and NSL-KDD datasets are generated and are tested using various machine learning classifiers. The results conclude that when adversarial examples are included in the training set, the performance of the machine learning classifiers especially the ensemble classifiers, improves. The adversarial attack examples generated are detected by these classifiers. Thus, using LAD we can create adversarial data which when included in the training of intrusion detection model enhances the chances of detection of attacks. This could prove helpful in mitigating the chances of adversarial attacks on IDS models. The experiment discussed in the paper opens a discussion: exploring Logical Analysis of Data further in the domain of generative adversarial net. Further, the adversarial data generated by LAD could also be used to evaluate robustness of deep learning classifiers. In order to improve our results, we can use different datasets to generate adversarial data.

References

1. Almuallim, H., Dietterich, T.G.: Learning Boolean concepts in the presence of many irrelevant features. Artif. Intell. **69**(1–2), 279–305 (1994)
2. Boros, E., Hammer, P.L., Ibaraki, T., Kogan, A.: Logical analysis of numerical data. Math. Program. **79**(1–3), 163–190 (1997)
3. Boros, E., Hammer, P.L., Ibaraki, T., Kogan, A., Mayoraz, E., Muchnik, I.: An implementation of logical analysis of data. IEEE Trans. Knowl. Data Eng. **12**(2), 292–306 (2000)
4. Breiman, L.: Bagging predictors. Mach. Learn. **24**(2), 123–140 (1996)
5. Breiman, L.: Random forests. Mach. Learn. **45**(1), 5–32 (2001)
6. Breiman, L., Friedman, J., Stone, C.J., Olshen, R.A.: Classification and Regression Trees. Chapman and Hall, Boca Raton (1984)
7. Chauhan, S., Gangopadhyay, S.: Design of intrusion detection system based on logical analysis of data (LAD) using information gain ratio. In: Dolev, S., Katz, J., Meisels, A. (eds.) CSCML 2022. LNCS, vol. 13301, pp. 47–65. Springer, Cham (2022). https://doi.org/10.1007/978-3-031-07689-3_4
8. Chen, T., Guestrin, C.: XGBoost: a scalable tree boosting system. In: Proceedings of the 22nd ACM SIGKDD International Conference on Knowledge Discovery and Data Mining, pp. 785–794. ACM (2016)
9. Cover, T., Hart, P.: Nearest neighbor pattern classification. IEEE Trans. Inf. Theory **13**(1), 21–27 (1967)
10. Cox, D.R.: The regression analysis of binary sequences. J. Royal Stat. Soc. Ser. B (Methodological) **20**(2), 215–242 (1958)
11. Crama, Y., Hammer, P.L., Ibaraki, T.: Cause-effect relationships and partially defined Boolean functions. Ann. Oper. Res. **16**, 299–325 (1988)
12. Das, T.K., Gangopadhyay, S., Zhou, J.: SSIDS: semi-supervised intrusion detection system by extending the logical analysis of data. CoRR abs/2007.10608 (2020). https://arxiv.org/abs/2007.10608

13. Freund, Y., Schapire, R.E.: A decision-theoretic generalization of on-line learning and an application to boosting. J. Comput. Syst. Sci. **55**(1), 119–139 (1997)
14. Friedman, J.H.: Greedy function approximation: a gradient boosting machine. Ann. Stat. **29**(5), 1189–1232 (2001)
15. Goodfellow, I.J., Shlens, J., Szegedy, C.: Explaining and harnessing adversarial examples. arXiv preprint arXiv:1412.6572 (2014)
16. Hammer, P.: Partially defined Boolean functions and cause-effect relationships. In: International Conference on Multi-attribute Decision Making Via OR-based Expert Systems. University of Passau, Passau, Germany (1986)
17. Kantartopoulos, P., Pitropakis, N., Mylonas, A., Kylilis, N.: Exploring adversarial attacks and defences for fake twitter account detection. Technologies **8**(4), 64 (2020)
18. Lejeune, M., Lozin, V., Lozina, I., Ragab, A., Yacout, S.: Recent advances in the theory and practice of logical analysis of data. Eur. J. Oper. Res. **275**(1), 1–15 (2019). https://doi.org/10.1016/j.ejor.2018.06.011
19. Mintz, M., Bills, S., Snow, R., Jurafsky, D.: Distant supervision for relation extraction without labeled data. In: Proceedings of the Joint Conference of the 47th Annual Meeting of the ACL and the 4th International Joint Conference on Natural Language Processing of the AFNLP, pp. 1003–1011 (2009)
20. Papadopoulos, P., Thornewill von Essen, O., Pitropakis, N., Chrysoulas, C., Mylonas, A., Buchanan, W.J.: Launching adversarial attacks against network intrusion detection systems for IoT. J. Cybersecurity Priv. **1**(2), 252–273 (2021)
21. Pitropakis, N., Panaousis, E., Giannetsos, T., Anastasiadis, E., Loukas, G.: A taxonomy and survey of attacks against machine learning. Comput. Sci. Rev. **34**, 100199 (2019)
22. Ratner, A.J., De Sa, C.M., Wu, S., Selsam, D., Ré, C.: Data programming: creating large training sets, quickly. In: Advances in Neural Information Processing Systems, vol. 29 (2016)
23. Rumelhart, D.E., Hinton, G.E., Williams, R.J.: Learning representations by backpropagating errors. Nature **323**(6088), 533–536 (1986)
24. Treder-Tschechlov, D., Reimann, P., Schwarz, H., Mitschang, B.: Approach to synthetic data generation for imbalanced multi-class problems with heterogeneous groups. BTW 2023 (2023)
25. Van, N.T., Thinh, T.N., et al.: An anomaly-based network intrusion detection system using deep learning. In: 2017 International Conference on System Science and Engineering (ICSSE), pp. 210–214. IEEE (2017)

Emotion Extraction from Likert-Scale Questionnaires
– An Additional Dimension to Psychology Instruments –

Renata Magalhães, Francisco S. Marcondes, Dalila Durães(✉),
and Paulo Novais

ALGORITMI Centre/LASI, University of Minho, Braga, Portugal
pg34321@alunos.uminho.pt, francisco.marcondes@algoritmi.uminho.pt,
{dad,pjon}@di.uminho.pt

Abstract. Sentiment analysis tasks are used in various domains, including education. Likert-scale questionnaires are often used to gain insights into the respondents' views in various contexts. However, these questionnaires can allow for more information than they are designed for. This research paper explores an emotion classification technique for extracting emotional information from likert-scale questionnaires. A case study is presented in which a tailored questionnaire was employed to gather students' opinions on school-related matters, such as learning importance, academic performance and family and peer involvement and support. The students (n = 845) answered the questionnaire using a scale from totally disagree to totally agree. Through this questionnaire-based approach, data on students' emotions was collected.

Keywords: sentiment analysis · emotion classification · natural language processing · likert scale questionnaires · education

1 Introduction

Likert-scale questionnaires are used in a variety of fields, such as marketing, education, medicine, nutrition, and nursing, but mostly in psychology [1]. It aims to measure responses to a statement within a scale of agreement from totally disagree to totally agree.

This research hypothesis is that it is possible to extract additional features from already existing Likert-Scales questionnaires, which would provide additional analysis dimensions. For this paper, the objective is to explore that hypothesis within the scope of emotion classification. Thus, given a Likert-scale questionnaire, the emotion of the respondent is to be assessed.

This paper was developed within the scope of a project aiming to prevent school failure and dropout. A Likert-scale questionnaire was developed as a supporting tool and applied to 845 students in northern Portugal. It was upon that questionnaire that this proposal is relying on, yet generalizable for any Likert-scale questionnaire with similar features. As a remark, the questionnaire

P. Quaresma et al. (Eds.): IDEAL 2023, LNCS 14404, pp. 166–176, 2023.
https://doi.org/10.1007/978-3-031-48232-8_16

is restrained due to copyright. Overall, the understanding that comes from emotion classification allows for improved interventions and strategies as well as support systems to address emotional well-being of the students.

Ethical Statement. The questionnaire utilized was designed by a team of educational psychologists. Considering that the participants in this study were underaged, parents were asked to sign an informed consent form. Researchers were only in charge of data analysis.

2   Related Work

Sentiment analysis, also known as opinion mining, is an area of research in the field of text mining that extracts and classifies sentiment, emotions, and attitudes expressed in text data [2]. Techniques such as natural language processing (NLP) and machine learning are used to enable the identification and analysis of emotions expressed in written or spoken language [3]. This is widely applied in various fields such as healthcare [4], customer feedback analysis [5], social media [6], and in education [7].

Emotion classification is an area of natural language processing, which allows the detection of emotions through data retrieved from the target user [8]. Current literature finds a varity of models for emotion classification performed either by text, audio, image or video [9]. Emotion lexicons are available for use, either manually annotated (*e.g.* NRC Emotion Lexicon and LIWC) or generated automatically using machine learning algorithms (*e.g.* WordNet Affect and SentiWordNet). These were not adquate for the context of this project as it is necessary to have spontaneous and genuine text in order to obtain accurate results. It is not the case as the aim is to assess the emotions that might be articulated on the questionnaire respondent given a question written by another person.

A questionnaire-approach, the AEQ, was developed in 2011 [10] to measure emotions in students' learning and performance. The framework used for defining emotions was the control-value theory of achievement, and different emotions were studied (enjoyment, hope, pride, relief, anger, anxiety, hopelessness, shame, and boredom). Overall, the AEQ contains 24 scales that tap these nine emotions occurring in three different academic achievement settings: class-related emotions, learning-relating emotions and test emotions. Students had to respond to this questionnaire with a 5 point Likert scale (1 = completely disagree, 5 = completely agree). The AEQ is not adequate for the context of this paper as it aims to specifically assess student's emotions, whereas the goal is not to extract and classify emotions directly, but to investigate the possibility of extracting emotions as additional feautres from a questionnaire that was not designed for that purpose.

3   The Emotion Dimension of Likert-Scale Questionnaires

Likert-scale questionnaires present statements to which the respondent has to reply to within a scale. This could be a scale of agreement (totally disagree to totally agree), a satisfaction scale (very dissatisfied to very satisfied), a likelihood scale (very unlikely to very likely), a good to bad scale (very bad to very good) or a frequency scale (never to always). This is usually presented in a five point scale, in which 1 represents the lowest score, 3 represents neutrality and 5 represents the highest score [11]. Scores 1 and 5 are usually variations of intensities of either the positive or negative poles.

Through Likert-Scale questionnaires it is possible to measure responses to a statement, that is, personal opinions and they are often used in correlation to emotional states. Psychologists have been struggling to find agreement in regards to a definition of emotion, yet the notion that it is complex is consensual. Plutchik's model of emotion is perhaps the most popular model for dimensional models [12] and is widely used in a varitety of research, including this study. Plutchik defines emotions as

a multifaceted and inferred response to a stimulus, including subjective feelings, cognitions, impulses to actions and behaviour [13, p. 551]

Following Plutchik's definition of emotion as a response to a stimulus, considering a Likert-scale questionnaire inquiry a stimulus, it is possible to say that, in addition to responding to a question, the respondent is also articulating an emotion that is dimmed by the intensity of the response. Therefore, combining the emotion associated to the question text with the reply to that question, it is possible to suggest the emotion that the respondent may be articulating.

Consider, for a reference, the Theory of Brain Functions and Behaviour *cf.* [14]. In short, it is a 4-point Likert scale questionary for assessing Behavioral Inhibition System (BIS), which regulates anxiety responses to stimuli associated with punishment, absence of reward, and novelty, while the Behavioral Approach System (BAS) governs appetitive motivation and responds to cues indicating rewards, nonpunishment, and escape from punishment. As a self-report questionnaire it has been used in various contexts, including for the elicitation of students' emotional states [15] in university examinations.

Looking at BIS-BAS, the first statement reads "A person's family is the most important thing in life". If the respondent responds with "very true for me", it means that the respondent values and cherishes family. Emotionally speaking, this statement articulates feelings of anxiety, submission, and sentimentality. However, if the respondent responds with "very false for me", it does not necessarily mean that the person hates their family. Emotionally speaking, this statement articulates a combination of feelings of outrage, contempt, and morbidness or derisiveness.

According to Plutchik's circumplex model, outrage is a combination of surprise and anger, contempt is a combination of disgust and anger and morbidness or derisiveness are a combination of disgust and joy. This suggests a complex combination of contrasting feelings of frustration, energy, distrust, disinterest and contempt.

Plutchik's three-dimensional circumplex model illustrates the interconnections between emotions using a color wheel analogy. It incorporates a vertical dimension for intensity, a circular dimension for similarity, and eight sectors representing primary emotions, with the empty spaces representing dyads, which are mixtures of two primary emotions [13]. The model's eight primary emotions dimensions can be arranged in four pairs of opposites: joy and sadness, fear and anger, anticipation and surprise, disgust and trust, as shown in Fig. 1.

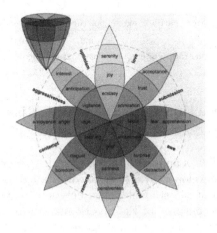

Fig. 1. Plutchik's circumplex model of emotions [13, p.349].

Plutchik states that emotions can combined in the same way that blue and red make purple, and, for instance, mixing joy and acceptance produces love; disgust and anger produces hatred or hostility.

It is of extreme importance to assess a whole questionnaire before proceeding to its emotional classification, to make sure there are no assumptions but evidence of emotion articulation. That is, by observing the presence of specific emotions in several statements of the questionnaire, that emotion can be accurately extracted, whereas if it is just present in one statement, it could not necessarily mean that specific emotion is being articulated by the respondent as he is just marking an answer on a scale but there is no reason or explanation as to why that answer was chosen. If a pattern or frequent presence is found, however, it is plausible to extract and classify that emotion.

This means that emotion classification is performed by classifying each statement on the questionnaire, with a score being assigned to each emotion present on that statement. For this paper, a lexical approach *cf.* [16] was used, for an instance refer to Table 1a. Thus, when the respondent reads the statement, it receives a stimulus, meaning that an emotion is being articulated. However, it is also necessary to consider the answer in order to verify its suitability. Each emotion articulated in Table 1a is then multiplied by the response weight presented in Table 1b and added to the respondent score. Since each emotion has its inverse, therefore, if the respondent chose "fully disagree" for the statement in Table 1a does not mean, for an instance, `anxiety = -0.4` but `hate = 0.4`.

Table 1. Demonstration of the proposed approach.

statement	anxiety	sentiment.	submission
A person's family is the most important thing in life	0.4	0.6	0.7

(a) Example of dataset for emotion classification proposed in this study.

response	fully agree	partially agree	not sure	partially disagree	fully disagree
emotion weight	1.0	0.5	0.0	−0.5	−1.0

(b) Demonstration of emotion classification calculation.

4   Emotion Extraction on a Drop-Out Risk Assessement Instrument

The proposed approach was applied to a Likert-scale questionnaire of about 50 statements which was developed by a team of educational psychologists and was administered to a total of 845 students, aged between 10 and 17 years old from northern Portugal. The purpose of the questionnaire was to evaluate multiple aspects including personal motivation, engagement in school, task management, perseverance, beliefs and attitudes towards the educational environment, active participation, family support, interpersonal relationships, and students' perceptions of teachers and the school environment. Participants were requested to indicate their level of agreement or disagreement to each statement, in which the responses available are "totally disagree", "disagree", "I am not sure", "agree", and "totally agree".

Examples of statements include "It is important to me to learn as much as I can", "I forget important deadlines" and "My family/parents are there when I need them" and, therefore, each section of the questionnaire articulates specific emotions within the respondent.

The first step was to identify emotions that were of interest in the questionnaire. Then, researchers and psychologists engaged in discussions to identify the emotions that would make sense in the context of the project, as according to current emotion detection practices [17].

Psychologists reinforced the importance of studying emotions that usually occur in an educational context and that are comprised in different spectrums of Plutchik's circumplex model.

Thus, the following emotions were analyzed within the study's specific context: happiness (also known as joy), trust, optimism, interest, boredom, anxiety, distraction and shame. Within Plutchik's Wheel of Emotions, joy and trust are recognized as two of the foundational primary emotions. Additionally, both optimism and interest are accounted for within the wheel. Optimism is characterized as a combination of anticipation and joy, while interest represents a lower intensity form of anticipation and vigilance. Anxiety, on the other hand, emerges from a combination of anticipation and fear, shame is rooted in a fusion of fear and disgust, boredom is a lower intensity of loathing and disgust, and distraction is a lower intensity of amazement and surprise, as well as a direct opposite of interest.

Each section contained emotional inferences for all emotions, but different emotions were more prevailing in different sections. The predominant emotions in the importance of learning and academic performance section are anxiety, interest and shame, in task management, preserverance and self-regulation are distraction and boredom, in beliefs and attitudes related to the educational environment is optimism, in attitudes and behaviours related to active participation and engagement in school is interest, in family support and involvement in school-related matters are happiness and anxiety, in friendship and peers relationships within the school context are happiness, trust and interest and in perceptions of students regarding their teachers and the overall school environment is trust.

A dataset was created that contains all the statements in the questionnaire, as well as a column for each of the chosen emotions. For each statement in the questionnaire a score was assigned to each emotion bearing in mind which emotions are articulated to the respondent when agreeing with the statement. This annotation was based on the intensity, typical sensations, similar words and utility of that specific emotion present in Plutchnik's Wheel. Assuming that the student fully agreed with that statement, a specific emotion or combination of emotions would take place, that is, a certain emotional inference would be produced. Subsequently, the scores for each emotion based on the student's response was calculated as proposed.

As illustration, an excerpt of the questionnaire classification can be seen in Table 2. Once the questionnaire was filled out, a normalized value between -1 to 1 for each emotion, per student, is calculated as illustrated in Table 3.

Table 2. Emotion Classification for the Questionnaire.

question	happiness	trust	optimism	interest	boredom	anxiety	distraction	shame
Q1_1	0.3	0.2	0.5	0.8	0	0.05	0	0
Q1_2	0.2	0.2	0.5	0.8	0.1	0.05	0	0.05
Q1_3	0.2	0.1	0.3	0.4	0	0.3	0.05	0.1

Table 3. Emotion Classification results.

student	happiness	trust	optimism	interest	boredom	anxiety	distraction	shame
1	0,70	0,70	0,66	0,66	0,5	0,55	0,47	0,56
2	0,76	0,80	0,76	0,75	0,83	0,75	0,84	0,80
3	0,54	0,50	0,51	0,45	0,47	0,66	0,45	0,66

The emotion classification model developed in the context of this study allowed for the gathering of students emotional scores on happiness, trust, optimism, interest, boredom, distraction, anxiety and shame, which were then combined into a dataset containing other important features for analysis such as school year and gender, for example.

The dataset utilized for this study contained a total of 845 students, from which 415 are female and the remaining 430 male. In terms of the school year, distribution is as follows: 212 from grade 5, 210 from grade 6, 138 from grade 7, 155 from grade 8 and 130 from grade 9, therefore showing an equitable gender and school year distribution.

After careful analysis of the results obtained, it was found that the emotional state that was most present among the eight classified was anxiety, followed by shame, boredom, distraction, interest, happiness, trust and optimism, at last. The mean value for anxiety is 0.59 while for optimism is 0.48. All mean values can be seen distribution can be visualized in the Table 4.

Table 4. Mean Values for all Emotions.

Anxiety	Shame	Boredom	Distraction	Interest	Happiness	Trust	Optimism
0.59	0.57	0.53	0.53	0.50	0.50	0.49	0.48

While anxiety was the emotion that contained the highest mean value, it is worth mentioning that the highest scores were found in grade 5 students. Keeping in mind that this questionnaire was administered in the beginning of the school year, this can be explained due to the fact that students move to a new school at grade 5. Anxiety is the combination of anticipation and fear, meaning that it is characterized by feelings of being alert, stressed and scared. This transition, which often includes students integrating a new class with new teachers and staff members at a different school can take a toll on students' anxiety scores. It is also worth mentioning that overall anxiety results were higher for male students. In fact, male students obtained a higher score of happiness, trust, boredom, anxiety, distraction and shame, whereas female students achieved higher levels of interest and optimism. It was also observed that male students report higher scores of anxiety in grades 5, 7 and 8, whereas female students report higher anxiety scores at grade 6 and 9. However, the difference of scores in grade 6 is almost non-existential.

Grade 5 students obtained higher values of all emotions in comparison to other school years. Overall, happiness, trust, optimism, and interest show a gradual decline as students progress from 5th grade to 9th grade. This could suggest a potential decrease in emotional well-being during the school years.

The transition from 6th grade to 7th grade appears to show the first signs of a notable impact on emotional experiences. Happiness, trust, and optimism experience a sharp decline, while boredom and distraction increase. This transition period may pose challenges for students' emotional well-being as it is a time of more responsibility and workload. In Portugal, students' transition to the 7th grade is known for being challenging as students start having more subjects than they did in the 5th and 6th grade.

However, lowest results occur at grade 8, specifically for happiness, interest, trust and optimism. There is a slight increase of these emotions from grade 8 to grade 9. Although reaching their lowest score also at grade 8, shame and anxiety scores overall remain more consistent throughout all school years in comparison to other emotions. Positive emotions register a significant decrease at grade 8, while anxiety, shame, boredom and distraction remain more consistent. This could indicate that grade 8 is an emotionally demanding year for students. While female students also report a small decrease in positive emotion scores at grade 8, this steady decrease is more remarkably noticed in male students.

Figure 2 shows the distribuiton of emotion scores per school year.

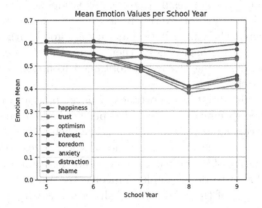

Fig. 2. Emotion Mean Distribution Per School Year.

Out of the eight emotions, shame is the one that remains more constant in different school years. With a slighty bigger mean value for male students, shame seems to remain consistent in the 5th and 6th grade, decreasing slightly in the 8th grade and experiencing a small increase by the 9th grade. Shame is a combined emotion, meaning that it occurs with a combination of fear and disgust. Typical sensations of shame include feeling bitter, and stressed. Male students report higher scores of shame at every school year, except for the 9th grade.

5   Discussion and Future Work

The proposed metholody shows that additional features can be extracted from likert-scale questionnaires. The case study presented shows that this questionaire-based emotion classification approach enabled an exploratory analysis of students' emotional states, taking place at the beginning of the school year, which contributes to the body of knowledge in sentiment analysis and emotion classification.

It was found that anxiety scores were the highest. This result was not expected. In contrast, optimism scores were found to be the lowest. Distraction and boredom are higher at grade 7 than 6, whereas interest, happiness, trust and optimism are found to be the opposite, perhaps due to the fact that students' workload is much higher in grade 7 and students usually have new teachers at this time. The lowest scores for all positive emotions were found at grade 8, which was expected due to the reputation that grade 8 holds for being a demanding year in Portugal.

An unexpected finding is that average anxiety results were slightly higher for male students. Previous studies report that female students usually achieve higher anxiety scores than male students [10,18]. Though only a small difference was found, this result can be paired with shame, for which male students also reported a slightly higher result than female students. However, answers to the questionnaire indicate that overall male students are more worried about improving their scores, showing others that they are good with school work and overall maintaing a good image among their peers, therefore explaining higher anxiety scores and validating the results obtained from the proposed emotion classification approach.

This research highlights the importance of developing and refining emotion detection techniques specifically tailored to specific domains and languages. As sentiment analysis continues to expand beyond English [2], it is crucial to invest in the development of accurate and extended emotion lexicons in other languages, as the majority of current lexicons are mostly available for the English language and are found to be incomplete, incorrectly translated, or simply containing transliterations of the original English terms, as is the case of the NRC Emotion Lexicon (also known as EmoLex). This presents a challenge for sentiment analysis in languages other than English. This research does not encounter that limitation as individual classification should be made for each statement in the questionnaire, performed in the questionnaire's language.

Multiple studies indicate that support and guidance from both teachers and parents are crucial factors in fostering student motivation and interest in school [19]. Taking this into account, the questionnaire used in this study pinpoints all the important factors in terms of academic success, ranging from the students' own views on their educational goals to their peers and family support, therefore creating an non-intrusive, undemanding, and reliable form of gaining insights into students' emotional states. Sentiment analysis and more specifically emotion detection can be used as a monitoring as well as predictive tool, creating advantages for the educational system at a low cost [20]. A future work hypotheses would be to have students repeat the whole questionnaire at the beginning and end of the school year to compare results and therefore making it possible to monitor students' emotional state throughout the school year.

This is just an example of the utlization of the proposed approach. When applied to questionnaires of different fields of study, this metholody can have the same impact, allowing for specific interventions in different domains.

6   Conclusion

The methodology proposed by this research shows that by classifying a likert-scale questionnaire for the purpose of emotion classification will result in a fine-tuned model that can be applied to any questionnaire of any field of study.

A case study was presented in which this approach was applied to a tailored likert-scale questionnaire administered to a sample of 845 K-12 students at the beginning of the school year. This questionnaire, designed by a team of educational psychologists, presents statements in relation to the importance of learning and academic performance within the school context and students had to answer within a scale of agreement and disagreement. It was possible to extract insights of aspects of students' experiences within the educational environment, and emotional levels for happiness, trust, optimism, interest, anxiety, distraction, boredom and shame.

In conclusion, this study showcases a new emotion classification model, implemented through likert-scale questionnaires. In an educational context it can aid in detecting students' emotional levels and allowed for a timely intervention or prevention of school failure. Though, this method can be applied to questionnaires of different areas of study, allowing for tailored interventions wihtin different contexts.

Acknowledgements. This work has been supported by FCT âĂŞ Fundação para a Ciência e Tecnologia within the R&D Units Project Scope: UIDB/00319/2020.

References

1. Pimentel, J.L.: A note on the usage of Likert scaling for research data analysis. USM R&D J. **18**(2), 109–112 (2010)
2. Medhat, W., Hassan, A., Korashy, H.: Sentiment analysis algorithms and applications: a survey. Ain Shams Eng. J. **5**(4), 1093–1113 (2014)
3. Cui, J., Wang, Z., Ho, S.-B., Cambria, E.: Survey on sentiment analysis: evolution of research methods and topics. Artif. Intell. Rev. 1–42 (2023)
4. Milne-Ives, M., et al.: The effectiveness of artificial intelligence conversational agents in health care: systematic review. J. Med. Internet Res. **22**(10), e20346 (2020)
5. Iqbal, A., Amin, R., Iqbal, J., Alroobaea, R., Binmahfoudh, A., Hussain, M.: Sentiment analysis of consumer reviews using deep learning. Sustainability **14**(17), 10844 (2022)
6. Bibi, M., et al.: A novel unsupervised ensemble framework using concept-based linguistic methods and machine learning for twitter sentiment analysis. Pattern Recogn. Lett. **158**, 80–86 (2022)
7. Dake, D.K., Gyimah, E.: Using sentiment analysis to evaluate qualitative students' responses. Educ. Inf. Technol. **28**(4), 4629–4647 (2023)
8. Li, W., Hua, X.: Text-based emotion classification using emotion cause extraction. Expert Syst. Appl. **41**(4), 1742–1749 (2014)
9. Jayalekshmi, J., Mathew, T.: Facial expression recognition and emotion classification system for sentiment analysis. In: 2017 International Conference on Networks & Advances in Computational Technologies (NetACT), pp. 1–8. IEEE (2017)

10. Pekrun, R., Goetz, T., Frenzel, A.C., Barchfeld, P., Perry, R.P.: Measuring emotions in students' learning and performance: the achievement emotions questionnaire (AEQ). Contemp. Educ. Psychol. **36**(1), 36–48 (2011)
11. Shaikh, S., Doudpotta, S.M.: Aspects based opinion mining for teacher and course evaluation. Sukkur IBA J. Comput. Math. Sci. **3**(1), 34–43 (2019)
12. Hung, L.P., Alias, S.: Beyond sentiment analysis: a review of recent trends in text based sentiment analysis and emotion detection. J. Adv. Comput. Intell. Intell. Inform. **27**(1), 84–95 (2023)
13. Plutchik, R.: The nature of emotions: human emotions have deep evolutionary roots, a fact that may explain their complexity and provide tools for clinical practice. Am. Sci. **89**(4), 344–350 (2001)
14. Carver, C., White, T.: Behavioral inhibition, behavioral activation, and affective responses to impending reward and punishment: the BIS/BAS scales. J. Pers. Soc. Psychol. **67**, 319–333 (1994)
15. Krupić, D., Corr, P.J.: Individual differences in emotion elicitation in university examinations: a quasi-experimental study. Pers. Individ. Differ. **71**, 176–180 (2014)
16. Jurafsky, D., Martin, J.H.: Speech and Language Processing, 3rd edn. (2023). https://web.stanford.edu/jurafsky/slp3/
17. Mohammad, S.M.: Best practices in the creation and use of emotion lexicons. arXiv preprint arXiv:2210.07206 (2022)
18. Hosseini, L., Khazali, H.: Comparing the level of anxiety in male & female school students. Procedia Soc. Behav. Sci. **84**, 41–46 (2013)
19. Affuso, G., et al.: The effects of teacher support, parental monitoring, motivation and self-efficacy on academic performance over time. Eur. J. Psychol. Educ. **38**(1), 1–23 (2023)
20. Huberts, L.C.E., Schoonhoven, M., Does, R.J.M.M.: Multilevel process monitoring: a case study to predict student success or failure. J. Qual. Technol. **54**(2), 127–143 (2022)

Recent Applications of Pre-aggregation Functions

G. Lucca[1]([⊠])(ID), C. Marco-Detchart[2](ID), G. Dimuro[3](ID), J. A. Rincon[2,4](ID),
and V. Julian[2,4](ID)

[1] Programa de Pós-Graduação em Engenharia Eletrônica e Computação,
Universidade Católica de Pelotas, Gonçalves Chaves, Pelotas 96015-560, Brazil
giancarlo.lucca@ucpel.edu.br
[2] Valencian Research Institute for Artificial Intelligence (VRAIN),
Universitat Politècnica de València (UPV), Camino de Vera s/n,
46022 Valencia, Spain
{cedmarde,vjulian}@upv.es, jrincon@dsic.upv.es
[3] Centro de Ciências Computacionais, Universidade Federal do Rio Grande,
Av. Itália km 08, Campus Carreiros, Rio Grande 96201-900, Brazil
gracalizdimuro@furg.br
[4] Valencian Graduate School and Research Network of Artificial Intelligence
(VALGRAI), Universitat Politècnica de València (UPV), Camí de Vera s/n,
46022 Valencia, Spain

Abstract. In recent years, a hot topic that has emerged is the concept
of pre-aggregation functions. This kind of function respects the same
property as an aggregation function, however, with a directional increase.
Taking this into consideration, we have performed a literature analysis
on Scopus digital library to select the most recent applications. From the
analysis, we have selected and analyzed 6 different studies from different
research fields.

Keywords: pre-aggregation functions · aggregation functions ·
Choquet integral · literature analysis

1 Introduction

An aggregation function [3,7] is a function responsible for combining several
inputs into a single one that can represent the aggregated information. We can
highlight, for example, some well-known aggregations that are used daily: The
mean, the maximum, the product, and the sum.

It is easy to notice that the usage of aggregation function is widely used
in many different systems. However, it is important to take into consideration
the role that these aggregations have. That is, the system can provide different
performances depending on the considered aggregation method.

An important function in the field of aggregation operators is the Choquet
integral [5]. This operator is able to combine the inputs taking into consideration

© The Author(s), under exclusive license to Springer Nature Switzerland AG 2023
P. Quaresma et al. (Eds.): IDEAL 2023, LNCS 14404, pp. 177–185, 2023.
https://doi.org/10.1007/978-3-031-48232-8_17

the relations that the data have. The concept behind this idea is possible due to the fuzzy measures [14], also known as capacity.

An innovative concept was stated in the literature called pre-aggregation function [11]. This kind of function has the same boundary properties as a common aggregation function. However, the required monotonicity is not so restrictive. The authors have shown that a construction way of a pre-aggregation is by replacing the product considered in the discrete Choquet integral by any triangular norm (t-norm) [9].

After the theoretical insertion of the pre-aggregation functions having the Choquet integral as construction method, different generalizations of the Choquet integral have been proposed, namely: CC-integrals [12], C_F-integrals [13], C_{F1F2}-integrals [10], dC_F-integrals [18] and dXC-integrals [17]. It is important to mention that each family of generalization has its properties and characteristics.

Considering the pre-aggregation functions that are Choquet based, this article aims to respond to the following:

Which are the most recent applications of the pre-aggregation functions?

To do so, we have conducted a revision of the literature aiming to find different articles that deal with this question and that were published in the last few years. For each one, we provide a brief discussion aiming at the copped problem and the role of the pre-aggregation.

This paper is organized as follows. Section 2 introduces the main concepts of the study. After that, in Sect. 3 the methodology is described showing the selected studies. The analysis of the applications is done in Sect. 4. In the end, the main conclusions are made.

2   The Role Choquet Integral as (pre-) aggregation

The main principles used in this paper are related to the Choquet integral and the notion of (pre-) aggregation functions. In this section, we define them.

A function $A : [0,1]^n \to [0,1]$ is an aggregation function (AF) [3] if:

(A1) A is increasing in each argument: for each $i \in \{1, \ldots, n\}$, if $x_i \leq y$, then

$$A(x_1, \ldots, x_n) \leq A(x_1, \ldots, x_{i-1}, y, x_{i+1}, \ldots, x_n);$$

(A2) A satisfies the boundary conditions:
  (i) $A(0, \ldots, 0) = 0$ and (ii) $A(1, \ldots, 1) = 1$.

Now, we recall the concept of directional monotonicity [4]. Let $r = (r_1, \ldots, r_n)$ be a real n-dimensional vector such that $r \neq 0 = (0, \ldots, 0)$. A function $F : [0,1]^n \to [0,1]$ is said to be r-increasing if, for all $x = (x_1, \ldots, x_n) \in [0,1]^n$ and $c > 0$ such that $x + cr = (x_1 + cr_1, \ldots, x_n + cr_n) \in [0,1]^n$, it holds $F(x + cr) \geq F(x)$. Similarly, one defines an r-decreasing function.

The idea of directional monotonicity was used by Lucca et al. [11] to develop the theory of pre-aggregation functions, that is, a function $PA : [0,1]^n \to [0,1]$ is said to be a pre-aggregation function (PAF) if the following conditions hold:

(PA1) PA is directional increasing, for some $r = (r_1, \ldots, r_n) \in [0,1]^n$, $r \neq 0$;
(PA2) PA satisfies the boundary conditions:
  (i) $PA(0, \ldots, 0) = 0$ and (ii) $PA(1, \ldots, 1) = 1$.

We call F an r-PAF whenever it is a PAF with respect to a vector r.

Given that this paper focuses on the Choquet integral, it is necessary to introduce the fuzzy measure, that is, a function $m : 2^N \to [0,1]$ that, for all $X, Y \subseteq N$, has these two properties:

(m1) m is increasing, that is, if $X \subseteq Y$, then $m(X) \leq m(Y)$;
(m2) m satisfies the boundary conditions, $m(\emptyset) = 0$, $m(N) = 1$.

This measure is a non-additive measure; that is, it is not required to hold the additive property, only an increasing one [14].

Definition 1. *[5] Let $m : 2^N \to [0,1]$ be a fuzzy measure. The discrete Choquet integral (dCI) is the function $\mathfrak{C}_m : [0,1]^n \to [0,1]$, defined, for all of $x \in [0,1]^n$, by*

$$\mathfrak{C}_m(x) = \sum_{i=1}^{n} \left(x_{(i)} - x_{(i-1)} \right) \cdot m \left(A_{(i)} \right), \tag{1}$$

where $\left(x_{(1)}, \ldots, x_{(n)} \right)$ is an increasing permutation on the input x, that is, $0 \leq x_{(1)} \leq \ldots \leq x_{(n)} \leq 1$, with $x_{(0)} = 0$, and $A_{(i)} = \{(i), \ldots, (n)\}$ is the subset of indices corresponding to the $n - i + 1$ largest components of x.

3 Methodology

This section provided the detailed methodology used in the search for the analyzed papers. Also, the details of the adopted inclusion criteria of the study are discussed.

The search is done in the Scopus digital library[1], one of the most used sources for scientific research. Then, to perform the analysis, we have selected the articles that cited the article that originated the concept of pre-aggregation [11].

Initially, more than 135 different studies have cited the original pre-aggregations. To reduce this scope, some refinements are done, precisely:

- Publication Year: 2023, 2022, 2021, 2020
- Document type: Article and Conference Paper
- Publication stage: Final
- Keyword: Aggregation functions, Choquet integral and Pre-aggregation

It is important to mention that no refinements were done per author, subject area, source title, affiliation, funding sponsor, country/territory, source type, or language. This is to produce a more generalist search and consequently provide a more precise and concise analysis. The final generated search query is defined as:

---

[1] https://www.scopus.com/home.uri.

LIMIT-TO (PUBSTAGE, "final") AND (LIMIT-TO (PUBYEAR, 2023) OR LIMIT-TO (PUB-
YEAR, 2022) OR LIMIT-TO (PUBYEAR, 2021) OR LIMIT-TO (PUBYEAR, 2020)) AND
(LIMIT-TO (DOCTYPE, "ar") OR LIMIT-TO (DOCTYPE, "cp")) AND (LIMIT-TO (EXAC-
TKEYWORD, "Aggregation Functions") OR LIMIT-TO (EXACTKEYWORD, "Choquet Inte-
gral") OR LIMIT-TO (EXACTKEYWORD, "Pre-aggregation")) AND (LIMIT-TO (LAN-
GUAGE, "English")) AND (LIMIT-TO (SRCTYPE, "j"))

As a consequence of the adopted refinement, a total of 35 different documents
were found. From these, we can observe in Fig. 1a the relation of the number
of documents per year. From this figure, we can observe that 2022 was the
year having the largest number of studies related to pre-aggregation functions.
Tanking into consideration this year, 2023, it is noticeable that the number of
documents presents an interesting number of citations, 5.

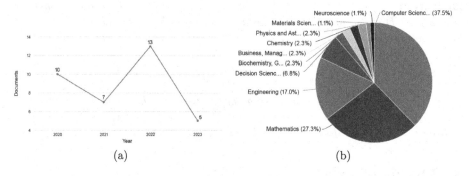

Fig. 1. Relation of documents per year (a) and subject area (b).

Figure 1b provides a graph with an analysis considering the subject area of
the documents that cite the pre-aggregations. Firstly, it is interesting to notice
that many different areas are using the concept of pre-aggregations. In fact,
for 10 different fields, namely: Computer Science (33 documents), Mathematics
(24), Engineering (15), Decision Sciences (6), Biochemistry (2), Genetics and
Molecular Biology (2), Business, Management, and Accounting (2), Chemistry
(2), Physics and Astronomy (2), Material Science (1) and Neuroscience (1).

3.1 Studies Selection

The initial selection method of studies returned a total of 35 different documents
classified into 10 different subject areas. In order to reduce the scope of the
analysis to produce a more specific evaluation of the studies, some exclusion
criteria were considered. This subsection aims to describe them.

The first considered exclusion criteria is performed by reading the abstract,
methodology, and results. If the paper does not include a complete and concise
application, we exclude it. This case intends to remove theoretical studies or the
ones that could perform a simple revision. After, a second exclusion criterion

is applied to produce a more focused analysis. Precisely, we have removed the articles that do not provide a direct usage of the pre-aggregations. This paper includes studies having references and no proper usage. From the application of both criteria, 21 articles were removed.

To reduce even more the scope of the study (14 remaining articles), a third exclusion criterion was adopted. It consists of eliminating those studies that have three or more authors from the original pre-aggregation study [11]. This is to include more diversity in the analysis and concentrate on those studies that have been done by different research groups. It is important to mention that the inclusion of this criteria has removed 8 articles. Therefore, the final analysis is performed over 6 different articles. The relation of the selected articles is provided in Table 1. In it, for each study, we provide the Reference (Ref), title, authors, year of publication, and the related journal (source).

Table 1. The selected papers after the exclusion process.

Ref	Title	Authors	Year	Source
[2]	Constructing multi-layer classifier ensembles using the Choquet integral based on overlap and quasi-overlap functions	Batista, T. ; Bedregal, B. and Moraes, R	2022	Neurocomputing
[6]	Motor-Imagery-Based Brain-Computer Interface Using Signal Derivation and Aggregation Functions	Fumanal-Idocin, J. et al.	2022	IEEE Transactions on Cybernetics
[8]	The Bonferroni mean-type pre-aggregation operators construction and generalization: Application to edge detection	Hait, S.R. et al.	2022	Information Fusion
[15]	Application of Fuzzy and Rough Logic to Posture Recognition in Fall Detection System	Pekala, B. et al.	2022	Sensors
[16]	A framework for generalized monotonicity of fusion functions	Sesma-Sara, M. et al.	2023	Fuzzy Sets and Systems
[19]	CT-Integral on Interval-Valued Sugeno Probability Measure and Its Application in Multi-Criteria Decision-Making Problems	Yang, H. ; Shang, L. and Gong, Z	2022	Axioms

4 Analysing the Recent Application of Pre-aggregation Functions

This section provides an analysis of the 6 selected papers (See Sect. 3). Firstly, for each paper, a discussion of the study is performed, and after that, a general analysis is made considering the tools provided by Scopus and bibliometrix [1].

The first article, [16] presents a framework that encompasses all the different forms of monotonicity of fusion functions found in the literature and enables the introduction of additional forms of monotonicity (dilative monotonicity). The authors discussed an application of sentiment analysis (text classification) problem in a dataset that is often used as a benchmark for sentiment analysis: the IMDb dataset. They have considered the word embedding technique to encode the meaning of words and then used different fusion functions to generate a document-level feature vector that is used as input into a feedforward neural network to classify the polarity of the document.

A new hybrid system, based on fuzzy and rough sets, has been developed to tackle the posture detection problem was done in [15]. In the presented study, different rule sets are investigated, the first one created based on domain knowledge and the second induced based on the rough set theory; also, two inference aggregation approaches are considered with and without knowledge measure. These measures, together with various aggregation methods are used to evaluate the accuracy of the classification of rule sets in the decision-making process.

In, [2], the CQO-Integral was introduced. The discrete Choquet integral (See Eq. 1) generalized by Quasi-Overlaps. The authors provided a method that combines this new approach in an ensemble to deal with classification problems. Precisely, 21 different datasets were selected from the UCI dataset repository, showing that their approach is superior when compared with other methods.

As stated in [11], a construction way of pre-aggregation functions is by replacing the product operator of the standard discrete Choquet integral with different T-norms. This is also called C_T-integral. In, [19], the discrete C_T-integral on the interval-valued Sugeno probability measure was analyzed. The authors have considered this approach to cope with a multi-criteria decision-making problem of end-of-life (EOL) strategy determination for a refrigerator component.

Another interesting application that considers pre-aggregation concepts is related to the Brain-Computer Interface (BCI). The research field concerns the communication between the human brain and external devices. In [6] proposed a new BCI framework called enhanced motor fusion framework (EMF). In each step of this framework, different aggregation functions were used, showing that a different combination of functions led to an improvement of the system.

Our last considered study presents an application to edge detection [8]. The authors investigate the Bonferroni mean-type (BM-type) pre-aggregation operators, proposing the construction methodology of the BM-type pre-aggregation operators. The study has utilized the proposed operator in the step of extracting the feature map of the image from a grey-scale image, and by using the BM-type pre-aggregation function, the intensity values of the feature image are acquired. The results are compared against classical extractors found in the literature and provide interesting discoveries.

In what follows, we provide in Table 2 the citation overview of the selected documents. This allows an analysis of the impact that the selected articles have had so far. The table is sorted according to the publication year (Pub. year) and shows the citations of the articles in the considered period. The last row is related to the total of citations that these documents have in the Scopus base.

From the citation overview, it is noticeable that the most cited paper has 16 citations, while 2 papers have 0 citations. However, observe that one of them was recently published. Also, it is possible to notice that this year, 2023, the

Table 2. Citation overview of the selected documents.

Reference	Pub. year	<2019	2019	2020	2021	2022	2023	Total
		0	0	0	3	14	11	
[2]	2022			1		1		2
[6]	2022		3		7		6	16
[8]	2022			3		3		6
[15]	2022			3		1		4
[16]	2023							0
[19]	2022							0

number of citations is almost the same as the complete 2022, which can indicate an increasing trend.

We provide in Fig. 2 a word cloud generated from the 6 selected articles. This allows us to analyze the terms that are used and investigate possibilities and future new research lines.

Fig. 2. Generated word cloud from the selected studies.

From the word cloud, it is possible to highlight terms related to monotonicity (generalized and directional), which is a central point in the pre-aggregations. Also, Choquet integral and discrete Choquet integrals are a way to construct pre-aggregations. Considering applications, we can highlight some interesting points such as biomedical signal processing, electrophysiology, detection systems, edge detection, feature extraction, and others.

5   Conclusion

Aggregation functions are methods that combine several values into a single one that can represent the input values. Several important applications use this kind of function.

An important question is whether aggregation functions are restrictive in terms of the monotonicity property. In this sense, the topic of pre-aggregations emerged, functions that respect the same property of aggregations, however, with a directional increase. Taking this into consideration, this paper intends to analyze the last applications related to pre-aggregation functions. To do so, we have performed a literature review in the Scopus base.

The results showed that 35 articles have cited the original pre-aggregation studies in the last 3 years. From these, we have selected 6 studies for a deeper analysis. The considered articles are related to different problems, such as Classification, Brain-Computer Interface, Edge detection, Posture Recognition, sentiment analysis, and multi-criteria decision-making.

Future works can follow the path of the generalizations of the Choquet integral and their applications. Also, the application of different functions in the selected studies can provide an increase in terms of performance.

Acknowledgements. This work was partially supported with grant PID2021-123673OB-C31 funded by MCIN/AEI/ 10.13039/501100011033 and by "ERDF A way of making Europe", Consellería d'Innovació, Universitats, Ciencia i Societat Digital from Comunitat Valenciana (APOSTD/2021/227) through the European Social Fund (Investing In Your Future) and grant from the Research Services of Universitat Politècnica de València (PAID-PD-22). The authors also would like to thank the Fundação de Amparo á Pesquisa do Estado do Rio Grande do Sul - FAPERGS/Brazil (Proc. 23/2551-0000126-8) and National Council for Scientific and Technological Development - CNPq/Brazil (3305805/2021-5, 150160/2023-2).

References

1. Aria, M., Cuccurullo, C.: bibliometrix: an R-tool for comprehensive science mapping analysis. J. Informetrics **11**(4), 959–975 (2017)
2. Batista, T., Bedregal, B., Moraes, R.: Constructing multi-layer classifier ensembles using the choquet integral based on overlap and quasi-overlap functions. Neurocomputing **500**, 413–421 (2022)
3. Beliakov, G., Pradera, A., Calvo, T.: Aggregation Functions: A Guide for Practitioners. Springer, Berlin (2007). https://doi.org/10.1007/978-3-540-73721-6
4. Bustince, H., Fernandez, J., Kolesárová, A., Mesiar, R.: Directional monotonicity of fusion functions. Eur. J. Oper. Res. **244**(1), 300–308 (2015)
5. Choquet, G.: Theory of capacities. Annales de l'Institut Fourier **5**, 131–295 (1954)
6. Fumanal-Idocin, J., Wang, Y.K., Lin, C.T., Fernandez, J., Sanz, J., Bustince, H.: Motor-imagery-based brain-computer interface using signal derivation and aggregation functions. IEEE Trans. Cybern. **52**(8), 7944–7955 (2022)
7. Grabisch, M., Marichal, J., Mesiar, R., Pap, E.: Aggregation Functions. Cambridge University Press, Cambridge (2009)
8. Hait, S., Mesiar, R., Gupta, P., Guha, D., Chakraborty, D.: The Bonferroni mean-type pre-aggregation operators construction and generalization: application to edge detection. Information Fusion **80**, 226–240 (2022)
9. Klement, E.P., Mesiar, R., Pap, E.: Triangular Norms. Kluwer Academic Publisher, Dordrecht (2000)

10. Lucca, G., Dimuro, G.P., Fernandez, J., Bustince, H., Bedregal, B., Sanz, J.A.: Improving the performance of fuzzy rule-based classification systems based on a nonaveraging generalization of CC-integrals named $C_{F_1 F_2}$-integrals. IEEE Trans. Fuzzy Syst. **27**(1), 124–134 (2019)
11. Lucca, G., et al.: Pre-aggregation functions: construction and an application. IEEE Trans. Fuzzy Syst. **24**(2), 260–272 (2016)
12. Lucca, G., et al.: CC-integrals: Choquet-like copula-based aggregation functions and its application in fuzzy rule-based classification systems. Knowl.-Based Syst. **119**, 32–43 (2017)
13. Lucca, G., Sanz, J.A., Dimuro, G.P., Bedregal, B., Bustince, H., Mesiar, R.: CF-integrals: a new family of pre-aggregation functions with application to fuzzy rule-based classification systems. Inf. Sci. **435**, 94–110 (2018)
14. Murofushi, T., Sugeno, M., Machida, M.: Non-monotonic fuzzy measures and the Choquet integral. Fuzzy Sets Syst. **64**(1), 73–86 (1994)
15. Pekala, B., Mroczek, T., Gil, D., Kepski, M.: Application of fuzzy and rough logic to posture recognition in fall detection system. Sensors **22**(4), 1602 (2022)
16. Sesma-Sara, M., et al.: A framework for generalized monotonicity of fusion functions. Information Fusion **97**, 101815 (2023)
17. Wieczynski, J., etal.: d-xc integrals: on the generalization of the expanded form of the Choquet integral by restricted dissimilarity functions and their applications. IEEE Trans. Fuzzy Syst. **30**(12), 5376–5389 (2022)
18. Wieczynski, J., et al.: dc_F-integrals: generalizing c_F-integrals by means of restricted dissimilarity functions. IEEE Trans. Fuzzy Syst. **31**(1), 160–173 (2023)
19. Yang, H., Shang, L., Gong, Z.: CT-integral on interval-valued Sugeno probability measure and its application in multi-criteria decision-making problems. Axioms **11**(7), 317 (2022)

A Probabilistic Approach: Querying Web Resources in the Presence of Uncertainty

Asma Omri[1]([⊠]), Djamal Bensliamne[2], and Mohamed Nazih Omri[1]

[1] MARS Research Laboratory, University of Sousse, Sousse, Tunisia
omri.asmaaa@gmail.com, mohamednazih.omri@eniso.u-sousse.tn
[2] University Claude Bernard Lyon 1, LIRIS Research Laboratory,
Villeurbanne, France
djamal.benslimane@univ-lyon1.fr

Abstract. Uncertainty in data naturally arises in various applications, such as data integration and Web information extraction. A few examples are the following. When information from different sources is conflicting, inconsistent, or simply presented in incompatible forms the result of integrating these sources necessarily involves uncertainty as to which fact is correct or which is the best mapping to a global schema. Data uncertainty is often ignored, or modeled in a specific, per-application manner. This may be an unsatisfying solution in the long run, especially when the uncertainty needs to be retained throughout complex and potentially imprecise processing of the data. In this paper, we study the basic activities of web resources that are affected by uncertainty, more specifically, modeling, programming and evaluation. We propose a probabilistic approach that treats uncertainty in all these activities.

Keywords: Web · Uncertainty · Probabilistic · ressources Web

1 Introduction

The data extracted from the Web is full of uncertainty: it may contain contradictions or result from processes that [1,2] are inherently uncertain, such as data integration or extraction. In this paper, we present the notion of uncertain web resource probabilistic, how they can be used to represent data from the Web, and the complexity of evaluating queries within this type of resources. Actual data (eg World Wide Web) is often uncertain due to the inherent uncertainty of the data itself or the data collection, integration and management processes. The main objective of this paper is to propose a theoretical framework for describing, manipulating, and evaluating uncertain data on the Web [6]. We propose an algorithm that can help answer the requests and determinate of degree of uncertainty of this answer. We present a model to define and interpret uncertain Web resources. We define an interpretation model and an algebra to compute uncertainty in the context of classical hypertext navigation and in the context of data query evaluation. This paper is structured as follows: Sections 2 and 3

© The Author(s), under exclusive license to Springer Nature Switzerland AG 2023
P. Quaresma et al. (Eds.): IDEAL 2023, LNCS 14404, pp. 186–195, 2023.
https://doi.org/10.1007/978-3-031-48232-8_18

describes our uncertainty model and interpretation. Sections 4 and 5 presents our operators used that can respond well to the queries. Sections 6 and 7 explains how we evaluate a query over an uncertain resource. Finally, Sect. 8 concludes and presents some future research studies

2 Uncertain Web Resources

2.1 Definition

The semantics of uncertain web resources can be explained on the basis of the possible world theory. In this case, an uncertain resource can be characterized by several possible representations that can potentially and individually be interpreted as varied. These possible representations can be interpreted as a set of possible worlds (PW1, ..., PWn) where each of them is characterized by a probability value (PWi) which represents the probability of having this world possible. In our work, we will assign to these possible worlds a special notation that will be used throughout this paper. Therefore, as already mentioned, these possible worlds are called possible webs. The data contained in these webs are considered certain data. Based on the definition of web resource, we propose a specific representation which helps represent uncertain resources. It is defined by the following equation:

$$\tilde{R} = \{uri_r, \langle rep_r, P_r \rangle \,|rep_r = \{\langle Sub - rep_i, P_i \rangle, ..., \langle Sub - rep_n, P_n \rangle, i \in [1..n], P_i \in [0..n]\}\} \quad (1)$$

Where repi are the possible representations of R. Since multiple representations of a resource cannot coexist at the same URI, they are mutually exclusive, therefore, we have $Pi \ln [0..1]$. The probabilities are not part of representations but they are meta-data provided by the server (Fig. 1).

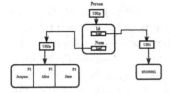

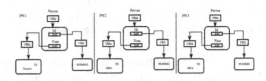

Fig. 1. Example of Uncertain Web Resources.

Fig. 2. Possible Webs of Person's resource.

Using the principle of possible worlds, as an example, this figure shows this representation of our Person resource that generates three webs, each of which identifies a certain representation. These possible worlds are represented by the Fig. 2.

This kind of representation explains all possible representations in the definition of a resource. If one has a resource R1 characterized by three attributes, from the start it is the user that defines all the possibilities (2 power 3). To solve this problem, we will use the principle of what is called probabilistic XML ProTB [3] which is a probabilistic XML model proposed by Nieman and Jagadish in 2002.This model also provides a convenient mix of probabilistic and non-probabilistic data.This approach differs from previous ones in developing the probabilistic relational systems in that we construct a probabilistic XML database. This design is motivated by application needs that involve data that are not readily available to a relational representation.

Applying this model, we will distinguish two types of nodes; independent type nodes and mutually exclusive nodes. We thus obtain this figure:

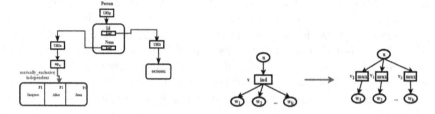

Fig. 3. Uncertain resource.   **Fig. 4.** Translating ind to mux.

In Fig. 3, it is noted that each resource is identified by a single URI and a single rep representation. This representation is uncertain since it is composed of elements which may be uncertain and in this case each element is characterized by a probability value. Using this model, we have two cases. The first represents an independent resource where each URI identifies a single resource independently from the other resources i.e a single representation is true in a given time. In this case, the model specifies that (1) possible resource representations are disjoint and (2) interpretations of a resource are independent.Figure 2 shows how we interpret uncertain resources as a set of possible representations with a probability of generating possible webs that have true and unique.

As already indicated an uncertain resource is identified by a single URI and a single uncertain representation. In the case where the node type is independent. We can have more than a representation that characterizes the resource. To solve this problem, we based on the principle used in the approach of [4] which allows to translate an independent node into several nodes mux this principle is presented by the Fig. 4.

This situation is shown in particular in the case of a node of type $< Dist >$ [mutuallyexclusive]. If one has the probability of the sum of the different possible worlds is less than 1 i.e. p < 1 this indicates that a part of the resource is not represented i.e considered as unknown. In our work, representations of unknown resources are noted.

3   Programmatic Representation of Uncertain Resources

In order to manage the uncertain resources, we propose a formalism to represent them physically.To do this, we will use Java Script Object Notation (JSON), which is a light data exchange format easy to read or write for humans and easily analyzable by machines. It is based on a subset of the Java Script programming language.

To adequately represent these uncertain resources, programmatic representation mechanisms have been proposed which may include all possible representations of the web, where each of them is characterized by its probability value.These mechanisms also make it possible to distinguish between two types of nodes: certain and uncertain nodes. They also make it possible to distinguish between the two categories of distributions: independent and mutually exclusive.

The two Figs. 5 and 6 below represent (show) some representations of uncertain resources using the JSON language.

```
{
  "Person": {
    "PersonId": " 09200001 ",
    "Name": {
      "Dist": {
        "-type": "mutualy-exclusive",
        "Val": [
          {
            "-Prob": "0.2",
            "#text": " Jean "
          },
          {
            "-Prob": "0.6",
            "#text": " Alice "
          },
          {
            "-Prob": "0.1",
            "#text": " Jackes "
          }
        ]
      }
    }
  }
}
```

```
{
  "Book": {
    "Title": " T1 ",
    "Author": {
      "Dist": {
        "-type": "mutualy-exclusive",
        "Val": [
          {
            "-Prob": "0.7",
            "#text": " A1 "
          },
          {
            "-Prob": "0.3",
            "#text": " A2 "
          }
        ]
      }
    }
  }
}
```

Fig. 5. Example 1: Programmatic representation of resource Person.

Fig. 6. Example 2: Programmatic representation of resource Book.

To manage uncertainty at the level of web resources, one must assume that the client is aware of the nature of these resources. Therefore, in this part of the article we will introduce the notion of uncertainty at the level of the client who is able to manipulate and treat these uncertain resources.

In our approach, we will try to adapt our proposition to each client and all types of clients. To achieve our goal, we must respect the principles of the web. For this purpose, we rely on the principle of content negotiation.

Content negotiation is a mechanism that defines an HTTP specification and offers the possibility of proposing different versions of the same document as a more general resource for the same URI. By using content negotiation, the HTTP client can inform the server of its preferences in terms of data format or language. It also implements some features to handle smart queries from navigation that send incomplete trading information.

In this paper, we distinguish between two types of GET queries: we have the classical GET and the uncertain GET. To define the uncertain GET, we start by proposing a special notation to differentiate it from the classical operation. The uncertain $\widetilde{GET}$ should also provide a specific rating that represents the uncertain web resource. This uncertain resource is represented by $\widetilde{R}$. Let $\widetilde{R}$ be an uncertain resource deployed by URIr, we have defined the behavior of this uncertain resource by the following formula:

$$\widetilde{GET}(uri_r) = \{P | rep_r = \langle Sub - rep_1, P_1 \rangle, ..., \langle Sub - rep_n, P_n \rangle\} \tag{2}$$

In the case where the client executes a request (GET) on a certain resource, the response to this request gives a representation with a probability value which is equal to 1. In our proposal, the $\widetilde{GET}$ does not define a new HTTP method. This uncertain query $GET]]$ works as the standard GET query. The only difference is that the $\widetilde{GET}$ is characterized by a specific HTTP header: $X - Accept - Uncertain : true$. This heading is explained in next section. We choose to define a specific header to avoid any interference with the standardized use of the accept header, which is the classical header used for the content negotiation, because it is used to specify an expected mine type for resource representation. To implement our approach, the practical part consists in proposing a specific adhoc heading to respect the standards of the web and of HTTP.

4   Composing Uncertain Web Resources

To respond to a user's request, it is sometimes necessary to compose several web resources. In a resource composition, each combination of representations of possible resources generates a possible new web Pwx. Every possible web is characterized by a probability value that represents the degree of probability of having this web.

To calculate this degree of probability, we used the probabilistic XML principle which is based in particular on the conditional probability. The probability of an element can be found by multiplying the conditional probabilities of resources throughout the path from the element to the root. Each probability will be valid in the traveled path to find the result. The probability of response is calculated by the following equation:

$$P(response) = P(Sub - rep) * ... * P(actual_node) * ... * P(root) \tag{3}$$

WhereP $(PWi) \in Card(PWx) and Card(PWx)$ represent the set of representations involved in Pwx. As already mentioned, there are certain representations of a resource that are not known if we have the sum of the *probabilityvalues* < 1. Thus, to calculate the probability of the unknown representations of a resource, we compute Ri by the following formula:

$$Prob(rep_a^x) = 1 - \sum_{i=1}^{n} Prob(rep_a^i) \tag{4}$$

Where *repia* are the different representations of a resource Ra.

In the next section, we will explain how to respond interpret and calculate a query within an uncertain resource.

5   The Operators

To compute the final result, we have to aggregate the data returned by these resources and calculate the values of result probability. To compute these values, we developed a set of algebraic operators that enable us to compute the degree. Each probability of a resource is allocated by conditioning the existence of the parent element. Thus, we used the conditional probability which is presented by the following example: $A \rightarrow B \rightarrow C$

$$Prob(B|A) = \frac{Prob(A|B) * Prob(B)}{Prob(A))} \tag{5}$$

$$Prob(B) = Prob(B|A) * Prob(A) \tag{6}$$

$$Prob(C) = Prob(C|B) * Prob(B|A) * Prob(A) \tag{7}$$

These operators are presented as follows:

- **Conjunction:** To calculate a simple conjunctive query, one must simply browse the ancestor chain of each node of the query without repeating the same path.
- **Disjunction:** $Prob(A_1 \vee A_2) = Prob(A_1) + Prob(A_2) - Prob(A_1 \wedge A_2)$.
- **Negation:** $Prob(\neg(A)) = 1 - Prob(A)$.

6   Evaluation of a Request Within an Uncertain Resource

In this section, we explain how to respond to a query that requires the composition of several certain and uncertain resources. We also show how to aggregate the response to a query from an uncertain composition based on hypertext navigation.

Actually, a query within a composition is defined as a set of ordered subqueries of resources presented in that composition. We should follow the same path through the various possible generated webs. In our approach, each possible web generates a result (response to a subquery) accompanied by its probability value. This result must be aggregated with all possible responses from other webs to provide the final response to the initial query. Generating each of these possible websites, i.e. combining and storing each combination in memory to meet the query and calculating the degree of probability in each of them, is a language task that takes a lot of Time and consumes a lot of memory. Therefore, we have to propose an approach that enables to aggregate these results directly without having to generate all possible webs. We have chosen to use probabilistic XML ProTDB [5] to reduce the consumption of the memory and the time and improve the result type. In our case, we have two types of uncertain nodes and some and in this way, we improve the nature of response.

Figure 7 shows an example of a request path. The query we are going to execute is "What are the names of the ChiefOfState requested by a user?". To execute this query, we will follow the path used in a classical composition of RESTful which is well detailed in the figure below. To follow this path, we assume that the resource representations specify the necessary semantics on their contents. For example, when searching for a name for a ChiefOfState resource. In this case the ChiefOfState functional property is required to complete. In our example, the representations that we semantically manipulate and the proposed changes are represented in the JSON-LD format.

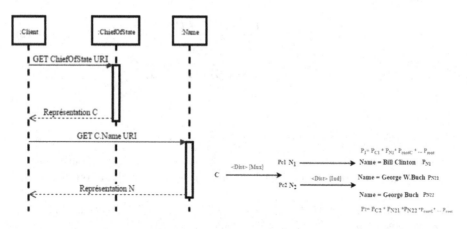

Fig. 7. Query answering in RESTful compositions.   **Fig. 8.** Generating tree pattern while navigating resources.

When we deal with uncertain resources, we go through our query path by representing the possible resources. This approach helps us to create a model that corresponds to a tree of possibilities composed of possible webs characterized by their probability values.

Figure 8 shows the tree model created from our path scenario to respond to the query. To improve our approach, we try to find a solution so as not to be obliged to generate all the possible worlds. As a solution, we propose an algorithm to calculate the resulting degrees of probability without being obliged to go through the stage of generating all the possible webs. This algorithm implements an operator that we labeled GET^{Prob} which functions as a router, since it follows step by step the tree of possible representations. It even enters inside this tree and across all the nodes that satisfy the sought query.

This algorithm GET^{Prob} takes as input parameters a list of URIs and returns the representations of possible resources that meet the user's needs. The operator GET^{Prob} executes the required sequence of the query http on URI data, applies the probability formula that has already been explained, and returns the set of representations accompanied by a single probability value that indicates the possibility value to have this result.

Algorithm 1. GET^{Prob}

1: For all (uri_i, Sub_{rep_r}).
2: Calculate probability values.
3: Test if we have traversed the node.
4: **if** TRUE **then**
5:     Prob= $Prob_i * ProbSub_r$.
6: **else**
7:     Prob =Prob.
8: **end if**
9: Test if we have traversed the node..
10: **if** $Representation \in results$ **then**
11:     Results.Add(Representation).
12: **else**
13:     Results.Update (Representation)
14: **end if**
15: AResults.ADD(Prob).
16: Return (Results).
17: End.

As an example, we have a list of URIs names extracted from the possible web set of ChiefOfState that is characterized by its probability value. The Get gives us the opportunity to search for the representation of each name. It enables to return a result with a probability value. It works according to two types of nodes and calculates the probability value by using the multiplication which is the basis of conditional probability.

7 Experimental Study and Analysis of the Results

All experiments are performed on a computer with a Windows 7 operating system and the Intel Core i5 processor and 6 GB of RAM. In order to analyze our approach and study its performance, we considered the execution time. We will present in this section some results and compare our uncertain approach with the certain semantic approach. The purpose of this experiment is to show that despite the great complexity of our uncertain invocation and web resources composition algorithm, queries can still be resolved in a reasonable amount of time. In this sense, we compare the performances of the certain composition algorithm which considers that wev ressources with certain semantics, with our proposition while increasing the number of web ressources, and varying the number of nodes of type ind per web resources. In this section we present two parts

Experiment 1: Number of Web Resources. In this section, we investigate the effect of changing the number of services on the execution time. In this series of experiments, we will measure the execution time required for the composition of the services by implementing the two approaches, with and without probability calculation. These experiments help us evaluate the cost incurred by calculating the probability of composition made to respond to a user's request. For this reason, we set up 16 Resfull services on an uncertain database, 100 MB in size, storing synthetic data on students, teachers and universities. We measured the execution time necessary for the composition with and without the calculation of the degrees of probability. These results are illustrated in Fig. 9:

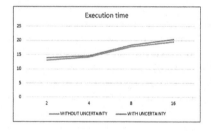

Fig. 9. Performance results: in terms of execution time with and without probability calculation.

Figure 9 shows the results in terms of execution time carried out through two approaches by varying the number of involved services to respond to a query. The result of this experiment shows that the time needed to return the responses to a query, which is estimated at 14,53 ms, is negligible. However, when the number of services is 2, the necessary execution time set by the certain approach is 14,17 ms. In all cases, the time required for the composition of services in our approach is considered negligible compared to that of uncertain approach. Based on these results, we can deduce that our approach gives better results since it enables us to manage the uncertainty of the data without consuming more time.

8   Conclusion

In this paper, we proposed a method for the treatment of uncertain resources, i.e. we proposed a formalism to physically represent them. Besides, we have defined an uncertain operation that works as the standard Get with a specific header. Moreover, we added two more headers to improve the uncertain representation with uncertain capabilities such as: Informing the client that the resource has an uncertain capacity, the client becomes able to accept uncertain queries by adding: X-accept-uncertain and put the probability of the representation chosen by adding: X-accept-Value. And finally, we proposed algebra for the composition of uncertain resources and explained how to interpret a composition.

References

1. Maleshkova, M., Pedrinaci, C., Domingue, J.: Investigating Web APIs on the World Wide Web. In: 8th IEEE European Conference on Web Services (ECOWS 2010), 1-3 December 2010, Ayia Napa, Cyprus, 2010, DBLP:conf/ecows/2010 (2017). https://doi.org/10.1109/ECOWS.2010.9
2. Halevy, A.Y., Rajaraman , A., Ordille, J.J.: Data integration: the teenage years. In: Proceedings of the 32nd International Conference on Very Large Data Bases, Seoul, Korea, September 12-15, 2006, DBLP:conf/vldb/2006, http://dl.acm.org/citation.cfm?id=1164130, dblp computer science bibliography, https://dblp.org
3. Abiteboul, S., Kanellakis, P.C., Grahne, G.: On the representation and querying of sets of possible worlds. Theor. Comput. Sci. **78**, 159–187 (1991). https://dblp.org/rec/bib/journals/tcs/AbiteboulKG91. DBLP computer science bibliography
4. Parag, A., Omar, B., Das, S.A., Chris, H., Shubha, N., Tomoe, S., Jennifer, W.: Trio: a system for data, uncertainty, and lineage. In: Proceedings of the 32nd International Conference on Very Large Data Bases (2006)
5. Nierman, A., Jagadish, H.V.: ProTDB: probabilistic data in XML. In: Proceedings of the 28th VLDB Conference. Springer (2002)
6. Benslimane, D., Sheng, Q.Z., Barhamgi, M., Prade, H.: Concepts, Challenges, and Current Solutions, TOIT, The Uncertain Web (2016)

Domain Adaptation in Transformer Models: Question Answering of Dutch Government Policies

Berry Blom[ID] and João L. M. Pereira[✉][ID]

University of Amsterdam, Amsterdam, The Netherlands
berry96@live.nl, j.p.pereira@uva.nl

Abstract. Automatic answering questions helps users in finding information efficiently, in contrast with web search engines that require keywords to be provided and large texts to be processed. The first Dutch Question Answering (QA) system uses basic natural language processing techniques based on text similarity between the question and the answer. After the introduction of pre-trained transformer-based models like BERT, higher scores were achieved with over 7.7% improvement for the General Language Understanding Evaluation (GLUE) score.

Pre-trained transformer-based models tend to over-generalize when applied to a specific domain, leading to less precise context-specific outputs. There is a marked research gap in experiment strategies to adapt these models effectively for domain-specific applications. Additionally, there is a lack of Dutch resources for automatic question answering, as the only existing dataset, Dutch SQuAD, is a translation of the SQuAD dataset in English.

We propose a new dataset, PolicyQA, containing questions and answers about Dutch government policies and use domain adaptation techniques to address the generalizability problem of transformer-based models.

The experimental setup includes the Long Short-Term memory (LSTM), a baseline neural network, and three BERT-based models, mBert, RobBERT, and BERTje, with domain adaptation. The datasets used for testing are the proposed PolicyQA dataset and the existing Dutch SQuAD.

From the results, we found that the multilanguage BERT-model, mBert, outperforms the Dutch BERT-based models (RobBERT and BERTje) on the both datasets. By introducing fine-tuning, a domain adaptation technique, the mBert model improved to 94.10% of F1-score, a gain of 226% compared to its performance without fine-tuning.

Keywords: Natural Language Processing · Question answering · Transformers · Domain adaptation · Dutch

1 Introduction

Question answering is concerned with automatically answering questions posed by humans in natural language. We can distinguish question answering into

P. Quaresma et al. (Eds.): IDEAL 2023, LNCS 14404, pp. 196–208, 2023.
https://doi.org/10.1007/978-3-031-48232-8_19

open-domain question answering, questions about any topic; and closed-domain question answering (questions under a specific domain). An example of an open-domain question would be "What did Albert Einstein win the Nobel Prize for?". This question is based on broad unrestricted knowledge and general ontologies. An example of a closed-domain question would be "What are my rights and obligations with a purchase agreement?", this question is asked by a citizen to the Dutch government about its policy. The responses to closed-domain questions are limited in terms of text availability, and are from a particular narrow domain. Additionally, there are two approaches to create an answer: the Generative Question Answering (GQA) generates text based on the context and the Extractive Question Answering (EQA) extracts the correct answer (a passage) from the context.

This study concentrates on closed-domain EQA concerning Dutch policy data. The importance of EQA lies in its ability to efficiently navigate large data volumes, pulling verifiable context-specific information directly from the source text. This capability is particularly vital in policy analysis, where precision and transparency are paramount. Recently, transformer based models which take raw text without almost no pre-processing and uses an attention mechanism for context, has led to advances in natural language processing tasks such as EQA [10]. Despite their advantages, pre-trained transformers models are prone to overfitting when applied to specific domains due to a large number of parameters [21]. Also, pre-training biases can result in erroneous model decisions [9].

This research aims to adapt three BERT-based models (BERTje [18], RobBERT [3], mBert [13]) by exploring domain adaptation techniques (e.g., fine-tuning) to Dutch policy data for EQA on a Dutch government policy dataset, an unexplored domain. The code and data used is made publicly available[1].

Our central Research Question (RQ) and Sub-Research Questions (SRQ) are:

RQ1. How do three BERT-based models (BERTje, RobBERT, mBert), a mix of multilingual and Dutch transformer models, and domain adaptation techniques perform in answering questions of Dutch government policies and questions translated from SQuAD dataset when compared to the Long Short-Term Memory (LSTM), a baseline model?

SRQ1.1 How does the baseline model (LSTM) effectively perform in answering Dutch questions?
SRQ1.2 How do the three BERT-based models effectively perform in answering Dutch questions?
SRQ1.3 How do the three BERT-based models with fine-tuning, a domain adaptation technique, effectively perform in answering Dutch questions?
SRQ1.4 What effect do fine-tuning the three BERT-based models using different learning rates per model layer have on the performance of answering Dutch questions?

The paper structure is as follows: Sect. 2 presents the related work in question answering; Sect. 3 details the new PolicyQA dataset; Sect. 4 outlines the experi-

---

[1] https://github.com/berryxmas/domain-adaptation-transformers-forQA.

mental setup; Sect. 5 presents the results; Sect. 6 presents the discussions of the results and new findings; and finally, Sect. 7 summarizes the main conclusions of this work and avenues of future improvements.

2 Related Work

This section explores the related work in Question Answering. Pre-trained language models have proven to be successful at the task of Extractive Question Answering (EQA), however, generalizability remains a challenge for most of the models. Pearce et al. [11] show that the BERT model [17] performs best on the English SQuAD 2.0 dataset [14] since the context, questions, and answers are all straightforward and the answers are purely extractive.

Before transformers, the approach used for Question Answering was Long Short-Term Memory (LSTM) networks [19], which are a type of recurrent neural network that are able to learn order dependence in sequence prediction problems and is the precursor of the Transformer model. Unlike normal feedforward neural networks, LSTM has feedback connections. This way, the network can process entire sequences of data.

Wang and Jiang [20] introduced an end-to-end neural architecture for answering questions of the English SQuAD dataset. The architecture is based on a match-LSTM model. This model goes through the tokens sequentially. At each position, a weighted vector representation is obtained. The weighted vector is then combined with the current token and fed to an LSTM.

One of the first QA systems for the Dutch language was SimpleQA [5] which was capable of answering Dutch questions where the answer was a location or a person. This QA system consisted of six steps to answer Dutch questions on which the answer type is a person or a location. The question was analyzed, rewritten, retrieved with the Google API, and the best-ranked answer was picked.

Araci [1] performed research in sentiment analysis, in the financial domain. A challenging task is to perform financial sentiment analysis due to the specialized language. Araci [1] hypothesized that pre-trained language models could be used for this problem because they require fewer labels and can be further specialized towards a domain using a domain-specific corpus. He introduced FinBERT to tackle NLP tasks in the financial domain [1].

Hazen et al. [4] compared a BERT-QA model on two QA datasets: the English SQuAD dataset 2.0 [14] and the BMW automobile manual training. An interesting observation was that the model trained on the English SQuAD dataset and tested on the Auto dataset gave a lower score than when trained only on the Auto dataset.

Isotalo [6] constructed a Dutch Question Answering dataset from reading comprehension exams for Dutch secondary school students. mT5, a large pretrained text generation model was used and that resulted in low scores even when trained on the same dataset.

Rouws et al. [16] created a new dataset, Dutch SQuAD, which is a machine-translated version of the original SQuAD v2.0 English dataset [16]. The research

Table 1. Summary of the related work in domain adaptation with domain-specific datasets.

Related Work	Year	Language	Domain	Task
Araci et al. [1]	2019	English	Financial context	Sentiment Analysis
Hazen et al. [4]	2019	English	BMW Automobile manual	Question Answering
Isotalo et al. [6]	2021	Dutch	Reading comprehension	Question Answering
Rouws et al. [16]	2022	Dutch	Labour Agreements (CAO)	Question Answering

demonstrates how to improve QA models with domain adaptation, by comparing pre-trained Dutch models, such as BERTje [18] and RobBERT [3], versus multilingual models like mBert [13].

Table 1 summarizes the existing related work in extractive QA with domain-specific datasets.

3 PolicyQA: A Dutch Government Policies Question and Answers Dataset

There is a lack of Dutch NLP resources, especially for EQA. For this reason, a new Dutch dataset is specifically designed for EQA. Government policies are long documents that are hard to read for users. PolicyQA is a challenging dataset with actual utility for the real world.

Government Policies. Dutch citizens can ask questions to the Dutch government about government policies. This dataset contains the most commonly asked questions by citizens. It appeared in 2016 and the amount of questions is subject to change. The dataset is maintained by the Ministry of General Affairs and updated weekly by domain experts from the Dutch government and is publically available through an API[2].

We use the API of the Dutch government to get the fields introduction and content. The introduction is the official answer, which is constructed by domain experts from the Rijksoverheid[3], which is the Dutch government. The content is related to the introduction and gives complementary information. The introduction is expected to always be present in the content text. However, by looking at Fig. 1, we can see this is not the case. The answer is not part of the content and in this case is even incorrect. To solve this problem, we merged both fields into one called context because the introduction field is not always correct and provides a longer answer. The collected text is pre-processed as all characters are set to lower-case and the HTML tags (including non-alphanumeric characters) are removed.

Additionally, to support extractive question answering, we added an extra field with the short answer to the question, which is a substring of the context field text. To fill the answer field, we manually labeled the Policy QA Dataset

---

[2] https://www.rijksoverheid.nl/opendata/vac-s.
[3] https://www.government.nl/#governmentnl.

> **Question:** Does a bank have to make and keep a copy of my passport?
> **Annotated Answer:** not obliged to make a copy or a scan of your ID.
> **Introduction:** Banks and financial service providers are not obliged to make a copy or a scan of your ID. They are, however, obliged to check and record the details of your proof of identity. This is stated in the Money Laundering and Terrorist Financing Prevention Act (WWFT).
> **Content:** Financial institution conducts mandatory customer due diligence. Financial institutions are required to carry out customer due diligence in certain cases. For example, if you become a customer of a bank or an insurer. Therefore, the institution checks and records your identity. The bank or insurer often makes a copy of your passport or your European identity card. (...)

Fig. 1. This question (translated to English) is about whether a bank has to make a copy of your ID or not. The answer is only in the Annotated Answer, which says "not obliged to make a copy of your ID". The Content contains a contradiction and says "The bank or insurer often makes a copy of your passport or your European identity card".

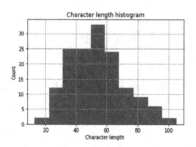

Fig. 2. Character Length for question in PolicyQA dataset

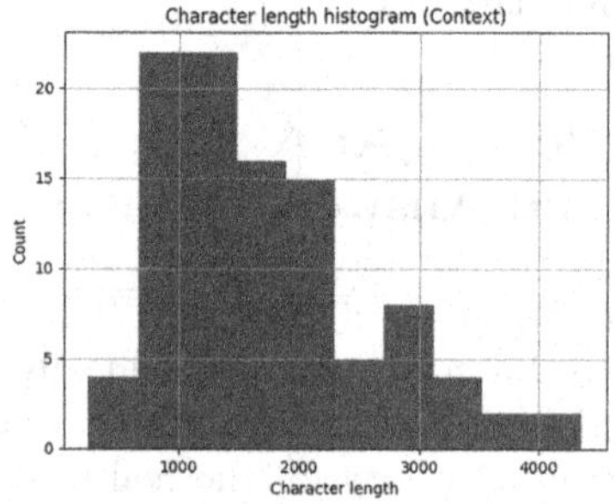

Fig. 3. Character Length for context in PolicyQA dataset

for the first 500 questions and answers. We specified a short answer, two to ten words from the context field to be able to perform extractive question answering.

Dataset Statistics. PolicyQA Dataset contains 1980 questions and contexts. The questions are relatively short, the amount of characters ranges from 10 to 110 characters. The Context is longer, the amount of characters ranges between 300 and 4400 characters and generally, it is between 800 and 1500 characters. The context contains one outlier with over 9000 characters.

Two charts with the length of characters are shown in Fig. 2 for questions and Fig. 3 for context.

4   Experimental Setup

The PolicyQA dataset, which was introduced in Sect. 3, is used to test the models. As well as the Dutch SQuAD Dataset, that was obtained by machine translating the original SQuAD v2.0 dataset from English to Dutch. We divide the experiments in two parts, extractive question answering and domain adaptation.

The experiments ran inside Google Colab, which has a Tesla T4 Graphics Processing Unit (GPU).

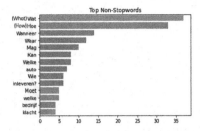

Fig. 4. Top non-stopwords for Policy QA questions

4.1 Datasets

This research uses two separate datasets: (i) the PolicyQA dataset containing government policy data and (ii) the Dutch SQuAD2. The data is preprocessed and fed to four QA models: an LSTM model, Dutch RobBERT model, Dutch BERTje model and mBert.

PolicyQA: The data is from the Dutch government, as described in Sect. 3, supplementary labels were annotated manually leading to 500 answers available to train and test extractive question answering. One example of the final and cleaned dataset is shown in Fig. 1. This question is about whether a bank is obliged to make and keep a copy of your passport. The short answer is derived from the context and says that a bank is not obliged to make and keep a copy of your passport. Also, the answer start character is set to 42 because the short answer begins from the 42nd character.
To get insights of the types of questions in the training set, we count the questions that begin with a few common start words. As we can see in Fig. 4 the most common non-stopword is What ("Wat") and a close second non-stopword is How ("Hoe").

Dutch SQuAD: The Dutch SQuAD is a translation of the original SQuAD 2.0 and contains 104.348 answers. The original SQuAD 2.0 [15] is a reading comprehension dataset which consists of questions posed by crowdworkers on Wikipedia articles where the answer is a segment of the text. The dataset also contains unanswerable questions, this is used to test if the system is capable of determining when no answer could be given. Figure 6 shows the character length of the context in the Dutch SQuAD dataset. The histogram shows that context range from 150 characters to 2300 characters and generally, it is between 600 and 1100 characters. We can observe, in Fig. 5, that most common questions begin with "How", "What", and "Is".

The Dutch SQuAD dataset is split into a training set and a test set, this is a 90/10 split. This split was chosen because it was the original split from the source data. The PolicyQA dataset is also split into a training set and a test set. The training set contains 80% of the samples and the test set contains 20% of the samples. This split was chosen arbitrarily.

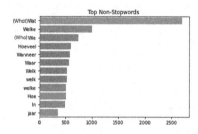

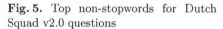

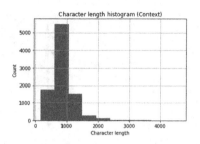

Fig. 5. Top non-stopwords for Dutch Squad v2.0 questions

Fig. 6. Frequency of question types

4.2 Extractive Question Answering Techniques

Baseline. In these experiments, we use match-LSTM, which is an adjusted version of the LSTM model. The difference between match-LSTM and LSTM is that match-LSTM contains an extra layer, which is called an answer pointer. This layer selects a set of tokens as the answer, this way used for extractive question answering. The match-LSTM sequentially aggregates the matching of each token to the weighted premise and uses the aggregated matching result to make a final prediction [20]. We also use word embeddings from GloVe to initialize the model. GloVe [12] functions as global vectors for word representation.

For the implementation of the LSTM model, we use a 3-layer bidirectional LSTM with h = 128 hidden units for both context and question encoding. The data is tokenized and lemmatized, also the training examples are sorted by length of the context and divided into batches of 32. All the hidden units of LSTM and word embedding have a dropout of p = 0.3 as in [2].

BERT-Models. We use the available BERT-models that support Dutch: mBert [13] is a multilanguage model trained on a diverse set of languages including Dutch; BERTje [18] and RobBERT [3] are both trained only on Dutch texts. The main difference is that BERTje uses BERT [17] base model and RobBERT uses Roberta [8], which comparatively with BERT is a larger neural network (i.e., contains more parameters) and is not trained using the next sentence prediction task. In addition to these models, we also utilized a pre-trained mBert model that was specifically trained on the Dutch SQuAD dataset. This choice was motivated by the expectation that a model pre-trained on a similar task in the same language (Dutch) would have an enhanced understanding of the language's nuances, thus potentially improving performance in our specific task. We use the Huggings Faces' PyTorch implementations[4] of the three models and the mBert trained on Dutch SQuAD.

---

[4] https://huggingface.co/docs/transformers/v4.20.1.

4.3 Domain Adaptation

Transformer models can suffer from performance and instability, this is often the case with large models and small datasets. Therefore, in our experiments, we further train the pre-trained BERT-based models in domain data, a process called fine-tuning. To help this process, we apply hyperparameter search on the weight decay and learning rate parameters using Adam optimizer to find the set of hyperparameters that resulted in the best model performance. When we refer to fine-tuned in our experiments, it refers to fine-tuning in domain data with hyperparameter search.

Moreover, since domain adaptation tasks have benefitted from setting higher learning rates in the top layers, we add the Layer-wise Learning Rate Decay (LLRD) [22] technique as an additional step in our fine-tuning process. LLDR sets a different learning rates for each layer in the model, by decreasing its values from top to bottom layers.

The default learning rate used for all BERT-based models was 5.5 and the calculated learning rate is 3.6e-06. The default weight decay was set to 0.0 and the calculated weight decay was 0.01. However, applying these parameters in practice presented challenges. For instance, we found that the lower learning rate significantly increased the time taken for our models to converge, requiring more computational resources than we initially planned.

4.4 Evaluation Metrics

To evaluate the quality of the extracted answers, we apply the following commonly used evaluation metrics in EQA:

F1-score: is a commonly used metric that by a harmonic mean combines precision and recall into a single metric. For EQA it compares the tokens between the true and the extracted answer. Precision is the number of correct tokens in the extracted answer (i.e., that appear in the true answer) divided by the total number of tokens in the extracted answer. Recall is the number of correct tokens in the extracted answer divided by the total number of tokens in the true answer.

Exact Match (EM): is calculated for a model by averaging over the individual answers. The score is either 1 or 0 per answer. If all characters of the extracted answer exactly match all characters of the true answer, then EM is 1, otherwise is 0.

5 Results

In Table 2, we report the F1-score and the EM score calculated per dataset and per model. The baseline model, LSTM, is used on the Dutch SQuAD dataset and PolicyQA. Also, the three BERT-based models and mBERT trained on Dutch

Table 2. Results for EQA on Dutch SQuAD and PolicyQA dataset using LSTM and three BERT-based models (mBert, BERTje, RobBERT).

		PolicyQA		Dutch SQuAD	
		F1	EM	F1	EM
LSTM		22.77	1.24	30.25	17.80
BERTje	pre-trained	15.44	0.00	59.80	54.30
	fine-tuned	57.25	26.00	61.23	55.43
	fine-tuned with LLRD	27.00	6.89	60.54	55.00
RobBERT	pre-trained	14.94	0.00	47.90	39.57
	fine-tuned	56.71	20.00	52.40	43.60
	fine-tuned with LLRD	30.45	5.38	50.30	40.92
mBert	pre-trained	16.07	0.00	64.67	61.26
	fine-tuned	61.20	29.00	77.29	69.20
	fine-tuned with LLRD	36.90	8.43	68.60	64.66
mBert	pre-trained	28.88	11.00	77.29	69.20
Dutch	fine-tuned	**94.10**	**83.50**	**79.28**	**72.38**
SQuAD	fine-tuned with LLRD	85.70	78.93	78.37	71.55

SQuAD are described pre-trained without any additional domain training, fine-tuned in same domain data, and fine-tuned using the Layer-wise Learning Rate Decay (LLRD) technique described in Sect. 4.3.

Testing the baseline model, LSTM, on the PolicyQA dataset resulted in an F1 score of 22.77% and an EM of 1.24. The multilingual BERT (mBert) pre-trained on the Dutch SQuAD dataset achieved the highest score with an F1 of 94.10 and an EM of 83.50. In this case, mBert was pre-trained on the Dutch SQuAD dataset and trained on the training data of the PolicyQA dataset.

Another observation we made was the increase in F1 score and EM score when the number of annotated samples increased. This was tested on for the highest performing model identified in Table 2, mBert trained on Dutch SQuAD. For every 100 samples, an 80/20 split was chosen arbitrarily for the train set and the test set. For example, for 100 samples 80 samples were used for training and 20 samples were used for testing. The samples for testing came from the last 100 samples which were never used for training the model. Interestingly, our experiments indicated that the F1 score increased linearly by approximately 3% for each additional set of 100 training samples. It is important to note that the testing samples were consistently drawn from the last 100 samples, ensuring they were never used in model training.

By fine-tuning the BERT-based models (mBert, BERTje, RobBERT) using the government policies, the F1 score improved. Also, the Layer-wise Learning Rate Decay (LLRD) technique was used during training. This technique did not result in an improvement of the F1 score.

The Dutch SQuAD was used as a second dataset. For this dataset, the LSTM model achieved an F1 of 30.25% and an EM of 17.80. The highest F1 and EM scores were obtained with mBert (F1 of 69.67% and EM of 66.26). In this approach, the model was trained on the Dutch SQuAD dataset without LLRD.

6   Discussion

Baseline - LSTM. By comparing the baseline approach with our results, we can see there is a big difference in F1 score as well as EM. In the research about the approach with Match-LSTM [20], which was described in the baseline section, a higher F1 score (69%) was achieved. However, this F1 score was achieved with slightly different data, this approach was based on the SQuAD v1.0 Dataset. The main difference between the SQuAD v1.0 dataset and the SQuAD v2.0 dataset (that originated Dutch SQuAD) is that over 50000 questions shouldn't be answered, in such case, the answer should be empty, but Match-LSTM always provides an answer. When we look at the F1 score for the PolicyQA dataset (22.77%), we can see the score is lower than 28.88. Moreover, the F1 score for the Dutch SQuAD dataset (50.23%) from our results is lower than the score obtained in the original English SQuAD V1.0 dataset [20] (77%). A possible reason for the lower score is the machine translation. When the English SQuAD dataset was translated to Dutch, the translation was not perfect [16]. This influences the output of the model.

BERT-Based. When we compare our results for the BERT-based models with the state-of-the-art approaches, we can see that Dutch SQuAD had a similar performance as previous research. For example, Rouws et al. [16] achieved similar scores using mBert on the Dutch SQuAD dataset [16]. With an F1 score of 71% that is only 10 points higher than our F1 score of 61%. Thus, we can assume our approach on mBert with training is successfully executed since the scores are similar. Since the results for the Dutch SQuAD are similar to previous research [16] we can presume the F1 score of 94.10% and EM of 83.50% for the PolicyQA dataset are also as expected. Mainly because of the similar results to the known dataset, the PolicyQA dataset is a new benchmark that can be used for future work.

Generally, a pre-trained large language model performs better on more data. Even for 100 samples in the PolicyQA dataset, the model scored high (79% of F1-score). Thus, we can confirm that by training on the PolicyQA dataset, a high score can be achieved with little data.

7   Conclusions and Future Work

This research focused on the lack of Dutch resources in the field of QA and the generalizability of pre-trained large language models. This paper provides a solution to the lack of Dutch resources, namely the PolicyQA dataset, and creates new insights on domain adaptation.

In this research, we tested and evaluated a baseline for extractive question answering in order to investigate the generalizability of pre-trained large language models and examine to what extent we can make a contribution to the field of extractive question answering.

SRQ1.1: By comparing the scores to existing research, we found that LSTM scored low (F1 of 22%) on the PolicyQA dataset. One possible problem is that the supplementary annotations in the data are not enough to train a model from scratch, these lead to bias and thus lower scores. Also, LSTM scored low on the Dutch SQuAD dataset. A possible reason for this is the machine translation from English to Dutch.

SRQ1.2: We tested the three BERT-based models without any domain adaptation technique (e.g., fine-tuning) to assess their performance in answering Dutch questions. In general, the results were low (below 17% of F1-score) for any of the models.

SRQ1.3: To understand the effect of fine-tuning the models on domain data, all three BERT-based models (BERTje, RobBERT, mBert) were evaluated with and without fine-tuning. The results show that fine-tuning leads to significant improvements in performance, e.g., with fine-tuning, mBert improves from 16% to 61% of F1 score in the PoliciQA data. Moreover, if fine-tuning is performed first in the same task and language but in a different domain (i.e., Dutch SQuAD) and then on domain data, we verify also considerable improvements using the mBert model to 94% of F1 score in PoliciQA. This is a high score, according to Lipton et al. [7]. Thus, we can conclude that the BERT-based models with fine-tuning adapt well to the Dutch government policy domain.

SRQ1.4:
We investigated if further improvements are obtained by fine-tuning with a different learning rate per layer. For that purpose, we conducted experiments with the Layer-wise Learning Rate Decay (LLRD) technique. We verify that for all models, LLDR resulted in considerable score decreases, at least 10% less of F1 score when fine-tuned without LLDR. So, for this domain adaptation task, which involves a drastic change of domain from Wikipedia (SQuAD) to Policy writings, the bottom layers required higher learning rates potentially because specific linguistic cues learned on those layers have become harder to train.

In conclusion, we evaluated how three BERT-based models (BERTje, RobBERT, mBert) perform in answering questions of Dutch government policies compared to an LSTM model. And we found that all three BERT-based models outperformed the baseline model, LSTM, with significant scores on both the Dutch SQuAD dataset and the PolicyQA dataset. We also showed that by training mBert on the Dutch SQuAD dataset and the PolicyQA dataset higher F1 scores and EM scores were achieved and that the use of LLRD did not improve the performance.

Compared to previous research, this research adds a new domain dataset, namely Dutch government policies, to the field of extractive question answering.

For future work, we suggest increasing the annotated texts of PolicyQA because we observed that the increase in the number of samples positively influences the performance. Also, other ways of domain adaptation like tuning other hyperparameters or adding top domain-specific layers without affecting pre-learned representations can be investigated using the PolicyQA dataset.

References

1. Araci, D.: FinBERT: financial sentiment analysis with pre-trained language models. Master's thesis, University of Amsterdam, The Netherlands (2019)
2. Chen, D., Fisch, A., Weston, J., Bordes, A.: Reading Wikipedia to answer open-domain questions. In: ACL (2017)
3. Delobelle, P., Winters, T., Berendt, B.: RobBERT: a dutch RoBERTa-based language model. In: Findings of ACL: EMNLP (2020)
4. Hazen, T.J., Dhuliawala, S., Boies, D.: Towards domain adaptation from limited data for question answering using deep neural networks. arXiv:1911.02655 (2019)
5. Hoekstra, A., Hiemstra, D., van der Vet, P., Huibers, T.: Question answering for dutch: simple does it. In: BNAIC (2006)
6. Isotalo, L.: Generative question answering in a low-resource setting. Master's thesis, Maastricht University, The Netherlands (2021)
7. Lipton, Z.C., Elkan, C., Naryanaswamy, B.: Optimal thresholding of classifiers to maximize F1 measure. In: Calders, T., Esposito, F., Hüllermeier, E., Meo, R. (eds.) ECML PKDD 2014. LNCS (LNAI), vol. 8725, pp. 225–239. Springer, Heidelberg (2014). https://doi.org/10.1007/978-3-662-44851-9_15
8. Liu, Y., et al.: RoBERTa: a robustly optimized BERT pretraining approach. arXiv:1907.11692 (2019)
9. Nadeem, M., Bethke, A., Reddy, S.: StereoSet: measuring stereotypical bias in pretrained language models. In: ACL IJCNLP (2021)
10. Pasch, S., Ehnes, D.: StonkBERT: can language models predict medium-run stock price movements? arXiv:2202.02268 (2022)
11. Pearce, K., Zhan, T., Komanduri, A., Zhan, J.: A comparative study of transformer-based language models on extractive question answering. arXiv:2110.03142 (2021)
12. Pennington, J., Socher, R., Manning, C.: GloVe: global vectors for word representation. In: EMNLP (2014)
13. Pires, T., Schlinger, E., Garrette, D.: How multilingual is multilingual BERT? In: ACL (2019)
14. Rajpurkar, P., Jia, R., Liang, P.: Know what you don't know: unanswerable questions for squad (2018)
15. Rajpurkar, P., Zhang, J., Lopyrev, K., Liang, P.: SQuAD: 100,000+ questions for machine comprehension of text. In: EMNLP (2016)
16. Rouws, N.J., Vakulenko, S., Katrenko, S.: Dutch SQuAD and ensemble learning for question answering from labour agreements. In: Leiva, L.A., Pruski, C., Markovich, R., Najjar, A., Schommer, C. (eds.) BNAIC/Benelearn 2021. CCIS, vol. 1530, pp. 155–169. Springer, Cham (2022). https://doi.org/10.1007/978-3-030-93842-0_9
17. Vaswani, A., et al.: Attention is all you need. In: NIPS (2017)
18. de Vries, W., van Cranenburgh, A., Bisazza, A., Caselli, T., Noord, G.V., Nissim, M.: BERTje: a dutch BERT model. arXiv:1912.09582 (2019)

19. Wang, D., Nyberg, E.: A long Short-Term memory model for answer sentence selection in question answering. In: ACL IJCNLP (2015)
20. Wang, S., Jiang, J.: Machine comprehension using match-LSTM and answer pointer. arXiv:1608.07905 (2016)
21. Yu, Y., Zuo, S., Jiang, H., Ren, W., Zhao, T., Zhang, C.: Fine-tuning pre-trained language model with weak supervision: a contrastive-regularized self-training approach. In: NAACL (2021)
22. Zhang, T., Wu, F., Katiyar, A., Weinberger, K.Q., Artzi, Y.: Revisiting few-sample BERT fine-tuning. arXiv:2006.05987 (2021)

Sustainable On-Street Parking Mapping with Deep Learning and Airborne Imagery

Bashini K. Mahaarachchi[1]([✉]) [ID], Sarel Cohen[2] [ID], Bodo Bookhagen[3] [ID],
Vanja Doskoč[4] [ID], and Tobias Friedrich[4] [ID]

[1] Institute of Computer Science, University of Potsdam, Potsdam, Germany
bashinim2011@gmail.com
[2] The Academic College of Tel Aviv-Yaffo, Tel Aviv, Israel
sarelco@mta.ac.il
[3] Institute of Geosciences, University of Potsdam, Potsdam, Germany
bodo.bookhagen@uni-potsdam.de
[4] Hasso Plattner Institute, University of Potsdam, Potsdam, Germany
vanja.doskoc@hpi.de, tobias.friedrich@hpi.de

Abstract. In large cities, a considerable amount of traffic congestion is caused by drivers actively looking for *unoccupied* parking spots, leading to reduced average velocities on the streets, reduced air quality, and an increase in sound pollution. A *parking space map* helps drivers to find vacant parking spaces more easily and helps minimising search time, fuel and energy consumption. We introduce a machine learning framework to detect *occupied* and *empty* parking spaces using airborne images. Our approach is based on three steps. First, we present novel ways to include observations from outside the visible spectrum (RGB) to delineate vegetation in urban environments. We remove roads on the images to minimise interference from moving vehicles and use a mask-RCNN to detect vacant parking spaces on the processed images. Third, we propose to use a logistic regression model (LRM) as a post-processing step in order to better filter true positives. From our experiments we conclude that Red-Green-NIR (RG-NIR) spectrum contains the most important information for training Faster-Region-based Convolutional Neural Network with Feature Pyramid Network (RCNN FPN) for vehicle detection. Our empirical studies on aerial images of Berlin, Germany, show that our framework can be employed successfully to create trustworthy parking space maps. Our best model obtain an accuracy of 97.7% and an F1-score of 0.645 for detecting both occupied and vacant on-street packing.

Keywords: Parking Space Map · Remote Sensing · Object Detection · NIR (near-infrared) · RCNN · FPN

1 Introduction

In the European Union urban traffic makes up 8% of the total emissions [3]. Approximately 30% of that traffic is the result of drivers searching for vacant parking spots [16,20]. The anticipation of parking space availability can significantly enhance traffic management, reduce congestion, and contribute to cleaner and quieter cities. To

P. Quaresma et al. (Eds.): IDEAL 2023, LNCS 14404, pp. 209–221, 2023.
https://doi.org/10.1007/978-3-031-48232-8_20

achieve these objectives, we require precise data on parking space locations and their occupancy status. *Parking space maps* containing this information are essential assets in the development of smart city systems that efficiently guide drivers and manage parking resources.

Various sources of data can be utilised in order to obtain such maps. In this work, we focus on the following two sources: Aerial images and vector-based road network[1]: OpenStreetMap (OSM, [11]). Aerial images are now routinely collected in most cities worldwide at annual time steps and usually contain various spectra of information. These images are often financed by the taxpayers and are publicly available. The high spatial resolution of 10 to 30 cm and their high repetition cycle make them an ideal target for automated approaches.

We use airborne images of Berlin-Germany, provided by the Berlin Senate Department for Urban Development and Housing, Germany [14]. Please refer to Sect. 3 for more information on the dataset. We combine this dataset with OpenStreetMap data [11] to incorporate information about the road system.

We use Detectron2's Mask RCNN FPN model [19] (as main model) and LRM (in a post-processing step) to detect and classify occupied and vacant parking spots on the aerial images.

Combining the mentioned sources of data and machine learning models, we build a framework to obtain reliable parking space maps. To facilitate future researchers, the annotations and the scripts have been made publicly available [2].

Structure of this Paper: Section 2 provides an overview of previous research conducted in the field of parking lot detection. In Sect. 3 we discuss the initial dataset which we use and modify throughout the work. In Sect. 4 we introduce our approaches and explain how to obtain the derived datasets we use. Afterward, we empirically analyse the parking space maps created in Sect. 5. Lastly, we conclude our work and discuss possible future work in Sect. 6.

1.1 Our Contribution

The objective of our research is to identify and map public on-street parking spaces while also providing insight into their occupancy rates. We are not aiming to do real-time parking location detection. Our focus is to build a method to do that using airborne images to identify the on-street parking areas, including on-street parking areas that reside below trees which might be more difficult to find.

In addition, publicly available satellite images, such as those provided by Google, do not consistently offer high-resolution images, and the frequency of their updates can be unreliable. The airborne images we use have higher resolution and they contain additional information such as NIR, enabling us to better locate parking spaces.

Our contributions can be summarised as follows:

1. We present a deep learning framework tailored to the creation of parking space maps using aerial imagery, with a particular emphasis on detecting both occupied and

---

[1] A vector-based road network refers to a digital representation of road infrastructure in geographic information system (GIS) or cartographic formats, where roads are represented as a series of interconnected vector elements.

vacant parking spaces. The widespread availability of aerial images, often publicly accessible, serves as a solid foundation for our approach.

2. We propose a post-processing step that enhances parking space detection accuracy by employing a LRM to filter out false positives, that is, parking spaces that were detected falsely, from the model's prediction. This way we increase the performance of our model.

3. We apply different image editing, image augmentation, and fusion techniques to create a variety of datasets from the original airborne images. This way, we study the effects of our changes to the predictive performance of the detection model we use, that is, of Detectron2. In particular, we apply removal methods, where we remove roads and trees from the images, and use different additional information encoded in the images, such as NIR information or height information from lidar point cloud data.

4. We publish the annotations we did for a dataset that is publicly available [13].

5. Overall, our research integrates geo-referenced data, applies advanced techniques, and produces geo-referenced predictions. These predictions are merged to construct comprehensive parking space maps. For a detailed exploration of our work, please refer to the subsequent sections.

2 Related Work

Identifying and segmenting vehicles and other objects in aerial images is a well-studied problem within the computer vision field generating many real-world applications. Ahrnbom et al. [1] propose a method using fast integral channel features and machine learning methods to decide if a region contains a vehicle or not. Their main task is to distinguish between vehicles and non-vehicles. Luo et al. [10] use the YOLOv3 framework to detect the presence of vehicles in different datasets. These datasets contain aerial images of parking garages. They have adopted image augmentation techniques such as random rotation, mirror flip, colour dithering, and adding gaussian noise to the images which have improved the model performance.

Koga et al. [7] apply correlation alignment domain adaptations and adversarial domain adaptations to a region-based vehicle detector that does not require labeled training data to detect vehicles in a target area.

Ways to detect vacant parking spots in bounded areas, such as underground garages and outdoor parking structures, have been proposed. Jones et al. [5] carried out experiments to build a system that makes use of existing infrastructure found in parking garages (surveillance cameras) to detect vacant parking spots. Due to the difficulty of detection under different lighting conditions, they limited the parking scenarios to parking structures where lighting conditions are constant. Singh et al. [17] used transfer learning methods to improve parking space detection accuracy, in different environmental conditions.

Seo et al. [15] devise a self-supervised learning algorithm that automatically generates a set of parking spot templates to learn the appearance of a parking lot and estimates the structure of the parking lot from the learned model. Li et al. [8] propose a vacant parking slot detection method using VPS-Net.

Kabak *et al.* [6] study the detection of *vacant* parking spots in aerial images of cities using support vector machines. They use image augmentation techniques for feature extraction.

In these studies, the methods rely on additional information given by the known surrounding, for example, well-visible marks on the ground or the topology of the parking lot and constant environment conditions. However, our approach, utilising both our method and dataset, eliminates the necessity for these constraints. This is achieved by removing trees from the images and employing the NIR channel as a fusion technique.

3 Dataset

In this work, we primarily use the *Digital Orthophotos Berlin 2020* (DOP) that records data in four wavelength regions: red, green, blue (0.4 to 0.7 μm), and NIR (0.7 μm to 1 μm). The DOP dataset is provided by the Berlin Senate Department for Urban Development and Housing, Germany [14]. The data are collected at 10 cm spatial resolution (DOP10), but resampled to 20 cm and made available to the general public (DOP20) [13]. They generally include NIR information to better elucidate vegetation cover and are often distributed as orthorectified images as a so-called *digital orthophotomosaic*.

An exemplary image of a DOP image is shown in Fig. 1. The dataset covers the entire city of Berlin and is tiled into 1×1 km^2 area with a positional accuracy of ± 0.4 m. There are 404 such image tiles covering the entire city out of which we use 15 image tiles for the annotations. We use the TIFF^2 files of the images to extract different colour bands which is explained in Sect. 4.1.

Fig. 1. Part of an exemplary image of the DOP dataset in Berlin, Germany. This covers a ground surface area of 85×85 square meters.

---

^{2} A TIFF, which stands for Tag Image File Format, is a computer file used to store raster graphics and image information.

4   Data Processing and Detection Framework

In this section we discuss our approach to create parking space maps from aerial images. Our method consists of three main steps. In a pre-processing step, we use the DOP dataset to create a variety of new datasets for our models. This step is described in Sect. 4.1. In Sect. 4.2, we discuss the annotation of the data as well as the deep learning model we use for detecting parking spaces. Finally, we specify the post-processing step, where we use a LRM to refine the models predictions, in Sect. 4.3.

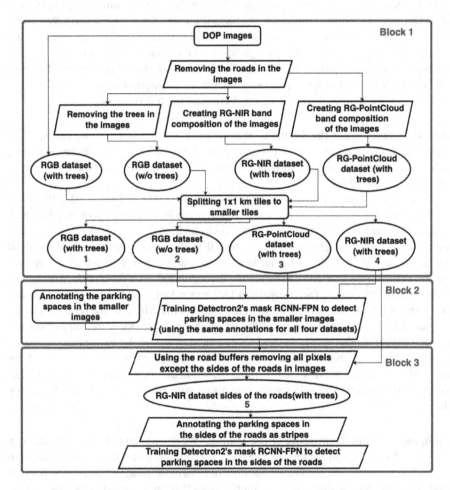

Fig. 2. This chart illustrates the dataset creation process using a flowchart divided into three blocks. The first block involves steps to create the initial four datasets, which include roads and trees removal and color composition adjustments. The second block covers the annotation process. Block 3 explains the creation of the roadside dataset based on Red-Green-NIR channels. (Color figure online)

4.1 Data Pre-Processing Step

In this section, we discuss the pre-processing step. In particular, we describe how to create various new datasets from DOP. An overview of the procedure can be seen in Fig. 2.

Removing Roads from the Images. Distinguishing between moving and parked vehicles in airborne images poses a challenge. To address this, we utilise road data from OSM [11] to remove the road network from images, eliminating moving vehicles typically located near road centers. We classify roads from OSM into two groups: *main roads* (motorways, trunks, primaries, secondaries, and related types) and *other roads* (footways, residentials, services, and steps). These roads are represented as lines, and we apply varying buffer widths to each group: 3 m on each side for main roads (resulting in a 6-meter-wide mask) and 1.3 m on each side for other roads (yielding a 2.6-meter-wide mask). This road removal process applies to all subsequent datasets.

Note that we remove the roads this way from all the following datasets.

Removing Trees from the Images. Trees often cover the sides of the roads. Their large crowns hide parked cars or parking spaces beneath them. This makes detecting such spaces on aerial images difficult. In this section, we discuss how trees can be removed from the images.

Modern aerial photographs record blue (B), green (G), red (R), and NIR wavelengths. The visible part of the spectrum (RGB) provides true-color photos, while NIR data are used to better identify vegetation due to high reflectance of vegetation in that wavelength. Removing pixels with chlorophyll in the images helps to locate parked vehicles under the trees. In particular leaves containing chlorophyll reflect a high percentage of the NIR signal and can be readily distinguished from other surface materials. A commonly used index, the Normalised Difference Vegetation Index (NVDI), relies on the spectral properties of leaves: high reflection in the NIR and low reflection in the red part of the spectrum. The ratio of these two channels provide a measure of vegetation density [4]. In this work, we apply a threshold of 0.3 NDVI to separate vegetated (≥ 0.3) and non-vegetated pixels (< 0.3). We select this threshold by looking at the image tiles for different NDVI and picking the value that best separate trees and other objects in the images. We only remove the pixels with NDVI (≥ 0.3) from the image tiles.

Changing the Band Composite of the Images. We discuss how the NIR channel can be used to detect trees in the images in Sect. 4.1. Here, we do not remove the trees from the images, but rather differentiate the texture of trees in the images using different color band mixes. We look at three channel combinations that capture most of the information in the images.

The correlation matrix of a randomly selected image tile (see Fig. 3) shows that green and blue channels are highly correlated to each other compared to other channel pairs. Therefore, we remove one of them, namely the blue channel, and add the NIR channel to the images to capture the maximum possible information in the data. We create one dataset with red, green, and NIR channel combinations.

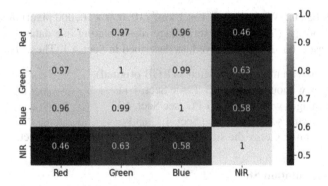

Fig. 3. The correlation matrix of red, green, blue, and NIR channels of a DOP image. Note the high correlation among the visible bands (red, green, blue) and the generally low correlation of the visible bands to the NIR band. This indicates that the NIR band contains information that are not included in the visible bands. (Color figure online)

Using only the Road-Side Information. Instead of removing only the roads (see Sect. 4.1), we remove everything but the side of the roads. This way, we focus only on the roadsides where we expect the vacant or occupied parking spaces to occur. Thus, when training the model on these images it only sees a limited number of objects in the images which then are mostly vehicles, remaining parts of the roads, and trees.

We apply this to the RG-NIR dataset (see Sect. 4.1). We choose this dataset as the Mask RCNN FPN model shows a superior behavior on it in initial experiments. Finally, we obtain the RG-NIR-RS data, that is, data containing NIR information but only displaying the side of the roads.

Adding Height Information from Lidar Point Cloud Data to the Images. Furthermore, we create another dataset using point cloud data, which have been acquired using airborne *light detection and ranging* (lidar, [9]) in 2020. Lidar data are acquired in several year intervals over all European and North American cities and most of European countries are covered by publicly available lidar data. The pulsed laser light in the NIR uses time of flight for distance calculation and generates a point density of $\sim 15 \text{ pts/m}^2$. We grid point-height data to a similar spatial resolution as the airborne RGB-NIR imagery (10 cm) and use the highest point return in each $20 \times 20 \text{ cm}^2$ cell to generate a Digital Surface Model (DSM) and upsample it to $10 \times 10 \text{ cm}^2$. Similar to Sect. 4.1, we substitute the blue channel with the DSM to obtain three channels with red, green, and height information of the image.

Although the lidar data have a high point density, they do not allow for a sampling distance of 10 cm and require interpolation for some empty grid cells. This leads to significant noise in the fused images. In addition, both datasets (airphotos and lidars) were not acquired at the same time, which leads to differences on roads and other dynamic surfaces. Therefore, the fused images of red-green-DSM
contain more noise than the fused images using red-green-NIR wavelength.

The Final Datasets. We split each 1×1 km^2 ($10,000 \times 10,000$ pixel) image tile into smaller 50×50 m^2 (500×500 pixel) image tiles, creating five datasets used in this work. All datasets have roads removed, as described in Sect. 4.1. These datasets are:

1. RGB images with trees (abbreviated as RGB original),
2. RGB images without trees (RGB-T, see Sect. 4.1),
3. RG-DSM images with trees (RG-PC, see Sect. 4.1),
4. RG-NIR with trees (RG-NIR, see Sect. 4.1) and
5. RG-NIR containing only the side of the roads (RG-NIR-RS, see Sect. 4.1).

4.2   Image Annotation Step

We use the Labelme [12, 18] tool to annotate the parking spaces as polygons, (see Fig. 4), particularly after removing roads as described in Sect. 4.1. However, some moving cars may still appear in the images due to imperfect alignment with OSM data or instances of partially visible cars, and we refrain from annotating instances where certainty about parked vehicles is lacking. To expand our annotated data, we employ a semi-supervised learning approach. We train Detectron2's Mask RCNN FPN [19] on the manually annotated data and refine its predictions on new data. We alter these annotations where necessary and add the missing annotations. This iterative process yields a substantial set of annotated training images, totaling 905 instances of occupied parking spaces and 522 instances of empty parking spaces.

Fig. 4. In (a), we illustrate annotations for occupied (red) and vacant (green) parking spaces. Image (b) displays the RG-NIR-RS dataset that contains only roadside areas, we mark vacant spaces in green and occupied spaces in red. In (a), individual spaces are annotated separately, while in (b), continuous stripes are marked as one polygon. (Color figure online)

4.3   Post-processing Steps

After we get the prediction masks from the Mask RCNN FPN We apply post-processing step to detect the false positive predictions generated by the model. We apply a LRM to identify spots wrongly predicted as parking spaces.

To apply LRM, we choose a random 1×1 km^2 image tiles and manually mark each of the predictions whether they are true or false positives. The image tile we choose contains 1282 instances of parking space predictions. Afterwards, within this image tile, we calculate the *area of each predicted parking space* and *its distance to the nearest parking space prediction*. These two are the input variables of the LRM.

We set the LRM parameters such as alpha, solver and max_iter that give the highest precision value since we aim to minimise false positives. After that, we mark each parking spot as true positive or false positive depending on the output of the LRM.

We notice that parking spaces that are categorised as false positives by the LRM but actually are true positives are mostly small and isolated parking spaces. These may belong to private households or are spots inside of buildings, see Fig. 5. Since our main objective is not to miss any of the public parking spaces, losing some of the small private parking spaces is acceptable. After removing the false positives we end up with one space map as seen in Fig. 5 (c).

4.4   Creating the Parking Space Map

Finally, to create a parking space map we save the model predictions as polygons together with their geo-reference coordinates and merge them according to the geo-reference coordinates. This results in a parking space map of our area of interest. Figure 5 shows an exemplary parking space map.

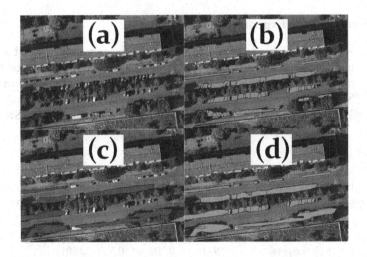

Fig. 5. In (a), we present an unaltered DOP image. In (b), we display the model predictions of the parking spaces of Image (a). Blue polygons indicate true positives, while red ones represent false positives. Image (c) shows the parking space stripes predicted by the model when it was trained with the roadside parking spaces dataset, and in (d), we merge parking spaces from (b) and (c) after eliminating false positives. (Color figure online)

5   Experiments

5.1   Experimental Setup

In this section we empirically evaluate our approaches. To conduct our experiments, we use *Google Colaboratory Pro*[3]. It offers an Intel(R) Xeon(R) @ 2.30 GHz CPU, a Nvidia Tesla T4 GPU and 25 GB RAM.

For our final predictions we use the Detectron2 Mask RCNN FPN's model (eventually extended by the LRM). To evaluate it we calculate the *accuracy, precision, recall,* and *F1 score* to compare the model's prediction results on our datasets. Thereby, the performance on the RBG original serves as a benchmark. To calculate the matrices we use the predictions on image tiles that cover an area of 1×1 km^2. We separate the total number of annotations of both classes mentioned in Sect. 4.2 into training and test sets, maintaining a ratio of 2:1.

We use Detectron2 Mask RCNN FPN model's base configuration (hyperparameters) to train the models except we set images per batch = 2, base learning rate = 0.0025 and maximum iteration = 700 in the model run across all the datasets. We tested different values for the hyperparameters mentioned and selected above as they gave us good quantitative results and more importantly they gave us best qualitative results when looking at the predicted bounding boxes overlaid on the images tiles.

5.2   Evaluation of the Experiments

We collect our results in Table 1. To calculate the performance metrics we use the area of the prediction polygons and true label polygons. We see that the approaches combining NIR information and post-processing using the LRM outperform the performance of Detectron2's Mask RCNN FPN's alone on the dataset. We apply LRM only on the datasets RG-NIR and RG-NIR-RS because these are the datasets that MASK RCNN has the best accuracy and precision values.

Table 1. Evaluation of the Mask RCNN model on the discussed datasets, see Sect. 4.1. Applying the LRM to the results of Mask RCNN model (as described in Sect. 4.3) is indicated as "+ LRM". The highlighted results refer to the dataset on which the model has the highest performance.

Dataset	Accuracy	Precision	Recall	F1 score
RGB original	0.941	0.306	0.580	0.401
RGB-T	0.965	0.493	**0.778**	0.604
RG-NIR	0.976	0.639	0.643	0.641
RG-PC	0.972	0.596	0.560	0.578
RG-NIR-RS	0.963	0.709	0.415	0.463
RG-NIR + LRM	**0.977**	0.673	0.619	**0.645**
RG-NIR-RS + LRM	0.949	**0.710**	0.580	0.638

---

[3] https://colab.research.google.com/.

The performance of Detectron2's Faster RCNN FPN continuously improves from red-green-blue images containing trees, red-green-blue images without trees, red-green and point cloud height information (instead of blue) images, and red-green and NIR images, see Table 1. This indicates that we do not necessarily have to remove trees from the images to improve the model performance, but should rather provide additional wavelength information in the images themselves. Therefore, we conclude that the approach of adding NIR channel data to the RGB aerial images can be used to improve the model's ability to differentiate different types of objects (e.g., trees, buildings, and vehicles) visible on the images.

Adding point cloud height data to the RGB images does not improve the model's performance as expected. We believe this is due to: the resolution of the point cloud data is 10 times lower than the resolution of DOP images, the noise contained in point cloud data, and DOP images and point cloud data being acquired at different times. Therefore, when we fused red and green channel data with point cloud height data, the final fused image could contain contradicting information (see Sect. 4.1). We suggest for future work to find ways to better bridge these modality gaps.

We get the highest precision for the dataset with images containing NIR information, see Table 1. We conclude that these (RG-NIR dataset) contain the most important information for training Detectron2's Faster RCNN FPN.

We get a better recall result for RGB-T dataset than the RG-NIR dataset. In RGB-T dataset, NIR channel is already used to remove tree pixels in the images. Low precision and high recall are observed because in RGB-T dataset pixels/area is less. There is lesser area available to detect parking spaces. From the available area, we detect many parking spaces but most of the predicted labels are false positives.

6 Conclusion and Future Work

When it comes to identifying parking spaces in airborne images, there are some obstacles that can affect the performance. Not only they contain trees, but also distinguishing driving cars, that is, cars which are not parked but on the road, from actually parked cars is difficult. In this work, we study different solutions to such problems in order to obtain a reliable parking space map. In particular, we use different methods to remove moving cars, such as removing roads from the images using OSM data, or detect non-parking cars. Furthermore, we study different modes of representation regarding the images. In particular, we add NIR information and point cloud height data to the aerial images. Lastly, we also apply a post-processing step to reduce the number of false positives in the models predictions.

Our method can also be applied to aerial images that do not contain NIR or spatial information. In these cases, one could follow the image prepossessing steps of the RGB original dataset creation and then apply the post-processing steps (see relevant steps in Fig. 2).

Future work may focus on extending this research. For example, one could rely on data obtained at different points in time. If the images are taken in timely proximity, moving cars may be detected as they would be at a slightly different position in the images. With the recent advent in airborne image acquisition techniques, one can expect high-resolution images in the near future that allow parking-spot detection from space. In addition, image acquisitions at various seasons, for example during the winter season, will help to detect parking spots in partly vegetated terrain. Further future work may focus on applying a similar approach to a real-time implementation of parking space maps. For example, similar image-analysis and filtering approaches can be carried out on unmanned aerial vehicles (UAV, aka drones) images with higher repetition rates.

References

1. Ahrnbom, M., Åström, K., Nilsson, M.G.: Fast classification of empty and occupied parking spaces using integral channel features. In: IEEE Conference on Computer Vision and Pattern Recognition Workshops (CVPR), pp. 1609–1615. IEEE Computer Society (2016)
2. Mahaarachchi, B.: Mapping on street parking from airborne images (2023). https://github.com/Bashinim/Mapping_On_Street_Parking_from_Airborne_Images
3. Cleverciti Global: Reduce unnecessary emissions - Clever parking for a healthy climate (2021). https://www.cleverciti.com/en/why-cleverciti/reduce-unnecessary-emissions
4. Fensholt, R., Sandholt, I., Stisen, S.: Evaluating MODIS, MERIS, and VEGETATION vegetation indices using in situ measurements in a semiarid environment. IEEE Trans. Geosci. Remote Sens. 44(7), 1774–1786 (2006)
5. Jones, M., Li, L.: Vacant parking spot detection (2012)
6. Kabak, M.O., Turgut, O.: Parking spot detection from aerial images. Stanford University, Final Project Autumn 2010, Machine Learning class (2010)
7. Koga, Y., Miyazaki, H., Shibasaki, R.: A method for vehicle detection in high-resolution satellite images that uses a region-based object detector and unsupervised domain adaptation. Remote. Sens. 12(3), 575 (2020)
8. Li, W., Cao, L., Yan, L., Li, C., Feng, X., Zhao, P.: Vacant parking slot detection in the around view image based on deep learning. Sensors 20(7), 2138 (2020). https://doi.org/10.3390/s20072138
9. Liu, X.: Airborne lidar for dem generation: some critical issues. Prog. Phys. Geogr. 32(1), 31–49 (2008)
10. Luo, X., et al.: Fast automatic vehicle detection in UAV images using convolutional neural networks. Remote. Sens. 12(12), 1994 (2020)
11. OpenStreetMap contributors: Openstreetmap (2017). https://www.openstreetmap.org
12. Russell, B.C., Torralba, A., Murphy, K.P., Freeman, W.T.: LabelME: a database and web-based tool for image annotation. Int. J. Comput. Vis. 77(1–3), 157–173 (2008)
13. Senate Department Berlin: Dop-d - digital orthophotos Germany (2020). https://www.berlin.de/sen/sbw/stadtdaten/geoportal/landesvermessung/geotopographie-atkis/dop-digitale-orthophotos/
14. Senate Department Berlin: Senate Department for Urban Development and Housing (2020). https://www.stadtentwicklung.berlin.de/index_en.shtml
15. Seo, Y., Ratliff, N.D., Urmson, C.: Self-supervised aerial image analysis for extracting parking lot structure. In: Boutilier, C. (ed.) IJCAI 2009, Proceedings of the 21st International Joint Conference on Artificial Intelligence, Pasadena, California, USA, July 11–17, 2009, pp. 1837–1842 (2009). https://ijcai.org/Proceedings/09/Papers/305.pdf
16. Shoup, D.: Pricing curb parking. Transp. Res. Part A: Policy Pract. 154, 399–412 (2021)
17. Singh, C., Christoforou, C.: Detection of vacant parking spaces through the use of convolutional neural network. In: The International FLAIRS Conference Proceedings, vol. 34 (2021)
18. Wada, K.: labelme: image polygonal annotation with python (2016). https://github.com/wkentaro/labelme

19. Wu, Y., Kirillov, A., Massa, F., Lo, W.Y., Girshick, R.: Detectron2 (2019). https://github. com/facebookresearch/detectron2
20. Yan, Q., Feng, T., Timmermans, H.: Investigating private parking space owners' propensity to engage in shared parking schemes under conditions of uncertainty using a hybrid random-parameter logit-cumulative prospect theoretic model. Transp. Res. Part C: Emerg. Technol. **120**, 102776 (2020)

Hebbian Learning-Guided Random Walks for Enhanced Community Detection in Correlation-Based Brain Networks

Roberto C. Sotero[1]([✉]) [iD] and Jose M. Sanchez-Bornot[2] [iD]

[1] Department of Radiology, and Hotchkiss Brain Institute, University of Calgary, Calgary, AB, Canada
roberto.soterodiaz@ucalgary.ca

[2] Intelligent Systems Research Centre, Ulster University, Derry Londonderry, UK

Abstract. Community detection in complex signed networks is a significant challenge, traditionally addressed using the Louvain method directly applied to the correlation matrix. This study introduces a two-tier approach that integrates a Hebbian learning rule within an adaptive signed random walk (ASRW) framework, then applies the Louvain method to the final weight matrix. This approach refines the network analysis process, providing a new tool for exploring community structure. Tested extensively on synthetic signed networks with defined community structures, our methodology consistently outperformed the traditional Louvain approach, particularly when communities were less clearly demarcated. Further application to resting-state functional MRI data from the ABIDE Preprocessed Initiative highlighted functional connectivity differences between neurotypical individuals and those diagnosed with Autism Spectrum Disorder (ASD). Our approach found key areas of significant difference, including several cerebellum regions, consistent with existing ASD literature. Our findings underscore the potential of the proposed technique to advance community detection in correlation-based networks.

Keywords: Community Detection · Hebbian Learning · Random Walks · Brain Networks · Autism Spectrum Disorder

1 Introduction

The identification of communities within brain networks is a pivotal task in neuroscience, providing insights into the functional and structural organization of the brain [1]. Communities, or modules, can elucidate the segregation and integration of information in the brain, offering crucial understanding into the brain's normal functioning [2] and potential abnormalities seen in neurological and psychiatric disorders [3]. A variety of methods have been developed for detecting communities in networks, ranging from optimization-based to dynamics-based methods [4]. However, finding a reliable, efficient, and robust community detection method remains a challenging problem, especially in the context of brain networks [5].

© The Author(s), under exclusive license to Springer Nature Switzerland AG 2023
P. Quaresma et al. (Eds.): IDEAL 2023, LNCS 14404, pp. 222–232, 2023.
https://doi.org/10.1007/978-3-031-48232-8_21

The conventional approach to community detection in brain networks often relies on the absolute value of the functional connectivity matrix, encapsulating statistical associations between various brain regions, commonly computed via correlation or coherence measures [6]. The absolute value is typically employed to yield a non-negative graph, thereby fulfilling the prerequisites for numerous community detection algorithms [4]. However, such a practice risks oversimplifying the nuanced neurobiological processes underpinned by positive and negative connections, thus potentially discarding valuable information about brain organization [7]. Recognizing this limitation, several adaptations to the modularity measure, a prevalent metric in community detection, have been presented to accommodate both positive and negative connections. For example, Rubinov and Sporns [8] introduced a general framework of modularity that is compatible with networks bearing both positive and negative connections. They contended that this approach offers a superior representation of the structure of complex brain networks compared to the conventional modularity measure. Rubinov and Sporns' method considers positive connections as constituents of communities and negative connections as the formation of 'anti-communities.' In another example, Gomez et al. [9] introduced an extension to the modularity measure, accommodating both positive and negative connections in networks, and thus providing a more nuanced perspective on community structure. However, their methodology requires a priori knowledge of the number of communities and does not offer an interpretation of the implications of negative weights in brain connectivity.

This study presents a unique approach to community detection, designed specifically for correlation-based brain networks. We enrich the adaptive signed random walk (ASRW) algorithm [10] by considering not just the strength but also the sign of the connections. This nuanced strategy allows for a more comprehensive examination of network structures. ASRW dynamically adjusts in response to the movements of random walkers within the network. By incorporating a Hebbian learning-inspired rule, we further amplify the effectiveness of our method. This rule suggests that the transition of a walker from one node to another influences the strength of their connection, effectively modifying its weight according to a learning rate, denoted as α. Consequently, for positive weights, a walker's transition from one node to another amplifies the connection between the two nodes. In contrast, for negative weights, such a transition diminishes the connection. The resulting weight adjustment influences future groupings of nodes, echoing the Hebbian principle of "neurons that fire together, wire together." After these dynamic adjustments, we apply the Louvain method to the final weight matrix. We validated our methodology using synthetic signed networks with predefined community structures and contrasted our results with the conventional approach that applies the Louvain method to the correlation matrix [8]. Furthermore, we applied our method to resting-state fMRI data from neurotypical individuals and those diagnosed with Autism Spectrum Disorder (ASD). By identifying functional connectivity differences between these two groups, we offer potential insights into the neurobiology of ASD while demonstrating the potency of our method in decoding complex network data.

2   Methods

2.1   Adaptive Signed Random Walk (ASRW)

In the ASWR model [10] transition probabilities are computed based on the functional connectivity matrix weights (C) and a similarity measure between the activities of the source and destination nodes. The primary concept is to encourage transitions between nodes exhibiting similar activities, taking into account whether network weights are positive or negative. Node activity is defined as the number of walkers visiting the node. Initially ($t = 0$), each of the K walkers is assigned to one of the N brain areas, yielding an initial activity $A_i(0)$ for each node i. Given these activities, the subsequent transition probabilities $P_{ij}(t + 1)$ and activities $A_i(t + 1)$ are computed as follows:

1. For each node i, we calculate the adjusted weights for all its neighboring nodes j as follows:
   a. We compute the similarity measure $S_{ij}(t)$ between the activity of node i and node j at time t:

$$S_{ij}(t) = 1 - \left| \frac{A_i(t) - A_j(t)}{max(A_i(t), A_j(t))} \right| \tag{1}$$

   b. If the weight C_{ij} is positive, the adjusted weight is $\left(\tilde{C}_{ij}(t) \geq 0 \right)$:

$$\tilde{C}_{ij}(t) = C_{ij} S_{ij}(t) \tag{2}$$

   If the weight C_{ij} is negative, the adjusted weight is:

$$\tilde{C}_{ij}(t) = |C_{ij}| \left(1 - S_{ij}(t) \right) \tag{3}$$

3. The transition probability $P_{ij}(t + 1)$ of a walker moving from node i to each neighboring node j in the next time step $t + 1$ is computed as:

$$P_{ij}(t + 1) = \tilde{C}_{ij}(t) / \sum_{j=1}^{N} \tilde{C}_{ij}(t) \tag{4}$$

   Thus, a walker is more likely to travel through a positive connection C_{ij} if the activities in nodes i and j are similar, while it is more likely to take a negative connection if the activities are dissimilar.
4. The random walkers move based on the computed probabilities $P_{ij}(t + 1)$, and the next activites $A_i(t + 1)$ are the new number of walkers at each node.
   The entire simulation process is iterated for T time steps.

2.2   Extending the ASRW With a Hebbian Learning-Inspired Strategy

In this work we incorporate a Hebbian learning rule into the ASRW implemented on a correlation-based brain network. As the random walkers move in the network, we update the network weights based on the Hebbian learning principle stating, "neurons that fire together, wire together" [11]. In the context of our model, the transition of a walker strengthens the connection between two nodes, effectively changing its weights. This weight change is governed by a learning rate, symbolized as α. Hence, the weights updates are defined as follows:

For positive weights, if a walker transits from node i to j:

$$C_{ij}(t) = C_{ij}(t) + \alpha \tag{5}$$

For negative weights, if a walker transits from node i to j:

$$C_{ij}(t) = C_{ij}(t) - \alpha \tag{6}$$

This increases highly transited positive weights and makes highly transited negative weights more negative. Upon conclusion of the walkers movement, the resulting weight matrix is normalized within the range of $[-1,1]$ to facilitate comparison with the initial correlation matrix.

The application of the Hebbian learning rule allows the weights of the network to evolve dynamically, reflecting the actual paths taken by the walkers and the patterns of activity in the network. This adaptive process enhances the strengths of relevant connections that are frequently used and weakens the connections that are not, leading to a more meaningful representation of the functional relationships between the nodes. The resulting final weight matrix, therefore, captures not only the static correlations but also the dynamic interplay and co-activation patterns between different nodes in the network. It emphasizes functionally relevant paths and connectivity patterns that may not be apparent in the initial correlation-based network.

2.3 Community Detection Based on the Final Weight Matrix

The final weight matrix $C_{ij}(T)$, derived from our ASRW-Hebbian methodology, serves as an optimal input for the Louvain community detection method. By emphasizing not only the nodes with a high correlation but also those that show persistent co-activation over time, the matrix presents a more accurate representation of the network's modular structure. These aspects are critical to defining functional communities. The Louvain method, renowned for optimizing network modularity [8], unveils intricate community structures when applied to our dynamically refined network, which may have been concealed in the initial correlation-based network. In our analysis, we utilize the Louvain method for signed networks, specifically as implemented in the Brain Connectivity Toolbox (BCT) [12]. It's important to note that the modularity function in this method contains a resolution parameter, γ. Conventionally, γ is set to 1, however, it could technically assume any positive value. The optimal choice of γ remains an open question, as there is no universally accepted selection criterion [5]. In light of this, we have conservatively selected $\gamma = 1$ for all analyses presented in this paper.

Upon the partition of nodes into communities or modules, we calculate the participation coefficient for each node. This measure describes the distribution of a node's connections across communities [13]. A participation coefficient nearing 1 implies a uniform distribution of a node's connections over communities, whereas a value close to zero indicates that the majority of a node's connections are within its own community. We utilize the code provided by the BCT to compute the participation coefficients.

3 Results

3.1 Synthetic Benchmarks

Our methodology was evaluated using synthetic signed networks with a predetermined community structure. These networks were constructed such that communities, or groups of nodes, are densely interconnected by positive edges, while negative edges predominantly exist between different communities. This design reflects the premise that most positive edges are found within communities, while most negative edges span between distinct communities [8]. To generate the networks we manipulated six parameters: the number of nodes in the network (N), the number of communities (k), the intra-community probability of a positive edge (p_{intra}^+), the intra-community probability of a negative edge (p_{intra}^-), the inter-community probability of a positive edge (p_{inter}^+), and the inter-community probability of a negative edge (p_{inter}^-). Performance assessment of our methodology was carried out by comparing the generated community partitions to the ones estimated by our method. This comparison used the normalized mutual information (MI), following the implementation provided by the BCT.

In our initial test (Fig. 1), we evaluated the proposed methodology under specific conditions: $N = 100$, $k = 5$, the number of time steps $T = 1000$, and the number of random walkers $K = 10^4$, $p_{inter}^+ = 0.1$, $p_{intra}^- = 0.1$. We varied p_{inter}^- in the range of 0.1 to 1 with increments of 0.1, p_{intra}^+ within the same range and increments and the learning rate α was set to one of [0.1, 0.25, 0.5, 1]. For each parameter combination we generated 100 synthetic networks with predetermined community structure and applied our Hebbian-learning ASRW. Subsequently, we used the Louvain method on the final weight matrix, averaging the results over the 100 iterations. The normalized MI between the known and estimated community partitions for each parameter combination are presented in Fig. 1A. For comparative purposes, we applied the Louvain method to the original networks and computed the MI (denoted as MI_0) as shown in Fig. 1B. The percentage change between our method and the standard Louvain approach ($100(MI/MI_0 - 1)$) is shown in Fig. 1C. Non-significant values were set to 0 and are displayed in white. Our findings reveal that when p_{intra}^+ is high (approaching 1) applying the Louvain method to both the modified and original weight matrices yield comparable results. In fact, in some cases, the original matrix performs slightly better. This outcome is expected as high p_{intra}^+ values indicate well-defined community structures. This is reflected in the close-to-one normalized MI values obtained by both methods (Fig. 1B and Fig. 1B). For very low p_{intra}^+ values (0.1–0.3), the community structures are loose and both approaches yield similarly low MI values. Nevertheless, for p_{intra}^+ values between 0.4 and 0.7, our methodology significantly improves community detection over the standard approach, with the greatest improvement (61.5%) being observed at the lowest $p_{inter}^+ = 0.1$. These results underscore the superiority of our methodology over the standard approach when community structures are not strongly defined.

In a subsequent experiment (Fig. 2), we increased the number of communities in the synthetic network to 10 and carried out a simulation analogous to that described for Fig. 1. This led to qualitatively similar results; however, the mutual information (MI) values were somewhat lower. The maximum improvement achieved by our method over the conventional approach was 35.2%. An increase in the number of communities introduces greater complexity and potential divisions for the algorithm to navigate, which can lead to misclassifications and a resultant decrease in MI score. This is further compounded if communities are loosely defined or if there is considerable overlap or noise in the network data. Smaller communities, which may become more prevalent as the total number of communities increases, may be particularly challenging to detect accurately, further contributing to the reduction in MI score.

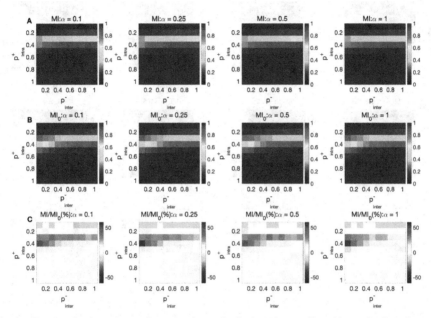

Fig. 1. Community detection performance in synthetic signed networks with varying learning rate and community structure. A) The normalized Mutual Information (MI) between 5 known community partitions and those estimated by applying the Louvain method to the final weight matrix, averaged over 100 network iterations. B) For comparison, the MI (denoted as MI_0) achieved by applying the Louvain method directly to the original network. C) The percent change in MI (MI/MI0) between our proposed method (A) and the standard Louvain approach (B). Non-significant values have been set to 0 and are displayed in white.

Ultimately, we found no distinct correlation between the learning rate and the achieved MI score, hinting at a complex interaction between the learning rate and the parameters that define the community structure.

3.2 Correlation-Based Functional Connectivity Networks Estimated from fMRI Data

In this study, resting-state fMRI (rs-fMRI) data from 948 subjects (457 with ASD and 491 neurotypical individuals) were acquired from the ABIDE Preprocessed Initiative [14]. To avoid potential biases introduced by custom preprocessing pipelines, we utilized data preprocessed via the C-PAC pipeline provided by ABIDE; the specifics of this preprocessing procedure are detailed elsewhere [15]. Pearson correlation coefficients were then calculated between each pair of time series from the 116 brain regions defined in the Automated Anatomical Labeling (AAL) atlas, generating functional connectivity matrices. Connections with False Discovery Rate (FDR)-adjusted p-values exceeding 0.05 were deemed non-significant and set to zero.

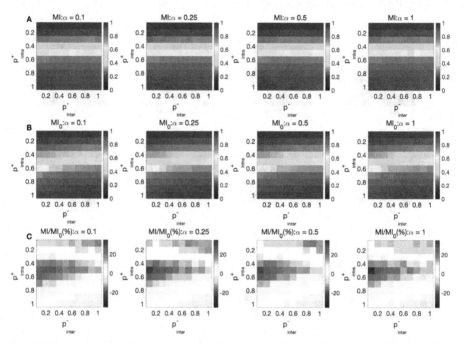

Fig. 2. Community detection performance with an increased number of communities. The simulation was performed similarly to Fig. 1, but with the number of communities increased to 10.

With the functional connectivity matrices in hand, we applied the community extraction methodology proposed in this paper to each subject's data, setting parameters as follows: $T = 1000$, $K = 10^4$, and $\alpha = 0.5$. This process was repeated 40 times to produce an ensemble of 40 community partitions. A consensus partition was then established using the algorithm proposed by Lancichinetti and Fortunato [16], as implemented in BCT. We utilized this consensus partition to calculate the participation coefficients from positive (Ppos) and negative (Pneg) weights.

Figure 3A and B illustrate the calculated Ppos values for each of the 116 brain regions in neurotypical and ASD subjects, respectively. Figure 3C highlights the 15 areas where significant differences in Ppos values were observed between neurotypical and ASD subjects. The most pronounced difference was noted in the 'Heschl R' region, an area associated with auditory processing, which has previously been reported to exhibit connectivity variations in individuals with autism [20], potentially explaining the sensory processing abnormalities frequently observed in ASD. In a similar manner, Fig. 4 presents the calculated Pneg values. In this analysis, 27 brain regions exhibited significant

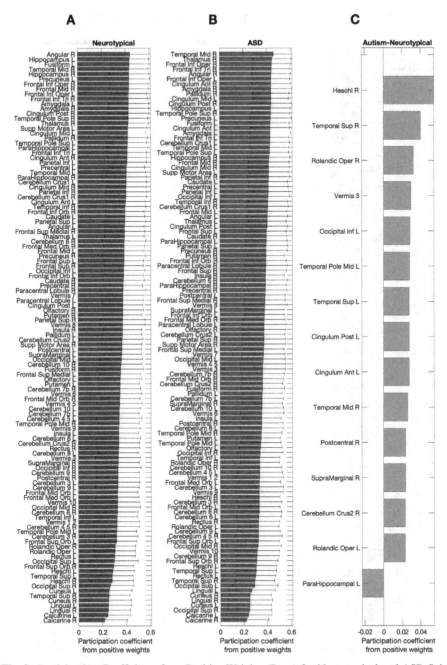

Fig. 3. Participation Coefficients from Positive Weights (Ppos) for Neurotypical and ASD Subjects. (A) The calculated Ppos for each of the 116 brain regions in neurotypical subjects. (B) The calculated Ppos for each of the 116 brain regions in ASD subjects. (C) The 15 brain regions demonstrating significant differences in Ppos values between neurotypical and ASD subjects. 'L' and 'R' denote left and right hemispheres, respectively.

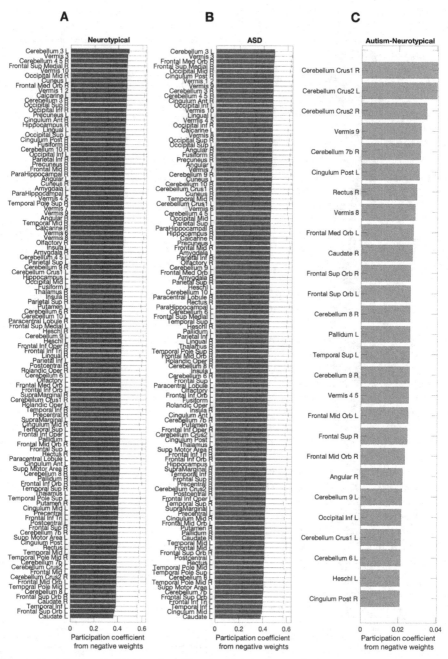

Fig. 4. Participation Coefficients from Negative Weights (Pneg) for Neurotypical and ASD Subjects. (A) The calculated Pneg for each of the 116 brain regions in neurotypical subjects. (B) The calculated Pneg for each of the 116 brain regions in ASD subjects. (C) The 27 brain regions demonstrating significant differences in Pneg values between neurotypical and ASD subjects. 'L' and 'R' denote left and right hemispheres, respectively.

differences between neurotypical and ASD individuals. One such region was 'Heschl L'. Notably, the most marked differences were found in two regions of the cerebellum. Of the 27 brain regions demonstrating significant differences, nine were in the cerebellum. This aligns with numerous studies reporting both structural and functional connectivity variations between individuals with autism and neurotypical counterparts [17, 18].

4   Conclusions and Future Work

In this study, we presented a novel technique for community detection in correlation-based brain networks, introducing a Hebbian learning rule into an ASRW [10]. In contrast to traditional methods, our approach enhances the Louvain method for signed networks [8] by dynamically adjusting connection weights based on node similarities and random walker transitions, applying it to the final, rather than the initial (correlation matrix), weight matrix. This methodology enables a more intricate and accurate assessment of complex network structures.

We rigorously evaluated our method on synthetic signed networks with controlled community structures, revealing marked improvements in community detection over the standard Louvain method, particularly for less clearly defined communities. As the number of communities increased, introducing greater complexity into the network structure, community detection algorithms faced additional challenges, reflected in lower normalized mutual information scores. This finding emphasizes the need for sophisticated community detection techniques, such as our proposed method, for accurately mapping complex network structures. Notably, our approach was tested with a modularity resolution of $\gamma = 1$ in the Louvain method, leaving room for future exploration at different resolution levels. We also applied our method to real-world data from resting-state functional MRI scans, demonstrating its practical relevance. By comparing functional connectivity in neurotypical individuals and those diagnosed with ASD, our approach effectively highlighted significant connectivity differences in key regions, aligning with existing ASD literature.

In conclusion, our approach represents a significant stride in community detection within complex signed networks. By incorporating Hebbian learning principles into the ASRW framework, we offer a powerful tool for dissecting intricate network structures. Our method proved successful in highlighting significant functional connectivity differences in real-world neuroscience data, potentially revealing novel insights into neurological conditions like ASD. Future efforts will refine and expand this methodology and apply it to other complex systems, further probing its versatility and robustness.

Acknowledgements. This work was supported by Grant 222300868 from the Alberta Innovates LevMax program, and by RGPIN-2022-03042 from Natural Sciences and Engineering Research Council of Canada.

References

1. Sporns, O.: Networks of the Brain | mitpressbookstore. 02 Dec 2016. https://mitpressbook store.mit.edu/book/9780262528986. Accessed 08 Jun 2023

2. Bassett, D.S., Bullmore, E.: Small-world brain networks. Neuroscientist **12**(6), 512–523 (2006). https://doi.org/10.1177/1073858406293182
3. Stam, C.J.: Modern network science of neurological disorders. Nat. Rev. Neurosci. **15**(10), 683–695 (2014). https://doi.org/10.1038/nrn3801
4. Fortunato, S.: Community detection in graphs. Phys. Rep. **486**(3), 75–174 (2010). https://doi.org/10.1016/j.physrep.2009.11.002
5. Betzel, R.F., et al.: The modular organization of human anatomical brain networks: accounting for the cost of wiring. Netw. Neurosci. **1**(1), 42–68 (2017). https://doi.org/10.1162/NETN_a_00002
6. Fornito, A., Zalesky, A., Bullmore, E.: Fundamentals of brain network analysis. In: Fundamentals of Brain network Analysis, pp. xvii, 476. Elsevier Academic Press, San Diego, CA, US (2016)
7. Kazeminejad, A., Sotero, R.C.: The importance of anti-correlations in graph theory based classification of autism spectrum disorder. Front. Neurosci. **14**, 676 (2020). https://doi.org/10.3389/fnins.2020.00676
8. Rubinov, M., Sporns, O.: Weight-conserving characterization of complex functional brain networks. Neuroimage **56**(4), 2068–2079 (2011). https://doi.org/10.1016/j.neuroimage.2011.03.069
9. Gómez, S., Jensen, P., Arenas, A.: Analysis of community structure in networks of correlated data. Phys. Rev. E **80**(1), 016114 (2009). https://doi.org/10.1103/PhysRevE.80.016114
10. Sotero, R.C., Sanchez-Bornot, J.M.: Exploring Correlation-Based Brain Networks with Adaptive Signed Random Walks. bioRxiv, p. 2023.04.27.538574 (2023). https://doi.org/10.1101/2023.04.27.538574
11. "Hebb, D. O. The organization of behavior: A neuropsychological theory. New York: John Wiley and Sons, Inc., 1949. 335 p. $4.00. Science Education **34**(5), 336–337 (1950). https://doi.org/10.1002/sce.37303405110
12. Rubinov, M., Sporns, O.: Complex network measures of brain connectivity: Uses and interpretations. Neuroimage **52**(3), 1059–1069 (2010). https://doi.org/10.1016/j.neuroimage.2009.10.003
13. Guimerà, R., Nunes Amaral, L.A.: Functional cartography of complex metabolic networks. Nature **433**(7028), 895–900 (2005). https://doi.org/10.1038/nature03288
14. Cameron, C., et al.: The Neuro Bureau Preprocessing Initiative: open sharing of pre-processed neuroimaging data and derivatives. Front. Neuroinform. Conference Abstract: Neuroinformatics 2013 (2013). https://doi.org/10.3389/conf.fninf.2013.09.00041
15. Kazeminejad, A., Sotero, R.C.: Topological properties of resting-state fMRI functional networks improve machine learning-based autism classification. Front. Neurosci. **12**, 1018 (2019)
16. Lancichinetti, A., Fortunato, S.: Consensus clustering in complex networks. Sci. Rep. **2**(1), 336 (2012). https://doi.org/10.1038/srep00336
17. Stanfield, A.C., McIntosh, A.M., Spencer, M.D., Philip, R., Gaur, S., Lawrie, S.M.: Towards a neuroanatomy of autism: a systematic review and meta-analysis of structural magnetic resonance imaging studies. Eur. Psychiatry **23**(4), 289–299 (2008). https://doi.org/10.1016/j.eurpsy.2007.05.006
18. Khan, A.J., Nair, A., Keown, C.L., Datko, M.C., Lincoln, A.J., Müller, R.-A.: Cerebro-cerebellar resting state functional connectivity in children and adolescents with autism spectrum disorder. Biol. Psychiatry **78**(9), 625–634 (2015). https://doi.org/10.1016/j.biopsych.2015.03.024

Extracting Automatically a Domain Ontology from the "Book of Properties" of the Archbishop's Table of Braga

José Pedro Carvalho[1] , Orlando Belo[1(✉)] , and Anabela Barros[2]

[1] ALGORITMI Research Centre / LASI, University of Minho, 4710-059 Braga, Portugal
zpscarvalho2@gmail.com, obelo@di.uminho.pt
[2] CEHUM, Centre for Humanistic Studies, University of Minho, 4710-059 Braga, Portugal
aldb@elash.uminho.pt

Abstract. During the last years, many researchers used ontology-based technologies for developing ontology-learning processes for acquiring knowledge included in textual information. In our case, we focused such processes to study the contents of Portuguese manuscripts from the 17th century, aiming to explore the formalities of the language and characterize the elements contained in the texts. In particular, we worked over a very specific manuscript. In this paper, we present and describe the implementation of a semi-automatic ontology learning process for extracting the knowledge contained in the "Book of Properties". This manuscript includes a detailed inventory of assets belonging to the Archbishop's Table of Braga at the beginning of the 17th century. Our goal is to provide a fundamental computational instrument for studying and learning the geography, culture, agriculture, history, and genealogy in Portugal at that time.

Keywords: Ontology Learning · Linguistic Ontologies · Extracting Knowledge from Textual Data · Natural Language Processing · Machine Learning

1 Introduction

During the last decade the domains of information retrieval [1], natural language processing [2] and machine learning [3] have evolved immensely. They contribute to rise new models, methods and techniques that are fundamental for implementing analy-sis processes for textual data sources, and providing expeditious forms for identifying, characterizing and representing data elements in such sources, revealing their knowledge. These techniques are very adequate to analyse large volumes of texts, which allows for the extraction of concepts, properties, relationships and axioms in a very effective way. Thus, ontology learning from texts [4, 5] had a big boost and developed very sharply over the last few years. The dissemination of these advances and the regular and systematic use of ontology learning processes in other application fields have contributed immensely to researchers in the field of semantic analysis and extraction of knowledge models from textual sources, whether historians, linguists or archaeologists, to start paying attention to the models and techniques of ontology learning from texts. We were no exception.

© The Author(s), under exclusive license to Springer Nature Switzerland AG 2023
P. Quaresma et al. (Eds.): IDEAL 2023, LNCS 14404, pp. 233–244, 2023.
https://doi.org/10.1007/978-3-031-48232-8_22

Over the past few years, some research work was done in the study and analysis of a Portuguese manuscript from the 17th century, entitled "The Book of Properties" [6, 7]. This book contains a very interesting inventory of assets belonging to the Archbishop's Table of Braga at the beginning of the 17th century, having a detailed description of all the rural and urban properties in four regions, mostly located in the north of Portugal, but extending to the Galicia and Santarém regions. In order to reduce the time of analysis of the texts and the identification and characterization of the various data elements related to the properties described in the manuscript, we decided to develop in a supervised way a specific ontology, using automatic means of natural language processing and machine learning. The goal was to get a knowledge domain ontology with the ability to characterize the knowledge expressed in texts of the "Book of Properties", incorporating a large diversity of properties' elements that can be used for the studying and learning the geography, culture, agriculture, economy, architecture, religion, history, and genealogy in Portugal at the 17th century.

In this paper, we present and discuss the implementation of a (semi)automatic ontology learning system we conceived, for extracting the knowledge about the properties of the inventory recorded in the "Book of Properties". The implementation of a (semi)automatic process for learning an ontology from the texts of the book implies its study and methodical preparation, actions that are commonly supported by models, techniques and tools of natural language processing, followed by text intensive analysis tasks, carried out through very specific machine learning processes. The remaining part of this paper is organized as follows. Section 2 exposes some related work about ontology learning processes, Section 3 presents our application case, Section 4 presents and discusses the ontology learning process implemented for constructing an ontology for the "Book of Properties", and, finally, Section 5, presents some conclusions and research lines for future work.

2 Background

Ontologies [8, 9] are not new in Computer Science. Yet, their definition stills often under discussion. The works [10–12] provided very interesting definitions for an ontology, justifying them generally as a mean for representing and sharing knowledge about a given application domain, which involves the conceptualization of a semantic network of entities and their relationships. Currently, we can find ontologies in many application domains, acting as useful instruments supporting knowledge bases of many problem-solving systems. In the field of biology, for example, [13] warned the importance of ontologies in structuring large amounts of information so that it can be consulted efficiently, especially when it is increasing at a high rate. In another domain, Health, [14] showed how ontologies can be used to help automate processes in health institutions. Finally, in Management and Business, [15] stated that the use of ontologies is necessary to interconnect different pieces of knowledge that can be used to represent the same information, preventing eventual communication failures and making procedures more efficient.

The manual construction of an ontology usually raises several problems, especially when knowledge domains are complex and wide-ranging, or the interpretation of its sources, in particular the textual ones, may generate different interpretations [16]. In these cases, knowledge acquisition and modelling become quite difficult, given the large number and diversity of the concepts that we may need to cover [17], implying the use of specific technologies, such as information retrieval, natural language processing or machine learning [18]. The systematic use of these technologies rose a new domain of work, recognized today as Ontology Learning. In the case of an automatic process of Ontology Learning [19, 20], it is necessary to use efficient machine learning methods and techniques for identifying the various ontological elements integrated in the sources of a given application domain. This kind of process still is not easy to accomplish, since it continuous requiring some kind of human intervention (supervision). Today, cases of application of fully automatic ontology learning are still rare. The process of Ontology Learning from texts is even more difficult to automate, since it requires the conjugation of natural language processing with machine learning techniques. Linguistic approaches in Ontology Learning often rely on the sophistication and versatility of natural language processing tools, especially in performing tasks of semantic and syntactic analysis of texts to make the discovery of representations of concepts and relationships [21].

As we saw, it is possible to approach the construction and development of an ontology in several distinct ways. It depends significantly from the knowledge domain, the information sources, and the knowledge, expertise and experience we have with ontology learning processes and applications. In a conventional ontology learning process, we use to accomplish a basic set of processing and analysis tasks, which were establish and documented a few years ago in [22], and refined later in [23] and [20]. Thus, to plan, develop and implement the ontology learning process for the "Book of Properties", we adopted and followed the recommendations of such authors in almost all the stages for constructing an ontology from text [24]. Before exposing and explaining the ontology learning process we developed, let us reveal a few things about our knowledge source: the "Book of Properties"

3 The "Book of Properties"

At the beginning of the 17th century, the properties of the Archbishop's Table densely filled the regions of Minho and Trás-os-Montes in Portugal, extending out of its religious domains, to the bishopric of Porto, even reaching Santarém, and climbing the borders of the kingdom towards Galicia, Spain. The inventory carried out at the time – the "Book of Properties" [6, 7] – recorded with great detail and precision, in 1288 A3 size pages, all the properties of the Archbishop's Table, rustic and urban, located in Valença, Vila Real, Chaves and Braga, as well as specify the payments due for the leasing of the properties. The "Book of Properties" is a codex of remarkable size and weight, of sturdy binding in wood and leather, with heavy decorative irons on both sides and leather and metal fitting clasps, housing within it 644 folios. The size, resistance and quality of the volume reveal the importance of its content and its legal reference and patrimonial value. It contains a clean copy of the inventory (or tomb) made by royal order of 1601, initialed on all folios and duly closed and signed in 1607. Its texts displays dates between the fifteenth and seventeenth centuries, corresponding to the term titles that in each locality were being presented to the Archbishop's Table, made by various notaries and signed by different archbishops and their procurators. Although mainly related to land, and secondly to houses or other buildings, these properties include other goods, such as boats. Its folios include abundant references to types of land, accidents of the terrain, and other geographical references, names of streets, places, rivers, settlements, and owners, biographical and genealogical notes, or products sown. It also includes references to types of existing trees, descriptions of houses and their characteristics, and many other things – we can see a small example of a property's description in Table 1.

Table 1. An excerpt of a description of a property included in the "Book of Properties".

Version	Description
Portuguese (original version)	*(…) umas casas colmaças que correm de norte a sul, e parte delas são sobradadas, e partem do Nascente e Norte com caminho, e das mais partes com terras deste casal; têm de comprido nove varas e de largo cinco. E pegado a estas casas estão três cortes e um palheiro, tudo colmaço, que tem de comprido dezanove varas e de largo quatro. E junto à dita casa, para a parte do sul, está um corrume de casas colmaças que têm a serventia para o Norte (…)*
English	*(…) some thatched houses that run from north to south, and part of them are two-story, and leave from the East and North with a path, and from the other parts with land of this couple; they are nine rods long and five rods wide. And attached to these houses are three courts and a haystack, all thatched, which is nineteen rods long and four wide. And next to the said house, towards the south, there is a cluster of thatched houses that have access from the north (…)*

4 Learning the Ontology

The construction and instantiation of an ontology for the "Book of Properties" will allow for revealing the numerous terms used in this relation of properties, which may include all the references and elements referred previously. In particular, the names of the lands, and respective contractors or cultivators, are very important elements for knowing the genealogy of the families of the regions approached in the inventory. Given the specific characteristics of the texts contained in the "Book of Properties", we adjusted the ontology learning process doing some of the tasks using other processing approaches required by some of the terms used in the book. Next, we present and discuss each of the tasks we performed in the ontology learning process we designed and implemented (Fig. 1).

Term Extraction. The ontology process begins with the extraction of the terms considered relevant to the domain ontology we wanted to construct. This is a very important task. It determines the terms that the ontology will incorporate. This task involved two

Fig. 1. The ontology learning process.

operations, namely: the extraction of the most common terms, and the extraction of the most relevant terms. To extract the most common terms, we used the OS Library [25] and OS.Path Library [26] to get all the words integrated in the texts. In this operation, the context of the terms was not taken into account. We removed all the punctuation characters contained in the text, such as commas or periods, so that they did not influence the results. Note that they have a strong presence in the texts. Then we removed stop words using the NLTK library [27]. A stop word is a word that is used in a common way, which does not add any value to a text in terms of context, such as determinants, personal pronouns and some prepositions. As such, these words should be ignored. After the removal of the stop words, we created a dictionary, containing all the common terms identified and the number of times they appeared in the analyzed texts. The extraction of the most relevant terms was carried out using YAKE [28]. It allows for extracting the most relevant words according to a given set of parameters, for example, the language we used (Portuguese), the number of words of the terms to be extracted (one or two) or the number of terms to extract. In addition, in the vast majority of cases, YAKE also made it possible to identify the stop words present in the texts, not including them in the process, even when they were in classical Portuguese. Then we moved on to the stage of identifying synonyms.

Synonym Detection. An ontology cannot have different elements representing a same concept. As such, it was necessary to analyze the texts to make this verification, and choose which of the identified terms represented the concept in question more appropriately. To do that, we used a public API of the Online Dictionary of Portuguese. For each of the identified elements, the API provided us with a list of synonyms. For example, for the term 'terra' (land), the API has given us the synonym 'terreno' (terrain). All the elements provided by the API were quite relevant to our process. However, during its use, we identified some less positive aspects. Namely, it was not possible to find synonyms for some of the terms contained in the texts, since their spelling is different from the ones in current Portuguese. This problem was not solved, because we did not have a dictionary (or an API) available that worked with the Portuguese of the 17th century.

Concept Definition. In this task, we separated the terms obtained previously, and reduced by the task of detecting synonyms, into distinct elements of the ontology, giving particular attention to the identification of concepts. First, we counted the most common terms to get the most relevant concepts of ontology. This type of elements tends to have its own specificities, something that Hejl [29] identified as being, for the most part, composed of common names or nouns. In this task, we applied Part Of Speech Tagging (POS Tagging) in order to determine the common names of all the terms resulting from the previous steps, using the Polyglot Tagger tool [30]. This tool is capable of handling multiple languages. From the results provided by Polyglot, we created a dictionary with the 100 most common terms, ordered from the most to the least frequent term. This dictionary included terms such as: 'parte' (part), 'terra' (land), 'semeadura' (seeding), 'norte' (north), 'alqueires' (bushels), 'caminho' (path), or 'terras' (lands), among many others. Through the analysis of these results, we could verify that some of the terms from the first stage were not considered as stop words, for example, 'hua', 'meo', 'he' or 'en', due to the differences in spelling between the classical Portuguese of the original texts and the modern Portuguese. Consequently, we removed these terms from the valid term list. Next, we counted the terms got previously using YAKE. As in the previous count, we used also Polyglot to perform the POS Tagging of the extracted terms. The algorithm considered the existence of terms with more than one word (multiword terms), since many of the terms extracted by YAKE were binomials. Often, we represent instances, such as names of people or organizations, using proper names. Thus, if POS Tagging resulted in a proper name, the algorithm considered the word as an instance, or a word of a possible instance, when it is a multiword term. When a term was multiword, it will be a concept if all of its constituent words have as a grammatical class a common name. For the term to be considered an instance, it would be constituted exclusively by proper names. If there were a word in an excerpt of a text that did not correspond to a common name or a proper name, the process would move to the next expression. We ignored excerpts of text that had a combination of proper names and common names for two reasons. First, most of the names that appear in binomials appear in other excerpts of text, which means that their removal does not result in a loss of information. Second, binomials that contain proper names and common names cannot be safely classified as concepts or instances, since we cannot give priority to any of the classes. At this stage, we found that a high percentage of the binomials that were considered as instances are names of people, mostly owning the properties of the Archbishop's Table. After separating the terms extracted by YAKE into concepts and instances, we proceed to verify the existence of synonyms. At this stage, we still detected terms that are not concepts. Some of them are only present due to the spelling differences we mentioned before. However, the results we got are positive. Table 2 shows a sample of these results.

Table 2. A small sample of the concepts extracted with the two methods.

Most common terms		Terms obtained with YAKE	
Portuguese	English	Portuguese	English
parte	part	norte	north
terra	land	sul	south
semeadura	seeding	poente	west
norte	north	semeadura	seeding
varas	sticks	largo	wide
alqueires	bushels	campo	field
caminho	path	caminho	path
terras	lands	terra	Earth
casal	couple	varas	sticks
(...)	(...)	(...)	(...)

To understand better which concepts are the most relevant to ontology, it was necessary to verify their definitions, how they could be linked together, and how they can be incorporated into the ontology, as concepts or as any other ontological element. However, to obtain or verify the definitions of the concepts obtained, consulting current Portuguese dictionaries does not work, not only due to the semantic evolution suffered by the words, but also due to changes in the definitions of the words themselves. To overcome such difficulties, we decided to consult the Bluteau's dictionary prepared and published by Rafael Bluteau [31]. The definitions in this dictionary are a more reliable representation of the meaning of the elements belonging to the list of concepts we have constructed. After consulting the definitions, we obtained some very valuable information for the definition and characterization of the terms. Let's look at three examples, 'semeadura' (seeding), 'norte' (north) and 'terra" (land), to illustrate the process of categorizing the terms. The term 'semeadura' refers to the cultivation of land. In the texts, this term is regularly used in the expression 'levara de semeadura' (seeding land), in order to indicate the amount of seed (or cereal), that is necessary to sow the land. Thus, 'semeadura' (seeding) or 'semente' (seed), as it is also used in the book, translates the value of the land, its size and productivity. From this, we can conclude that this term should not be considered as a concept, but rather as a characteristic of a land. In turn, the term 'norte' (north) refers to an orientation used to indicate confrontations (or borders) between different properties. This information allows us to conclude that terms like this are more suitable to represent relationships, establishing links with other properties, based on the existing confrontations between them. The same is also true to the terms 'sul' (south), 'nascente' (east), and 'poente' (west). Finally, we have the term 'terra' (land). This term represents a land belonging to the Archbishop's Table. It has its own essence. It is not a feature of a concept nor an indication of a relationship between terms. As such, we considered it as a concept.

From the information collected in the Bluteau's dictionary, we only identified one term as being a concept relevant to ontology, namely 'land' (land) or 'propriedade' (property). These terms means the same concept. However, after a new analysis of the texts, we identified two more concepts: 'título' (title) and 'emprazador' (contractor). The term 'título' represents a large set of properties of the Archbishop's Table employed by a small number of architects, while the term 'emprazador' designates a person to whom the Archbishop's Table leases one or more lands for cultivating them to subsist, in exchange for a payment of a forum to the Archbishop's Table. In this task, we found the relevant concepts of the ontology, as well as their attributes and the relative terms of the relations between the concepts. These results made this task the most relevant of the whole process carried out.

Relationship Validation. After obtaining the essential terms and concepts, we identified the relationships between the different concepts found. The analysis of the terms obtained at the end of the previous task allowed us to have a notion of the concepts that are relevant to the ontology, as well as of the possible relationships that occur between them and their characteristics. We only validated the relationships established between concepts already defined. To perform this task, we used an association rules algorithm, the Apriori algorithm [32]. This algorithm allows us to calculate the strength of the

Table 3. An excerpt of the relationships established by the Apriori algorithm.

Relationship		Lift value
1st term	2nd term	
casal	propriedades	4.828087167070217
caminho	casas	2.9540740740740743
campo	terras	2.739010989010989
campo	largo	2.3205320033250207
largo	varas	2.304099664631525
campo	terra	2.115980024968789
alqueires	semeadura	2.0770833333333334
casas	varas	2.0691251596424007
herdade	terra	1.9837312734082395
(...)	(...)	(...)

relationships between two terms, using a lift function to measure the quality and degree of interest of the relationship, combining the support and trust functions [33].

In this task, we started by separating all the sentences from the texts of the book and, for each of them, we removed all the phrasal elements that did not contain alphanumeric characters and discarded the words independent of the context using the Pandas library [34]. Next, we applied the Apriori algorithm to all the content of the texts, storing in a dictionary the relationships established between the different terms, ensuring that the terms were identified in the previous task. We only considered relationships in which the terms appeared in the same sentence at least 20 times. Then we defined the support, trust and lift functions, and saved the results in another dictionary. The resulting pairs of terms were then reordered based on their lift value, from the strongest to the least strong pairs (Table 3). From these results, we selected only the pairs of terms that had a lift value greater than 2.0, because we found that below this value, it was notorious that the results did not indicate direct relationships.

Obtaining Instances. Having concepts, relations and their respective attributes, we proceeded to obtain their instances. Firstly, we proceed to get properties. We separated the titles and information relating to their properties, the types of land that could be associated with each land, and the confrontations of the various properties. Then, we checked whether the properties belonged to any of the titles previously obtained. If this happened, we established the respective links between the several instances involved. After this, we looked at the different confrontations that a property would have. We detected four different possibilities of confrontations: 'norte' (north), 'sul' (south), 'nascente' (east), and 'poente' (west). In this verification, we extracted information about the confrontations of other orientations related to the confrontation that was being analyzed. We also checked to see if there was any information about how the land was cultivated, since the expression used to introduce the cultivation of a land was always 'levara de semeadura' (taken from seeding). Initially, we used this expression to make the division between the last confrontation and the cultivation done.

Table 4. A small set of elements of the ontology.

Element		Type
emprazador	(contractor)	Class
título	(title)	Class
vinha	(vineyard)	Subclass
vinha (is_a 'property')	(vineyard)	Taxonomic relationship
propriedade (is_part_of 'título')	(property)	Non taxonomic relationship
nome	(name)	Simple attribute
tipo	(type)	Multivalue attribute
semeadura	(seeding)	Simple attribute
confrontações	(confrontations)	Composite attribute
(...)	(...)	(...)

All the verifications were carried out according to the various expressions we detected, covering numerous aspects related to the properties, such as the types of land and the way they were cultivated, the number of men needed to prepare the land or, simply, the names of the properties and their owners or contractors. We proceed to the instantiation of the ontological elements defined during the process. Finishing this task, we have the ontology learning process and consequently the definition of ontology for the "Book of Properties" finished.

Just a final note. As for possible axioms for ontology, we chose to do their detection manually, not including in the process a specific task for their detection. It should be note, as a simple curiosity, that we can only found a single axiom: 'Sempre que um

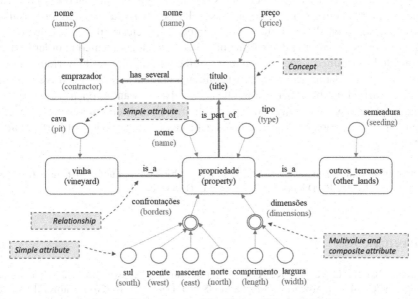

Fig. 2. The ontology of the "Book of Properties".

terreno tem uma cava, este tem que possuir vinha' (Whenever a piece of land has a pit, it must have a vineyard.). We end this section presenting in Table 4 a list of some of the elements that integrate the ontology we constructed, and in Fig. 2 an illustration of the ontology for the "Book of Properties".

5 Conclusions and Future Work

In this paper we presented and described the process of building an ontology for the "Book of Properties", which followed a standard development methodology but approached in a slightly different way, given the experience and knowledge acquired by the authors in other similar projects. The ontology we got from this process, it is just a "first" version. However, even so, we already recognize its value to identify and characterize the various elements of the properties inventoried in the "Book of Properties", as well as to analyze the characteristics of the entities referred in the texts, in a totally separate, effective and autonomous way. A concrete example of the availability of this knowledge, and of what we can already have using the ontology, beyond the simple direct consultation of ontological elements, is to make some concrete calculations on. For example, it is already possible to know the total area of a plot of land, the amount of cereal that could be produced on the land, or how many men would be needed to dig in the identified vineyards, just to name a few. Aspects like these, allow us to highlight the power of the ontology already defined. In our point of view, this is quite interesting. It offers a very rich field of research for students, professors, researchers and individual users for studying and learning some aspects of geography, culture, and agriculture, among many others, of Portugal at the 17th century. The semi-automatic process carried out, had the ability to discover the elements that allow us to reveal concretely properties of the inventory and their characteristics, as well as to make calculations such as those referred above. The balance of this work is, in our view, clearly positive. However, we believe that is necessary to review some of the tasks we implemented in order to refine the knowledge acquired, especially in terms of the instantiation of the various ontological elements, validating again their structure and content through intensive querying processes carried out by experts in the field. Finally, we want to open the ontological system to all who wish to study and learn the enormous and very interesting knowledge containing in the "Book of Properties" and get an extraordinary view of Portugal at the 17th century.

Acknowledgements. This work has been supported by FCT – Fundação para a Ciência e Tecnologia within the R&D Units Project Scope: UIDB/00319/2020.

References

1. Tonny James, N., Kannan, R.: A survey on information retrieval models, techniques and applications. Int. J. Adv. Res. Comput. Sci. Softw. Eng. **7**(7), 16 (2017). https://doi.org/10.23956/ijarcsse.v7i7.90

2. Sharma, A.: Natural language processing and sentiment analysis, in international research. J. Comput. Sci. **8**(10), 237 (2021). https://doi.org/10.26562/irjcs.2021.v0810.001

3. Zhang, L., Wang, S., Liu, B.: Deep learning for sentiment analysis: a survey. WIREs Data Min. Knowledge Discov. **8**(4), e1253 (2018). https://doi.org/10.1002/widm.1253

4. Biemann, C.: Ontology learning from text: a survey of methods. LDV Forum **10**, 75–93 (2005)

5. Lourdusamy, R., Abraham, S.: A Survey on methods of ontology learning from text. In: Jain, L.C., Peng, S.-L., Alhadidi, B., Pal, S. (eds.) Intelligent Computing Paradigm and Cutting-edge Technologies: Proceedings of the First International Conference on Innovative Computing and Cutting-edge Technologies (ICICCT 2019), Istanbul, Turkey, 30–31 Oct 2019, pp. 113–123. Springer International Publishing, Cham (2020). https://doi.org/10.1007/978-3-030-38501-9_11

6. Barros, A.: Apontamentos lexicais sobre o Livro das Propriedades ou Tombo da Mitra Arquiepiscopal de Braga: designações de terras e outros aspetos das propriedades. In: Maia, C.A., Santos, I.A. (eds.) Estudos de linguística histórica: mudança e estandardização, pp. 393–428. Imprensa da Universidade de Coimbra, Coimbra (2019)

7. Barros, A.: A edição do Livro das Propriedades ou Tombo da Mitra Arquiepiscopal de Braga. In: Abreu, P., et al. (eds.) Os sete castelos: D. Rodrigo de Moura Teles, pp. 183–218. Instituto Universitário da Maia & Instituto de História e Arte Cristã, Maia (2020)

8. Guarino, N., Oberle, D., Staab, S.: What is an ontology? In: Staab, S., Studer, R. (eds.) Handbook on Ontologies, pp. 1–17. Springer, Berlin, Heidelberg (2009). https://doi.org/10.1007/978-3-540-92673-3_0

9. Keet, M.: An Introduction to Ontology Engineering, University of Cape Town. https://people.cs.uct.ac.za/~mkeet/OEbook/ (2018). Last accessed 8 Jul 2023

10. Uschold, M., Gruninger, M.: Ontologies: Principles, methods and applications Ontologies: Principles, Methods and Applications Mike Uschold Michael Gruninger AIAI-TR-191 February 1996 To appear in Knowledge Engineering Review Volume 11 Number 2, June 1996 Mike Uschold Tel: Mi. Knowledge Engineering Review (1996)

11. Borst, W.: Construction of engineering ontologies for knowledge sharing and reuse. Twente (1997)

12. Gruber, T.R.: Toward principles for the design of ontologies used for knowledge sharing? Int. J. Human-Comput. Stud. **43**(5–6), 907–928 (1995). https://doi.org/10.1006/ijhc.1995.1081

13. Kim, J., Caralt, J., Hilliard, J.: Pruning bio-ontologies. In: Proceedings of the Annual Hawaii International Conference on System Sciences (2007). https://doi.org/10.1109/HICSS.2007.455

14. Kiong, Y., Palaniappan, S., Yahaya, N.: Health ontology system. In: Proceedings of the 7th International Conference on Information Technology in Asia: Emerging Convergences and Singularity of Forms (2011). https://doi.org/10.1109/CITA.2011.5999506

15. Fan, S., Zhang, L., Sun, Z.: An ontology based method for business process integration. In: Proceedings of the International Conference on Interoperability for Enterprise Software and Applications, IESA (2009). https://doi.org/10.1109/I-ESA.2009.31

16. Belhoucine, K., Mourchid, M.: A Survey on Methods of Ontology Learning from Text, pp. 113–123 (2020). https://doi.org/10.1007/978-3-030-38501-9_11

17. Maedche, A., Staab, S.: Ontology learning for the semantic web. IEEE Intell. Syst. **16**(2), 72–79 (2001). https://doi.org/10.1109/5254.920602

18. Drumond, L., Girardi, R.: A survey of ontology learning procedures. In: Proceedings of the 3rd Workshop on Ontologies and their Applications, Salvador, Bahia, Brazil (2008)

19. Cimiano, P., Mädche, A., Staab, S., Völker, J.: Ontology learning. In: Staab, S., Studer, R. (eds.) Handbook on Ontologies. International Handbooks on Information Systems IHIS, pp. 245–267. Springer, Heidelberg (2009). https://doi.org/10.1007/978-3-540-92673-3_11

20. Asim, M., Wasim, M., Khan, M., Mahmood, W., Abbasi, H.: A survey of ontology learning techniques and applications. Database **2018**, bay101 (2018). https://doi.org/10.1093/database/bay101

21. Wong, W., Liu, W., Bennamoun, M.: Ontology learning from text: a look back and into the future. ACM Comput. Surv. **44**, 20 (2012). https://doi.org/10.1145/2333112.2333115

22. Brewster, C.: Ontology learning from text: methods, evaluation and applications paul buitelaar, philipp cimiano, and bernado magnini (editors) (DFKI Saarbrücken, University of Karlsruhe, and ITC-irst), Amsterdam: IOS Press (Frontiers in artificial intelligence and appl. Comput. Linguist. **32**(4), 569–572 (2006). https://doi.org/10.1162/coli.2006.32.4.569

23. Tiwari, S., Jain, S.: Automatic ontology acquisition and learning. Int. J. Res. Eng. Technol. **03**(26), 38–43 (2014). https://doi.org/10.15623/ijret.2014.0326008.

24. Buitelaar, P., Cimiano, P., Magnini, B.: Ontology Learning from Text: An Overview, in Ontology Learning from Text: Methods, Evaluation and Applications/Frontiers in Artificial Intelligence and Applications volume 123, pp. 1–10. Paul Buitelaar, Philipp Cimiano, Bernardo Magnini Editors, IOS Press (2005)

25. OS Library. https://docs.python.org/3/library/os.html. Last accessed 9 Jul 2023

26. OS.Path Library. https://docs.python.org/3/library/os.path.html. Last accessed 9 Jul 2023

27. NLTK – Natural Language Toolkit. Homepage, https://www.nltk.org/. Last accessed 9 Jul 2023

28. Campos, R., Mangaravite, V., Pasquali, A., Jorge, A., Nunes, C., Jatowt, A.: YAKE! Keyword extraction from single documents using multiple local features. Inform. Sci. **509**, 257–289 (2020). https://doi.org/10.1016/j.ins.2019.09.013

29. Hejl, L.: Evolution of the Conception of Parts of Speech. Thesis, Palacky University in Olomouc, Faculty of Arts, Department of English and American Studies (2014)

30. Polyglot. https://polyglot.readthedocs.io/en/latest/. Last accessed 10 Jul 2023

31. Bluteau, R.: Vocabulario Portuguez e Latino. Dictionary (1712-28)

32. Agrawal, R., Srikant, R., Fast algorithms for mining association rules. In: Proceedings of the 20th International Conference on Very Large Data Bases, VLDB, pp. 487–499. Santiago, Chile (1994).

33. Harun, N., Makhtar, M., Aziz, A., Zakaria, Z., Abdullah, F., Jusoh, J.: The application of Apriori algorithm in predicting flood areas. Int. J. Adv. Sci. Eng. Inform. Technol. **7**(3), 763–769 (2017). https://doi.org/10.18517/ijaseit.7.3.1463

34. Pandas Library. https://pandas.pydata.org. Last accessed 9 Jul 2023

Language Models for Automatic Distribution of Review Notes in Movie Production

Diego Garcés[1,2(✉)], Matilde Santos[3], and David Fernández-Llorca[4,5]

[1] Computer Science Faculty, Complutense University of Madrid, Madrid, Spain
digarces@ucm.es
[2] Skydance Animation Madrid, Madrid, Spain
diego.garces@skydance.com
[3] Institute of Knowledge Technology, Complutense University of Madrid,
Madrid, Spain
msantos@ucm.es
[4] Computer Engineering Department, University of Alcalá de Henares, Madrid, Spain
[5] European Commission, Joint Research Centre, Seville, Spain
david.fernandez-llorca@ec.europa.eu

Abstract. During the several years of production of an animated movie, review meetings take place daily, where supervisors and directors generate text notes about fixes needed for the movie. These notes are manually assigned to artistic departments for them to fixed. Being manual, many notes are not properly assigned and are never fixed, lowering the quality of the final movie. This paper presents a proposal for automating the distribution of these notes using multi-label text classification techniques. The comparison of the results obtained by fine-tuning several transformer-based language models is presented. A highest mean accuracy of 0.776 is achieved assigning several departments to each of the review notes in the test set with a BERT Multilingual model. A mean accuracy of 0.762 was reached in just 10 epochs and 10 min of training on an RTX-3090 with a DistilBERT transformer model.

Keywords: Film production · Text Mining · Large Language Models · Natural Language Processing · Deep learning · Text Classification

1 Introduction

Movie production is above all an iterative process. Every day during production, review meetings take place, where supervisors and directors give notes to improve the in-progress material presented in the review. Taking into account that the production of an animated movie takes several years, this means that thousands of these text notes are generated for each production. These notes have to be addressed by different artists, belonging to different departments. Production is in charge of distributing these notes among the departments concerned so

P. Quaresma et al. (Eds.): IDEAL 2023, LNCS 14404, pp. 245–256, 2023.
https://doi.org/10.1007/978-3-031-48232-8_23

that they can be taken into consideration. Despite animating movies being a very technical discipline, little or no technical help is provided for this part of the process, remaining as a highly manual task. Combined with the fast-paced nature of movie production, this means that many notes are not properly distributed and consequently not properly addressed, impacting the final quality of the released movie.

These notes take the form of free text. They are written by the Production Department during the review meeting. Distributing these notes among the different departments is a text classification problem.

Text classification and its application in several industries has been widely studied [6,13]. Tasks such as ticket assignment [3], news categorization [12], topic classification [16] or sentiment analysis [21] have made use of these techniques.

Nevertheless, the application of Natural Language Processing (NLP) techniques for text classification on some fields is completely new. Movie production is an industry which is yet to adopt them.

Animated movie production is a highly specialized process. A Movie Studio is composed of many departments. Each one performs a very specific task. Examples of departments found in a typical animation studio are Modeling, Surfacing, Rigging, CFX, Animation, Layout, Final Layout, FX or Lighting. They are often grouped in departments working in pre-production (creating assets like characters or props) and others working in production (shots of the movie).

An individual note usually provides feedback for several aspects at the same time, like the animation and hair of a character, which are the responsibility of two different departments (Animation and CFX). The problem of assigning department to notes becomes a multi-label text classification problem [22]. Multi-label classification has been studied in the past in the context of text categorisation [1].

It is important to differentiate the nature of these movie review notes to the ones contained in publicly available datasets, used, among others, for sentiment analysis tasks [21], that classify movie reviews as positive or negatives. These reviews are comments about the complete final movie. The ones used in this research are specific about a particular aspect of the movie, like a shot. They analyze in-progress material of the movie production and they propose ways to improve the quality of the reviewed material. Usually they are very specific and technical. Table 1 shows some examples of notes taken during these reviews. Only a segment of the complete text has been added for brevity.

Table 1. Review notes and target departments.

Note text	Destination Departments
Rain should be more dense and generate a greater bokeh	FX and Lighting
Movement should be exaggerated more. Hair looks too stiff considering the movement of the head	Animation and CFX
There is some texture stretching in the shoulders	Rigging and Surfacing
Tree pattern in the street is too obvious, distribute them	Final Layout

To our knowledge there are no recent publications about applying NLP text classification techniques for review notes[1] in movie production. These datasets contain highly confidential information related to important intellectual property. That is why the dataset used in this research cannot be released to the public.

This proposal builds on top of our recent previous work [5] where we presented preliminary results using DistilBERT embeddings and a logistic regression classifier. Here, we extend our work by comparing several transformer-based language models, including their own embeddings, that are fine-tuned to perform multi-label text classification.

2   Proposed Method

Figure 1 depicts a complete overview of the whole process developed for the distribution of the movie review notes among the studio's departments. The diagram shows how data are transformed and which processes are executed to achieve the final result.

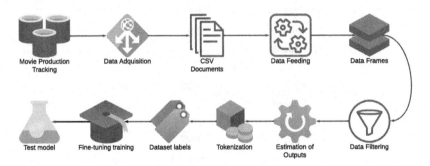

Fig. 1. Flow chart of the proposed approach for automatic distribution of review notes in movie production.

2.1   Label Estimation

The notes extracted from the Movie Production Tracking software are not properly labeled to be used in Machine Learning processes. Manual labeling is not possible due to the high number of notes available and the time it would take. However, the notes, apart from the note text, contain other fields that can help to estimate which departments were interested in those notes. There is a list of artists in the note for the purpose of sending notifications. Also, a list of tasks performed related to the note is registered within the note. With this extra

---

[1] It is important to differentiate the review notes of the production process from the reviews of finished movies carried out by users, such as, the Large Movie Review Dataset [11].

information, labels can be estimated with enough certainty to allow the use of training algorithms with them.

The automatic estimation will be performed only using the note text, as these other fields will not be available. Artists and tasks are added to the note when fixing the note, a process than usually happen several days after the review.

The labeled dataset is then loaded into data frames to be processed efficiently by the next steps.

Each department works with different workflows. Some have more continuous internal reviews, like Animation, than others. Some need more iteration with the Director. This means that the labeled dataset will be unbalanced. Some departments will have many more notes labeled for them, such as Animation or Lighting, than others, like Crowds or Layout. Table 2 shows the distribution of amount of notes per label.

Table 2. Distribution of review notes and target departments.

Department	Labeled notes
Animation	32.5%
CFX	14.1%
Compo	0.9%
Crowds	2.5%
Final Layout (FL)	11%
FX	5.5%
Layout	2.6%
Lighting	12.3%
Matte	0.9%
Modeling	9.8%
Rigging	0.9%
Surfacing	6.6%

2.2 Tokenization

Transformer models cannot be trained directly on free text data. Is it necessary to transform this text into a numerical format that can serve as input to the model. This numerical data takes the form of tensor data, representing the whole note text.

There are several tokenization strategies available. When training a transformer model for classification, its corresponding tokenizer is used to maximize compatibility. BERT and DistilBERT tokenizers use the WordPiece algorithm [18], while RoBERTa tokenizer uses the Byte-pair enconding (BPE) algorithm [4]. Both algorithms belong to the Subword tokenization strategy. This strategy improves over word and character based tokenizers. These algorithms do not split the frequently used words into smaller subwords. Instead, they split rare words into smaller meaningful subwords.

2.3 Classification

The objective of the classification step is generating predictions about which departments would be interested on a particular review note. Using only the note text as input, the output will be the a list of departments. These departments will be notified, so they can create the appropriate tasks to fix the note given by the Supervisor or Director.

When working with transformer models for classification, there are usually two options to train the model. The first one consists on using the model to extract features, called hidden states, to use them as inputs for a custom classifier, for example Logistic Regression. This method maintains the transformer model in its pre-trained state. This option architecture can be seen in Fig. 2. This is the approach used in our previous work with DistilBERT [5].

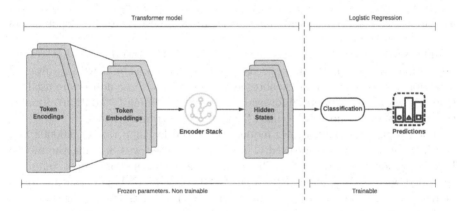

Fig. 2. Architecture of Transformer model with Logistic Regression.

The second one trains the whole transformer model end-to-end. The training process updates all the parameters of the pre-trained model. This process is called fine-tuning and its architecture can be seen in Fig. 3.

3   Experiments and Results

In order to compare the performance, several transformer-based language models have been tested. BERT [2], DistilBERT [17], RoBERTa [9] and BERT multilingual [15] models have been used and fine-tuned. A DistilBERT model used to obtain the hidden states with a Logistic Regression classifier has been trained as

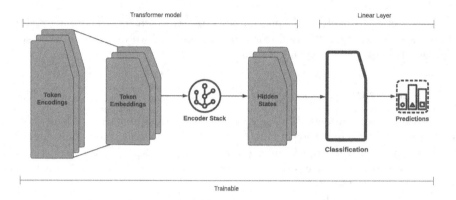

Fig. 3. Architecture of Transformer model with fine-tuning.

the feature extraction option for classification. The solver for the Logistic Regression classifier was Limited-memory Broyden-Fletcher-Goldfarb-Shanno (lbfgs) from the SciKit-learn library. Since this is a multilabel classification problem, a One vs Rest [7, 20] strategy had to be used. This strategy consists on training one classifier for each class. The classes in this dataset are the departments interested on the note.

The same exact dataset has been used in all the training so that results are comparable. For performance reasons, the dataset was built with a subset of 5000 notes from the total movie notes available. 80% of those notes were used for training, while 10% were used for evaluation while training (validation). Finally, the rest 10% of the notes were used for testing the trained model.

For training the transformer models, an initial learning rate of 0.00002 and a weight decay of 0.01 were used. The batch size used was 8, using the metric F1-score to load the best model. The training was performed on a local computer equipped with a RTX 3090 GPU with 24 GB of VRAM. For measuring tools, SciKit-learn have been used to compare the different methods and models.

Precision [10] measures the amount of True Positives among all the Positives detected. It shows how reliable the system is when labeling a note with a particular department. It doesn't take into account notes that have not been labeled with a particular department.

$$\text{Precision} = \frac{\text{True Positives (TP)}}{\text{True Positives (TP)} + \text{False Positives (FP)}} \tag{1}$$

Precision for all the models by label can be seen in Fig. 4. For clarity, only the labels with the highest number of notes are represented in the graph.

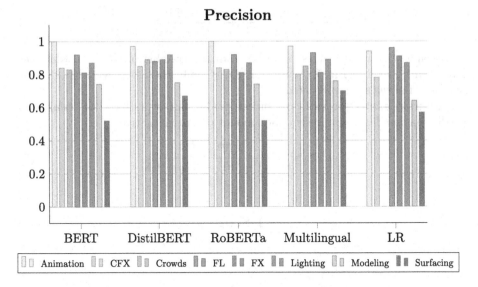

Fig. 4. Precision per label for the different language models.

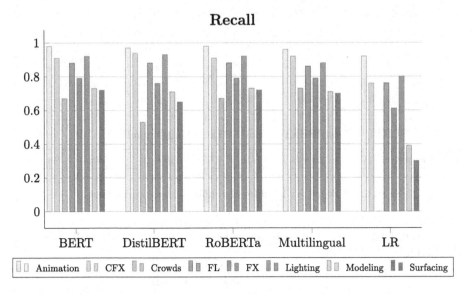

Fig. 5. Recall per label for the different language models.

Recall measures how many positives have been detected by the system from the total positives contained in the dataset. It is a measure of the model ability to find all the notes that have to be labeled with a particular department. It does not take into account notes labeled with a department that should not contain that label.

$$\text{Recall} = \frac{\text{True Positives (TP)}}{\text{True Positives (TP)} + \text{False Negatives (FN)}} \qquad (2)$$

The graph with recall values resulting from the training of all the models per label can be seen in Fig. 5. Again, for clarity, only the labels with the highest number of notes are represented.

Accuracy shows the ratio of correct predictions over the total number of predictions made. TN represent the amount of True Negatives.

$$\text{Accuracy} = \frac{\text{TP} + \text{TN}}{\text{TP} + \text{FP} + \text{TN} + \text{FN}} \qquad (3)$$

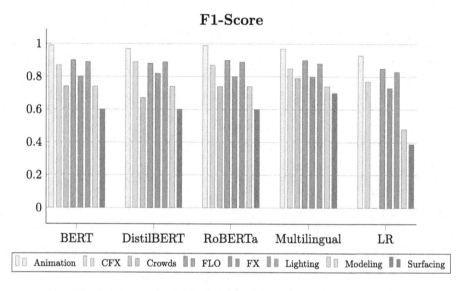

Fig. 6. F1-Score per label for the different language models.

Finally, F1-Score is used to balance between precision and recall, following this formula:

$$\text{F1-Score} = 2 \cdot \frac{\text{precision} \cdot \text{recall}}{\text{precision} + \text{recall}} \qquad (4)$$

The F1-Score values of all the models per label can be seen in Fig. 6.

The averaged accuracy and F1-Score taking into account all the labels are presented in Fig. 7. The accuracy and F1-Score have been averaged using the weighted strategy. It calculates a mean taking into account the label support within the dataset.

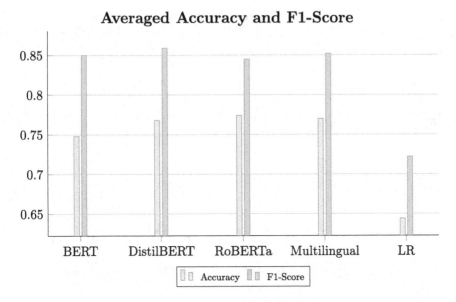

Fig. 7. Averaged Accuracy and F1-Score for the different language models.

With these data, it can be seen that the fine-tuning methods improve significantly the feature extraction method compare with the use of Logistic Regression on the hidden state values. In the labels with fewer samples the improvement is very high. For example, Crowds has a F1-Score of 0 with Logistic Regression. It goes up between 0.74 and 0.79 with the transformer fine-tuning methods.

The measurements of the different fine-tuning methods are very similar. *Surfacing* category has a higher F1-Score with the Multilingual BERT which is probably due to the fact that *Surfacing* in this show was very international and notes can contain fragments or words from other languages. Some labels even reach a precision of 1.0, meaning that all notes labeled with that department were correctly predicted.

The averaged accuracy and F1-Score show an improvement of over 0.12 in accuracy and 0.13 in F1-Score in the fine-tuned transformer models over the method that only uses a pre-trained DistilBERT for feature extraction and Logistic Regression for classification.

The F1-score obtained from the evaluation dataset after each training epoch is shown in Fig. 8.

All language models converged rapidly in about 10 epochs, reaching good values above 0.83. F1-score measure oscillates during training, especially between epochs 10 and 40. Finally, the training gets very stable after 40 epochs.

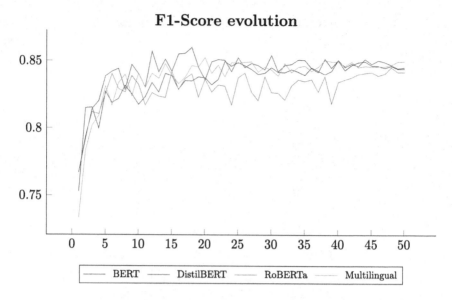

Fig. 8. Eval F1-Score evolution over 50 epochs

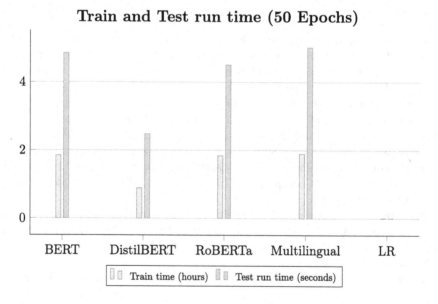

Fig. 9. Training and test time for the different models trained over 50 epochs.

Figure 9 shows that smaller models, like DistilBERT, provide much better performance in train time and test run time than larger models like BERT or RoBERTa. They have similar accuracy and F1-Score ratios, so a model like DistilBERT would be preferable in this situation.

The confusion matrix showing the DistilBERT fine-tuning model compared against the one for the feature extraction method with Logistic Regression is shown on Table 3. The confusion matrix represents the amount of True Negatives (TN), False Positives (FP), False Negatives (FN) and True Positives (TP) found while running the classifier with the test split of the dataset. They are represented as percentages over the total number of notes present in the test dataset.

Table 3. Confusion Matrix for DistilBERT (grey) vs Logistic Regression (white)

	TN	FP	FN	TP	TN	FP	FN	TP
Animation	49.5	0.1	0.8	31.7	48.5	2	2.7	29.9
CFX	66.6	2.3	0.8	13.2	65.9	3	3.3	10.8
Crowds	80.3	0.2	1.2	1.3	80.2	0.3	2.5	0
FLO	70.7	1.3	1.3	9.6	71.8	0.3	2.7	8.3
FX	77	0.5	1.3	4.2	77.2	0.3	2.2	3.3
Layout	79.9	0.5	2	0.7	80.2	0.2	2.5	0.2
Lighting	69.7	1	0.8	11.4	69.3	1.5	2.5	9.8
Modelling	70.9	2.3	2.8	7	71.1	2.2	6	3.8
Surfacing	74.2	2.2	2.3	4.3	74.9	1.5	4.6	2

4   Conclusions and Future Work

The conclusion from this research project is that fine-tuned models provide much better performance than using the transformer model in its pre-trained state just for feature extraction. Training the whole model with the specific dataset of the movie review notes gives a significant improvement in the results of the predictions made by the system. This makes the proposed fine-tuning system a good approach to distribute movie review notes among artistic departments in a movie studio.

As future works, some of the limitations of this approach could be addressed, such as the use of a limited and unbalanced dataset, and the lack of a human input to verify the classification results. Besides, more hyperparameter optimization could be carried out to improve the accuracy.

Exploring different Large Language Models (LLM) like GPT-4 [8], Falcon [14] or Llama [19] and comparing their performance, both in results and resource consumption, is a future line of research.

References

1. Chen, X., et al.: A survey of multi-label text classification based on deep learning. In: Sun, X., Zhang, X., Xia, Z., Bertino, E. (eds.) ICAIS 2022. LNCS, vol. 13338, pp. 443–456. Springer, Cham (2022). https://doi.org/10.1007/978-3-031-06794-5_36

2. Devlin, J., Chang, M.W., Lee, K., Toutanova, K.: Pre-training of deep bidirectional transformers for language understanding. In: naacL-HLT, pp. 4171–4186 (2019)
3. Feng, L., Senapati, J., Liu, B.: TaDaa: real time ticket assignment deep learning auto advisor for customer support, help desk, and issue ticketing systems. arXiv:2207.11187 (2022)
4. Gage, P.: A new algorithm for data compression. C Users J. **12**(2), 23–38 (1994)
5. Garcés, D., Santos, M., Fernández-Llorca, D.: Text classification for automatic distribution of review notes in movie production. In: García Bringas, P., et al. (eds.) SOCO 2023. LNNS, vol. 749, pp. 3–12. Springer, Cham (2023). https://doi.org/10.1007/978-3-031-42529-5_1
6. Gasparetto, A., Marcuzzo, M., Zangari, A., Albarelli, A.: A survey on text classification algorithms: from text to predictions. Information **13**(2), 83 (2022)
7. Goštautaitė, D., Sakalauskas, L.: Multi-label classification and explanation methods for students' learning style prediction and interpretation. Appl. Sci. **12**(11), 5396 (2022)
8. Liu, Y., et al.: Summary of ChatGPT/GPT-4 research and perspective towards the future of large language models (2023)
9. Liu, Y., et al.: RoBERTa: a robustly optimized BERT pretraining approach. arXiv:1907.11692 (2019)
10. Llorella, F.R., Iáñez, E., Azorín, J.M., Patow, G.: Binary visual imagery discriminator from EEG signals based on convolutional neural networks. Rev. Iberoamericana Autom. Inform. Ind. **19**(1), 108–116 (2022)
11. Maas, A.L., Daly, R.E., Pham, P.T., Huang, D., Ng, A.Y., Potts, C.: Learning word vectors for sentiment analysis. In: 49th Annual Meeting of the Association for Computational Linguistics: Human Language Technologies, pp. 142–150 (2011)
12. Martin, T.: The reuters dataset (2017). https://martin-thoma.com/nlp-reuters
13. Minaee, S., Kalchbrenner, N., Cambria, E., Nikzad, N., Chenaghlu, M., Gao, J.: Deep learning-based text classification: a comprehensive review. ACM Comput. Surv. **54**(3), 1–40 (2021)
14. Penedo, G., et al.: The RefinedWeb dataset for falcon LLM: outperforming curated corpora with web data, and web data only. arXiv:2306.01116 (2023)
15. Pires, T., Schlinger, E., Garrette, D.: How multilingual is multilingual BERT? (2019)
16. Rahman, M., Akter, Y.: Topic classification from text using decision tree, K-NN and multinomial naïve bayes. In: 1st International Conference on Advances in Science, Engineering and Robotics Technology (ICASERT) (2019)
17. Sanh, V., Debut, L., Chaumond, J., Wolf, T.: DistilBERT, a distilled version of BERT: smaller, faster, cheaper and lighter. arXiv:1910.01108 (2019)
18. Schuster, M., Nakajima, K.: Japanese and Korean voice search. In: 2012 IEEE International Conference on Acoustics, Speech and Signal Processing (ICASSP), pp. 5149–5152 (2012)
19. Touvron, H., et al.: LLaMA: open and efficient foundation language models. arXiv:2302.13971 (2023)
20. Tsoumakas, G., Katakis, I.: Multi-label classification: an overview. Int. J. Data Warehouse. Min. **3**, 1–13 (2009)
21. Zhang, L., Wang, S., Liu, B.: Deep learning for sentiment analysis?: a survey. WIREs Data Min. Knowl. Discov. **8**(4), e1253 (2018)
22. Zhang, M.L., Zhou, Z.H.: A review on multi-label learning algorithms. IEEE Trans. Knowl. Data Eng. **26**(8), 1819–1837 (2014)

Extracting Knowledge from Incompletely Known Models

Alejandro D. Peribáñez, Alberto Fernández-Isabel[(✉)], Isaac Martín de Diego,
Andrea Condado, and Javier M. Moguerza

Data Science Laboratory, Rey Juan Carlos University, c/Tulipán s/n, 28933
Móstoles, Spain
{alejandro.delgado,alberto.fernandez.isabel,isaac.martin,andrea.condado,
javier.moguerza}@urjc.es
http://www.datasciencelab.es

Abstract. In the age of Artificial Intelligence (AI), the use of trained
models is a common practice to solve a huge amount of different problems. However, it is highly complicated to understand the decision-
making process of these models and how the used training data affect
them during the production phase. For this reason, multiple techniques
related to model extraction have appeared. These techniques consist of
analyzing the behavior of a (sometimes partially) unknown model and
generating a clone that reacts similarly. This process is relatively simple with basic models, but it becomes arduous when complex models
must be analyzed and replicated. This paper tackles this issue by presenting the Neural NetwOrk Models (VENNOM) system. It is a general framework architecture to extract knowledge and provide explainability to unknown models. The proposed approach uses low-capacity,
high-explainability neural networks to produce flexible and interpretable
models. The proposed framework offers several advantages, particularly
in terms of obtaining explanations through visual and textual content
for models that are otherwise opaque. The framework has been tested
on tabular data sets, demonstrating its performance and potential.

Keywords: Model cloning · Knowledge extraction · Machine
Learning · Model understanding · Explainability

1 Introduction

The rise of (AI) has become a revolution in all senses in our daily days. Different
previously trained models can be found in multiple places where they make real-
time decisions, and humans are unaware of that. However, these models have
used data to reach their abilities, but it is difficult to understand what their
acquired knowledge is in the context, or what variables are enough relevant to
modify their behavior [19].

This issue leads to evaluating existing models put in production and trying
to extract their knowledge to produce clones. These new models can provide

© The Author(s), under exclusive license to Springer Nature Switzerland AG 2023
P. Quaresma et al. (Eds.): IDEAL 2023, LNCS 14404, pp. 257–268, 2023.
https://doi.org/10.1007/978-3-031-48232-8_24

information about the original unknown model or malicious details to attack infrastructures. This process is called model extraction, and it is only necessary to know information about the data the original model manages [5].

On the other hand, existing unknown models present behavior that could be difficult to predict. Therefore, the replication in another flexible and explainable model could solve questions about concept drift issues or incorrect classifications that cannot be understood easily. This fact is particularly relevant in critical situations where human beings use models to provide support (e.g., healthcare domain or military actions).

Thus, the formal problem definition is to use and understand Machine Learning (ML) models when only partial information is available.

In this context, decision trees or logistic regression models are some of the most easily interpretable by humans [7]. However, they present a lack of flexibility that makes them inadequate to produce model extractions generically. This issue has moved to select neural networks as the ideal candidate to clone unknown models. These models have a high capability to solve extremely difficult problems using a huge number of parameters and hidden layers (i.e., Deep Learning techniques [12]). As a consequence, the scientific community has put into the spotlight the possibility of generating techniques to evaluate and interpret neural network models.

This paper presents the Versatile Explanations through Neural NetwOrk Models (VENNOM) system. It is a novel complete framework to tackle the model extraction from a given dataset as input and produces different explanations through the generation of a neural network model. The extraction is achieved by consulting the original unknown model with several data. These data are obtained from an original dataset which can be not significant enough to train a new model. Therefore, the framework incorporates an automatic data augmentation module specifically designed for that task. The neural network model is developed by another module that analyzes the type of data and selects the most appropriate architecture automatically. Then, modifications over the neural network model are considered, increasing the number of neurons or layers, and later reducing them (i.e., a distillation process [25]) to compact the outcome. When the obtained result obtains enough similar results in the response to the unknown model, a third module enters the scene to provide explainability using Layer-wise Relevance BackPropagation (LRP) techniques to provide information about the most relevant variables. All these features locate VENNOM as a system that overcomes previous developments usually focused on standard model extraction issues and on producing basic neural network structures through genetic algorithms. Moreover, typically, these approaches do not include explainability tasks to show the strengths and weaknesses of the models.

Some experiments have been achieved at this point of development, showing its viability and capability to solve the provided issues.

The rest of the article is organized as follows. Section 2 describes the three main foundations of the framework: data augmentation, model extraction techniques, and explainability in neural networks. Section 3 presents the three different modules that integrate the proposal, while Sect. 4 focuses on a complete experiment with tabular data specifically developed to validate the prototype. Finally, Sect. 5 concludes and proposes possible future guidelines.

2 Related Work

This section tackles the foundations of the framework. It revisits the different perspectives related to data augmentation, model extraction techniques, and the explainability of neural network models.

In the first case, data augmentation techniques are addressed illustrating their ability to produce synthetic data (see Sect. 2.1). In the second case, multiple approaches focused on knowledge transferring, model extraction attacks, and model behavior replication are presented (see Sect. 2.2). Finally, explainability techniques are detailed highlighting how they are organized (see Sect. 2.3).

2.1 Data Augmentation

Unbalance data is a common problem in the ML domain. It is a situation where there are observations belonging to a majority class and at least one minority class (i.e., sets of observations with significantly fewer labels than the others) [21]. The problem is addressed through two main perspectives: undersampling and oversampling techniques. The first perspective uses different techniques focused on randomly deleting labeled observations from the majority class, establishing an equilibrium between the classes. On the other hand, the second perspective has the purpose of generating non-previously existing data through different techniques to level the number of observations of different classes. In this scenario, data augmentation techniques appear as a solution.

In general, there are three well-known techniques that can be used for a big amount types of data: random production, Synthetic Minority Oversampling TEchnique (SMOTE) and similar approaches, and generative adversarial neural networks.

Random production is the most simple strategy. It consists of generating new observations by making random modifications to some of the variables of the dataset. It could produce strange situations and also non-sense observations. It is typical to consider one-class-based algorithms to establish a frontier between useful points and outliers to reduce possible problems.

SMOTE related approaches are based on producing new observations measuring the similarity between observations and selecting neighbors. Then, intermediate points between neighbors are created. This task generates several observations very similar to the original ones. This process can be completed by introducing some limitations following the random production strategy in the form of one-class approaches. Notice that SMOTE usually produces duplicated data, therefore it is relevant to include some filters or enhanced versions of the algorithm. Some of these versions are Borderline SMOTE and k-means SMOTE.

Generative adversarial neural networks complete a system where both networks work together to produce new synthetic data. This data is created by introducing modifications in the original observations. If the modified instance is admitted by the system, then that observation can be included as an element of the original dataset. On the contrary, if the system does not admit the modified observation, it is considered an outlier.

In the proposal, VENNOM includes a module to address the data augmentation process using SMOTE and one-class Support Vector Machine (SVM).

2.2   Model Extraction

Model extraction is a process where relevant information about a (completely or partially complete) unknown model is replicated [10]. This cloning task is desirable to be independent of the architecture and parameters of the original model, and only the data accepted for this latter is relevant.

Data is used to make several requests to the original model to obtain the associated label or value (depending if it is a classification or regression model) in response. The data accepted by the original is the minimum requirement to know to start the replication process (i.e., the original model is considered a black box architecture). Alternatively, the features of the original model could be another starting point. In this case, the replication process is easier to achieve.

Delving into the new model, it could be developed mainly under two paradigms: cloning behavior (also called cloning knowledge) or cloning response measures [13]. The first perspective is focused on obtaining a similar response (a label if it is a classification problem, or a value if it is a regression problem) to each observation evaluated. On the other hand, the second perspective accepts possible differences in behavior fixing the general quality indicators like accuracy, precision, or recall. Therefore, this latter is commonly used only in classification problems where the quality of the model is easy to measure.

Regarding the context where this technique is commonly used, it appears mainly in two areas: privacy and cyber attacks [20], and legacy model cloning. In the first domain, the main usage consists of extracting the behavior of the original model with the goal of producing a similar one to hack the protected resource. These attacks are very typical, and the way of protecting against them is basically the limit of requests accepted by the original model (e.g., bot attacks). The second domain pursues the replication of the original model that is usually in production. The objective of this replication is usually to upgrade the architecture (changing the type of model and its configuration) [4], for explainability purposes [3], or to obtain a model that can be retrained [2]. This latter occurs when the model in production is frozen or it cannot be accessed.

In the case of VENNOM, it is a complete framework oriented to legacy model cloning where the model extraction is achieved for explainability purposes. Therefore, it is focused on cloning the behavior of the original model using a more flexible architecture to provide generalization capability and resilience to the approach.

2.3 Explainability in Neural Networks

Explainable Artificial Intelligence (XAI) is one the most relevant branches in AI. It consists of understanding the behavior of a model, knowing how it reacts to some data, and how it produces the expected result [1].

In general, these models can be classified as white boxes, black boxes, and transparent boxes [16]. The first models are those that provide information about parameters and architecture, but they are usually enough complex to understand them easily. The second models are those that present a hidden architecture and also unknown parameters. Neural networks are included between them because the deep architecture opaques how they work. The last models are ideals, as their design is relatively easy to explain, their parameters are known, and they usually follow a well-known algorithm.

Delving into explanations, they are mainly developed for black-box models (i.e., deep neural networks in general). They can be organized into qualitative and quantitative evaluations. The first provides information related to explainable methods from the perspective of human understanding and usability. The second evaluates the outcomes of the model for a specific task or problem.

Methods used in these explanations can be classified into two different types: ante-doc methods and post-hoc methods.

The ante-doc methods are model-specific and they are focused on a reduced set of similar models [18]. They are oriented to models considered white boxes and explain their content and behavior specifically. Well-known instances of these methods are: Bayesian Case Model (BCM), Generalized Additive Model (GAM), Bayes Rule List (BRL), and Neural Additive Model (NAM). The first two are qualitative while the other two are quantitative respectively.

The post-hoc methods are algorithms that look inside the model to interpret the decision path [14]. There are two categories: model-specific and model-agnostic [9]. The first category refers to explanations of a particular model and its learning process. Well-known instances are LRP, and Automatic Conceptbased Explanation (ACE). LRP provides quantitative explanations, while (ACE) is focused on qualitative explanations. The second category refers to methods that generate explanations independent of the internal logic used to achieve predictions. Well-known instances are Local Interpretable Model Explanation (LIME) and SHapley Additive ExPlanation (SHAP). LIME tackles quantitative explanations, while SHAP provides qualitative explanations respectively.

In the case of VENNOM, the explainability system produces quantitative evaluations based on LRP to calculate the relevance of the variables.

3 Proposed Framework

VENNOM has as its main purpose to extract knowledge from an incompletely known model. To achieve this task, it uses input data accepted by the original model to replicate, making several requests to gather the response (the output of the original model) [22]). Then, it replicates this model using neural networks

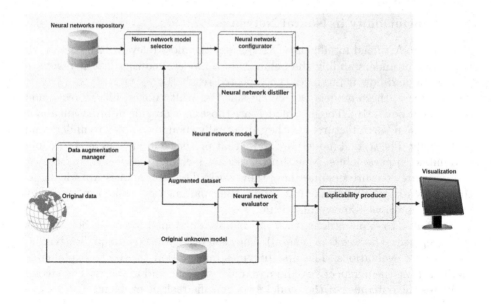

Fig. 1. Overview of the architecture of the framework. Each specific module is represented with a color. (Color figure online)

and tries to interpret and explain the observed behavior through the detection of the most relevant variables coming from the data.

This prototype can be seen as an aggregation of three modules (see Fig. 1). As neural networks work better with several observations, a data augmentation module is included to produce new synthetic data when it is needed. Otherwise, neural networks present different architectures according to the problem to tackle and the complexity of the data. For this reason, a module able to produce specific neural networks that can be modified easily is introduced. Finally, to interpret and explain the resulting models and the observed behavior, another module based on the LRP technique has been joined.

Notice that in the next sections, more details about these systems and how they interact in the complete prototype are provided.

3.1 Data Augmentation Module

This part of the framework addresses the generation of new data using a previously created dataset as starting point. It produces synthetic information, developing a vast set of similar observations and evaluating if they could be included in the original data as new items.

The module follows a specific generative methodology to produce new information. It starts using the original dataset as input and produces a one-class model ignoring the possible labels (i.e., response variable). Therefore, all the data is considered with the same label (i.e., it happens to be that each item has the label of being part of the original data).

A one-class SVM [15] has been selected to develop the one-class model. Then, all the original observations that are inside the decision frontier of the SVM are considered for the next step. Then, SMOTE is used to increment data creating new synthetic samples. SMOTE is used after reducing the samples of each one of the classes presented in the dataset. A parameter α establishes the percentage of the reduced elements. Next, these samples are filtered deleting duplicates to avoid possible inefficiency in the process. And finally, they are validated against the frontiers established by the one-class SVM. Samples are labeled as useful (they are similar enough to the original set of items) or outliers (they are not similar enough to the original set of items). Only useful samples are included in the final result. The process concluded when the algorithm reaches a predetermined number of samples.

Regarding the architecture of the *Data augmentation manager* module (see the blue area in Fig. 1), it is a simple structure where the outcome is the new dataset that includes the original and the synthetic data.

When the system has generated the augmented data, the next step consists of training the neural network model that fits better to the issue.

3.2   Neural Network Generation Module

This module is in charge of producing the resulting neural network that emulates the behavior of the original unknown model. It uses a set of specialized modules which work together in sequential steps to develop the final model.

Regarding the architecture, the components are the *Neural network model selector*, the *Neural network configurator*, the *Neural network distiller*, and the *Neural network evaluator* (see the pink area in Fig. 1). They are complemented with three items: the original unknown model, the dataset with synthetic data previously built, and a repository with different predefined neural networks. The outcome is the neural network that fits better with the specifications provided (i.e., the model that emulates the behavior of the original unknown model according to a measure of performance). In this case, the selected measure is General Performance Score (GPS) [6].

The *Neural network model selector* component uses the repository with several well-known architectures of neural networks [17]. It also processes the augmented data previously obtained to understand the possible type (e.g., tabular data, image processing, or time series). This procedure leads to selecting the most appropriate architecture.

The *Neural network configurator* produces variations in the selected architecture of the neural network according to the results obtained in the models that evaluate it. The addition of neurons in a specific layer, the insertion of complete new layers, the modification of connection through adapting the dropout, the substitution of the activation functions for other options, or the selection of the optimizer are some of the most typical tasks that are used following intelligent rules. These produce the desired flexibility in the architecture to adapt the neural network model to the expected performance and a well-built structure. This component also follows a knowledge preservation strategy. It consists of storing the weights calculated in a previous iteration with a more simple neural network and initializing the next adaptation fixing these weights to the existing neurons. This procedure avoids the complete random initialization of the neural network models, producing an incremental development [23].

The *Neural network distiller* component reduces possible oversized neural networks previously developed. It is important to simplify and compact the resulting model, to promote explainability and interpretation of the obtained outcomes. The distillation process can erase unnecessary neurons, connections, and also in weird cases, complete layers [25]. A low-capacity neural network with high explainability is the desired outcome at this point.

The *Neural network evaluator* component makes comparisons between the performance of the unknown original model and the selected neural network model. Notice that the idea behind this module is to compare the behavior of models according to the same input. To achieve it, the component selects a set of the augmented data (non-evaluated by either of the models yet) and observes the output. Then, both outputs are compared, and if the neural network model obtains enough similarity in results to the original unknown model the process finishes. Otherwise, new modifications in the neural network are included.

When this module reaches a satisfactory result, the model is processed by the next one to explain its features. This result is controlled by a parameter β that specifies the GPS value provided by the neural network model.

3.3   Explainability Module

This part of the framework follows the guidelines of the XAI approach [8]. Therefore, it is focused on explaining relevant information about the built model with neural networks. It can describe the most relevant variables of the model, illustrating which ones have more impact on the decisions. It is particularly interesting to understand the unknown model that was replicated.

The explanations generated by the system are provided through visual and textual content using LRP techniques to detect the relevance of the variables. XAI systems usually present a lack of details in this context, and the inclusion of a Natural Language Generation (NLG) engine eases human comprehension. The engine selected has been Generative Pre-trained Transformers (GPT) in release 3.5 turbo [24]. It allows consulting an API to make the process transparent to the rest of the module, providing the context and returning the outstanding explanation.

Regarding the architecture of this system, it comprehends a global module with several components inside (see the green area in Fig. 1). These components are in charge of the different described operations. It allows compartmentalizing the functionalities making them transparent to the user and between them.

When the system concludes, the final result of the framework and the three modules are shown to the users via a graphical interface. There, users can introduce modifications or visualize specific details about the produced neural network model, the augmented data, or the provided explanations.

4   Experiments

The proposal has been evaluated through a complete experiment to validate its viability and the quality of the generated outcome. Thus, the three modules of VENNOM are evaluated following its predefined working process.

The experiment starts by selecting a dataset. In this case, a well-known tabular dataset related to the detection of spam in emails was the chosen one [11]. It consists of a set of 57 continuous variables and a nominal class label. The number of observations is 4, 601, labeled as spam 1, 813 (i.e., 39.4%).

Then, a simple decision tree is used as the unknown model as it promotes the explanations of variables and their relevance. This fact is useful for achieving comparisons with the results of the cloned model in future steps. Note that, in a real situation, information regarding the unknown model will not be available.

The data are split into 80% for training purposes and 20% for testing purposes. For this experiment, the decision tree is built using 5 relevant variables, 4 levels of depth in the tree, and fixing the node split in at least 25 observations. This model produces a GPS equal to 0.86.

Once the starting point is completed, it is time to achieve the evaluation of the complete VENNOM system. To do that, the workflow of the framework is followed by evaluating the different modules in independent processes.

Firstly, the framework uses the data augmentation system to produce synthetic data. This decision is automatic and it comes from the reduced number of observations presented in the original dataset for training a neural network architecture. Thus, the system uses SMOTE and the one-class SVM to generate 4, 365 new observations after discarding several duplicates and outliers. To achieve this task, the parameter α (which proportion of data with the corresponding label selected to be used by SMOTE) is fixed to 0.2. Thus, for class 0 (i.e., labeled no spam), 20% of the data was used, and the same proportion for class 1 (i.e., labeled as spam).

After that, for testing purposes, another decision tree model was trained to ensure the quality of the new synthetic data, obtaining very similar accuracy values. Therefore, it can be said that the generation was satisfactory.

Secondly, the framework analyzes the spam dataset detecting its type. It decides that it is tabular data as expected and selects a basic multi-layer neural network architecture to work. Next, the parameter β is fixed to 0.8. Notice that this parameter is the threshold that establishes the acceptable GPS to stop the process that increases the complexity of the neural network (i.e., adding neurons and hidden layers). GPS is calculated using the response given by the unknown model and the response given by the neural network model, measuring the similarity between models.

Next, for evaluating the explainability module, the 5 most relevant variables from the decision tree are extracted in order of importance (see Table 1).

Coming up next, the generation step of the neural network is achieved 100 times (producing 100 different neural networks). It allows reducing the influence of random generations of neural networks and non-typical cases. All networks have an input layer with 5 neurons (one per variable) and a neuron as output.

Table 1. Arrangement of the variables in the unknown model (decision tree).

Position	Variable
1	char_freq_$
2	word_freq_remove
3	char_freq_!
4	capital_run_length_average
5	word_freq_hp

Table 2. Selections for each variable in the corresponding position for 100 neural networks.

Position	char_freq_$	word_freq_remove	char_freq_!	capital_run_length_average	word_freq_hp
1	**46**	30	11	7	6
2	19	**30**	15	19	17
3	14	14	**26**	23	23
4	10	14	20	**29**	27
5	11	12	27	22	**28**

The number of hidden layers is configurable by VENNOM, but they use Relu activation functions and a Sigmoid function in the output layer. Notice that hidden layers are modified by adding neurons or are included new ones when the GPS is significantly less than the β value.

Finally, the system analyzes each obtained neural network trying to reduce its architecture through a distillation process. Given the simplicity of the networks built, the distillation does not drop out any neuron or layer in this experiment.

For each neural network, the LRP technique is used. The obtained results indicate that most of the neural networks consider very similar relevance to the variables, maintaining the arrangement in most of the cases (see Table 2). It can be seen how the first two most relevant variables (*char_freq_$* and *word_freq_remove*) are selected with a higher margin than the others in the first positions, while the last two relevant variables (*capital_run_length_average* and *word_freq_hp*) are almost tied with other more relevant in the last positions. Moreover, the absolute mean relevance (LRP could produce negative relevance values) has been estimated by obtaining the next LRP values: *char_freq_$* equals to 3155.77, *word_freq_remove* equals to 2539.60, *char_freq_!* equals to 1834.94, *capital_run_length_average* equals to 844.59, and *word_freq_hp* equals to 625.76. This confirms the arrangement according to the relevance of the variables. Therefore, the neural network successfully detects the most relevant variables.

In the last step, the system uses these results to make requests to GPT 3.5 turbo to produce a final explanatory text easily understandable by humans.

Once the experiment is completed, it can be concluded that the prototype of VENNOM produces satisfactory results. The model extraction process is accurate and the behavior is very similar to the unknown model in most cases. The neural networks generated are adequate for the selected problem, and the most relevant variables are usually found and provided in a very similar arrangement of importance.

5 Conclusions

This paper has proposed the VENNOM system. It is a novel framework to extract knowledge from incompletely unknown models. It uses neural network models to produce clones of unknown models and explanations related to the clones through the LRP technique.

VENNOM uses three different modules. The first module tackles the data augmentation task. The second module analyzed the type of data and selected the best option for the neural network model. Then, the built model is simplified by applying distillation techniques. The third module addresses the explainability task, evaluating the neural network model to obtain the relevance of the variables and consulting GPT to produce the final textual explanation.

Regarding the evaluation of the framework, a complete experiment using tabular data has been developed. There, VENNOM has shown its ability to produce adequate results automatically. The data augmentation process worked perfectly, while the explainability task showed a good performance obtaining similar results to the unknown model in the detection of the most relevant variables. However, different experiments and scenarios should be developed to continue the validation of the proposal.

In the future, the framework could be upgraded to address different types of data, and also develop a specific distillation methodology. Finally, more complex explainable methods could be developed to evaluate the importance of the variables, and if the model breaks some rule or regulation related to some type of discrimination.

Acknowledgments. This work has been partially supported by the Spanish MICINN under the XMIDAS project (PID2021-122640OB-I00), the VAE project: TED2021-131295B-C33 funded by MCIN/AEI/ 10.13039/501100011033, and by the "European Union NextGenerationEU/PRTR", and donation of the Titan V GPU by NVIDIA.

References

1. Arrieta, A.B., et al.: Explainable artificial intelligence (XAI): concepts, taxonomies, opportunities and challenges toward responsible AI. Inf. Fusion **58**, 82–115 (2020)
2. Attaoui, M., Fahmy, H., Pastore, F., Briand, L.: Black-box safety analysis and retraining of DNNs based on feature extraction and clustering. ACM Trans. Softw. Eng. Methodol. **32**(3), 1–40 (2023)
3. Bastani, O., Kim, C., Bastani, H.: Interpretability via model extraction. arXiv preprint arXiv:1706.09773 (2017)
4. Cánovas Izquierdo, J.L., García Molina, J.: Extracting models from source code in software modernization. Softw. Syst. Model. **13**, 713–734 (2014)

5. Chang, C.C., Pan, J., Xie, Z., Hu, J., Chen, Y.: Rethink before releasing your model: ML model extraction attack in EDA. In: 28th Asia and South Pacific Design Automation Conference, ASPDAC 2023, pp. 1–6 (2023)
6. De Diego, I.M., Redondo, A.R., Fernández, R.R., Navarro, J., Moguerza, J.M.: General performance score for classification problems. Appl. Intell. **52**(10), 12049–12063 (2022)
7. Ding, W., Abdel-Basset, M., Hawash, H., Ali, A.M.: Explainability of artificial intelligence methods, applications and challenges: a comprehensive survey. Inf. Sci. **615**, 238–292 (2022)
8. Dwivedi, R., et al.: Explainable AI (XAI): core ideas, techniques, and solutions. ACM Comput. Surv. **55**(9), 1–33 (2023)
9. Ghorbani, A., Wexler, J., Zou, J.Y., Kim, B.: Towards automatic concept-based explanations. In: Advances in Neural Information Processing Systems, vol. 32 (2019)
10. Holzinger, A.: Introduction to machine learning & knowledge extraction (make) (2019)
11. Hopkins, M., Reeber, E., Forman, G., Suermondt, J.: UCI spambase data set (1999). https://archive.ics.uci.edu/ml/datasets/Spambase
12. Janiesch, C., Zschech, P., Heinrich, K.: Machine learning and deep learning. Electron. Mark. **31**(3), 685–695 (2021). https://doi.org/10.1007/s12525-021-00475-2
13. Junejo, K.N., Goh, J.: Behaviour-based attack detection and classification in cyber physical systems using machine learning. In: Proceedings of the 2nd ACM International Workshop on Cyber-Physical System Security, pp. 34–43 (2016)
14. Molnar, C., König, G., Bischl, B., Casalicchio, G.: Model-agnostic feature importance and effects with dependent features: a conditional subgroup approach. Data Min. Knowl. Discov. 1–39 (2023)
15. Razzak, I., Zafar, K., Imran, M., Xu, G.: Randomized nonlinear one-class support vector machines with bounded loss function to detect of outliers for large scale iot data. Futur. Gener. Comput. Syst. **112**, 715–723 (2020)
16. Saleem, R., Yuan, B., Kurugollu, F., Anjum, A., Liu, L.: Explaining deep neural networks: a survey on the global interpretation methods. Neurocomputing **513**(7), 165–180 (2022)
17. Sharkawy, A.N.: Principle of neural network and its main types. J. Adv. Appl. Comput. Math. **7**, 8–19 (2020)
18. Srihari, S.: Explainable artificial intelligence: an overview. J. Wash. Acad. Sci. **106**(4), 9–38 (2020)
19. Sullivan, E.: Understanding from machine learning models. Br. J. Philos. Sci. **73**(1) (2022)
20. Tramèr, F., Zhang, F., Juels, A., Reiter, M.K., Ristenpart, T.: Stealing machine learning models via prediction APIs. In: USENIX Security Symposium, vol. 16, pp. 601–618 (2016)
21. Wang, L., Han, M., Li, X., Zhang, N., Cheng, H.: Review of classification methods on unbalanced data sets. IEEE Access **9**, 64606–64628 (2021)
22. Wu, B., Yang, X., Pan, S., Yuan, X.: Model extraction attacks on graph neural networks: taxonomy and realisation. In: Proceedings of the 2022 ACM on Asia Conference on Computer and Communications Security, pp. 337–350 (2022)
23. Yao, X., Liu, Y.: Towards designing artificial neural networks by evolution. Appl. Math. Comput. **91**(1), 83–90 (1998)
24. Ye, J., et al.: A comprehensive capability analysis of GPT-3 and GPT-3.5 series models. arXiv preprint arXiv:2303.10420 (2023)
25. Zhang, L., Bao, C., Ma, K.: Self-distillation: towards efficient and compact neural networks. IEEE Trans. Pattern Anal. Mach. Intell. **44**(8), 4388–4403 (2021)

Threshold-Based Classification to Enhance Confidence in Open Set of Legal Texts

Daniela L. Freire[1(✉)] , Alex M. G. de Almeida[2] , Márcio de S. Dias[3] ,
Adriano Rivolli[4] , Fabíola S. F. Pereira[5] , Giliard A. de Godoi[4] ,
and Andre C. P. L. F. de Carvalho[1]

[1] University of Sao Paulo, Sao Paulo, Brazil
{danielalfrere,andre}@icmc.usp.br
[2] Ourinhos College of Technology, Ourinhos, Brazil
alex.marino@fatecourinhos.edu.br
[3] Federal University of Catalan, Catalão, Brazil
marciodias@ufcat.edu.br
[4] Federal Technological University of Paraná, Curitiba, Brazil
rivolli@utfpr.edu.br, giliardgodoi@alunos.utfpr.edu.br
[5] Federal University of Uberlândia, Uberlândia, Brazil
fabiola.pereira@ufu.br

Abstract. Machine Learning has revolutionized the categorization of vast legal documents, minimizing costs and improving evaluations. However, conventional models struggle with unseen data categories in real-world scenarios, a challenge termed Open Set Classification. Our study tackles the issue faced by the Court of Justice in São Paulo, Brazil, to identify recurring lawsuit themes from texts, as manual sorting is inefficient. We introduce a method to enhance confidence in text classification using an open dataset by converting multiclass challenges into binary ones with four confidence tiers. By testing various techniques, we found that combining doc2vec with the Support Vector Machine classifier delivers trustworthy results and robust performance. Ultimately, our method offers an effective solution for classifying legal texts confronting Open Set Classification issues in the legal sector.

Keywords: Machine learning · Text classification · Open Set Classification · Threshold-based Classification · Multiclass dataset · Natural language processing

1 Introduction

Machine Learning (ML) is revolutionizing the legal realm, classifying a high volume of unstructured textual documents as one of its most pertinent applications [2]. Much scientific research in the legal domain seeks to automate the task of reviewing those documents to reduce the cost of performing this task manually [15]. Properly categorized with relevant tags, these legal documents can improve evaluations by legal experts, automated recommendation systems, or other data-centric applications [1].

Traditional machine learning models are often trained under the closed-set assumption, where training and test samples are from the same distribution. However, in real-world scenarios, it can only obtain data samples representing some possible classes to compose the training set used in an ML model. However, all classes can be encountered in the test dataset, including ones the classifier has not trained. This scenario is called Open Set Classification (OSC), where the train data belongs to K classes, where K represents the total number of classes and, subsequently, during the testing phase, the model faces N distinct classes, out of which $N - K$ classes were not encountered during the training process. Regrettably, models can potentially assign deceptive confidence levels to unseen test samples [8, 14, 16, 17].

In a real-world problem, Brazil's Court of Justice of São Paulo (in Portuguese, Tribunal de Justiça de São Paulo - TJSP) seeks to identify repetitive lawsuit themes using document text of legal decisions. Repetitive themes denote clusters of appeals that share identical legal arguments from akin points of law. Within the TJSP, public officials encounter the everyday obstacle of manually perusing numerous extensive lawsuit verdicts to ascertain their classification as repetitive. Automating the identification of repetitive themes can accelerate the procedure, preserving time and streamlining the prompt resolution of lawsuits. In the training dataset, all samples of document text of legal decisions are labelled into three themes-nevertheless, new themes can be created. So, the test dataset has samples that do not belong to any of the themes for which the model was trained. There are far fewer processes belonging to the repetitive themes than there are. If a lawsuit is classified as a repetitive theme, this lawsuit has a different procedure in the court of justice. Therefore, a wrong classification can be harmful.

This paper proposes a strategy to provide more confidence in the multiclass text classification problem predictions with an open text dataset. We used the One-vs-rest (OvR) heuristic method associated with a threshold scale. OvR method splits the multiclass dataset into multiple binary classification problems. The threshold scale defines four confidence levels for the predictions: certainty, similarity, doubt, and unknown. We applied our strategy to the TJSP problem. We compared different learning technique combinations, varying word embedding techniques, and ML algorithms in the classification task of repetitive themes. Our goal was to point out which of the combinations provides more confidence in its predictions. The doc2vec with Support Machine Vector combination was best in predictions in production scenarios, avoiding wrong classifications of lawsuits in repetitive themes, besides it achieved good performance metrics at validation time (balanced accuracy = 0.9894, precision 0.9850, recall = 0.9840, f1-score = 0.9842, geometric mean = 0.9893, kappa = 0.9653). The remainder of this paper is organized as follows. The proposal tool is discussed in Sect. 2, and the experimentation with a real case study is reported in Sect. 3. Finally, Sect. 4 reports the main conclusions and discusses future research directions.

2   Proposed Strategy

This section proposes a strategy to enhance confidence in multiclass text classification with an open text dataset. We describe the motivation to use the One-vs-rest method and our proposal of threshold scale in classifying legal texts.

2.1   Motivation for One-vs-Rest

Text classification consists of assigning each text its corresponding class. A binary classification corresponds to a classification with only two classes, and a multiclass classification corresponds to one with more than two classes. There are two approaches to performing multiclass classification. The former is to directly apply multiclass techniques such as decision trees and obtain the predicted class. The latter is to decompose the multiclass problem into several binary problems that are easier to solve. One decomposition way is One-vs-rest (OvR for short, also referred to as One-vs-All or OvA)One-vs-rest (OvR for short, also referred to as One-vs-All or OvA), a heuristic method that splits the multiclass dataset into multiple binary classification problems, then a binary classifier is trained on each binary classification problem and predictions are made using the model that is the most confident [9].

Since OvR requires a model for each class, it is typically not recommended for large datasets with millions of samples, slow models such as neural networks, or problems with hundreds of classes. However, for open datasets, the OvR technique can help to avoid a new sample belonging to an unseen class being incorrectly classified in one of the trained classes.

OvR can be implemented using the OneVsRestClassifier class within the Python programming language's scikit-learn library [10]. This class offers the flexibility to apply the OvR method with various classifiers, such as LR, SVM, RF, LGBM, and CAT.

2.2   Threshold Scale

In the classification of open sets, a threshold-based approach is often employed. This approach involves rejecting or classifying input samples into distinct classes with well-defined training samples based on thresholds that are determined empirically [3,4,13]. Therefore, the threshold plays a critical role in the decision-making process.

Typically, a confidence threshold is established; when the confidence surpasses this threshold, the sample is accepted (considered positive). Otherwise, it is rejected (deemed harmful). Consequently, discrimination effectively becomes a binary classification problem, distinguishing between acceptance and rejection options. However, the difficulty in this problem arises from the scarcity or lack of representativeness in the harmful data. In simpler terms, obtaining the rejected samples proves challenging, and even if collected using a specific strategy, they may not accurately represent future samples that require rejection.

When dealing with real-world problems, it is essential to be mindful of the potential consequences of a false-positive classification. We often define a high threshold for positively classifying a sample to mitigate this risk. At the same time, we aim to reduce the effort required for empirically analyzing extensive lawsuit texts. By carefully considering these factors, we can arrive at more accurate and efficient decision-making processes. So, for our confidence threshold scale, we define the three levels to acceptance and one to rejection, as illustrated in Fig. 1 and described as follows.

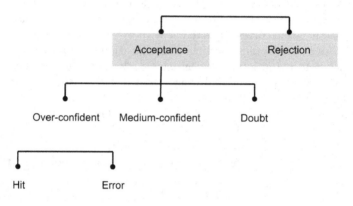

Fig. 1. Threshold Scale

- **Over-confident or O** - when the highest probability to belong to a given class is higher or equal to the maximal threshold. In training and validation datasets:
    - **Hit or H** - when the class with the highest probability is the actual class.
    - **Error or E** - when the class with the highest probability differs from the actual class.
- **Medium-confident or M** - when the highest probability to belong to a given class is higher or equal to the maximal threshold and lower to the medium threshold.
- **Doubt or D** - when the sum of the two highest probabilities to belong to a given class is higher than the maximal threshold and lower the maximal threshold and when the absolute value of the difference of the two most enormous probabilities is lower than the minimal threshold. This case occurs when there are two high and similar probabilities of belonging to a given class.
- **Rejection or R** - otherwise.

We can mathematically express a prediction according to Eq. 1.

$$prediction = \begin{cases} \mathbf{O} \; when \; (\hat{p} \geq \tau_{max}) \\ \mathbf{M} \; when \; (\tau_{max} \leq \hat{p} < \tau_{med}) \\ \mathbf{D} \; when \; ((\hat{p} + \hat{p}_2) > \tau_{max} \; and \; (|(\hat{p} - \hat{p}_2)|) < \tau_{min}) \\ \mathbf{R} \; Otherwise \end{cases} \quad (1)$$

We mathematically model the score to the results achieved by ML models. Then, the model with the highest score will be considered the most suitable for the problem addressed. To formulate this mathematical model, first, we assign a weight to each acceptance and rejection level, c.f. Table 1. After, we can mathematically express a prediction score to validation according to Eq. 2 and a prediction score to production according to Eq. 3. In these equation, the symbol '$|$ $|$' means *number of*. In Eq. 2, the total number of samples in the validation dataset is expressed as $|Tot| == |H| + |E| + |M| + |D| + |R|$. In Eq. 3, the total number of samples in the production dataset is expressed as $|Tot\_p| == |O\_p| + |M\_p| + |D\_p| + |R\_p|$.

$$
\begin{aligned}
\boldsymbol{val\_score} &= [|Acceptance| * \omega_{accep} - |Rejection| * \omega_{reject}]/|Tot| \\
&= [(|H| * \omega_{hit} + \#E * \omega_{error} + |M| * \omega_{medium} \\
&\quad + |D| * \omega_{doubt}) - |R| * \omega_{reject}]/|Tot|
\end{aligned}
\tag{2}
$$

$$
\begin{aligned}
\boldsymbol{prod\_score} &= [|Acceptance\_p| * \omega_{accep\_p} - |Rejection\_p| * \omega_{reject\_p}]/\#Tot\_p \\
&= [(|O\_p| * \omega_{over\_p} + |M\_p| * \omega_{medium\_p} + |D\_p| * \omega_{doubt\_p}) \\
&\quad - |R\_p| * \omega_{reject\_p}]/|Tot\_p|
\end{aligned}
\tag{3}
$$

Table 1. Confidence Level Weights

			Validation weight	Production weight
Acceptance	Over-confident	Hits	ω_{over_hit}	ω_{over_p}
		Errors	ω_{over_error}	
	Medium-confident		ω_{medium}	ω_{medium_p}
	Doubt		ω_{doubt}	ω_{doubt_p}
Rejection			ω_{reject}	ω_{reject_p}

The validation and production score sum, c.f, find the final score. Equation 4.

$$
\begin{aligned}
\boldsymbol{score} &= val\_score + prod\_score \\
&= [(|H| * \omega_{hit} + |E| * \omega_{error} + |M| * \omega_{medium} + |D| * \omega_{doubt}) \\
&\quad - |R| * \omega_{reject}] \\
&\quad + [(|O\_p| * \omega_{over\_p} + |M\_p| * \omega_{medium\_p} + |D\_p| * \omega_{doubt\_p}) \\
&\quad - |R\_p| * \omega_{reject\_p}]
\end{aligned}
\tag{4}
$$

3 Study Experimental

In this section, we use our proposed strategy in an experiment that aims to find an ML model that results in confident predictions in the training and production datasets, considering that the model is faced with new and unknown classes in the production environment. We test fifteen scenarios combining three-word embedding techniques (tf-idf, hash, and doc2vec) and ML algorithms, namely Logistic Regression (LR), Support Machine Vector (SVM), Random Forest (RF), Light Gradient-Boosting Machine (LGBM), and CatBoost (CAT) in the classification task of repetitive themes.

3.1   Dataset

The dataset for lawsuit decisions from Court São Paulo has been divided into three classes, described in Table 2 [5]. The training dataset contains 5,244 lawsuit decision document texts, with 3,989 samples of theme 1033, 798 of theme 1039, and 457 of theme 1101. The validation dataset has 752 lawsuit decision document texts, 90 samples of theme 1033, 128 of theme 1039, and 534 of theme 1101. Lastly, the production dataset includes 41,724 samples, of which only 21 belong to one of the classes known during training.

Table 2. Repetitive Theme Descriptions

Theme	Descriptions
1033	concerns the interruption of the statute of limitations to claim compliance with a collective sentence due to the filing of a protest action or collective execution by legitimized to propose collective demands
1039	deals with the setting of the initial term of the prescription of the indemnity claim against an insurer in the contracts, active or extinct, of the Housing Financial System
1101	addresses the final term of the incidence of remunerative interest in cases of collective and individual actions claiming the replacement of inflationary purges in savings accounts

3.2   Preprocessing

Firstly, we performed the preprocessing of document texts, deleting stop-words and punctuation, in addition to some transformations of some terms in the Brazilian legal domain, such as adding the word laws/norms/legal codes or sections of them (articles, paragraphs, items) with their numbering and abbreviation in full, for example: "article 42 of the CDC" is transformed into "article_42 of the Consumer Defense Code" (in Portuguese, Código de Defesa do Consumidor). We used the packages from the Natural Language Toolkit (nltk) and string libraries. The nltk library provides stop words set, and string library punctuation marks set. Then, we normalized the frequency of each word by creating a word-frequency table via tokenization, where each sentence has a score according to the words represented.

Word Embedding are a technique used to derive text features, enabling us to feed these features into an ML model for text data processing. Their purpose is to retain both syntactic and semantic information. These embeddings are valuable for reducing the dimensionality of a corpus, capturing inter-word semantics, and depicting or visualizing underlying usage patterns within the corpus used for their training. We use tf-idf [12] for text vectorization, limiting our vocabulary to 3,000 features, where each feature, or n-gram, represents a measurable piece of text that can be used for analysis. N-grams are contiguous sequences of words collected from a text sequence. We used n-grams with n from 1 to 5. Hash vectorizer [6] transforms a set of text documents into a matrix representing the frequency of tokens. This text vectorization method employs the hashing trick to establish the mapping between the token stream names and their corresponding integer feature indices. We used the same number of features (3000) and n-grams (1 to 5). The collection of doc2vec [7] utility scripts facilitates the training of doc2vec models, provided by the gensim Python library [11]. Once these models are trained, they allow for the extraction of document embeddings from the dataset or the inference of embeddings for different datasets. We will create a Doc2Vec model with 1024-dimensional vectors and train it on the training corpus for 40 iterations. We have set the minimum word count to 2 to filter out words with very few occurrences.

3.3 Environment and Supporting Tools

Regarding environment and supporting tools, the computational language used to develop the codes was Python version 3.8. The platform for executing these codes was the platform Google Colaboratory, which allows the execution of code in the Python language, using its computational resources, such as CPUvirtual, RAM, NVIDIA TESLA P100 GPUs (Graphics Processing Unit), and TPUs (Tensor Processing Units).

3.4 Training

We adopt the same execution pipeline, default hyperparameter setting, and OneVsRestClassifier class for all algorithms except LR and SVM. We use the SGDClassifier class in the Python programming language's scikit-learn library for LG and SVM. This class is utilized to apply the stochastic gradient descent (SGD) technique for handling classification problems. The SGDClassifier builds an estimator using a regularized linear model and SGD-based learning. Additionally, we used the CalibratedClassifierCV class for SVM because this algorithm does not directly produce predictions of probabilities, so a prediction of probabilities must be approximated. Some examples include neural networks, support vector machines, and decision trees. For CalibratedClassifierCV, the parameters used were estimator = ovr, cv = 5, and method = sigmoid'.

3.5 Threshold Scale Application

In the validation dataset, we seek an ML model that maximizes the total number of *Acceptance*, except by the errors and minimizes *Rejection* since all samples have known classes in validation. Nevertheless, we consider the hit number ($|H|$) as the most positive significant result; the error number $|E|$ and $|R|$ as the most negative significant results; $|M|$ as the second most positive significant result; and $|D|$ as the least positive considerable result. Whereas, in the production dataset, most samples have unknown classes, so our goal is the opposite, i.e., we seek a machine-learning model that minimizes *Acceptance* and maximizes the total number of *Rejection*. The same way, we consider the over-confidence number ($|O_p|$) as the most negative important result; $|R_p|$ as the most positive significant result; $|M_p|$ as the second most negative significant result; and $|D_p|$ as the least negative considerable result. Then, we assign the values to the weights as follows: $\omega_{hit} = 10$, $\omega_{error} = (-10)$, $\omega_{medium} = 5$, $\omega_{doubt} = 1$, $\omega_{reject} = (-10)$, $\omega_{over\_p} = (-10)$, $\omega_{medium\_p} = (-5)$, $\omega_{doubt\_p} = (-1)$, $\omega_{reject\_p} = 10$. Substituting the weight values in Eq. 2 and 3, we arrive at Eq. 5.

$$
\begin{aligned}
score &= [((|H| - |E|) * 10 + |M| * 5 + |D| - |R| * 10]/|Tot| \\
&\quad - [((|O\_p| * 10 + |M\_p| * 5 + |D\_p| - |R\_p| * 10]/|Tot\_p|
\end{aligned} \tag{5}
$$

3.6   Results

We measure the performance using traditional metrics, such as balance accuracy (bal_acc), precision, recall, and f1-score. Additionally, we use the geometric mean (g-mean) and Cohen's kappa (kappa), which performance metrics have been used to evaluate imbalance classification tasks. G-mean is calculated based on two parameters referred to as sensitivity and specificity. Sensitivity evaluates the positive class's performance, while specificity assesses the negative class's performance. On the other hand, kappa measures the agreement between the target and predicted classes, similar to accuracy but accounting for the possibility of change predictions. The ML community has embraced these metrics for comparing classifier performances. We also measure the training time (time).

According to Table 3, the models achieved performance above 94% in all performance metrics. The model of SVM∧doc2vec, 99.96% obtained the best bal_acc. The best precision was 99.02% obtained by the model of SVM∧tf-idf. The best recall was 98.94% obtained by models of SVM∧tf-idf and RF∧tf-idf. Model of SVM∧doc2vec, 98.96% obtained the best f1-score. The highest g-mean was 99.50% obtained by the model of RF∧tf-idf. The best kappa was 97.68% obtained by models of SVM∧tf-idf and RF∧tf-idf. Regarding training time, the LR and SVM algorithms were quickest when using doc2vec. The most extended training was with the CAT algorithm using the tf-idf, hash, and doc2vec.

We use our confidence threshold scale, and the detailing of the results obtained for the levels of Acceptance and Rejection in validation and production datasets are reported in Table 4. The scores of models in validation and production datasets are shown in Table 5. The model that achieved the best validation score LR∧doc2vec (9.62 points), followed by CAT∧tf-idf (9.57 points), and LGBM∧tf-idf (9.56 points). Regarding production, the model that achieved the best score SVM∧doc2vec (9.85 points), followed by RF∧hash (9.73 points) and LR∧hash (9.73 points). Table 6 reports the final ranking of scores, obtained

Table 3. Performance Metrics (%).

Algorithm	Word Embed.	Bal_acc	Precision	Recall	F1-score	G-mean	Kappa	Time
LR	tf-idf	99.13	98.39	98.14	98.19	99.12	95.98	49.74
	hash	98.69	98.11	97.87	97.93	98.69	95.42	25.23
	doc2vec	96.85	97.93	97.87	97.87	96.79	95.32	0.8
SVM	tf-idf	99.30	**99.02**	**98.94**	**98.96**	99.30	**97.68**	42.65
	hash	99.25	98.54	98.40	98.43	99.25	96.54	26.96
	doc2vec	**99.96**	98.50	98.40	98.43	98.94	96.53	2.1
RF	tf-idf	99.50	98.99	**98.94**	98.95	**99.50**	**97.68**	62.8
	hash	98.29	97.74	97.21	97.33	98.28	94.03	68.8
	doc2vec	98.29	97.65	97.21	97.31	98.28	94.02	30.1
LGBM	tf-idf	99.44	98.89	98.80	98.82	99.44	97.40	78.5
	hash	98.16	97.82	97.34	97.35	98.15	94.29	74.2
	doc2vec	97.63	97.52	97.34	97.38	97.62	94.27	15.9
CAT	tf-idf	99.19	98.49	98.27	98.32	99.18	96.26	646.9
	hash	98.16	97.82	97.34	97.35	98.15	94.29	615.1
	doc2vec	98.06	97.78	97.61	97.65	98.06	94.84	211.8

by the sum of validation and production scores, c.f. Eq. 4. The best model was SVM∧doc2vec (18.29 points), followed by LR∧tf-idf (18.26 points), and SVM∧tf-idf (17.71 points).

3.7   Discussions

According to the performance metrics results obtained in the validation dataset, the models trained with SVM and RF algorithms and tf-idf word embedding are the best options, c.f. Table 3. However, since all combinations achieved close and good values, it is necessary to apply statistical tests to verify if there is a significant difference between them, which was not addressed in this paper, remaining as a suggestion for future treatments.

Table 5 shows that the results in validation and production are pretty different. LR∧doc2vec, the first position in the validation score, got in the last position in the production score. On the other hand, SVM∧doc2vec was in the eleventh position in the validation ranking, but with a good score of 8.44 points, it was the best in the production score of 9.85. The final score pointed out SVM∧doc2vec as the model most appropriate to the classification of lawsuit decisions from Court São Paulo in repetitive themes, shown in Table 6. This score confirms the results found in the balanced accuracy metric to SVM∧doc2vec. The third (SVM∧tf-idf) and fifth RF∧tf-idf) position in the ranking also is not a surprise because of the good performance metrics results. LR∧tf-idf was not in first position in validation and production scores individually but achieved second in the final score.

Table 4. Model Results.

Algorithm	Word Embbeg.	Validation					Production			
		|H|	|E|	|M|	|D|	|R|	|O_p|	|M_p|	|D_p|	|R_p|
LR	tf-idf	669	1	41	0	41	297	581	3	40843
	hash	316	1	369	0	66	204	279	289	40952
	doc2vec	736	12	1	2	1	38562	558	1741	863
SVM	tf-idf	703	1	28	1	19	645	3358	157	37564
	hash	692	1	31	0	28	609	5859	892	34364
	doc2vec	653	1	54	0	44	159	153	90	41322
RF	tf-idf	641	1	66	0	44	679	5389	78	35578
	hash	412	1	195	0	144	146	104	611	40863
	doc2vec	336	1	214	0	201	38243	556	1726	1199
LGBM	tf-idf	729	7	8	1	7	32399	3298	308	5719
	hash	725	12	5	1	9	24999	6903	724	9098
	doc2vec	719	5	9	2	17	38401	648	1731	944
CAT	tf-idf	723	3	17	0	9	9507	9135	947	22135
	hash	704	13	14	0	21	18546	10503	972	11703
	doc2vec	692	3	26	1	30	3495	13178	945	24106

Table 5. Validation and Production Scores

Validation				Production			
Order	Algorithm	Word Embbeg.	Score_v	Order	Algorithm	Word Embbeg.	Score_p
1	LR	doc2vec	9.62	1	SVM	Rdoc2vec	9.85
2	CAT	tf-idf	9.57	2	RF	hash	9.73
3	LGBM	tf-idf	9.56	3	LR	hash	9.73
4	LGBM	hash	9.40	4	LR	tf-idf	9.65
5	LGBM	doc2vec	9.33	5	SVM	tf-idf	8.44
6	SVM	tf-idf	9.27	6	RF	tf-idf	7.72
7	SVM	hash	9.02	7	SVM	hash	7.37
8	CAT	hash	9.00	8	CAT	doc2vec	3.34
9	CAT	doc2vec	8.94	9	CAT	tf-idf	1.91
10	LR	tf-idf	8.61	10	CAT	hash	−2.92
11	SVM	doc2vec	8.44	11	LGBM	hash	−4.66
12	RF	tf-idf	8.36	12	LGBM	tf-idf	−6.80
13	LR	hash	5.76	13	RF	doc2vec	−8.99
14	RF	hash	4.85	14	LGBM	doc2vec	−9.10
15	RF	doc2vec	3.20	15	LR	doc2vec	−9.14

Table 6. Final Scores

Order	Algorithm	Word Embbeg.	Score_v	Score_p	Score
1	SVM	doc2vec	8.44	9.85	18.29
2	LR	tf-idf	8.61	9.65	18.26
3	SVM	tf-idf	9.27	8.44	17.71
4	SVM	hash	9.02	7.37	16.39
5	RF	tf-idf	8.36	7.72	16.08
6	LR	hash	5.76	9.73	15.49
7	RF	hash	4.85	9.73	14.58
8	CAT	doc2vec	8.94	3.34	12.28
9	CAT	tf-idf	9.57	1.91	11.48
10	CAT	hash	9.00	−2.92	06.08
11	LGBM	hash	9.40	−4.66	4.74
12	LGBM	tf-idf	9.56	−6.80	2.77
13	LR	doc2vec	9.62	−9.14	0.48
14	LGBM	doc2vec	9.33	−9.10	0.23
15	RF	doc2vec	3.20	−8.99	−5.78

4   Conclusion

The challenge of classifying extensive legal texts, especially in scenarios with the potential for unseen classes during testing, necessitates robust and dependable methods. The study we presented aptly addressed a pertinent challenge faced by the Court of Justice of São Paulo, Brazil, ensuring a more streamlined and accurate way to identify repetitive themes in legal lawsuits. We successfully transformed the multiclass dataset problem into manageable binary classification challenges by adopting a threshold-based classification strategy.

Our findings underscore the efficacy of using the One-vs-Rest heuristic combined with a threshold scale. Notably, the doc2vec with Support Vector Machine (SVM) classifier exhibited outstanding performance, significantly reducing classification errors. These results are particularly salient considering the practical implications in the legal domain: an erroneous classification can have direct and potentially harmful consequences on the court's procedures.

However, it is imperative to note that while our method has showcased a promising approach for addressing open-set classification challenges, it is full of potential improvements. One immediate avenue for future research is the exploration of model hyperparameters, mainly focusing on regularization techniques. Regularization could further enhance the model's performance, ensuring it stays within the training data while maintaining its generalization capabilities for unseen classes.

Our study contributes significantly to the growing body of research aimed at automating and refining text classification tasks in the legal sector. The proposed strategy is theoretically sound and beneficial, potentially revolutionizing how legal documents are processed and classified. By continually refining and tuning our models, we believe the future holds even more accurate and efficient methods for handling such challenges in legal texts.

References

1. Coelho, et al.: Text classification in the Brazilian legal domain. In: International Conference on Enterprise Information Systems, pp. 355–363 (2022)
2. Fernandes, W.P.D., et al.: Extracting value from Brazilian court decisions. Inf. Syst. **106**, 101965 (2022)
3. Geng, C., Chen, S.: Collective decision for open set recognition. IEEE Trans. Knowl. Data Eng. **34**(1), 192–204 (2020)
4. Geng, C., Huang, S.J., Chen, S.: Recent advances in open set recognition: a survey. IEEE Trans. Pattern Anal. Mach. Intell. **43**(10), 3614–3631 (2020)
5. de Justiça Secretaria de Jurisprudência, S.T.: Precedentes qualificados (2023)
6. Kanada, Y.: A vectorization technique of hashing and its application to several sorting algorithms. In: PARBASE, pp. 147–151 (1990)
7. Le, Q., Mikolov, T.: Distributed representations of sentences and documents. In: International Conference on Machine Learning, pp. 1188–1196. PMLR (2014)
8. Miller, D., Sunderhauf, N., Milford, M., Dayoub, F.: Class anchor clustering: a loss for distance-based open set recognition. In: Proceedings of the IEEE/CVF Winter Conference on Applications of Computer Vision, pp. 3570–3578 (2021)
9. Murphy, K.P.: Machine Learning: A Probabilistic Perspective. MIT Press, Cambridge (2012)
10. Pedregosa, F., et al.: Scikit-learn: machine learning in python. J. Mach. Learn. Res. **12**, 2825–2830 (2011)
11. Řehřek, R., Sojka, P., et al.: Gensim—statistical semantics in python. Retrieved from genism.org (2011)
12. Salton, G., Buckley, C.: Term-weighting approaches in automatic text retrieval. Inf. Process. Manag. **24**(5), 513–523 (1988)
13. Scheirer, W.J., Jain, L.P., Boult, T.E.: Probability models for open set recognition. IEEE Trans. Pattern Anal. Mach. Intell. **36**(11), 2317–2324 (2014)
14. Sun, X., Yang, Z., Zhang, C., Ling, K.V., Peng, G.: Conditional gaussian distribution learning for open set recognition. In: Proceedings of the IEEE/CVF Conference on Computer Vision and Pattern Recognition, pp. 13480–13489 (2020)
15. Wei, F., Qin, H., Ye, S., Zhao, H.: Empirical study of deep learning for text classification in legal document review. In: 2018 IEEE International Conference on Big Data (Big Data), pp. 3317–3320. IEEE (2018)
16. Yoshihashi, R., Shao, W., Kawakami, R., You, S., Iida, M., Naemura, T.: Classification-reconstruction learning for open-set recognition. In: Proceedings of the IEEE/CVF Conference on Computer Vision and Pattern Recognition, pp. 4016–4025 (2019)
17. Yu, Q., Aizawa, K.: Unsupervised out-of-distribution detection by maximum classifier discrepancy. In: Proceedings of the IEEE/CVF International Conference on Computer Vision, pp. 9518–9526 (2019)

Comparing Ranking Learning Algorithms for Information Retrieval Systems

J. Zilles[1] , E. N. Borges[1] , G. Lucca[2(✉)] , C. Marco-Detchart[3] ,
Rafael A. Berri[1] , and G. P. Dimuro[1]

[1] Centro de Ciências Computacionais, Universidade Federal do Rio Grande,
Av. Itália km 08, Campus Carreiros, Rio Grande 96201-900, Brazil
{eduardoborges,rafaelberri,gracaliz}@furg.br
[2] Programa de Pós-Graduação em Engenharia Eletrônica e Computação,
Universidade Católica de Pelotas, Gonçalves Chaves, Pelotas 96015-560, Brazil
giancarlo.lucca@ucpel.edu.br
[3] Valencian Research Institute for Artificial Intelligence (VRAIN), Universitat
Politècnica de València (UPV), Camino de Vera s/n, 46022 Valencia, Spain
cedmarde@upv.es

Abstract. The growth of machine learning applications in various fields
has enabled the advancement of Information Retrieval systems. As a
result of this evolution, it has become possible to solve the well-known
document classification problem. In the beginning, document positions
within a result were given from a score, where each document receives
an assigned value based on the terms used in the input query. The use
of machine learning in this field is known as Learning to Rank, which
allows the classification of documents to better meet user search requirements, taking into account aspects such as document preference, importance, and relevance. This paper presents a comparison of different algorithms for ranking documents using machine learning. It is observed that
RankSVM presents relatively satisfactory results in smaller datasets,
while algorithms that use Gradient Boosting obtain better results for
larger datasets.

Keywords: Machine Learning · Information Retrieval · Learning to
Rank

1 Introduction

The high increase in the number of document indexing databases makes it almost
impossible for users to find relevant information without seeking help from some
Information Retrieval (IR) system [5]. These systems were developed to make a
certain set of information available to users. It is expected that for a given search
term, the returned document order follows a logical and relevant sequence, with
the most relevant documents shown at the top and the least relevant at the
bottom of the resultant search query.

In this way, the ranking problem consists of sorting a document list using
criteria provided by the user to return the best answer at the top of the list.
This problem is considered the main question in the IR research field [1].

P. Quaresma et al. (Eds.): IDEAL 2023, LNCS 14404, pp. 281–289, 2023.
https://doi.org/10.1007/978-3-031-48232-8_26

Document ranking has three main characteristics, (i) relevance, (ii) Importance and (iii) Preference. The first produces a score for each document, the second the degree of importance of the document concerning the input and the last evaluates the behavior of the user who is searching for a document [3].

Once the user preference is known, it is possible to use Machine Learning (ML) techniques [10] to support the ranking problem solution. Precisely, is this preference that is used to improve the model, also known as the online model, since the model is continuously learning according to the new data input. The usage of ML to solve ranking problems is defined as Learning to Rank (LR) [6], which is a research field in the IR area.

The evaluation of a ranking system consists of the application of certain metrics, in a similar way that is used in ML. Then, a ranking generated by a model or IR algorithm is considered satisfactory if the obtained results are supported by the usage of such metrics. We highlight for example, the Mean Averaging Precision (MAP) [7], Discounted Cumulative Gain (DCG), and Normalized Discounted Cumulative Gain (NDCG) [4].

Having this in consideration, the objective of this paper is to analyze the training phase of different LR algorithms along with their evaluation by using practices applied in the literature. To do so, we consider the usage of public datasets in which we applied the algorithms and analyzed the results in terms of time consumption, performance, and quality.

This paper is organized as follows. The adopted methodology is described in Sect. 2 and the obtained results in Sect. 3. In the end, the main conclusions are drawn.

2   Methodology

This work starts with a dataset as input, which is converted to the format used by the algorithms. Then, the algorithm that will be used in training is picked. After that, once the algorithm is chosen, the training is made followed by the analysis of the results. If the obtained result is considered satisfactory, the model is up to be used in the prediction of the relevance of the documents. Otherwise, an adjustment of the hyperparameters is made and another training is done. In what follows the main points of the study are stated.

2.1   Considered Tools

In this paper, we make use of open-source tools to apply the considered methods. Precisely, we have adopted the LightGBM[1] and eXtreme Gradient Boosting (XGBoost)[2]. This is due to the fact that this tool allows development using different programming languages, as well as LR algorithms. Moreover, both implemented ML and combine it with LR with the Gradient Boosting [8]. The main characteristics of the tools are:

---

[1] https://lightgbm.readthedocs.io/en/latest/.

[2] https://xgboost.readthedocs.io/en/latest/index.html.

Table 1. Summary of the datasets considered in the study.

Dataset	Attributes	Queries	Train	Validation	Test	Documents
OHSUMED	45	106	9.219	3.538	3.383	16.140
TD2003	44	50	29.085	10.070	10.016	49.171
TD2004	44	75	44.156	15.012	15.002	74.170
MSLR-WEB10K	136	10.000	723.412	235.259	241.521	1.200.192
MSLR-WEB30K	136	30.000	2.270.296	747.218	753.611	3.771.125

- **LightGBM:** is considered as a Gradient Boosting for ML [9]. The tool provides decision-making based on a tree structure using a level wise growing. Also, it can be used in tasks such as: Classification, Regression, and ranking. Another interesting characteristic is that the tool allows the changing of the considered boosting technique, the standard one is Gradient Boosting Decision Tree (GBDT), to different alternatives like Random Forrest.
- **XGBoost:** is an efficient, fast, and scalable based on tree boosting [2]. It is used in different contests related to ML. This tool provides a set of hyperparameters to improve the model training. One of these parameters is the considered boosting technique (called as booster) which by default is the Gradient Boosting Tree. However, others can be considered like: Dropouts meet Multiple Additive Regression Trees (dart) and Linear (bglinear).

Another important software used in this study is the **FLAML** [11]. It consists of a solution based on Automated Machine Learning (AutoML), where given a training dataset and an error metric, it finds the best model with the fitted parameter sets in a short period. The training time is defined by the user, and at the same time, if the algorithm occupies 80% of the training time the tool suggests an increase in the given time to provide a better result.

2.2 Datasets

The datasets considered in this paper are publicly available in the LETOR[3] collection. This project is an alternative that provides a benchmark to be used in the comparison among LR algorithms. We summarize the datasets in Table 1, which is sorted by the total of documents. These documents are divided into train, validation, and testing. Representing a distribution of the records in 60, 20, and 20% respectively.

2.3 Parameter Setup

As mentioned before, we have into consideration different tools to execute the models. Thus, we provide in this subsection the parameters used by the methods in each one. The training parameters were generated by the FLAML, with

---

[3] For more information about the LETOR access – https://www.microsoft.com/en-us/research/project/letor-learning-rank-information-retrieval/.

Table 2. Parameter setup of the algorithms used by the LightGBM.

Parameter	lambdarank and rank_xendcg					Regression				
	OHSUMED	TD2003	TD2004	MSLR-WEB10K	MSLR-WEB30K	OHSUMED	TD2003	TD2004	MSLR-WEB10K	MSLR-WEB30K
max_bin	10	9	7	1023	255	6	8	7	1023	1023
learning_rate	0.28695	0.42311	0.61275	0.02527	0.06965	0.10241	0.71577	0.0487	0.02527	0.08477
num_leaves	4	4	5	1654	748	63	5	59	1654	132
colsample_bytree	0.86993	0.75614	0.28454	0.53523	0.64148	0.01	0.85394	0.48709	0.53523	0.64253
min_child samples	12	11	64	13	6	19	2	4	13	58
reg_alpha	0.01023	0.01107	0.01443	0.00097	0.00813	3.9119	0.42812	0.00653	0.00097	0.00706
reg_lambda	0.01197	0.00097	1.87352	1.88456	0.00946	0.05005	0.00373	0.02416	1.88456	30.481071

Table 3. Parameter setup of the algorithms used by the XGBoost.

Parameter	rank:ndcg					eg:squarederror				
	OHSUMED	TD2003	TD2004	MSLR-WEB10K	MSLR-WEB30K	OHSUMED	TD2003	TD2004	MSLR-WEB10K	MSLR-WEB30K
colsample_bylevel	0.71141	0.78529	0.50516	0.31869	0.77365	0.88701	0.29293	0.93842	0.76663	1
colsample_bytree	0.697	1	0.82459	0.52361	0.60049	0.85157	0.49427	0.58875	0.75828	0.77693
learning_rate	0.45659	0.42996	0.15021	0.01626	0.0415	0.29203	0.00658	0.07547	0.03898	0.09752
max_leaves	4	50	36	2101	10562	11	472	41	112	815
min_child_weight	28.19307	1.12332	4.34532	2.01377	28.41485	5.5726	0.31666	5.60004	25.75814	28.92609
n_estimators	14	16	65	5283	1394	13	6421	296	31891	579
reg_alpha	0.01266	0.02347	0.00097	0.00577	0.03808	0.00148	0.00097	0.02205	0.00097	0.01831
reg_lambda	0.04371	0.04668	0.01008	29.95845	78.68247	0.00212	7.42879	0.00109	0.00305	0.2085
subsample	0.85963	0.98278	0.77738	0.86704	1	0.52277	0.83634	0.94078	0.9064	0.7858

a maximum execution time of 10.000 s (approximately 2 h and 46 min)[4]. Concerning the parameters it was set the total number of iterations to 500 for the datasets MSLR-WEB10K and MSLR-WEB30K and 100 for the otter cases. This limitation is not applied to the RankSVM since this method does not present such parameters.

We provide in Tables 2 and 3 the parameter used by the algorithms (per rows) for the considered datasets (per columns), according to the tool used in the execution (LightGBM or XGBoost). We highlight that the RankSVM does not make use of the configuration of so many parameters to be trained. However, it has the parameter t, which is related to the kernel to be used (the linear is used by default), and c, which represents the trade-off between the error reached in training and the bound defined by the algorithm at the beginning of its execution.

3   Obtained Results

3.1   Time Consumption Analysis

In this subsection, we provide a run-time analysis of the execution of different algorithms over the considered datasets. Precisely, we show in Table 4, per rows, the adopted datasets, per columns, the different algorithms, and the values in each cell are related to the time consumption in seconds.

Considering the LightGBM algorithms, and the characteristics of the dataset (See Table 1), we can notice that the size of the dataset reflect in the consumption time, where the smallest run time is obtained in the OHSUMED, TD2003, and

---

[4] The configurations of the PC used in the experiments are: nim i5 8400, 24 Gb RAM with a gtx 1060 graphic card with 6 Gb.

Table 4. Execution time (in seconds) of the different algorithms.

Dataset	LightGBM			XGBoost	
	lambdarank	rank_xendcg	regression	rank:ndcg	reg:squarederror
OHSUMED	0.355	0.2629	0.505	0.745	0.8372
TD2003	0.454	0.375	0.5349	1.6019	1.1964
TD2004	0.632	0.654	1.557	1.689	1.576
MSLR-WEB10K	375.6719	325.6496	277.2962	40.1566	36.5476
MSLR-WEB30K	254.1841	172.7014	177.8121	102.6979	88.0698

TD2004. For the datasets MSLR-WEB10K and MSLR- WEB30K, this behavior was not reflected. Note that the times related to MSLR-WEB10K were greater than the ones obtained by the MSLR-WEB30K despite the latter having more records. Its also necessary to highlight that in the first three datasets 100 iterations were applied, while in the last two, it was realized 500. The datasets OHSUMED, TD2003, and MSLR-WEB30K obtained the shortest training time using the rank_xendcg algorithm. In the case of the TD2004, the best time was obtained using lambdarank and with MSLR-WEB10K it was a regression.

Taking into account the run time analysis of algorithms used by XGBoost, it is observable a similar situation to considering the LightGBM, that is, the larger the size of the dataset, the longer the training time. Therefore, we have that in the OHSUMED and TD2003 obtained the smallest times both if the algorithm considered is the rank:ndcg. The TD2004, MSLR-WEB10K, and MSLR-WEB30K datasets obtained the best time using reg:squarederror. The algorithm training time that took 1.71 min was assigned to the MSLR-WEB30K using the rank:ndcg algorithm.

3.2 Training Behavior

To provide a training behavior analysis, we provide in Fig. 1 the NDCG evaluation metric considering the 10th position of the document returned list (NDCG@10) during the training phase of each dataset. The plotted values are computed based on the validation base. The X axis is related to the number of iterations and the result obtained by the NDCG@10 is presented in Y axis. The first three datasets have been executed with 100 iterations, moreover, the remaining ones make use of 500 iterations.

In a closer look into the first graph, related to OHSUMED datasets, we have that the largest values are achieved at the beginning going to iteration 20, where the LambdaRank is at the top and rank:ndcg is the lowest. After this period, the results present a quality loss, followed by a stagnation in which no improvements are presented. The dataset also presents a behavior in the range of 0.4 and 0.5.

A more dispersive behavior is presented in the TD2003 dataset. The values vary from 0.15 and 0.45. The method that obtained the largest NDCG@10 in more iterations is reg:squarederror, however, it can be noticed at the end of

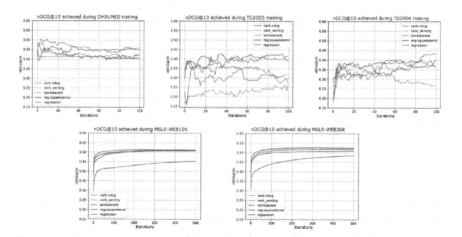

Fig. 1. Comparison among the algorithms during the training phase in the considered datasets.

training that rank:ndcg outperforms the reg:sqarederror (that stagnates in 0.4). Finally, the worst achieved result is obtained by rank_xendcg which does not present a significant improvement.

In the case of TD2004 dataset, there is a behavior with the variation of NDCG@10 from 0.1 to iteration 60 where the lambdarank, which was already getting the worst result among the other algorithms, presented a decrease while the others still show a positive result. In this dataset, in the end, the best result obtained is attributed to the regression.

In the end, for the MSLR-WEB10K and MSLR-WEB30K datasets, a similar behavior is noticed considering that in both cases the worst achieved NDCG@10 is related to the rank:ndcg algorithm (the remaining cases the algorithms presented similar results with a low variation). But, the best algorithm is not the same for both cases. In MSLR-WEB10K the regression presents a better behavior, while in MSLR-WEB30K the lambdarank that have superiority.

3.3   Evaluations

In this subsection, we analyze the results obtained for each dataset. To do so, we show them in Table 5. In it, for each dataset (rows) we present the results achieved by the considered evaluation metrics for the considered algorithms (columns). For the different datasets and algorithms metric we also provide the obtained Mean and the related standard deviation (Stdev). Finally, we highlight with **boldface** the largest obtained values for each metric.

It is possible to observ that the OHSUMED, which presents the lowest amount of registry, obtained the best results using the RankSVM with 0.3512 of MAP and 0.5025 of NDCG. Considering all the algorithms, this case also obtained a MAP and NDCG with 0.3242 and 0.4602 respectively.

Table 5. Obtainded results by the MAP and MeanNDCG evaluation metrics.

Algorithm	LightGBM						XGBoost						Mean		Stdev	
	lambdarank		rank_xendcg		regression		RankSVM		rank:ndcg		reg:squarederror					
Dataset	MAP	NDCG	MAP	NDCG	MAP	NDCG	MAP	NDCG	MAP	NDCG	MAP	NDCG	MAP	NDCG	MAP	NDCG
OHSUMED	0.324	0.4741	0.3122	0.4334	0.3233	0.4526	**0.3512**	**0.5025**	0.345	0.4952	0.2899	0.4036	0.3242	0.4602	0.0203	0.0346
TD2003	0.1292	0.3432	**0.234**	0.4225	0.1645	0.36	0.1652	0.4043	0.2198	**0.4454**	0.1815	0.4254	0.1823	0.4001	0.0353	0.0366
TD2004	0.3002	0.4833	0.3076	0.5079	0.3135	0.4912	0.3884	**0.5663**	**0.3978**	0.5434	0.311	0.4797	0.3364	0.5119	0.0403	0.0322
MSLR-WEB10K	0.3709	**0.5667**	0.3706	0.5666	**0.3714**	0.5653	0.2152	0.3826	0.3331	0.5286	0.3679	0.5643	0.3381	0.529	0.0566	0.0668
MSLR-WEB30K	0.3868	**0.589**	0.3842	0.586	**0.3874**	0.5865	*	*	0.3557	0.5563	0.382	0.5798	0.3792	0.5795	0.0119	0.0119
Mean	0.3022	0.4912	0.3217	0.5032	0.312	0.4911	0.28	0.4639	0.3302	0.5137	0.3064	0.4905	–	–	–	–
Stdev	0.0919	0.0866	0.0534	0.0667	0.0788	0.0815	0.0924	0.0743	0.0593	0.0398	0.0712	0.0711	–	–	–	–

*Not executed due to the lack of memory.

In relation to the TD2003 dataset, having into account the MAP metric, the best algorithm is the rank_xendcg (0.234) from LightGBM and rank:ndcg (0.4454) of MeanNDCG from XGBoost. In this dataset, the obtained mean was 0.1823 (MAP) and 0.4001 (MeanNDCG). As it is the lowest obtained mean, we can conclude that this dataset has the most difficult patterns to learn by the ML algorithms. In the TD2004 dataset, the rank:ndcg obtained the largest MAP (0.3978) and the RankSVM the best MeanNDCG (0.5663), where the achieved mean was 0.3364 and 0.5119 for MAP and MeanNDCG. The MSLR-WEB10K dataset presented with lambdarank the best MeanNDCG (0.5667) and for MAP the regression (0.3714). The mean MAP for the dataset was 0.3381 while for the MeanNDG was 0.529. We highlight that the MSLR-WEB30K was not evaluated to the RankSVM due to the lack of RAM. However, evaluating the remaining algorithms we have noticed that for MAP the dataset obtained the best results by using the regression, and for the other metric using the lambdarank.

We can observe that for the OHSUMED dataset, considering the MAP and NDCG, the best algorithm to be used is the RankSVM. The TD2003 presented the lowest performance, due to the lack of considered queries. In this case, the used algorithm will vary from the considered evaluation metric. By using MAP the chosen will be rank_xendcg and by using NDCG rank:ndcg. Concerning the TD2004 dataset, the algorithms to be used are rank:ndcg considering the MAP and regression otherwise. Finally, for MSLR-WEB10K and MSLR-WEB30K, which present a similar behavior, by considering MAP as an evaluation metric the regression algorithm is considered as a good option, while for the NDCG is the lambarank. This result can be considered similar since both datasets share more than 10k queries.

By following the same structure as the previous table, we provide in Table 6, the obtained values by the evaluation metrics in the positions 1 (P@1) and 10 (P@10). Then, considering the OHSUMED dataset for P@1, the best obtained result is related to rank:ndcg with a precision of 0.5, that is, for all executed queries, evaluating the top of the ranking, in 50% of times the retrieved document was a relevant one. Considering the P@10, the precision presents a decreasing, where the best result is obtained by the RankSVM (having a precision of 0.3773). The obtained mean for this dataset among all algorithms was 0.3939 and 0.3356 for P@1 and P@10 respectively.

Table 6. Comparative results between P@1 and P@10.

Algorithm	LightGBM								XGBoost							
	lambdarank		rank_xendcg		regression		RankSVM		rank:ndcg		reg:squarederror		Mean		Stdev	
Dataset	P@1	P@10	P@1	P@10	P@1	P@10	P@1	P@10	P@1	P@10	P@1	P@10	P@1	P@10	P@1	P@10
OHSUMED	0.4545	0.3	0.2727	0.3136	0.3636	0.3545	0.4545	**0.3773**	0.5	0.3591	0.3182	0.3091	0.3939	0.3356	0.0815	0.0291
TD2003	0.3	0.13	0.2	0.13	0.3	0.15	0.4	0.15	0.4	0.2	0.3	**0.17**	0.3167	0.155	0.0687	0.0243
TD2004	0.4	0.2	0.4667	0.18	0.4	0.16	0.4	0.22	**0.6**	**0.2267**	0.4667	0.2	0.4555	0.1977	0.0711	0.0226
MSLR-WEB10K	**0.559**	0.3768	0.524	0.3749	0.551	**0.3795**	0.212	0.1939	0.4855	0.3284	0.548	0.3765	0.4799	0.3383	0.1222	0.0669
MSLR-WEB30K	0.5728	**0.3954**	0.5568	0.3886	**0.5834**	0.3931	*	*	0.5387	0.3562	0.5726	0.3869	0.5648	0.384	0.0155	0.0142
Mean	0.4572	0.2804	0.404	0.2774	0.4396	0.2874	0.3666	0.2353	0.5048	0.294	0.4411	0.2885	–	–	–	–
Stdev	0.1017	0.1019	0.1417	0.1043	0.1094	0.1088	0.092	0.0857	0.0657	0.0673	0.1134	0.0891	–	–	–	–

*Not executed due to the lack of memory.

In the TD2003 dataset a tie between the methods RankSVM and rank:ndcg occurs when the P@1 is considered. However, for P@10 the best result is obtained by reg:squarederror (0,17). With the mean, we can observe a decreasing in P@10, almost half of the obtained in P@1 (0,3167), resulting in 0.155. The same algorithm, rank:ndcg, has obtained the best results in P@1 and P@10 for the TD2004 dataset. In the remaining datasets, MSLR-WEB10K and MSLR-WEB30K, the best result is achieved by using regression, with 0.3795 for the first and lambdarank with 0.3954.

Regarding the mean and the standard deviation related to the datasets, it is noticeable that for P@1 and P@10 the MSLR-WEB30K and TD2003 present the best and worst results respectively. Also, the obtained means and standard deviation in P@10 are inferior to P@1 in all cases. Concerning the algorithms, we have that at P@1 and P@10 the rank:ndcg obtained the largest mean.

4   Conclusions

In this paper, the goal is the application of different LR algorithms and perform different analyses among them. To do so, we have used two public libraries that implement LR algorithms that are publicly available: LightGBM and XGboost. The methods were applied to a collection of datasets that are also public and widely used, known as LETOR 3.0 and LETOR 4.0. We have compared the methods considering their run time, the behavior presented in the training phase, and different evaluation metrics.

Between the smallest datasets, which have executed 100 iterations, we can notice that rank_xendcg obtained the best run time in the datasets OHSUMED and TD2003. Considering the MSLR-WEB10K and MSLR-WEB30K, which have executed 500 iterations, the small execution time was obtained by the usage of the reg:squareerror.

The results show that different algorithms are performing better for different datasets. The exception is when the OHSUMED is considered. In this case, the best results are obtained by the RankSVM algorithm (which is available in the LightGBM library) in both considered metrics, MAP and NDCG. Additionally, the TD2003 dataset provided the worst results with these metrics (except for reg:squarederros).

In future works, we intend to use the RankLib as an option to evaluate the results obtained by the algorithms, once it has implemented many RL algorithms. The usage of TensorFlow is another interesting possible option.

Acknowledgements. This work was partially supported with grant PID2021-123673OB-C31 funded by MCIN/AEI/ 10.13039/501100011033 and by "ERDF A way of making Europe" and grant from the Research Services of UPV (PAID-PD-22). The authors also would like to thank the FAPERGS/Brazil (Proc. 23/2551-0000126-8) and CNPq/Brazil (3305805/2021-5, 150160/2023-2).

References

1. Baeza-Yates, R.A., Ribeiro-Neto, B.: Modern Information Retrieval. Addison-Wesley Longman Publishing Co., Inc., USA (1999)
2. Chen, T., Guestrin, C.: XGBoost: a scalable tree boosting system. In: Proceedings of the 22nd ACM SIGKDD International Conference on Knowledge Discovery and Data Mining, KDD 2016, pp. 785–794. Association for Computing Machinery, New York (2016). https://doi.org/10.1145/2939672.2939785
3. Harrag, F., Khamliche, M.: Mining stack overflow: a recommender systems-based model (2020). https://doi.org/10.20944/preprints202008.0265.v1
4. Kalervo, J., Jaana, K.: Cumulated gain-based evaluation of IR techniques. ACM Trans. Inf. Syst. **20**(4), 422–446 (2002). https://doi.org/10.1145/582415.582418, https://doi.acm.org/10.1145/582415.582418
5. Kowalski, G.: Information Retrieval Systems: Theory and Implementation. The Information Retrieval Series. Springer, USA (2007). https://books.google.com.br/books?id=hfT6hFXNT4sC
6. Li, H.: Learning to Rank for Information Retrieval and Natural Language Processing. Synthesis Lectures on Human Language Technologies. Morgan & Claypool Publishers (2011). https://doi.org/10.2200/S00348ED1V01Y201104HLT012
7. Manning, C.D., Raghavan, P., Schütze, H.: Introduction to Information Retrieval. Cambridge University Press (2008). https://doi.org/10.1017/CBO9780511809071, https://nlp.stanford.edu/IR-book/pdf/irbookprint.pdf
8. Ogunleye, A., Wang, Q.G.: XGBoost model for chronic kidney disease diagnosis. IEEE/ACM Trans. Comput. Biol. Bioinf. **17**(6), 2131–2140 (2020). https://doi.org/10.1109/TCBB.2019.2911071
9. Singh, S.P., Singh, P., Mishra, A.: Predicting potential applicants for any private college using LightGBM. In: 2020 International Conference on Innovative Trends in Information Technology (ICITIIT), pp. 1–5 (2020). https://doi.org/10.1109/ICITIIT49094.2020.9071525
10. Tan, P.N., Steinbach, M.S., Karpatne, A., Kumar, V.: Introduction to Data Mining, 2nd edn. Pearson, London (2019)
11. Wang, C., Wu, Q., Weimer, M., Zhu, E.: FLAML: a fast and lightweight AutoML library. In: Smola, A., Dimakis, A., Stoica, I. (eds.) Proceedings of Machine Learning and Systems, vol. 3, pp. 434–447 (2021). https://proceedings.mlsys.org/paper/2021/file/92cc227532d17e56e07902b254dfad10-Paper.pdf

Analyzing the Influence of Market Event Correction for Forecasting Stock Prices Using Recurrent Neural Networks

Jair O. González[✉], Rafael A. Berri, Giancarlo Lucca, Bruno L. Dalmazo, and Eduardo N. Borges

Centro de Ciêcias Computacionais, Universidade Federal de Rio Grande, Rio Grande, Brazil
jogonzalezc@furg.edu.co, {jogonzalezc,rafaelberri,giancarlo.lucca, dalmazo,eduardoborges}@furg.br

Abstract. Understanding financial behavior, particularly in the stock market, has become a crucial task in recent years due to its significant impact on the global economy. One of the areas that addresses the relationship between finance and computer science to generate prediction models in this field is known as stock market prediction. This area aims to forecast the behavior of different stocks in the financial market. One of the most well-known and used techniques is Deep Learning, which is composed of different structures of deep neural networks that enable the learning of non-linear models. In this study, we utilized open data from the largest companies in Brazil, namely Petrobras and Itaúsa, provided by BovDB, a stock quotes dataset of all companies in the Brazilian stock exchange between 1995 and 2020. The data from the considered stocks were processed by means of a recurrent neural network aiming to analyze the influence of a price correction method that takes into account the impact of previous market events, e.g. an Ex-bonus, on the training and validation results produced by the RNN model. The experimental results show a good affinity of the model with temporal data, as well as a positive influence on noise reduction and fewer prediction errors, achieving an error reduction up to 26% considering Petrobrás.

Keywords: Deep Learning · Time Series · BovDB · Brazilian Stock Exchange · RNN

1 Introduction

Investment market stocks are a capitalization instrument where profitability is variable, depending on the issuer company's trading results, reflected in the stock's buying or selling price [15]. Stock data is analyzed mainly for short-term and long-term trading objectives, presenting different profit patterns. These patterns tend to vary continuously due to external factors such as politics, economics, and breaking news, generating high uncertainty in the financial

© The Author(s), under exclusive license to Springer Nature Switzerland AG 2023
P. Quaresma et al. (Eds.): IDEAL 2023, LNCS 14404, pp. 290–302, 2023.
https://doi.org/10.1007/978-3-031-48232-8_27

world [19]. One critical factor affecting the stock market's unpredictability is its constantly changing nature. Before the advent of machine learning and deep learning, statistical models, such as the ARIMA [12] and GARCH [3], were developed to predict market values. However, these models were ineffective in handling temporal data as they assumed linearity, seasonality, and normality in their behavior [10]. Subsequently, more sophisticated models, such as machine learning, have emerged, with extensive studies conducted on their potential for predicting stock prices [18]. Examples of such models include logistic regression and Support Vector Machines (SVM), which provide highly accurate results [1]. With the rise of Big Data on the web [9], new methods have evolved to handle large quantities of data, with deep learning being the most well-known [13]. Deep learning is composed of different structures of deep neural networks that allow nonlinear learning of data [11], and utilize the parallel processing power of Graphic Processing Units (GPUs).

One of the most important applications of deep learning is time series prediction [4]. It has the ability to recognize patterns and identify irregularities in data, making it a highly sought-after technique in the financial market. Several techniques and methods have been proposed in the literature for financial market prediction, with a variety of studies on stock prediction [19]. The Recurrent Neural Network (RNN) model has emerged as one of the most relevant models in this area.

In this paper, we have used data from BovDB [5], a data set that was built from the Brazilian Stock Exchange (B3) from the year 1995 to 2020. Data behavior was evaluated by means of the RNN model. We have compared the results obtained from the model with and without the inclusion of the correction factor, which takes into account the effects generated by events in the market. Making use of this comparison we have assessed the effectiveness of the correction and determined its impact on the accuracy of the predictions.

This paper is organized as follows. Preliminaries concepts are introduced in Sect. 2, Sect. 3 presents the related work. Section 4 discuss in detail the methodology. Section 5 shows the obtained results. The conclusions and some future work are addressed in Sect. 6.

2 Preliminaries Concepts

This section presents the key concepts necessary to have a better understanding of the research. We emphasize the event market correction method applied to the data and the hyperparameters that fit the models.

2.1 Market Event Correction Factor

A cumulative variable known as "factor" was proposed by Cardoso et al. [5]. This approach takes into account past events to correct present stock prices. To illustrate the influence of the factor on the data, Fig. 1 is presented, where the mean and standard deviation (black line on top of each annual mean) are compared between the VALE3 stock with (orange bars) and without (blue bars) the factor. It can be observed that data without the factor (prices without any correlation) have a large deviation due to events that naturally adjust stock prices in their respective year. This adjustment means that old prices are not directly compared over time with current prices. On the other hand, the data adjusted with the factor present a lower deviation and a comparable mean (in quantity) throughout the entire time series.

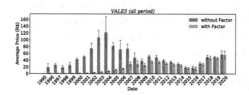

Fig. 1. Average and Standard Deviation of VALE3 between 1995 and 2020, with and without market event correction factor [5].

2.2 Recurrent Neural Network

An RNN model is an artificial neural network [13] that use their internal memory to process input data and to predict future data. The architecture of an RNN model typically consists of different numbers of layers and units in each layer, where one of the main characteristics is its nodes, which are cyclically connected along a sequence, helping it to exhibit dynamic temporal behavior.

2.3 Data Adaptability in Windowed Dataset

When working with neural networks and time series data, it's important to structure the data in a way that can be implemented in an RNN model. This representation can be implemented by a technique called "windowed dataset" [15]. To illustrate this technique, let's consider the task of predicting the value of a stock at time "t" based on its previous values. For example, consider the behavior of a generic stock value shown in Fig. 2. If we want to predict the value of the stock at time 1200, we have to consider the previous 30 values as input to predict this last value. Therefore, values from time 1170 to 1199 will determine the value at time 1200. In summary, the "windowed dataset" technique involves selecting a fixed window size and sliding it along the time series to create multiple input-output pairs that can be used to train the RNN model. This allows the model to capture the temporal dependencies present in the data and make accurate predictions.

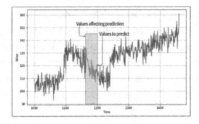

Fig. 2. Previous values impacting prediction in a windowed dataset [15].

3 Related Works

Recently, various Deep Learning models have been developed for predicting stock prices, which have shown to be innovative and relevant in the areas of finance and marketing. In previous years, for example, [17] developed tools for the financial industry. Support Vector Machine (SVM) has been one of the most widely used models in conjunction with Random Forest, Kernel Factory, AdaBoost, Neural Networks, K-Nearest Neighbors, and Logistic Regression [2]. A systematic review [17] shows one of the early Deep Learning models that performs well in the financial prediction area. Other models mentioned in the literature include Deep Multilayer Perceptron (DMLP), RNN, LSTM, CNN, Restricted Boltzmann Machines (RBM), DBN, Autoencoder (AE), and DRL [14].

In the literature, financial time series prediction is mainly considered a regression problem [17]. In [6], the RNN model was applied to predict stock prices and demonstrated excellent results. In the Brazilian market, research has been conducted in this area applying artificial intelligence models with very good results. For example, [8] compared the response of the Arima-Garch and RNA models in predicting prices in different economic sectors. Different from previous related work, in this paper, we have analysed 20 years of data from Brazilian market and used an RNN model to verify the influence of a market event correction method on the accuracy of the predictions. The main contribution is precisely the analysis of the influence of the correction factor on the model's predictive capacity.

4 Methodology

In this chapter, we present the methodology used in the development of this study, which applies the concepts mentioned previously. We discuss the types of data used and the necessary tools for analysis.

As data sources, we will use temporal series from the companies Petrobras (PETR4) and Itaúsa (ITSA4), which are recognized as the most important in the Brazilian stock market. Since extracting the data is a complex task and the influence of events on the stock prices is not straightforward, we obtained the data from the BovDB [5], database, which contains structured and preprocessed information for each stock.

The dataset we consider has been developed for applications such as stock price analysis and prediction, and it includes the following variables for each stock: trading date, opening price, maximum price on the day, minimum price on the day, closing price, volume, and the correction factor.

The code was developed using the Python programming language and the following libraries: Tensorflow[1], Pandas[2], sklearn[3], and Numpy[4].

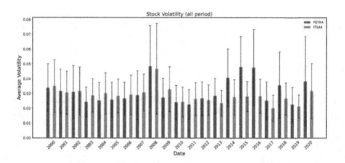

Fig. 3. Annual volatility for each Stock.

5   Obtained Results

In this section, we present the results obtained from our analysis. In order to identify the price variations and behavioral similarities of each stock, we plotted the mean volatilities of each stock for each year, as shown in Fig. 3, along with its standard deviation (black line), where the stocks exhibit similar behavioral values. In total, have a similar global behavior, indicating the complexity of their behavior. This suggests that the stocks exhibit similar levels of unpredictability regardless of the magnitude of the price changes. This complexity can be attributed to a variety of factors, including market conditions, company performance, and external events.

We evaluated the impact of the market event correction atribute, called "factor", on the price of Petrobras (PETR4) and Itaúsa (ITSA4) stocks during the training and validation of a recurrent neural network. To conduct this analysis, the stocks were divided into two sections: the first was used for model training, while the second was used for validating/testing the results. Specifically, the Petrobras stock has a total of 5078 movements, out of which 85% are used for training and the remaining 15% for validation. And For ITSA4, out of the 6425 records available, 80% of the data is used for training and the remaining 20% for validation. This ensures that the model is trained on a sufficient amount of data while also having a separate dataset to validate the model's performance.

---

[1] www.tensorflow.org.
[2] www.pandas.org.
[3] www.sklearn.org.
[4] www.numpy.org.

Figures 4 and 5 illustrate the behavior of the "closing price" variable for PETR4 and ITSA4 stocks, respectively, over the years. The blue line represents the training section, while the red line represents the validation section used in the RNN model. The figures also show the corrections made by the correction factor to deal with the impact of events on the stock prices. In the case of PETR4 stocks, a total of 72 events were recorded, including Ex-Dividends, Ex-Bonus, Ex-Interest, Ex-Stock S, Ex-Stock I, and Ex-Rights. These events occurred on different dates over the years, specifically in March 2003, August 2005, September 2004, December 2018, June 2000, and March 2001, respectively.

On the other hand, ITSA4 stocks experienced 201 events, among which are Ex-Divideds in February 2006, Ex-subs in November 2008, Full Rights in April 2007, Ex-Bonus in May 2009, Ex-Interest in June 2013 and Ex-Rights in September 2000. These events are relevant in the stock market and can have a significant impact on stock behavior and investor decisions. It's essential to stay informed about these occurrences to gain a better understanding of the dynamics of the financial markets, Being the factor variable capable of correcting this influence on the data and presenting us with a better understanding of the stocks.

5.1   The Learning Rate

The Learning Rate (Lr) hyperparameter is a crucial variable to fit a neural network model, as it controls the rate at which the model updates its weights during the training process. In this study, we plotted the loss function of the model obtained using different learning rates to select the optimal rate that provides the lowest loss value for the model. By systematically testing a range of learning rates, we were able to identify the rate that minimized the loss function, improving the accuracy and effectiveness of the model. This approach ensures that the neural network is trained to its maximum potential and provides a more reliable basis for making predictions.

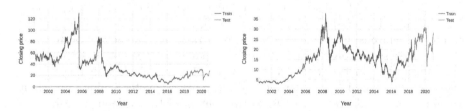

Fig. 4. The closing price of PETR4 without (left) and with correction factor (right).

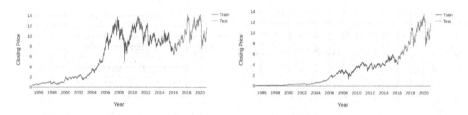

Fig. 5. The closing price of ITSA4 without (left) and with correction factor (right).

The Fig. 6 demonstrates how the model, when applied with the correction factor in the price, is capable of reducing noise in the training results and providing a good picture of the minimum loss point, helping to have a lower prediction errors. (WF) representa o modelo treinado sem factor e (F) com factor.

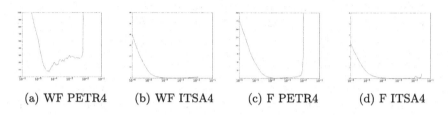

  (a) WF PETR4   (b) WF ITSA4   (c) F PETR4   (d) F ITSA4

Fig. 6. Loss versus Learning rate considering PETR4 and ITSA4.

5.2 Tunning Hyperparameters

The selection of hyperparameters is a critical step in the process of building a neural network model. Hyperparameters are the settings of the model that cannot be learned from data and must be manually tuned to achieve optimal performance. We considered several hyperparameters in the neural network model. These included the batch size, window size, the number of hidden layers, the number of neurons in each layer, the learning rate, and the activation function [7]. The process of choosing the optimal hyperparameters can be challenging and time-consuming. In this work, Keras Tunner [16] was used, with the random search method who is a popular method for hyperparameter tuning in machine learning and deep learning models. In the context of Keras, random search can be used to explore different hyperparameter values and find the optimal settings for a neural network model. The basic idea behind random search is to randomly sample hyperparameters from a specified distribution and evaluate their performance on a data set. To implement random search in Keras, we first define a search space by specifying the range of values for each hyperparameter and the training loss variable was used as the objective function. This because this variable represents the difference between the predicted output of the neural network and the actual output for the training data. The batch size is the number of samples processed in one iteration during training, while the window size refers to the number of time steps considered in the input sequence. The number of hidden layers and neurons determine the depth and width of the neural network architecture, respectively. Finally, the activation function determines the non-linearity of the model. The ranges for each hyperparameter were chosen based on previous studies and knowledge of the problem domain and were selected to cover a wide range of possibilities while avoiding extreme values that may cause convergence issues or overfitting.

Table 1. Best hyperparameters for each Stock

	Configuration	Units_1	Activation	Units_2	Activation	Batch_size	Score_Loss
PETR4	1	32	relu	128	tanh	64	**3,215**
	2	128	relu	128	relu	64	3,226
	3	32	relu	128	relu	32	3,233
ITSA4	1	128	tanh	32	tanh	64	**0,046**
	2	50	tanh	32	tanh	64	0,055
	3	50	tanh	50	relu	64	0,057

The first entry for each stock in Table 1 corresponds to the optimal configuration of the respective model. These configurations were also used in the models trained with the corresponding stock factor, allowing for a performance comparison between the two approaches.

5.3 Proposed Model

To evaluate the performance of the model, we used the mean squared error (MSE) due to its good performance in regression tasks. We also used the stochastic gradient descent (SGD) optimizer to update the weights of the network during training. The results of the model are presented below, with a comparison of the loss function applied to the training and validation data for each epoch. The goal of this comparison is to ensure that the model is not overfitting to the training data and is able to generalize well to new, unseen data.

Figure 7 shows a good training for both cases and Fig. 8 shows the validation data (blue line) together with the model prediction results (red line). It can be observed that the model applied with the factor presents better similarity to the data compared to the model without the factor. In addition, as a comparison, the error for both cases was calculated using the mean absolute error function, giving a result of 0.558 for the model without a factor and 0.530 for the model with a factor, demonstrating the improvement of the model when applying the data of this company when the market event correction method was applied.

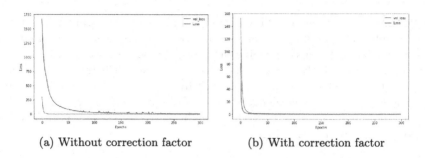

(a) Without correction factor   (b) With correction factor

Fig. 7. Loss Function vs Epoch for PETR4 Train (blue) and Valid (red) Data. (Color figure online)

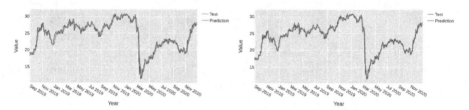

Fig. 8. Validation data (blue line) and model prediction (red line) for PETR4 without (left) and with (right) correction factor. (Color figure online)

The loss function obtained for the ITSA4 stock is presented in Fig. 9. It can be observed that both models with and without market event correction factor show a decreasing trend throughout each epoch, indicating that the model is improving with each iteration. However, in the model applied with the factor, it is noticeable that the loss function of the validation data is significantly higher compared to the loss function of the training data, indicating that it is not fitting well with this information. This happens when the model is learning the training data too well and it becomes too specific to the training data, thus not generalizing well to new, unseen data.

In the model applied without the factor in the Fig. 10 can be observed that the model manages to fit but presents some flaws in certain parts of the data behavior. On the other hand, the model applied with the factor initially presents a very good prediction, but after July 2016, it fails to have a good prediction. To validate this, the error obtained until this date was calculated in both cases. The model applied without the factor obtained an error value of 0.17453712, while the model applied with the factor had an error of 0.09779243, showing its positive influence. The problem that happens after this date is the difference in the training and testing closing price variation, where the first ranges from values 0.05 up to 4.97 and the second from 4.02 up to 13.51. As we are working with Recurrent Neural Network (RNN) models, these are capable of extrapolating data in their predictions to some extent, but their ability to do so depends on several factors. Therefore, since most of the validation values are outside the training range, it is quite difficult to make an accurate prediction.

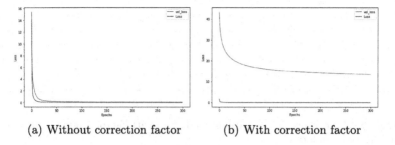

(a) Without correction factor   (b) With correction factor

Fig. 9. Loss Function vs Epoch for ITSA4 Train(blue) and Valid (red) Data. (Color figure online)

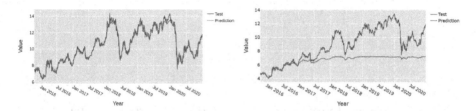

Fig. 10. Validation data (blue line) and model prediction (red line) for ITSA4 without (left) and with (right) factor. (Color figure online)

To mitigate the negative effect generated by the difference in the interval between the training and testing sections on the model's prediction, specifically for the ITSA4 stock applied with the factor, data partitioning was carried out. The stock price data was divided into 10 folds, as shown in Fig. 11, where the five blue data sets were used to train the model, while the other five red data sets were used for testing. This was done to generate a better distribution of behavior in both sections. In Fig. 12, the results obtained in terms of loss for each epoch are presented, showing a good behavior for both cases. For each training epoch, both the model applied with and without the factor decrease the prediction error, reaching low values. Figure 13 shows prediction results obtained in the experiment. Visually, it is evident that the model performs well with and without the factor, except for a significant failure observed between the years 2019 and 2020 in the model with factor, which can be attributed to the range of values. Specifically, the training interval in terms of the closing value was smaller than 12. but even with this, the total error of the model with factor was of 0.263 and the error while in the other case with the model without factor was of 0.548, demonstrating once again the effectiveness of its application.

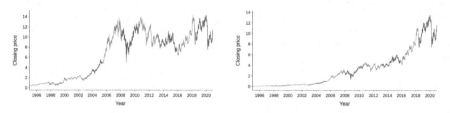

Fig. 11. ITSA4 stock prediction without (left) and with factor (right) using five folds for training and validation.

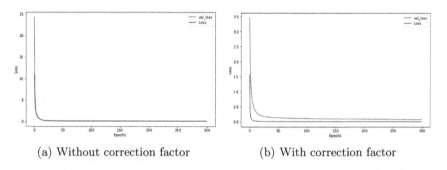

(a) Without correction factor   (b) With correction factor

Fig. 12. Loss Function vs Epoch for ITSA4 Training (blue) and Validation (red) Data, with folds. (Color figure online)

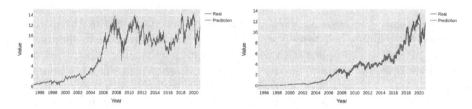

Fig. 13. Validation data (blue line) and model prediction (red line) for ITSA4 without (left) and with (right) factor, with folds. (Color figure online)

6   Conclusion

In conclusion, the field of stock market predictions is a complex area due to the volatility of stocks, making it challenging to generate a model that can adapt to all the variations that stocks experience on a day-to-day basis. Recurrent neural network (RNN) models have become popular in this field as they have been able to adapt and generate behavior patterns, helping to understand future stock behavior. On the other hand, correcting the influence of past events on stock prices significantly contributed to the model's training, leading to an improvement of up to 26% in its predictions, as was the case with Petrobras. However, as demonstrated in this study, when dealing with stocks like ITSA4 whose values increase consecutively over the years, starting with very old data with lower values does not contribute enough to predict future high values, because the RNN models are not extrapolative, which results in predictions with values far away from the real. The approach taken in this study to address the issue of range differences was to distribute the data into 10 folds, and it was demonstrated that the model, when applied to the values of ITSA4 and the range problem was addressed, was able to generate good predictions for both data sets with and without the factor, and it adapted well to the variations of the stocks. Thus, this method proved to be effective in mitigating the challenge of range differences in stock market prediction using RNN models. However, further research is needed to explore this along with other techniques that may address this issue and to apply this approach to a wider range of stocks to assess its effectiveness in different market conditions. Nonetheless, the positive impact of the market event correction method was qualitatively and quantitatively evidenced, where all errors calculated were smaller when using the correction factor. This demonstrates that taking into account events that have occurred over the years in a stock contributes to the model's ability to predict values better. Although there is still room for improvement in stock market prediction models, this study highlights the importance of considering all relevant events and trends when developing such models.

Acknowledgements. This research was funded by FAPERGS/Brazil (Proc.23/25510000126-8, CNPq/Brazil (305805/2021-5, 150160/2023-2).

References

1. Alpaydin, E.: Introduction to Machine Learning. MIT Press, Cambridge (2020)
2. Ballings, M., Van den Poel, D., Hespeels, N., Gryp, R.: Evaluating multiple classifiers for stock price direction prediction. Expert Syst. Appl. **42**(20), 7046–7056 (2015)
3. Bollerslev, T.: Generalized autoregressive conditional heteroskedasticity. J. Econom. **31**(3), 307–327 (1986)
4. Brownlee, J.: Deep Learning for Time Series Forecasting: Predict the Future with MLPs, CNNs and LSTMs in Python. Machine Learning Mastery (2018)
5. Cardoso, F.C., et al.: BovDB: a data set of stock prices of all companies in B3 from 1995 to 2020. J. Inf. Data Manage. **13**(1) (2022)
6. Chandra, R., Chand, S.: Evaluation of co-evolutionary neural network architectures for time series prediction with mobile application in finance. Appl. Soft Comput. **49**, 462–473 (2016)
7. Chollet, F.: Deep Learning with Python. Simon and Schuster (2021)
8. De Oliveira, F.A., Nobre, C.N., Zárate, L.E.: Applying artificial neural networks to prediction of stock price and improvement of the directional prediction index-case study of PETR4, Petrobras, Brazil. Expert Syst. Appl. **40**(18), 7596–7606 (2013)
9. Dietrich, D., Heller, B., Yang, B., et al.: Data Science & Big Data Analytics: Discovering, Analyzing, Visualizing and Presenting Data. Wiley, Hoboken (2015)
10. Fu, T., Chung, F., Luk, R., Ng, C.: Preventing meaningless stock time series pattern discovery by changing perceptually important point detection. In: Wang, L., Jin, Y. (eds.) FSKD 2005. LNCS (LNAI), vol. 3613, pp. 1171–1174. Springer, Heidelberg (2005). https://doi.org/10.1007/11539506_146

11. Goodfellow, I., Bengio, Y., Courville, A.: Deep Learning. MIT Press, Cambridge (2016)

12. Hyndman, R.J., Athanasopoulos, G.: Forecasting: Principles and Practice. OTexts (2018)

13. Jiang, W.: Applications of deep learning in stock market prediction: recent progress. Expert Syst. Appl. **184**, 115537 (2021)

14. LeCun, Y., Bengio, Y., Hinton, G.: Deep learning. Nature **521**(7553), 436–444 (2015)

15. Moroney, L.: AI and Machine Learning for Coders. O'Reilly Media, Sebastopol (2020)

16. O'Malley, T., et al.: Kerastuner (2019). https://github.com/keras-team/keras-tuner

17. Sezer, O.B., Gudelek, M.U., Ozbayoglu, A.M.: Financial time series forecasting with deep learning: a systematic literature review: 2005–2019. Appl. Soft Comput. **90**, 106181 (2020)

18. Wu, H., Zhang, W., Shen, W., Wang, J.: Hybrid deep sequential modeling for social text-driven stock prediction. In: Proceedings of the 27th ACM International Conference on Information and Knowledge Management, pp. 1627–1630 (2018)

19. Zhang, L., Aggarwal, C., Qi, G.J.: Stock price prediction via discovering multi-frequency trading patterns. In: Proceedings of the 23rd ACM SIGKDD International Conference on Knowledge Discovery and Data Mining, pp. 2141–2149 (2017)

Measuring the Relationship Between the Use of Typical Manosphere Discourse and the Engagement of a User with the Pick-Up Artist Community

Javier Torregrosa[(✉)], Ángel Panizo-LLedot, Sergio D'Antonio-Maceiras, and David Camacho

Department of Computer System Engineering, Universidad Politécnica de Madrid, Calle de Alan Turing, 28031 Madrid, Spain
javiertorregrosalopez@gmail.com

Abstract. The Manosphere movement and its subgroups have garnered the attention of researchers seeking deeper insights into their dynamics. This study specifically focuses on a particular subgroup within the Manosphere called Pick Up Artists (PUAs). PUAs concentrate on teaching heterosexual men sexual seduction techniques to attract women, often promoting a clear objectifying and sexist perspective. The primary goal of this research is to explore whether users who engage more with the PUA community tend to adopt the typical Manosphere discourse. To achieve this objective, we aim to measure the relationship between linguistic features and the number of relevant Pick-Up Artist accounts followed by a user. Four linguistic features were considered: *sexist language*, *misogyny*, *use of slang*, and *Manosphere topics*. To assess these linguistic features, we employ relevant state-of-the-art tools, including various classifiers to identify misogynistic and sexist content, topic extraction techniques like BERTopic, and dictionaries containing typical Manosphere slang. The results of our analysis indicate a moderate correlation between the presence of misogynistic/sexist content and the number of accounts followed. However, we did not find a similar correlation with the other variables. This could indicate that the amount of relevant accounts followed does not influence the discourse of an account. Nevertheless, further research is necessary to gain a better understanding of the typical discourse within the PUAs subculture of the Manosphere as the tools used in this study do not seem to clearly identify the expected discourse.

Keywords: Manosphere · Pick-Up Artist · Natural Language Processing · Social Platforms · Discourse Analysis

This research has been supported by the EU under *Malicious actors profiling and detection in Online Social Networks through Artificial Intelligence* (MARTINI) project (CHIST-ERA-21-OSNEM-004) and by the Spanish Ministry of Science and Education under FightDIS (PID2020-117263GB-100) and XAI-Disinfodemics (PLEC2021-007681) grants.

P. Quaresma et al. (Eds.): IDEAL 2023, LNCS 14404, pp. 303–310, 2023.
https://doi.org/10.1007/978-3-031-48232-8_28

1   Introduction

The "Manosphere", a cultural movement known for its opposition towards Feminism, and its subgroups (e.g. Men Right Activists, Involuntary Celibates, Men Going Their Own Way, and so on), discourse and ideology have attracted lot of popular and academic attention, especially due to their harassing style against some (e.g. feminists) or all women [10,13,15]. Studying the discourse and terminology of three Manosphere related groups in Reddit found that women tend to be "dehumanised and sexually objectified, negatively judged for morality and veracity", while "men were constructed as victims of female social actors and external institutions and, as a result, as unhappy and insecure" [13]. Terms related with victimisation from this movement towards women were also found after studying their communities, such as "misandry", or hate against men [14]. It was also found that misogynistic content and users with violent attitudes were common on these communities (on Reddit) [8].

The Pick-Up Artists (PUA) community, one of the subgroups of the Manosphere, is composed primarily by men with interest in sexual seduction techniques [12]. Some of those men act like "gurus", teaching other heterosexual men how to seduce women and improve their "masculinity" [4], with a clear objectifying and sexist vision. This subculture, with a main activity focused on online environments, benefits from different online platforms (e.g. Twitter, Reddit or, more recently, Telegram) to spread their ideology and "knowledge", mostly based on attracting as many women as possible and getting part in "confidence games" with a twisted perspective of how interactions between men and women work [4]. Those gurus create relevant profiles on Twitter that are constantly sharing pieces of advice regarding this ideology (so called "The Game"), which ultimately leads to creating communities that share that common ideology.

Regarding these communities, previous studies that have approached radicalisation and extremist ideologies [20] found evidence of a relationship between a user's level of interaction with an extremist group and their adoption of the group's typical discourse. In the same line, this study aims to explore whether users who engage more with the PUA community are more likely to adopt the typical Manosphere discourse. Specifically, the article investigates whether following more or less relevant PUA accounts (as a measure of engagement) is correlated with linguistic features related to the typical Manosphere discourse. To address this, we consider the general concepts present in that discourse [4,7,8] and propose three main objectives:

- To check the relationship between mysoginistic and sexist discourse and relevant PUA accounts followed.
- To check the relationship between the presence of typical slang from the manosphere and relevant PUA accounts followed.
- To check the relationship between the presence of specific conversational topics related to the manosphere and relevant PUA accounts followed.

2   Data and Resources

The dataset for the analysis was built using a snowball technique. Starting from 35 relevant accounts from the PUA community (from now on called *seeds*), their list of followers was extracted (N=128.329). For each follower, the last 300 tweets were obtained. With all this information, a series of filters were applied: 1) only original and retweet messages were selected, as it was seen that there isn't tools in this topic suited to handle an exchange of replies; 2) Tweets in a language other than English were discarded; 3) Only tweets from 2018 onward were included; 4) Seeds with less than 1000 followers were excluded; 5) Only accounts with 50 or more tweets published were included in the study; 6) Finally, to prevent bots from affecting the analysis, accounts with an unusually high number of tweets published (those considered outliers) were discarded, as well as accounts with less than 5% of original tweets, i.e. not retweets. The final corpus for analysis included 2.118.995 tweets (835.589 originals and 1.283.406 retweets) from 01 January 2018 to 14 November 2022; 15.122 accounts; and 32 of the 35 initial seeds.

It shall be mentioned that, while examining the variables in the dataset, we discovered a significant relationship between the number of seed accounts followed and the median number of tweets published by the users who followed that particular number of seed accounts (Spearman's rho $= 0.602$, $p < 0.001$). This relationship and its possible implications are considered in the following sections.

3   Methodology

As stated on the introduction, the PUA community has attracted less interest than other subgroups inside the Manosphere and, with some exceptions (see [6]), their discourse has not been studied especifically; it has only been approached as part of a "whole" Manosphere community [7,8]. Considering this lack of information, and to achieve the objectives of this work, four different variables were calculated, namely: *sexism*, *misogyny*, *use of slang*, and *manosphere topics*.

For the first two variables, two models extracted from the state of the art were applied to evaluate the probability of every tweet to include sexist or misogynist content. Both concepts shall be considered separately, as sexism is conceptually different from misogyny: the former has a justificatory component of the inequality, while the latter has a hostile component of the same inequality [18]. These models were: i) *Bertweet-large-sexism-detector*[1]: a fine-tunned model of BERTweet-large on a specific sexism detection dataset that reached an F1 score of 0.903 for English tweets [21]. ii) *Bert-base-uncased-ear-misogyny*[2]: a fine tuned model of bert-base-uncased on the Automatic Misogyny Identification dataset [9] that uses Entropy-based Attention Regularization to reduce overfitting of specific terms during training. It reached a F1 score of 0.918 [3].

---

[1] https://huggingface.co/NLP-LTU/bertweet-large-sexism-detector.
[2] https://huggingface.co/MilaNLProc/bert-base-uncased-ear-misogyny.

For the third variable, the number of slang terms used in each tweet was analysed. Since there is no specific set of slang dictionaries from the PUA community, we opted for a selection of commonly used slang dictionaries from the Manosphere community, as described by Farrell [8]. All the dictionaries proposed by Farrel et al. were considered: *belittling, flipping the narrative, homophobia, hostility, patriarchy, physical violence, racism, sexual violence* and *stoicism*.

For the final variable, tweets that discuss conversational topics that were considered theoretically relevant for this subculture were identified. For this task, BERTopic [11] was used to extract relevant topics from the dataset. Then, the topics where manually inspected to found those theoretically relevant for the Manosphere subculture. To configure BERTopic, the recommended settings were used, which include UMAP [17] for dimensionality reduction, HBDSCAN [16] for doing the clustering, and KeyBert [19] for summarising the topics. Finally, the most relevant hyperparameters of UMAP ($n\_components$ and $n\_neighbors$) and HDBSCAN($min\_cluster\_size$ and $min\_samples$) where tuned using the Optuna [1] framework with a random sample of 100.000 tweets and the Calinski Harabasz Score [5] as optimisation metric.

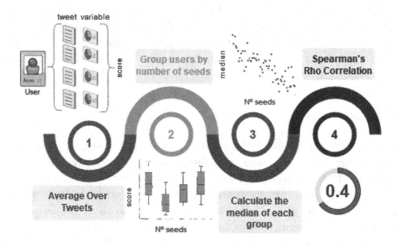

Fig. 1. Worflow for calculating the correlation between the number of seeds followed and the variables of interest

To measure the correlation between the different variables and the number of seeds a user follows, a score must be calculated for each user. It is important to note that the variables mentioned above are associated with tweets, and each user has a varying number of tweets published. Consequently, the average score across all tweets published by each user is used as the score for a particular variable concerning that user. Once the scores for each user are calculated, the correlation between the number of seeds followed and the variable of interest can be determined. However, it should be acknowledged that the method used for calculating these correlations, Spearman's Rho [2], is highly sensitive to noise.

Therefore, the input to the Spearman's Rho method consists of the median score obtained from users who follow a particular number of seeds. Figure 1 provides an overview of this procedure.

Finally, it was essential to investigate whether the number of tweets, which showed a positive correlation with the number of seeds, was influencing our experiments. To address this concern, we performed 100 simulations ($s = 100$) in which an equal number of tweets ($t = 100$) were randomly selected for each user. Subsequently, correlations were calculated between the number of seeds and each of the four variables to check whether the observed effects are preserved when the number of tweets is fixed.

4   Results

Table 1. Outcomes of the simulations and the correlations

Variable correlated	Simulations			Final score (all data)	
	Number of significant test (p <0.05)	Mean Spearman's Rho	Std Spearman's Rho	Spearman's Rho	p-value
Mysoginistic score	100	0.492	0.060	0.555	≈ 0.0
Sexist score	81	0.427	0.088	0.484	0.005
Average slang term	13	−0.067	0.206	−0.232	0.200
Manosphere topics	8	0.127	0.179	0.171	0.348

The simulations meant to check the influence of the number of tweets in the rest of correlations, together with the final correlations obtained using all the data from each variable (without considering the number of tweets), are displayed on Table 1. As can be inferred, the variables showing a significant correlation without controlling the number of tweets also correlate significantly when controlling them. Same happens when there is a lack of correlation. Therefore, the number of tweets does not seem to moderate the correlations.

Considering now the correlations per se, both mysoginistic and sexist score showed positive correlation with the number of seeds followed (in this case, a median-strong correlation, considering the Sperman's Rho score). However, no correlation was found between the average slang terms used by each user and the number of seeds followed. Neither was found when comparing the variable manosphere topics per user.

Regarding the slang terms, most of them were not suitable for conducting any correlation due to the lack of data when checking each dictionary separately. Only physical violence (Rho $= -0.017$, p $= 0.925$), hostility (Rho $= 0.065$, p $= 0.721$), and racism (Rho $= 0.179$, p $= 0.3261$) had enough tweets to allow unbiased comparisons. However, none of them showed any correlation with the number of seeds followed.

Moving on to the topics found, a total of $753,539$ tweets were assigned to 327 topics. After manual examination, 7 topics were selected as manosphere-related, which can be roughly classified as: *masculinity, quit porn, testosterone levels, naked ig bitches, alpha male affirmation, semen retention,* and *hoes.* Similar to the slang terms, only the *masculinity* topic had enough data to conduct a correlation. However, it also showed no correlation with the number of seeds followed (Rho $= 0.104$, p $= 0.570$).

5 Conclusion

The objective of this article was to check the relationship between four language variables expected to be relevant for the Manosphere subcultures (misogynistic and sexist content, the use of slang and talking about Manosphere related topics) and the number of seeds an account followed. Previous studies had shown that there seems to be a correlation between the relevance of a user in a radical community and the use of its specific slang and an aggressive tone [20]. However, based on the obtained results, we can conclude that there is no clear relationship between the engagement of an account with the PUA community, as measured by the number of relevant PUA accounts followed, and the adoption of the typical Manosphere discourse.

Only the presence of sexist and misogynist discourse in the tweets was found to be positively and significantly correlated with the number of seeds followed. It was observed that the presence of both types of discourse should be further analyzed, as the amount of tweets labeled as sexist or misogynous was very low. This could be attributed to the possible mixing of both terms when creating the classifiers. It is essential to remember that, while closely related, these terms do not represent the same type of discourse [18].

Neither the typical slang extracted nor the Manosphere related topics displayed any significant relationships. The limited amount of tweets including slang or relevant topics for this community (which was expected to be involved in talks about masculinity, critics towards women, pick-up strategies and references to terms typical from their "Game") appears to be a path to explore in the future. In fact, as mentioned earlier, there are few articles discussing this group, and even fewer addressing their typical discourse, with only support from the studies conducted by [7,8]. This suggests that the language used by both low and high engaged pick-up artists followers may differ from what is "expected" from a Manosphere follower. Investigating the specific language of this subculture and its distinctions from other subcultures within the Manosphere should be a focus of future research.

Disclaimer. The dehydrated dataset can be accessed at https://doi.org/10.5281/ zenodo.8335331. Unfortunately, the recent changes in the X (former Twitter) API limit the access. Those colleagues interested in this research and data, please feel free to contact the corresponding author.

References

1. Akiba, T., Sano, S., Yanase, T., Ohta, T., Koyama, M.: Optuna: a next-generation hyperparameter optimization framework (2019). https://doi.org/10.48550/arXiv. 1907.10902. https://arxiv.org/abs/1907.10902
2. Artusi, R., Verderio, P., Marubini, E.: Bravais-Pearson and spearman correlation coefficients: meaning, test of hypothesis and confidence interval. Int. J. Biol. Markers **17**(2), 148–151 (2002)
3. Attanasio, G., Nozza, D., Hovy, D., Baralis, E.: Entropy-based attention regularization frees unintended bias mitigation from lists. In: Findings of the Association for Computational Linguistics: ACL 2022, pp. 1105–1119. Association for Computational Linguistics, Dublin, Ireland (2022). https://doi.org/10.18653/v1/2022. findings-acl.88. https://aclanthology.org/2022.findings-acl.88
4. Banet-Weiser, S., Bratich, J.: From pick-up artists to Incels: Con (fidence) games, networked misogyny, and the failure of neoliberalism. Int. J. Commun. **13**, 1 (2019)
5. Caliński, T., Harabasz, J.: A dendrite method for cluster analysis. Commun. Stat. Theory Meth. **3**(1), 1–27 (1974)
6. Dayter, D., Rüdiger, S.: The Language of Pick-up Artists: Online Discourses of the Seduction Industry. Routledge, Milton Park (2022)
7. Farrell, T., Araque, O., Fernandez, M., Alani, H.: On the use of jargon and word embeddings to explore subculture within the Reddit's Manosphere. In: Proceedings of the 12th ACM Conference on Web Science, pp. 221–230 (2020)
8. Farrell, T., Fernandez, M., Novotny, J., Alani, H.: Exploring misogyny across the Manosphere in reddit. In: Proceedings of the 10th ACM Conference on Web Science, pp. 87–96 (2019)
9. Fersini, E., Nozza, D., Rosso, P.: Overview of the Evalita 2018 task on automatic misogyny identification (AMI). In: CEUR Workshop Proceedings, vol. 2263, pp. 1–9. CEUR-WS (2018)
10. Ging, D.: Alphas, betas, and Incels: theorizing the masculinities of the Manosphere. Men Masculinities **22**(4), 638–657 (2019)
11. Grootendorst, M.: BERTopic. https://github.com/MaartenGr/BERTopic
12. King, A.S.: Feminism's flip side: a cultural history of the pickup artist. Sex. Cult. **22**(1), 299–315 (2018)
13. Krendel, A.: The men and women, guys and girls of the 'Manosphere': a corpus-assisted discourse approach. Discourse Soc. **31**(6), 607–630 (2020)
14. Lawrence, E., Ringrose, J.: @ notofeminism,# feministsareugly, and misandry memes: how social media feminist humor is calling out antifeminism. In: Emergent Feminisms. Routledge (2018)
15. Marwick, A.E., Caplan, R.: Drinking male tears: language, the Manosphere, and networked harassment. Fem. Media Stud. **18**(4), 543–559 (2018)
16. McInnes, L., Healy, J., Astels, S.: HDBSCAN: hierarchical density based clustering. J. Open Source Softw. **2**(11), 205 (2017)
17. McInnes, L., Healy, J., Melville, J.: UMAP: uniform manifold approximation and projection for dimension reduction (2020). https://doi.org/10.48550/arXiv.1802. 03426. https://arxiv.org/abs/1802.03426
18. Richardson-Self, L.: Woman-hating: on misogyny, sexism, and hate speech. Hypatia **33**(2), 256–272 (2018)
19. Sharma, P., Li, Y.: Self-supervised contextual keyword and keyphrase retrieval with self-labelling (2019)

20. Torregrosa, J., Panizo-Lledot, Á., Bello-Orgaz, G., Camacho, D.: Analyzing the relationship between relevance and extremist discourse in an alt-right network on twitter. Soc. Netw. Anal. Min. **10**, 1–17 (2020)
21. Vaca-Serrano, A.: Detecting and classifying sexism by ensembling transformers models. In: IberLEF (2022). https://ceur-ws.org/Vol-3202/exist-paper3.pdf

Uniform Design of Experiments for Equality Constraints

Fabian Schneider[1]([✉]), Ralph J. Hellmig[2], and Oliver Nelles[1][ID]

[1] Institute of Mechanics and Control Engineering – Mechatronics,
University of Siegen, Siegen, Germany
{fabian2.schneider,oliver.nelles}@uni-siegen.de
[2] Institute of Material Science, University of Siegen, Siegen, Germany
ralph.hellmig@uni-siegen.de

Abstract. Design of Experiments (DoE) is essential for data-driven models. It is applied to create uniformly distributed designs for the model's input space. For good uniformity, the point density of the generated design shall be kept roughly constant, i.e., too close points and too big data holes shall be avoided. The generation of space-filling designs, even for the unconstrained design space, is challenging, especially with increasing dimensionality and design size. These difficulties increase in the presence of constraints that occurs in many engineering applications. Equality constraints are one group, i.e., mixture or volume constraints. Specialized approaches are required for such constraints. A point-distance-based optimization is not suitable. The computational effort for large designs is too high, and for higher-dimensional designs, too many points are placed close to the boundary of the design space. The proposed approach, instead, is based on the projection of uniformly distributed designs onto the constraint. Thus, this work presents the idea of using maximin Latin Hypercubes (LH) with their good space-filling properties and projecting them onto the constraint. The resulting design might be non-uniformly distributed so that large data holes could occur. To fill these holes, this projection of an LH onto the constraint is repeated for the remaining inputs by going over all inputs. This method is analyzed in a first two-dimensional case study. In a second case study, it is applied to a nonlinear constraint. A design has to be placed on a spherical surface. The results are compared to an existing state-of-the-art method.

Keywords: Design of Experiments · Design of Experiments for Constraints · Equality Constraints · Latin Hypercubes

1 Introduction

The quality of data-driven models depends highly on the available data used during training. Thus, several design of experiment (DoE) methods have been developed [1]. The model's design space (or input space) is a d-dimensional hypercube for the unconstrained case. Latin hypercubes [2] and Sobol' sequences [3] are frequently applied designs that achieve appropriate model accuracies [4].

In most cases, little prior knowledge of any specific desired distribution of the design data is available. Thus, the most common, robust, general-purpose approach is to require a space-filling design, i.e., an even or uniform distribution [1]. Consequently, no data holes or data clusters exist. Besides this, uniform marginals (1-D) are also desired.

If the design space is constrained, difficulties arise – the more severe, the stricter the constraints are. Generally, constraints can be divided into equality constraints and inequality constraints. Equality constraints are generally more severe than inequality constraints. Some ideas have been pursued addressing Latin hypercubes (LH) for inequality constraints [5–7]. A straightforward approach to addressing inequality constraints is to create an unconstrained design and remove all points that violate the constraints. One drawback is that, dependent on the severeness of the constraint, the ratio of feasible points to unfeasible points tends to approach zero, especially for higher-dimensional problems. However, even if this approach leads to usable designs for inequality constraints, it does not apply to equality constraints. It is improbable that a so-generated point does fulfill the constraint.

Specialized approaches have to be developed for DoE for equality constraints, and research has been done in the field of mixture designs. Mixture designs occur, for example, in chemistry or material science. A simple two-dimensional example would be a mixture problem of two substances or materials. So, we define a variable u_i for each compound, further referred to as input, in the range 0 to 1, where 1 denotes the pure component. Obviously, they have to obey the mixture rule $\sum_{i=1}^{n} u_i = 1$, where n is the number of inputs, here $n = 2$. So, the constraint $u_1 + u_2 - 1 = 0$, a linear equation, has to be fulfilled. Otherwise, a point is unfeasible. Developed DoE strategies are simplex-lattice designs [8], simplex-centroid designs [9] or extreme-vertices designs [10]. They have been developed to create designs for mixture constraints, particularly for polynomial models. Thus, they are not universally applicable, especially for general nonlinear data-driven models. These designs are not suitable for data-driven models since they are non-uniformly distributed. Uniform designs for mixture experiments have been also developed. One strategy is to transform a regular uniform design onto the constraint [11,12]. A second one is searching in interiors of simplex-lattices [13]. A further approach is to define a tolerance criterion to allow small violations of the constraint [14]. This simplifies the generation of points because the feasible space is increased. An approach based on the projection of points onto the constraint is presented in [14]. In [15], an algorithm is proposed to generate a design on Riemannian manifolds.

Besides mixture constraints, other equality constraints occur in applications. For example, a multiplicative one could occur instead of an additive connection of the variables, like $\prod_{i=1}^{n} u_i = 1$, or a combination of both types. In this case, the equation is nonlinear. A multiplicative constraint, for example, could result if a certain body volume, e.g., cuboid with the three variables u_1, u_2, and u_3, its length, width, and height, is requested. Such volume constraints also occur for more complex bodies in the forging or forming processes. So, if a body is parameterized to generate different geometries, the parameters have to fulfill the volume constraint. Methods to create space-filling designs for nonlinear equality constraints are rare. In [11], a uniformly distributed mixture design is used as a starting point to generate uniformly distributed points on a spherical surface.

The remaining work is organized as follows. In Sect. 2, Latin hypercubes are introduced. The proposed method is described in Sect. 3. The method is applied to two case studies. They are presented in Sect. 4, followed by a conclusion in Sect. 5.

2 Latin Hypercubes

Latin hypercubes (LH) are frequently applied to create uniformly distributed designs, especially for meta-models. LHs are a grid-based approach. An LH with N data points is based on an $N \times N \times \ldots \times N$ grid. The main features of LHs are:

1. Each grid line in each dimension is taken by exactly one data point.
2. As a consequence for equally spaced grids (standard case): The data distribution projected on each axis is inherently uniform, and each grid line also contains exactly one data point in the projection.
3. Design is space-filling if an optimization of the points is performed [2].

Without optimization, the designs may have poor space-filling properties [2]. For example, this is the case if LHs are created according to the originally proposed random combination of the available levels. To improve the design quality, many different optimization strategies have been developed. An overview is given in [2]. Those strategies differ in the objective function and also in the algorithm applied. A design with large nearest neighbor (NN) distances for all points is preferred. A common strategy to achieve large NN distances is to maximize the minimal NN, i.e., optimizing the maximin criterion. Here, only the critical point pair is considered. So it is non-smooth. So the frequently applied approximation of the maximin criterion has been developed, the ϕ_p-criterion [16]:

$$\phi_p(\mathbb{U}) = \left\{ \sum_{i=1}^{N-1} \sum_{j=i+1}^{N} \|\underline{u}_j - \underline{u}_i\|^{-p} \right\}^{1/p}, \qquad (1)$$

where $\mathbb{U}$ is the set of points of the design. $\underline{u}_i = [u_{i,1}\, u_{i,2}\, \cdots\, u_{i,d}]^{\mathrm{T}}$ defines a point, with d being the input space dimension. Frequently used algorithms are based on coordinate swapping. For example, the Extended Deterministic Local Search algorithm (EDLS) [17]. Besides this, genetic algorithms [18] or simulated annealing strategies [16] are applied to improve the designs.

The method developed in this paper requires space-filling designs. Santner [4] shows that optimized LHs are slightly superior in nonlinear modeling tasks compared to Sobol' sequences, proposed in [3]. Thus, LHs are selected and generated by the EDLS algorithm. The goal of the EDLS algorithm is to minimize the ϕ_p-criterion. Thus, a uniformly distributed design results. The designs are suitable for nonlinear modeling tasks [19], especially in combination with Local Model Networks [20,21]. The price to be paid for these advantages is a higher computational load compared to the Sobol' sequences.

3   Design of Experiments for Equality Constraints

The previously mentioned DoE methods for equality constraints are mainly developed to create uniformly distributed designs for linear constraints. In [11], where a design is generated for a nonlinear constraint, a uniformly distributed design for a mixture constraint is also used. After such a design is created as initialization, it is projected onto the nonlinear constraint. By contrast, the proposed method is directly applicable to linear and nonlinear constraints.

The design points should be uniformly distributed over the feasible design space. This is especially important if they are used to train nonlinear models. Besides this, uniform marginals are also desired. However, there exists a fundamental conflict between the desired space-filling property and the uniform projection property in the case of constraints. Since certain areas of the input space are not accessible, a uniform data point distribution (space-filling) in the original input space (high-D) yields non-uniform marginals (1-D). The priority is given to the space-filling property.

The developed approach is similar to the idea of slack variable models. They are one modeling technique for mixture experiments [22]. This approach is necessary since every equality constraint, for example, the mixture rule $\sum_{i=1}^{n} u_i = 1$, where n is the number of inputs, renders one variable redundant. One variable u_i can be expressed as a combination of all other variables, in the case of a mixture constraint according to $u_i = 1 - \sum_{j=1, j \neq i}^{n} u_j$. This leads to a failure of the least squares estimation of the model. For slack variable models, one variable, which is called slack variable, is dropped from the model. The auxiliary condition expresses this variable. For a two-dimensional mixture constraint, $u_1 + u_2 - 1 = 0$, the slack variable model is defined by dropping u_2 and obtaining the reduced model $\hat{y} = a_0 + a_1 \cdot u_1$ with the auxiliary condition $u_2 = 1 - u_1$. So the dimensionality of the model is reduced by one. Important is that the selection of the slack variable affects the model quality [22].

The slack variable model approach uses the reduced input space of the model to enable the least squares estimations. In analogy, this DoE method also defines a slack variable and the constraint, also called auxiliary condition in the context of mixture constraints. So a reduced design space of dimensionality $d - 1$ results. This reduced space is called support space. A uniformly distributed design, the support design, is created for this support space. The constraint is evaluated for each point, further mentioned as a support point. Both parts concatenated yield the design point. The selection of the slack variable is also crucial in this case, so the catch of the method is to use every variable once as slack variable. By this, the influence of the selection is reduced. Algorithm 1 describes this procedure in more detail.

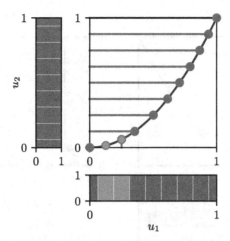

Fig. 1. Resulting two-dimensional design for the equality constraint $u_1^2 - u_2 = 0$ with $N = 11$ data points. Purple: the first input, u_1, is the slack variable. Orange: the second input u_2, is the slack variable. Only points with large NN distances are added to the design. (Color figure online)

Algorithm 1: Design of Experiments for Equality Constraints

Data: N_{sp}: number of support points, d: dimensionality, $\mathbb{C}$: constraints
Result: $\mathbb{U}$: generated design
for $i \leftarrow 1$ **to** d **do**
    Define slack variable u_i;
    $\mathbb{D}_{sp} \leftarrow \{1, 2, \ldots, d\} \setminus \{i\}$;   % Dimensions of the support points
    Create uniformly distributed support design $\mathbb{U}_{sp}$, a maximin LH
    of size N_{sp} for dimensions $\mathbb{D}_{sp}$;
    Evaluate constraint $\mathbb{C}$ for every support point of $\mathbb{U}_{sp}$ and
    attach dimension i to it;
    Remove all unfeasible points from $\mathbb{U}_{sp}$;
    if $i \geq 2$ **then**
        Calculate point selection criterion $\delta_{crit}(\mathbb{U})$;
        Remove all points from $\mathbb{U}_{sp}$ that violate δ_{crit};
    end
    $\mathbb{U} \leftarrow \mathbb{U} \cup \mathbb{U}_{sp}$;   % Merge remaining points to design
end

The constraint $u_1^2 - u_2 = 0$ for the $d = 2$ dimensional design space is used to explain the method and is visualized in Fig. 1. If u_1 is selected, for instance, as slack variable, $u_1 = \sqrt{u_2}$ results as constraint. The support design $\mathbb{U}_{sp}$ is a one-dimensional LH of size 9. Consequently, an equally spaced grid is defined over u_2. For each point, the constraint is evaluated. It is added to the design if the point is feasible and does not validate the selection criterion. Both iterations' points are visualized in Fig. 2. Here, it is obvious that the selection of the slack variable

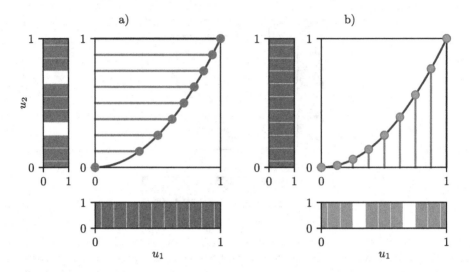

Fig. 2. Resulting designs for the constraint $u_1^2 - u_2 = 0$ with $N = 9$, if only one variable is selected as slack variable. a) The first input, u_1 is the slack variable. b) The second input u_2 is the slack variable.

influences the design quality. The points for the second case, Fig. 2b), are more uniformly distributed than the points of the first one. The distance to the NN depends on the slope of the constraint. For both choices of slack variables, a different part is covered with a finer resolution. While the constraint $u_1 = \sqrt{u_2}$ is more sensitive for small values of u_2, the second auxiliary condition $u_2 = u_1^2$ is more sensitive for larger values of u_1 and therefore larger values of u_2. The severity of this effect depends on the shape of the constraint.

Figure 1 illustrates that the second iteration is done to fill the holes that could not be filled during the first one. The orange points fill the large hole of the purple design. The selection is done according to all points' previously mentioned NN distances. In detail, a feasible point, the ith point, is only added if its NN distance $d_{NN,i}(\mathbb{U})$ to the design $\mathbb{U}$ is larger than the threshold $\delta_{crit}(\mathbb{U})$. This threshold is defined as

$$\delta_{crit}(\mathbb{U}) = \frac{1}{2N} \sum_{i=1}^{N} d_{NN,i}(\mathbb{U}), \qquad (2)$$

with the design size N of design $\mathbb{U}$ and the nearest neighbor (NN) distance $d_{NN,i}$ of point i to the points of design $\mathbb{U}$. Thus, a point is added if half the NN distance of it is at least as large as the mean of all NN distances of the design. So it is ensured that no points are too close to each other. This threshold is updated after each iteration. So, for every slack variable, a new threshold is calculated. By this procedure, the maximum possible design size is $N = d \cdot N_{sp}$, which has to be considered during the definition of the support design. Since the points of the subsequent iterations are only used to fill the holes, it is expected that this maximum design size is not reached by far. This depends on the nonlinearity of the constraint.

Besides the selection criterion, the support design $\mathbb{U}_{sp}$ affects the design quality. The dimensionality of the support design is $d_{sp} = d - 1$. Increasing the dimensionality leads to a severe growth in complexity and effort necessary to create designs, also known as the curse of dimensionality. While the one-dimensional grid is an appropriate support design for the analyzed example, transferring it straightforwardly to higher dimensional problems is unsuitable. If a one-dimensional grid with N levels is expanded to d dimensions, the number of points increases from N to N^d. If the size of the two-dimensional support design is increased to $N_{sp} = 25$, five levels for each dimension results. Note that the resolution along each axis is nearly halved compared to the previous example, although more than twice the support points are used. This is one effect of the curse of dimensionality and demonstrates its relevance. So, in consequence of the arguments of Sect. 2, Latin hypercubes are selected as support designs instead of grids to avoid this exponential increase of support points. Note that a one-dimensional LH is equivalent to a one-dimensional grid. The LH-based designs' level resolution is higher than the grid-based support designs. In each support design, no level is populated twice. The benefits of the LHs become essential with increasing dimensionality. Since a loop over all dimensions is performed, one additional dimension more requires one further iteration. This is generally benign behavior. The dimensionality of the support design also rises by one. The design size of a grid-based support design with l levels would increase from l^d to l^{d+1}. On the contrary, no additional support points are necessary for an LH-based support design if the number of levels l is not varied. Nevertheless, a slight increase of the design size N_{sp} is suggested because, otherwise, the mean point distance of the support points would increase. These properties indicate that the algorithm is well-suited for higher dimensional constraints.

Another important step is the calculation of the slack variables. Therefore, a constraint is necessary. For the discussed case, it is possible to derive this condition easily. Unfortunately, there are cases where this is not suitable or even impossible. This happens, for example, if an analytical derivation of each condition is too time-consuming. In such cases, the constraint can not be derived.

Instead of such a condition, an equation solver could obtain the solution of the ith slack variable u_i for every support point $\underline{u}_{sp}$. In this case, an equation of the type

$$u_i \overset{!}{=} f(\underline{u}_{sp}), \tag{3}$$

where $f(\underline{u}_{sp})$ describes the constraint, depending on all other dimensions, has to be solved. As an example, the function for the slack variable u_1 of constraint $u_1^2 - u_2 = 0$ could be defined as $f(\underline{u}_{sp}) = \sqrt{u_{sp,2}}$. The resulting point is added to the design if the equation solver finds a solution inside the design space, typically inside $[0, 1]$. If the solution lies outside the interval or no solution can be found, the support point $\underline{u}_{sp}$ is rejected.

Consequently, $d \cdot N_{sp}$ executions of the equation solver are necessary to create a design. Compared to an evaluation of the constraint, more effort is necessary. The increase of the computational cost depends significantly on the computational cost of the constraint evaluation. Nevertheless, this is why this method could be applied to a wider range of constraints.

This easy application to different constraints is one of the advantages of the method. Besides this, the proposed method has no hyperparameter and is deterministic. Compared to the proposed projection method in [14] where several inputs could be used for projection, only one input is used at a time. This is advantageous because the curse of dimensionality makes the optimization of points difficult [5]. The amount of points close to the limit rises with increasing dimensionality. The usage of the support designs prevents this. A drawback is that the possible constraints are limited. The method is not suitable to constraints for that more than one u_i fulfills (1).

4   Case Studies

Two case studies are presented in this section. The proposed method is used in the first one to analyze the already considered two-dimensional case. The second one shows an application of the method to a three-dimensional problem. Here, a design on a Pareto front is created.

4.1   Two-Dimensional Comparison

In this section, the method is analyzed in more detail. As mentioned above, the point distribution's non-uniformity is independent of the design size. The resulting points are non-uniformly distributed as long as the levels are uniformly distributed. Depending on the constraint, the levels could be arranged so that the resulting points are uniformly distributed. This means that the design points have the same NN distance. It is possible to create such a design in this simple case. The NN distances of all points could be calculated easily, and the levels could be placed accordingly. A more general approach is to approximate the NN distance via linearizing the constraint. In this case, the approximated distance $\tilde{d}_{l,m}$ of point $\underline{u}_m$ to point $\underline{u}_l$ is calculated according $\tilde{d}_{l,m} = (u_{m,1} - u_{l,1})\sqrt{1 + 2 * u_{l,1}}$, where $u_{m,i}$ is the ith dimension of the mth point. The error of this approximation depends on the linearization error. In Fig. 3, a design is shown, where levels of u_1 are placed so that the approximated distances are equal. So, this method aims to minimize the variance $\sigma^2_{d_{NN}}(\mathbb{U})$ of all NN distances of the design $\mathbb{U}$. To evaluate this, the NN distance of each point to the remaining design points is calculated, and the variance of these distances is determined. The results are visualized in Fig. 3. Here, the proposed method, where all variables are once the slack variable (DoE EQ), is compared with both cases where only u_1 or u_2 is the slack variable (Proj u_1 & Proj u_2) and the previously described non-uniformly spaced levels obtained by the linearization (Approx). The designs generated by linearization are the best for large design

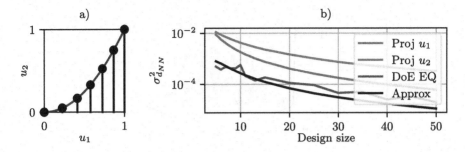

Fig. 3. a) Resulting designs for the constraint $u_1^2 - u_2 = 0$ with $N = 7$, if levels of u_1 are placed to obtain equal approximated distances. b) Variance of the NN distances (small values are better) for the constraint $u_1^2 - u_2 = 0$. The different methods are compared for different design sizes.

sizes. For small design sizes, the linearization error reduces the quality. The method proposed in this work achieves slightly poorer but nevertheless comparable results. Compared to both cases with only one slack variable, the method proposed is superior. For higher-dimensional problems, the adjustment of the levels is not suitable. For this, a grid with all its drawbacks has to be used as support design.

4.2 Three-Dimensional Pareto Front

This section applies the method to a three-dimensional problem and compares it to an existing one. In [11], a method is proposed to create points on nonlinear equality constraints. They applied their method to the spherical surface

$$u_1^2 + u_2^2 + u_3^2 = 1, \tag{4}$$

with $u_1 \geq 0$, $u_2 \geq 0$ and $u_3 \geq 0$ in Cartesian coordinate system. In the mentioned method, a uniform design for mixture experiments for three inputs, or variables, is generated. This design is projected on the surface in a second step. So, an initial design is modified to fulfill the constraint. To obtain suitable designs, they improved in an underlying step the applied uniform designs for mixture experiments [23] by the cutting method [24]. They evaluated the generated designs for the mixture constraint via the centered L_2-discrepancy of mixture experiments (CDM_2) [11,25]. Two designs with 66 points are generated, and the CDM_2 values are given for each. The CDM_2 of the standard method is 0.0389, while their improved version achieves a value of 0.0102. An equivalent design generated by this work's proposed algorithm is shown in Fig. 4 to benchmark the algorithm. It achieves a CDM_2 value of 0.0151. This lies between the results of the other methods and is closer to the improved one. Here, it should be noted that the method developed is not specialized for this particular constraint, and no optimization is performed apart from the optimization of the support designs via the EDLS algorithm. So a more universally applicable method with suitable results

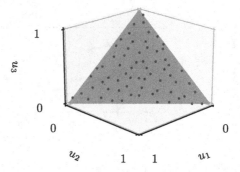

Fig. 4. Design for three-dimensional mixture constraint $u_1 + u_2 + u_3 - 1 = 0$. The design contains 66 points.

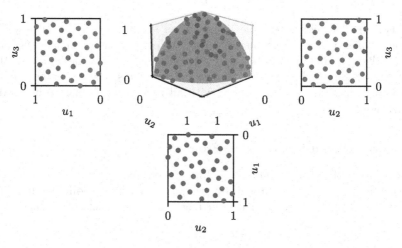

Fig. 5. Design on the spherical surface $u_1^2 + u_2^2 + u_3^2 = 1$. The color of each design point marks which slack variable was chosen to create the point. Slack variable: u_1 (purple), u_2 (orange), u_3 (blue). Unfeasible support points are marked gray. (Color figure online)

is developed. Besides this, the proposed algorithm is also applicable to a larger number of variables and large design sizes.

The design created by the proposed algorithm for the spherical surface is shown in Fig. 5. The points are spread over the entire surface. This example also illustrates the importance of selecting all variables as slack variables. Otherwise, important regions of the constraint are not well covered with points. Compared with the design in [11], the coverage of the regions close to the limits is superior.

5   Conclusion

This paper presents a new Latin-hypercube-based method for Design of Experiments for constrained design spaces. It is developed to create uniformly distributed designs for equality constraints. The target was to construct a universal method that can be applied to a wide range of equality constraints and is suited for powerful nonlinear modeling approaches, e.g., Gaussian process models.

An approach similar to the idea of slack variable models is proposed. By going over each dimension of the design space, each dimension is selected as slack variable. In each iteration, a uniformly distributed support design, a maximin Latin hypercube, is generated for the remaining dimensions. Each point of this support design is projected onto the constraint to calculate the slack variable. If the resulting point is feasible, it is added to the design. From the second iteration on, it is ensured that the new point is not too close to an existing one. Otherwise, it is not added. So, only the remaining data holes are filled. The two case studies demonstrate that this method generates suitable designs for different constraints. The curse of dimensionality affects this approach only weakly. The computational effort increases mildly with the dimensionality or the design size. This is accomplished by the application of Latin hypercubes as support design. The iteration over all dimensions leads to good coverage of the feasible design spaces for arbitrary processes and constraints with comparable results even to specialized methods for mixture constraints.

The method does not apply to constraints with more than one feasible solution for the projection of the support points. Therefore, further work will focus on this topic. Besides this, the selection criterion affects the results significantly. Additionally to the presented distance-based criterion, distribution-based criteria could improve the design quality.

References

1. Pronzato, L., Müller, W.G.: Design of computer experiments: space filling and beyond. Stat. Comput. **22**(3), 681–701 (2011)
2. Viana, F.: Things you wanted to know about the Latin hypercube design and were afraid to ask. In: Proceedings of the10th World Congress on Structural and Multidisciplinary Optimization, Orlando, FL, USA, 19–24 May 2013 (2013)
3. Sobol', I.: On the distribution of points in a cube and the approximate evaluation of integrals. USSR Comput. Math. Math. Phys. **7**(4), 86–112 (1967)
4. Santner, T.J., Williams, B.J., Notz, W.I.: The Design and Analysis of Computer Experiments. Springer, New York (2018). https://doi.org/10.1007/978-1-4939-8847-1
5. Schneider, F., Schüssler, M., Hellmig, R.J., Nelles, O.: Constrained design of experiments for data-driven models. In: Proceedings - 32. Workshop Computational Intelligence, Berlin, 1–2 December 2022 (2022)
6. Petelet, M., Iooss, B., Asserin, O., Loredo, A.: Latin hypercube sampling with inequality constraints. AStA Adv. Stat. Anal. **94**(4), 325–339 (2010)
7. Khan, S., Gunpinar, E.: An extended Latin hypercube sampling approach for cad model generation. Anadolu Univ. J. Sci. Technol.: Appl. Sci. Eng. **18**, 301–314 (2017)

8. Kayacier, A., Yüksel, F., Karaman, S.: Simplex lattice mixture design approach on physicochemical and sensory properties of wheat chips enriched with different legume flours: an optimization study based on sensory properties. LWT Food Sci. Technol. **58**(2), 639–648 (2014)
9. Scheffé, H.: The simplex-centroid design for experiments with mixtures. J. Roy. Stat. Soc.: Ser. B (Methodol.) **25**(2), 235–251 (2018)
10. Snee, R.D., Marquardt, D.W.: Extreme vertices designs for linear mixture models. Technometrics **16**(3), 399–408 (1974)
11. Hao, Z., Liu, Z., Feng, B.: Application of uniform design for mixture experiments in multi-objective optimization. In: 2014 IEEE International Conference on Progress in Informatics and Computing, pp. 350–354 (2014)
12. Borkowski, J.J., Piepel, G.F.: Uniform designs for highly constrained mixture experiments. J. Qual. Technol. **41**(1), 35–47 (2009)
13. Zhao, H., Li, G., Li, J.: Uniform test on the mixture simplex region. Symmetry **14**(7), 1371 (2022)
14. Stumpf, J., Naumann, T., Vogt, M.E., Duddeck, F., Zimmermann, M.: On the treatment of equality constraints in mechanical systems design subject to uncertainty. In: Balancing Innovation and operation. The Design Society (2020)
15. Li, H., Castillo, E.D.: Optimal design of experiments on Riemannian manifolds. J. Am. Stat. Assoc. 1–12 (2022)
16. Morris, M.D., Mitchell, T.J.: Exploratory designs for computational experiments. J. Stat. Plann. Inference **43**(3), 381–402 (1995)
17. Ebert, T., Fischer, T., Belz, J., Heinz, T.O., Kampmann, G., Nelles, O.: Extended deterministic local search algorithm for maximin Latin hypercube designs. In: 2015 IEEE Symposium Series on Computational Intelligence, pp. 375–382 (2015)
18. Bates, S., Sienz, J., Toropov, V.: Formulation of the optimal Latin hypercube design of experiments using a permutation genetic algorithm (2004)
19. Belz, J.: Fighting the curse of dimensionality with local model networks. Ph.D. thesis, Universität Siegen (2018)
20. Murray-Smith, R., Johansen, T.: Local learning in local model networks. In: 1995 Fourth International Conference on Artificial Neural Networks, pp. 40–46 (1995)
21. Nelles, O., Isermann, R.: Basis function networks for interpolation of local linear models. In: IEEE Conference on Decision and Control (CDC), pp. 470–475 (1996)
22. Javier, C.S.: Selecting the slack variable in mixture experiment. Ingeniería Invest. Tecnol. **16**(4), 613–623 (2015)
23. Ning, J., Fang, K.T., Zhou, Y.: Uniform design for experiments with mixtures. Commun. Stat.-Theory Methods **40**, 1734–1742 (2011)
24. Ma, C., Fang, K.T.: A new approach to construction of nearly uniform designs. Int. J. Mater. Prod. Technol. **20**, 115–126 (2004)
25. Ning, J.H., Zhou, Y.D., Fang, K.T.: Discrepancy for uniform design of experiments with mixtures. J. Stat. Plann. Inference **141**(4), 1487–1496 (2011)

Globular Cluster Detection in M33 Using Multiple Views Representation Learning

Taned Singlor[1], Phonphrm Thawatdamrongkit[1], Prapaporn Techa-Angkoon[1], Chutipong Suwannajak[2], and Jakramate Bootkrajang[1(✉)]

[1] Department of Computer Science, Faculty of Science, Chiang Mai University, Chiang Mai, Thailand
{taned_s,phonphrm_t,prapaporn.techaang,jakramate.b}@cmu.ac.th
[2] The National Astronomical Research Institute of Thailand (Public Organization), Chiang Mai, Thailand
chutipong@narit.or.th

Abstract. Globular clusters (GC) are crucial for understanding galaxy formation and evolution. However, identifying them in large imagery datasets is a time-consuming task. This prompts the development of an automated GC detection algorithm. Although GC detection is fundamentally an object detection problem, the state-of-the-art object detection algorithms are unable to produce accurate results. Motivated by how GCs are identified by astronomers, we propose a deep neural network that fuses multiple views of raw imaging data and learns a better representation of the input image. The proposed network is then combined with YOLO object detection algorithm resulting in YOLO for Globular Cluster detection (YOLO-GC) model. Experimental results based on a real catalog of GCs in the M33 Galaxy showed that the proposed multi-view representation learning technique helps improve detection performance.

Keywords: Globular Cluster Detection · Deep representation learning · Multi-view learning

1 Introduction

Globular clusters are systems of stars that formed early with galaxies. They contain hundreds of thousands of stars and are generally found in the halos of galaxies. Globular clusters can provide information on the formation and evolution of galaxies [1]. The classic approach to identifying GCs involves visually examining the images of galaxies. However, due to a large amount of astronomical data, experts can only inspect a limited number of images in a given amount of time. This can be difficult and time-consuming. Therefore, an automated technique to identify and analyze these GCs is needed to facilitate their studies. In general, astronomical object detection can be seen as an object detection task, for which the existing state-of-the-art algorithms such as YOLO [2], R-CNN

P. Quaresma et al. (Eds.): IDEAL 2023, LNCS 14404, pp. 323–331, 2023.
https://doi.org/10.1007/978-3-031-48232-8_30

[3] or SSD [4], should be readily applicable. However, the existing algorithm often misses out on subtle details of the object of interest rendering sub-optimal detection performance. We propose that the problem can be improved if we can incorporate some inductive biases into the model.

Previous studies on astronomical object detection incorporated additional inductive bias to improve the detection performance. The work by [5] presents a method for automatic galaxy detection and galaxy identification using deep learning and data augmentation. The method is based on a YOLO network. They converted raw FITS [6], a common data format of astronomical images, in igr bands to RGB images and also created five pixel-scaling/conversion methods to improve the overall galaxy detection. From their experimental results, it was found that non-linear scaling is more useful compared to linear scaling. They also found that the accuracy of the detection method is dependent on the quality of the conversion method used. In a related work by [7], the authors have developed a new galaxy cluster detection algorithm YOLO-CL, a modified version of the deep convolutional network for object detection named YOLO. The study showed that YOLO-CL performed well on Sloan Digital Sky Survey (SDSS) when compared to conventional methods. The authors demonstrated that deep neural networks possess a significant advantage compared to the traditional techniques for detecting galaxy clusters.

For GC detection task, astronomers usually makes a decision based on multiple views of the same input images[1] The reason is that some transformations or views can reveal the details of the relatively darker region but might loss some details in the brighter regions, while some may be able to retain details in the bright areas more. Motivated by the way an astronomer inspects multiple views of the input before concluding, in this work we incorporated such practice into an object detection pipeline. Our approach borrows an idea from multi-view learning to combine multiple non-linear transformations of input images. Specifically, we proposed a deep neural network for fusing multiple views, which are the results of applying non-linear functions, for example, `log, power, sinh, arcsinh, sqrt`, to the input image. We then stacked the deep representation learning network on an object detection algorithm. Here we adopted YOLO [2] as a backbone detector and named the proposed GC detector as YOLO-GC.

2   Dataset

2.1   Data Collection

In this work, we focused our interest on detecting GCs in the M33 galaxy. We manually collected data from the Barbara A. Mikulski Archive for Space (MAST) [8] in the form of FITS format, which is a standard format for storing and exchanging astronomical data. In order to obtain all the necessary information from the archive certain parameters namely, "science", "HST", "HAP",

---

[1] We note that in this study we use 'view' to refer to a result of a mathematical transformation applied to the pixel intensity of the input image.

"ACS/WFC", "F814W" and "Optical" were selected. The "science" parameter was selected so that the data contains scientific data gathered from various observations and experiments, providing a broad overview of the M33 galaxy. The "HST" (Hubble Space Telescope) parameter focuses specifically on data acquired from the Hubble Space Telescope. The "HAP" (High-Level Science Products) parameter lets us retrieve data that has undergone advanced processing and calibration. The "ACS/WFC" (Advanced Camera for Surveys/Wide Field Channel) parameter indicates that we get data that was taken using the Advanced Camera for Surveys on board the Hubble Space Telescope. This instrument captures wide-field images, allowing for a comprehensive examination of large areas within the M33 galaxy. The "F814W" parameter [9] refers to observations made in the F814W filter, which corresponds to a specific wavelength range in the optical spectrum. Optical data, as represented by the "Optical" parameter, encompasses observations conducted across various optical wavelengths, providing critical insights into the properties and characteristics of objects within the M33 galaxy.

2.2 Data Pre-processing

Once the FITS data is obtained, we used SAOImageDS9 (DS9) application to label all the globular clusters in each image according to the M33's globular clusters catalog [10]. We also divided the original image into smaller-sized images (1024 pixels by 1024 pixels). The division helps improve the accuracy and efficiency of object detection task, particularly when dealing with small objects within large image. Dividing a large image into multiple smaller images is also preferable as compared to resizing as the latter operation may render the loss of important nuances in the image. In total, we obtained 110 training images with an average of 9 GCs per image, while the test set contains 21 images with an average of 7 GCs in each test image.

2.3 Data Transformations

After the pre-processing step, we obtained 110 images for our training set and 21 images for our testing set. We then applied zero-one normalization to pixel values within the resulting images. Subsequently, we applied six non-linear transformations to each image. For each filter, the original pixel value $x \in [0, 1]$ is transformed according to the mathematical formulas given in Table 1 in order to obtain the respective views.

Table 1. Non-linear transformations employed in this study.

Original	Log	Power	Sqrt	Squared	Arcsinh	Sinh
x	$\frac{e^{-10x}-1}{e^{-10}-1}$	$\frac{1000^x-1}{1000}$	$\sqrt{x}$	x^2	$\frac{\text{arcsinh}(10x)}{3}$	$\frac{\sinh(3x)}{10}$

In this research, the focus is primarily on six specific transformations: Log, Power, Sqrt, Squared, Arcsinh, and Sinh, plus the original image. These transformations were chosen based on their relatively low correlations and their ability to complement each other. For instance, the Log accentuates low-intensity regions while the Linear preserves the original data values. The Sqrt emphasizes tiny intensity differences, whereas the Power emphasizes contrast. The Sinh and Arcsinh respectively highlight intermediate and extreme intensity values while the Squared filter emphasizes high-intensity zones. After the transformation, we standardized all the images once again using transformation-specific statistics, i.e., the means and the variances were computed from images undergone the same transformation. Figure 1 illustrated the outcome of some transformations and the distributions of pixel intensity as a result of some of the transformations.

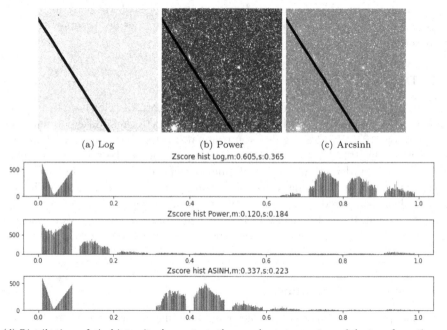

(d) Distributions of pixel intensity demonstrate the complementary nature of the transformations.

Fig. 1. Examples of the images and pixel distributions after each transformation.

3   The Proposed YOLO-GC Model

3.1   Problem Setting

Given a set of training data $\{x_i, y_i\}_{i=1}^N$ of size N, where $x_i \in R^{W \times H}$ represents a single channel image of size $W \times H$, and y_i is a tuple of coordinates indicating the locations of GCs in x_i. The task of GC detection is to infer a real-valued function $f : X \to Y$ using the training data so that it can be used to detect and output the coordinates of GCs in an unseen test image.

3.2   YOLO-GC Architecture

To enhance the efficiency of globular cluster detection, we present YOLO-GC, a multi-view learning approach that can be regarded as a representation learning technique as shown in Fig. 2. The network is composed of two primary components called the *Stack Block* and the scalable *Route Block*. These components serve as the core building blocks of the network.

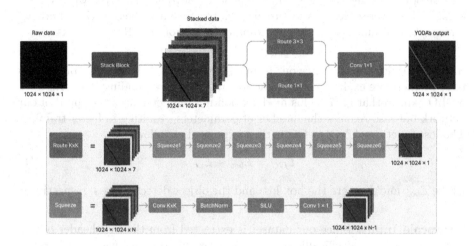

Fig. 2. The YOLO-GC Network's Architecture.

Stack Block. The process of stacking multiple views of input images in the Stack Block is motivated by the way astronomers and experts in the field inspect GCs. As mentioned earlier, astronomers utilize several image transformations while observing or visualizing globular clusters to enhance their perception and comprehension during their analysis. The transformations provide astronomers with various perspectives and representations of the globular cluster, allowing them to gain crucial details and characteristics that might not be visible from the raw data alone. To this end, seven views (six transformations of the raw image and the original raw image) are stacked along the depth dimension. Through this process, the extraction of significant features from the input data is expected. Each transformation is designed to complement each other and provides a distinct perspective and insights into specific aspects of GCs presented in the raw input data.

Route Block. The central learning block in YOLO-GC is the Route Block, which serves as a feature extractor. Within this block, the input consists of stacked data derived from the Stack Block. The primary objective of the Route Block is to extract essential information by simultaneously considering all seven views and subsequently merging and reducing the channels one at a time until only one channel remains. This reduction is accomplished by utilizing the Squeeze Block within the Route Block, which consists of convolutional layers with varying kernel sizes, customized for each Route Block. In this study, we use two Route Blocks with two kernel sizes: a 3×3 kernel and a 1×1 kernel. This approach diverges from the use of a single Route Block, providing us with the ability to capture different levels of granularity and increasing the Receptive Fields [11,12]. Following the operations performed within the Route Block, the last pivotal step involves the utilization of a Convolutional layer with a 1×1 kernel. This layer serves as a feature aggregator, enabling the selection of crucial information obtained from the preceding outputs from the Route Blocks.

Single Object Class. Since our primary objective is to identify globular clusters, we therefore exclude class prediction and the corresponding loss term from YOLO akin to that in [7]. This modification not only enhances the speed of our network but also reduces the number of parameters, resulting in faster training. The loss function of YOLO-GC is then expressed as follows

$$\mathcal{L}_{GC} = \mathcal{L}_{box} + \mathcal{L}_{obj} \tag{1}$$

where $\mathcal{L}_{box}$ and L_{obj} are the box loss and the object detection loss, respectively.

Grayscale Input. Since our dataset is extracted from the SCI header of raw FITS files, which specifically contains I filter, in terms of wavelength range images, with a single channel, it is necessary to process and treat them as grayscale images. The first layer of YOLO has to be adapted accordingly. There are various benefits to modifying the initial layer of YOLO to accept a single-channel image as opposed to the typical three-channel RGB format. First, by lowering the processing demands of the initial layer, it improves computational efficiency and speeds up inference. Second, by emphasizing intensity over color changes, the model becomes less dependent on particular color representations and more centered on structural information as needed in GC detection.

4   Experimental Results

We compare YOLO-GC, which uses YOLOv5 as a backbone, with the baseline YOLOv5. We trained both models until they are sufficiently converged with a batch size = 4, learning rate = 0.01 using a SGD optimizer. We then measured the detection performance using the mean average precision at 50 (mAP@50) [13], for which the predicted region is considered a hit when the Intersection over Union (IoU) between the predicted box and the ground truth box exceeds 50%. Such a degree of overlapping is generally sufficient for GC detection task. Also included for reference is a more stringent mAP@50:95 measure where we averaged mAPs over a range of IoU from 50% to 95% with 5% increment.

Table 2 summarizes the performances of the two models without any data augmentation. In the case where the detector network is YOLOv5 of size nano (n), we see that the proposed YOLO-GC outperformed baseline models (YOLOv5) with an mAP of 0.457 at IoU 50%, and an mAP@50:95 of 0.416. This is equivalent to roughly 12% and 20% increases in detection performance, respectively. However, when we employed YOLOv5 size s as a backbone detector, we observe a smaller improvement in mAP@50:95 over the vanilla YOLO. Interestingly, the baseline YOLO slightly outperforms YOLO-GC in terms of mAP@50 in this case. We speculate that this is due to the lack of training data.

Table 2. Performances of YOLO and YOLO-GC trained *without* data augmentation.

Size	Model	Number of parameters	mAP@50	mAP@50:95
n	YOLOv5n	1764118	0.401	0.332
	YOLO-GC	1765939	**0.457**	**0.416**
s	YOLOv5s	7020022	**0.491**	0.382
	YOLO-GC	7021843	0.484	**0.419**

To see if our speculation is indeed the case, we employed data augmentation methods such as mosaic, vertical flipping, and horizontal flipping to increase the size of our training data. The detection performances are summarized in Table 3. From the results, we observed that data augmentation indeed helps improve the performance of all the models. In the case of using YOLO nano as a backbone, the proposed YOLO-GC achieved an mAP@50 of 0.628, and an mAP@50:95 of 0.549 compared to mAP@50 of 0.604 and an mAP@50:95 of 0.523 by the vanilla YOLO. Similarly, when using YOLO size s as a backbone, YOLO-GC outperformed the baseline YOLO with roughly 17% improvement for mAP@50 and 10% improvement for mAP@50:95. These results highlight the benefit of having the multi-view representation learning added to the baseline YOLO model. We would like to note also that YOLO-GC only introduces marginal computational overhead over the standard YOLO as measured by the number of parameters.

Table 3. Performances of YOLO and YOLO-GC trained *with* data augmentation.

Size	Model	Number of parameters	mAP@50	mAP@50:95
n	YOLOv5n	1764118	0.604	0.523
	YOLO-GC	1765939	**0.628**	**0.549**
s	YOLOv5s	7020022	0.600	0.547
	YOLO-GC	7021843	**0.726**	**0.608**

To gain further insights, we employed GradCAM [14] to analyze and compare the decision-making process of both the baseline model (pure YOLOv5n) and our proposed method (YOLO-GC). Specifically, we utilize the objectness score

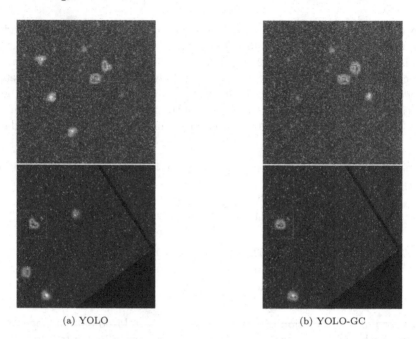

(a) YOLO   (b) YOLO-GC

Fig. 3. Activation maps for YOLO and YOLO-GC. The YOLO-GC seems to learn to focus on globular clusters better. Red box represents the ground truth. (Color figure online)

as the criteria for evaluation. This analysis provides information on how each model prioritizes certain areas of the input image. The results show that YOLO-GC enables the network to focus more on location where the object of interest is presented as can be seen in Fig. 3.

5   Conclusion

We proposed YOLO-GC, a deep neural network for multi-view representation learning for GC detection which used YOLO as a backbone detector. The multi-view learning part was inspired by the way astronomer examines multiple non-linear transformations of the input image to detect globular clusters. The proposed YOLO-GC demonstrated improvements over the baseline YOLOv5 models, both with and without data augmentation. The best-performing YOLO-GC model of size n achieved the highest mAP50:95 of 0.549, while the best YOLO-GC model of size s achieved the highest mAP50:95 of 0.608. The two models outperformed the YOLOv5 baseline models in almost all cases tested. The findings from our study not only support the effectiveness of our method but also offer essential information for the next astronomical research on feature extraction and object detection. In the future, we aim to employ additional imagery bands namely i, g, and r, into our stacking procedure. This strategy will facilitate better representation learning and will hopefully improve GC detection performance even further.

Acknowledgement. We thank the National Astronomical Research Institute of Thailand (Public Organization), the Department of Computer Science, Faculty of Science, and the Graduate School at Chiang Mai University for providing computing facilities and financial support.

References

1. Ashman, K.M., Zepf, S.E.: Globular cluster systems. Globular Cluster Systems (2008)
2. Redmon, J., Divvala, S., Girshick, R., Farhadi, A.: You only look once: unified, real-time object detection. In Proceedings of the IEEE Conference on Computer Vision and Pattern Recognition, pp. 779–788 (2016)
3. Ren, S., He, K., Girshick, R., Sun, J.: Faster r-cnn: towards real-time object detection with region proposal networks. In: Advances in Neural Information Processing Systems, vol. 28 (2015)
4. Liu, W., et al.: SSD: single shot multibox detector. In: Leibe, B., Matas, J., Sebe, N., Welling, M. (eds.) ECCV 2016. LNCS, vol. 9905, pp. 21–37. Springer, Cham (2016). https://doi.org/10.1007/978-3-319-46448-0_2
5. González, R.E., Munoz, R.P., Hernandez, C.A.: Galaxy detection and identification using deep learning and data augmentation. Astron. Comput. **25**, 103–109 (2018)
6. Pence, W.D., Chiappetti, L., Page, C.G., Shaw, R.A., Stobie, E.: Definition of the flexible image transport system (fits), version 3.0. Astron. Astrophys. **524**, A42 (2010)
7. Grishin, K., Mei, S., Ilic, S.: YOLO-CL: galaxy cluster detection in the SDSS with deep machine learning. arXiv preprint arXiv:2301.09657 (2023)
8. Marston, A., Hargis, J., Levay, K., Forshay, P., Mullally, S., Shaw, R. : Overview of the mikulski archive for space telescopes for the James Webb space telescope data archiving. In: Observatory Operations: Strategies, Processes, and Systems VII, vol. 10704, pp. 416–428. SPIE (2018)
9. Blakeslee, J.P., et al.: Surface brightness fluctuations in the hubble space telescope ACS/WFC F814W bandpass and an update on galaxy distances. Astrophys. J. **724**(1), 657 (2010)
10. Sarajedini, A., Mancone, C.L.: A catalog of star cluster candidates in M33. Astron. J. **134**(2), 447 (2007)
11. Ding, X., Zhang, X., Han, J., Ding, G.: Scaling up your kernels to 31x31: revisiting large kernel design in CNNs. In: Proceedings of the IEEE CVPR, pp. 11963–11975 (2022)
12. Luo, W., Li, Y., Urtasun, R., Zemel, R.: Understanding the effective receptive field in deep convolutional neural networks (2017)
13. Henderson, P., Ferrari, V.: End-to-end training of object class detectors for mean average precision. In: Lai, S.-H., Lepetit, V., Nishino, K., Sato, Y. (eds.) ACCV 2016. LNCS, vol. 10115, pp. 198–213. Springer, Cham (2017). https://doi.org/10.1007/978-3-319-54193-8_13
14. Selvaraju, R.R., Cogswell, M., Das, A., Vedantam, R., Parikh, D., Batra, D.: Grad-CAM: visual explanations from deep networks via gradient-based localization. Int. J. Comput. Vision **128**(2), 336–359 (2019)

Segmentation of Brachial Plexus Ultrasound Images Based on Modified SegNet Model

Songlin Yan[1] , Xiujiao Chen[2(⊠)] , Xiaoying Tang[3], and Xuebing Chi[4]

[1] Guilin Tourism University, Guilin 541006, China
yansl@sccas.cn
[2] The Second Affiliated Hospital of Guilin Medical University, Guilin 541199, China
chenxiujiao@glmc.edu.cn
[3] Southern University of Science and Technology, Shenzhen 518055, China
tangxy@sustc.edu.cn
[4] Computer Network Information Center, Beijing 100190, China
chi@sccas.cn

Abstract. The images automatic segmentation is an important technique for medical treatment. It can help doctor to relieve from heavy works of reading ultrasound images, especially for brachial plexus images. Moreover, the deep learning technology assists doctors in locating the catheters and improves the efficiency and accuracy of injection. However, the research of brachial plexus ultrasonic image segmentation is too few to satisfy the needs of medical application. In this paper, we used a novel modified SegNet to accurately segment brachial plexus. In the training stage, the original training set was divided into two parts randomly (training set 90% and validation set 10%), and the parameters of models were determined and optimized by adopting cross-validation method and data augmentation which can avoid over-fitting effectively. Computational results show that, the model significantly increases nerve segmentation accuracy with 96%; meanwhile, the model is scored 0.644 by Kaggle competition (The Kaggle competition uses CSV files containing the final results for scoring. Besides, the Kaggle competition does not require participants to provide open source code, and all participants' competition scores and rankings can be found on the website: https://www.kaggle.com/c/ultrasound-nerve-segmentation/leaderboard).

Keywords: Automatic segmentation · Modified SegNet model · Brachial plexus ultrasonic image

1 Introduction

In recent years, with the rapid development of deep learning, convolutional neural network (CNN) has made great achievements in image classification, target recognition and image segmentation. In terms of image classification, there are famous models such as AlexNet [1], GoogLeNet [2], and VGG19 [3], which achieved remarkable results in the ILSVRC (ImageNet Large Scale Visual Recognition Challenge) competition. Besides, in automatic image segmentation, a large number of classic models have been developed, including FCN [4], SegNet [5], DeconvNet [6] and DeepLab [7]. With many famous companies and universities designing more and more new models, these excellent models have also been widely applied in many downstream tasks.

P. Quaresma et al. (Eds.): IDEAL 2023, LNCS 14404, pp. 332–344, 2023.
https://doi.org/10.1007/978-3-031-48232-8_31

In terms of disease types and their risk prediction, the deep learning model could analyze various diseases, including diabetic retinopathy, nonvalvular atrial fibrillation and breast cancer. For example, Gao *et al.* [8] used a combination of CNN and recurrent neural networks to classify the severity of nuclear cataracts. Yang *et al.* [9] applied deep learning methods to model brain magnetic resonance images to classify cerebellar dyskinesia.

On the other side, computer-aided presurgical evaluation has become an essential treatment method now. Some works have many contributions to automatic segmentation of lesion tissues for presurgical preparation. Lee *et al.* [10] summarized the relevant progress in using computers to help preliminary analysis of cancer surgery. Nonetheless, the researches of the precise segmentation and localization for brachial plexus are rare. The brachial plexus innervates the pain conduction of the upper limb. To the best of our knowledge, brachial plexus block, as a local anesthesia method, is widely used in upper limb surgery. As a main strategy of brachial plexus block, the use of ultrasound image guided injection needles can improve the local anesthetic effect and reduce side effects. Unfortunately, compared with CT and MRI, ultrasound images have low resolution, high noise, blurry boundaries, and high real-time requirements. These lead to great challenges for manual operations. Therefore, this paper proposed a novel modified SegNet model for automatic segmentation of brachial plexus ultrasound images.

Overall, the modified SegNet for brachial plexus ultrasound image segmentation model mainly has the following contributions:

(1) The proposed ultrasound image segmentation for brachial plexus nerves based on an innovative model is a pioneering approach that can effectively assist doctors and nurses in identifying brachial plexus nerves and improve efficiency.
(2) Supervised deep learning requires a large number of training samples to achieve high accurate classification, and we have adopted some data augmentation methods to expand the dataset, and used cross validation and fine-tuning techniques to train the models.

Moreover, the database in this article is downloaded from the Kaggle competition database (https://www.kaggle.com/c/ultrasound-nerve-segmentation/data), and we evaluate our models using the website, ensuring the objectivity and authenticity of our model.

2   The Modified SegNet Model

2.1   The Structure of SegNet Model

SegNet [5, 11] is a new convolutional neural network model which can perform semantic pixel-wise image segmentation on images. The SegNet model is mainly composed of encoder, decoder and soft-max layers, Encoders and decoders appear in pairs for image feature extraction. Its architecture is illustrated in Fig. 1.

2.2   Our Improvements

SegNet is an end-to-end network, and its feature extraction is partially derived from VGG16. Existing VGG16 models can be used for fine-tuning. At the same time,

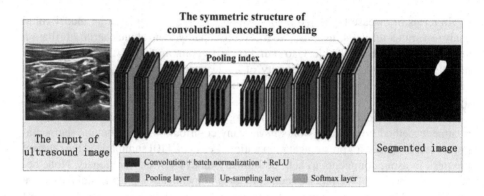

Fig. 1. An illustration of the SegNet model architecture

SegNet as a lightweight network has low requirements for the number of training data, and it has a fast inference speed. In the initial training of SegNet, the authors only used 367 training data and 233 test data[1] from the CamVid Road dataset.

Table 1. The change of the modified layers*

Convolution layer	VGG16	SegNet	Ours	Our padding
1	$3 \times 3 \times 64$	$3 \times 3 \times 64$	$7 \times 7 \times (64 \times 3)$	3
2	$3 \times 3 \times 64$	$3 \times 3 \times 64$	$7 \times 7 \times 64$	3
3	$3 \times 3 \times 128$	$3 \times 3 \times 128$	$5 \times 5 \times 64$	2
4	$3 \times 3 \times 128$	$3 \times 3 \times 128$	$5 \times 5 \times 64$	2
5	$3 \times 3 \times 256$	$3 \times 3 \times 256$	$3 \times 3 \times 64$	1
6	$3 \times 3 \times 256$	$3 \times 3 \times 256$	$3 \times 3 \times 64$	1
7	$3 \times 3 \times 256$	$3 \times 3 \times 256$	$3 \times 3 \times 64$	1
8	$3 \times 3 \times 512$	$3 \times 3 \times 512$	$3 \times 3 \times 128$	1
9	$3 \times 3 \times 512$	$3 \times 3 \times 512$	$3 \times 3 \times 128$	1
10	$3 \times 3 \times 512$	$3 \times 3 \times 512$	$3 \times 3 \times 128$	1
11	$3 \times 3 \times 512$	$3 \times 3 \times 512$	$3 \times 3 \times 128$	1
12	$3 \times 3 \times 512$	$3 \times 3 \times 512$	$3 \times 3 \times 128$	1
13	$3 \times 3 \times 512$	$3 \times 3 \times 512$	$3 \times 3 \times 128$	1

* Please note that we do not present the max-pooling layer in this table. And we only demonstrate the encoding part of the whole model.

But this model has not been used for medical images segmentation task. The original model is based on high-resolution indoor or outdoor scene images for segmentation, which are RGB tricolor images with good color gradient features. Unfortunately,

---

[1] https://github.com/alexgkendall/SegNet-Tutorial.

brachial plexus ultrasound images have lower resolution (580 × 420) and it is the 8-bit grayscale image. These are different from the common image recognition and classification problem. In order to achieve pixel level image segmentation with high accuracy, each pixel in the image needs to be classified and predicted. Therefore, the original model needs to be improved to meet the requirements of medical image processing.

Some of highlights are presented as follows:

Each convolution layer in the beginning increases the area of the receptive field and reduces the number of convolution kernels to obtain the feature information of a larger area in the image. At the same time, this structure also solves the over-fitting phenomenon caused by the small samples. The details of modified layers are depicted listed in Table 1.

Besides, we maintained the main structure of the encoding section, and it is logically similarity to the top 13-layers of VGG16. However, we not only removed its fully connected layers, but also modified the convolutional layers structure in the encoding part. And this structure also solves the over-fitting phenomenon caused by the small samples. Its parameters are as follows:

(1) the size of convolution kernels is from 7 × 7 to 3 × 3. The number feature maps is 64, 128, 192 in various stages, where all of them are $stride = 1$. Each neuron in the encoder adopts ReLU activation function;
(2) To maintain the classic structure, the pooling layer still uses the 2 × 2 size window, and $stride = 2$;

And the decoding section is also changed in our model, because of the structural similarity between the decoder and the encoder. The decoder part is consisting of a series of up-sampling layers and convolutional layers. The core of an up-sampling layer is its pooling layer labelling information, namely the pooling index. The pooling index decoding principle is shown in Fig. 2[2].

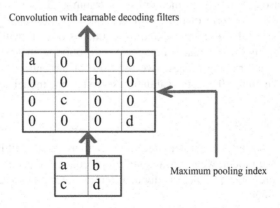

Fig. 2. An illustration of SegNet decoder(The function of max pooling indices)

---

[2] We did not change its decoding format, but modified its up-sampling index number in our model because of the different sizes with the original types.

The other innovations are as follows:

(1) Because of our input images are grayscale images, the number of convolutional kernels in the first convolutional layer is changed from 64 to 3×64, meeting the requirements for the number of decoders and final output channels;

(2) Adjust the number of neurons of the output layer to 243600×1. The number of output categories is the same as the required number of categories, here it is 2 classes.

3   Training Details

3.1   The Dataset of Brachial Plexus Ultrasound Images

All training and test data use the ultrasonic image data of brachial plexus provided by Kaggle website, in which there are 5635 training sets, as shown in Fig. 3[3].

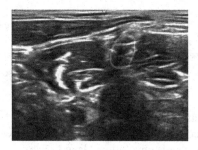

Fig. 3. A sample of brachial plexus ultrasonic image in the training set

The dataset also contains the same number of training labels manually marked by experts and saved in the form of binary image. Some of the training samples serve as negative samples in the training set, and these images do not contain the region of interest (ROI). The number of images in the test set is 5508, with a ratio close to 1 : 1 (training: test). In the training process, we use 90% of the entire training set as the training set, and the other 10% as the validation set to test the model accuracy.

3.2   Data Preprocessing

Although SegNet is a lightweight network, in order to avoid over-fitting and improve the robustness of the model, we use Matlab to preprocess our data. The preprocessing of the dataset before the experiment is divided into two steps: (1) data augmentation; (2) image enhancement.

The main strategies include:

---

[3] This image is a clip of the original dynamic images that we downloaded from: https://www.kaggle.com/chefele/ultrasound-nerve-segmentation/animated-images-with-outlined-nerve-area/code.(the red thin bounding line is the post-production mark.).

(1) Brightness adjustment (only for original images);
(2) Contrast adjustment (only for the original image);
(3) Flip left and right (original image and its corresponding label);
(4) Flip up and down (original image and its corresponding label);
(5) Random rotation of x degree (original image and its corresponding label), bilinear interpolation and clipping are also used in rotation to ensure that the image size remains unchanged, and x is a pseudo-random integer subject to discrete uniform distribution within the range of $[0, 360]$;
(6) Translation (original image and its corresponding label), with a translation distance (X, Y) of $[-50, 50]$ ranging from a pseudo random integer that follows a discrete uniform distribution;
(7) Randomly zoom in and out (original image and training label), with a scaling factor of $[0.5, 2.0]$ to obey uniformly distributed pseudo random numbers.

And then, the mask image needs to be transcoded from the binary image (black and white image) to a binary matrix only with the elements 0 and 1. We also use the script program written by Matlab to progressive scan the mask image line by line, and set the pixel value of the target area as 1, and other areas of the background area as 0, which digitalized the ROI or not in the mask image by pixel. After that, the file name and storage address of the image and its mask using the key-value pair form is saved in a text file. And the images and labels are converted into lmdb format through the script program provided by a deep learning framework - "Caffe" [12] as the experimental data source.

4   Experiments

Several experimental details will be demonstrated in this section.

4.1   Parameter Settings

In the network training process, the random gradient descent algorithm (SGD) is used to refresh the weights. The training data set is randomly selected when each batch of data is input, so the accuracy of each test model is different. The initial learning rate is manually adjusted by drawing the loss function curve, so as to find out the learning rate that makes the model have a faster rate of convergence.

To accelerate the training process and prevent over-fitting, we used the modified VGG16 model pre-trained by ImageNet for fine-tuning. However, since our data is a grayscale image and the original training data for VGG16 is a RGB color style, during fine-tuning, the first convolutional layer needs to be retrained, i.e., initialized by Gaussian distribution. And the other basic parameters of the model are still remained. In the experiments, the model after 30000 iterations can converge well at the initial learning rate $\lambda = 0.001$. Therefore, the initial learning rate in our training sets $\lambda = 0.001$. Since we used the fine-tuning strategy, in order to avoid oscillation and divergence, we adopt the fast decay learning rate strategy, namely, "step" learning rate strategy, and the step size is set to 10000.

When the modified SegNet model is working for segmenting each pixel, the model divides the corresponding category probability of each pixel in the image into belonging regions (one category represents a segmentation region), and assigns the label corresponding region with the highest probability to that pixel. Due to the fact that the model deals with image segmentation corresponding to two classification problems (dividing the image into ROI or not), we need to set different weights for each class to calculate the total loss of the network. Therefore, combined with cross validation methods, we upload the test results of models with higher training accuracy to the Kaggle website, and use their test data to comprehensively evaluate the accuracy and robustness of our model. The loss function adopts a weighted softmax loss function, which filters the model by adjusting the weights. The weighted softmax loss function is obtained by adding the weighted loss of each pixel as follow:

$$L(\Omega) = -\frac{1}{n}\left[\sum_{i=1}^{n}\sum_{j=1}^{m}\alpha_j \cdot 1\left\{y^{(i)} = j\right\} \cdot ln\frac{e^{\Omega_j^T X^{(i)}}}{\sum_{p=1}^{m}e^{\Omega_p^T X^{(i)}}}\right], \tag{1}$$

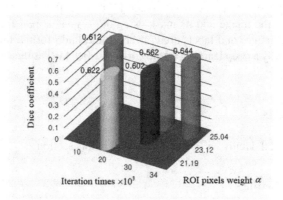

Fig. 4. Dice coefficients of our models with different parameters. They contain two core parts: 1. the values $\alpha = 23.12$ in our model generated from random initial value iteratively; 2. we manually set the weight $\alpha = 25.04$ and $\alpha = 21.19$ respectively.

where, n represents the number of samples per batch, m is the number of output categories, α_j denotes the average weight ratio of the ROI or not, $y^{(i)}$ shows the model output, $X^{(i)}$ indicates the input of the softmax layer, Ω definds the softmax layer weights, and $1\{\cdot\}$ depicts a truth function. Its criteria are:

$$1\left\{Expression : value = True\right\} = 1, \tag{2}$$

$$1\left\{Expression : value = False\right\} = 0. \tag{3}$$

For example, the value of an expression $1\left\{2 + 2 = 4\right\}$ is 1, but the value of $1\left\{1 + 1 = 5\right\}$ is 0.

Although there are only two classes of output, the weighted cross entropy loss function can be used to calculate the loss of each pixel, whereas in order to facilitate and keep the layer form consistent with the original SegNet model, the softmax loss function form is still used here. Therefore, in the training stage, another important external parameter is the weight ratio α of each type in the loss function. In the beginning, the default weight ratio value α_0 is assigned with the uniform distribution $U = (0, 50)$. When the network begins to iterate, the value α is generated by the proportion of the similarity between the output and its validation mask image. It can be expressed by:

$$\alpha = (\frac{S_T}{S_P} \times 0.5 + \zeta \times 0.5), \tag{4}$$

where, S_T represents the size of the ROI (measured by the number of pixels), S_p denotes the size of the original image (also measured by the number of pixels), and ζ is the similarity between the result and its mask image (calculated by the center distance and size). The real value α is updated after each validation epoch, and its strategy is:

$$\alpha_{i+1} = 0.5 \times \alpha_{i+1} + 0.5 \times \alpha_i, i = 1, 2, \ldots, k. \tag{5}$$

Due to the outputs in our model are only divided into two classes, the loss function is setting the weight of the non-ROI as 1 and the weight of the ROI as α by default, which means normalizing the weight of the non-ROI.

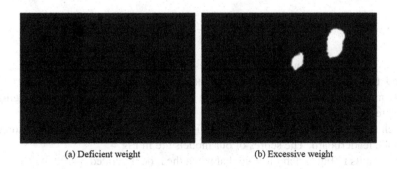

(a) Deficient weight   (b) Excessive weight

Fig. 5. The error results of unfitting weights

The other parameter settings are as follows:

(1) The modified model adopts a SegNet neural network with five sets of encoder-decoder pairs (maintaining the same convolutional layer as VGG16);
(2) Weight initialization method: Use $N(0, 1)$ Gaussian distribution to assign default weights to some non-fine-tuning layers of the network, including softmax layers;
(3) Regularization method: $L2$ (Ridge regression);
(4) Mini-batch size (the size of training data and validation data is the same): The number of images is 25. Due to the large number of layers in the SegNet network and the need for additional storage space to store pooling indexes, the batch size of input with the SGD optimization algorithm is smaller, ensuring sufficient GPU memory;

(5) Training method: Perform 20 iterations in each mini-batch but only update the network weights 10 times.

4.2 Implementation Details

We used Sugon W780-G20 server in the experiment, which equipped with 8 NVIDIA Tesla K80 and 2 Intel Xeon processors (2.4 GHz). The data preprocessing was processed by Matlab with CPU and the model training was executed simultaneously in 8 K80 GPUs by parallel.

During training phase, each epoch shuffles the data from the training set and sequentially extracts them to ensure that each image is only used once per epoch, and a batch of validation data is used to evaluate model accuracy after every 10 mini-batches. For calculating the loss, we set different weights for each class. And then, we trained several models in both random initial and manual setting of the weight α.

The average accuracy is obtained through 5-fold cross validation each time. In order to test network performance and avoid using only training accuracy and validation results for model evaluation, after cross validation stage, we test some selected models by using the Kaggle website. The test score of the Kaggle website is the benchmark to select our final model which has the highest score. This Kaggle contest uses the Dice coefficient to evaluate the similarity between the segmentation results and their masks. And the Dice coefficient function is:

$$D = \frac{2 \times |X \bigcap Y|}{|X| + |Y|} \tag{6}$$

where, X is the prediction result, Y represents the mask of the corresponding data (ground truth). D denotes the similarity between the prediction segmented and its labeled mask. The accuracy of the training model can be evaluated by averaging all image prediction results.

Each selected model will upload to the Kaggle competition, and will be scored on the public leaderboard. The scores of our models are in Fig. 4.

The results in Fig. 4 demonstrate that when the model iterated 34000, its Dice coefficient is 0.64421, and when it iterated 10000, its Dice coefficient is 0.61201.

On the other hand, we also tested some manually selected weights α that were deviated significantly from the classic values, as shown in Fig. 5.

In Fig. 5, its subplot (a) shows that when the weight sets too low $\alpha = 3.8536$, the background area pixels completely occupy the entire image. This result implies an abnormal phenomenon that the model could not distinguish the ROI correctly.

The subplot (b) is the reverse of the previous one. It depicts that when the weight is too high $\alpha = 34.6821$, the model prefers to the false ROI. Thus, the low weight means a higher probability of classifying pixels as background points, while the high weight infers that some parts that are just slightly similar to the ROI will also be misjudged into the target area.

From the results, we can see that the model scores obtained near random weights are relatively close, and as the number of iterations increases, the scores gradually increase. However, these random weights may not be the optimal solutions. Therefore, the initial

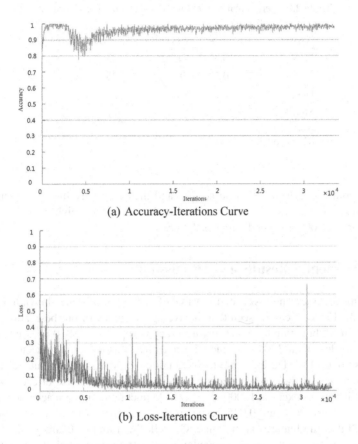

(a) Accuracy-Iterations Curve

(b) Loss-Iterations Curve

Fig. 6. The accuracy/loss curve of our best model

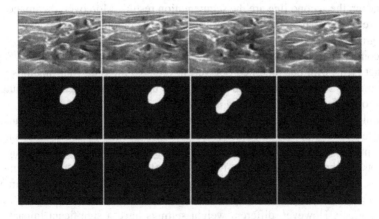

Fig. 7. An illustration of segmentation result on validation set

Table 2. Experimental results for the 4 segmentation algorithms

Networks	MSE	Accuracy(%)	Time/step(ms)
FCN	0.3185	71.20	33
SegNet	0.2796	78.01	20
Unet	0.2135	82.87	26
Modified Seg-Net (Our Model)	0.2204	83.40	21

weight α could be selected by grid search, and then manually fine-tune them to the optimal weight. From the experiments, we find that when the weight is $\alpha = 25.04$ and the iteration is 34000, our model makes the best.

5   Experimental Results and Discussion

The training accuracy and loss function value of our best model are shown in Fig. 6.

From the Fig. 6, it can be seen that the average accuracy of our best model is more than 96%. In the beginning, the training accuracy is approximately 0.81. And when the iterations are less than 3000, it depicted a high but illusory training accuracy, where the loss is about 0.18. During 3000 to 7500 iterations, the loss continued to decrease and the accuracy of the model decreased, mainly due to the noise effects. When the number of iterations exceeds 10000, there is little fluctuation in both accuracy and loss, with an average of 96% and 0.045, respectively. Therefore, we believe that the model has learned the true features at this time. Overall, the training accuracy curve and the loss curve converge fast. Some segmentation results from the validation dataset are demonstrated in Fig. 7.

The first row in Fig. 7 shows the original brachial plexus ultrasound images. And the images in the second line are the segmentation results. Moreover, the images in the bottom are the corresponding ground truth. It suggests that the modified SegNet model could segment the ROI very well.

We compared three other modes: "Fully Convolutional Networks" (FCN), "SegNet" and "Unet [13]" with our modified SegNet. In order to measure four methods reasonably, all models were used the same preprocessed dataset. The experimental outputs of our 4 various networks are illustrated in Table 2.

We only upload our best model to the Kaggle website and our scores and ranking on the Kaggle public or private test data set are demonstrated in Fig. 8. The results in Fig. 8 indicate that our model can effectively segment brachial plexus ultrasound images, and it also gets a rather good ranking on the private test data set. Meanwhile, the modified model can repeatedly adjust the weight α in the loss function through experiments to improve model accuracy. Increasing the number of iterations can also slightly improve model accuracy. However, different weight settings have a significant impact on the segmentation effect of brachial plexus ultrasound images.

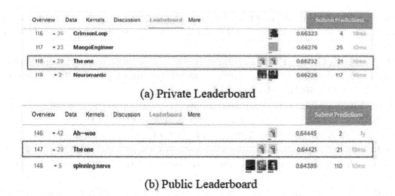

(a) Private Leaderboard

(b) Public Leaderboard

Fig. 8. Our ranking and scores on the public/private test data set

6   Conclusions

Our work is based on the modified SegNet model for semantic segmentation of brachial plexus ultrasound images, and analyzes the accuracy of the segmented area and network parameters. We can see the improved SegNet model successfully segments the brachial plexus neural structure in the ROI with high accuracy. These results will provide guidance for further improving work.

Overall, compared to traditional medical image processing methods, computer-aided diagnosis and treatment is not only an innovation in computer applications, but also a revolution in the field of medical diagnosis. Using the modified SegNet model to automatically segment images can simplify the doctor's treatment process. Therefore, the modified SegNet model has considerable feasibility in the application of ultrasound image segmentation, and we also hope that this method can provide new ideas for deep learning in medical image recognition.

References

1. Krizhevsky, A., Sutskever, I., Hinton, G.: ImageNet classification with deep convolutional neural networks. In: International Conference on Neural Information Processing Systems, vol. 25, no. 2, pp. 1097–1105 (2012)
2. Szegedy, C., Liu, W., Jia, Y.: Going deeper with convolutions. In: Proceedings of the IEEE Conference on Computer Vision and Pattern Recognition, vol. 22, no. 4, pp. 1889–1897 (2016)
3. Simonyan, K., Zisserman, A.: Very deep convolutional networks for large-scale image recognition. In: Proceedings of the International Conference on Learning Representations (ICLR), pp. 1–14 (2015)
4. Long, J., Shelhamer, E., Darrell, T.: Fully convolutional networks for semantic segmentation. In: Proceedings of IEEE International Conference on Computer Vision and Pattern Recognition, vol. 79, no. 10, pp. 3431–3440. IEEE Computer Society, Washington DC (2015)
5. Badrinarayanan, V., Kendall, A., Cipolla, R.: SegNet: a deep convolutional encoder-decoder architecture for image segmentation. IEEE Trans. Pattern Anal. Mach. Intell. **39**(12), 2481–2495 (2015)

6. Noh, H., Hong, S., Han, B.: Learning deconvolution network for semantic segmentation. In: Proceedings of the IEEE International Conference on Computer Vision, vol. 11, pp. 1520–1528 (2016)
7. Liang-Chieh, C., George, P., Iasonas, K., et al.: DeepLab: semantic image segmentation with deep convolutional nets, atrous convolution, and fully connected CRFs. IEEE Trans. Pattern Anal. Mach. Intell. **40**(4), 834–848 (2017)
8. Gao, X., Lin, S., Wong, T.Y.: Automatic feature learning to grade nuclear cataracts based on deep learning. IEEE Trans. Biomed. Eng. **62**(11), 2693–2701 (2015)
9. Yang, Z., Zhong, S., Carass, A., Ying, S.H., Prince, J.L.: Deep learning for cerebellar ataxia classification and functional score regression. In: Wu, G., Zhang, D., Zhou, L. (eds.) MLMI 2014. LNCS, vol. 8679, pp. 68–76. Springer, Cham (2014). https://doi.org/10.1007/978-3-319-10581-9_9
10. Lee, H., Chen, Y.P.: Image based computer aided diagnosis system for cancer detection. Expert Syst. Appl. **42**(12), 5356–5365 (2015)
11. Kendall, A., Badrinarayanan, V., Cipolla, R.: Bayesian SegNet: Model Uncertainty in Deep Convolutional Encoder-Decoder Architectures for Scene Understanding. arXiv:1511.02680v1 (2015). https://arxiv.org/abs/1511.02680
12. Jia, Y., Shelhamer, E., Donahue, J., et al.: Caffe: convolutional architecture for fast feature embedding. In: Proceedings of the 22nd ACM International Conference on Multimedia, pp. 675–678. ACM, New York (2014)
13. Ronneberger, O., Fischer, P., Brox, T.: U-Net: convolutional networks for biomedical image segmentation. In: Navab, N., Hornegger, J., Wells, W.M., Frangi, A.F. (eds.) MICCAI 2015. LNCS, vol. 9351, pp. 234–241. Springer, Cham (2015). https://doi.org/10.1007/978-3-319-24574-4_28

Unsupervised Online Event Ranking
for IT Operations

Tiago Costa Mendes[1,2](✉) ⓘ, André Azevedo Barata[1,2], Miguel Pereira[3],
João Mendes-Moreira[1,2]ⓘ, Rui Camacho[1,2]ⓘ, and Ricardo Teixeira Sousa[1,2]ⓘ

[1] LIADD/INESC Tec, R. Dr. Roberto Frias, 4200-465 Porto, Portugal
tiago.c.mendes@inesctec.pt
[2] Faculty of Engineering of the University of Porto, R. Dr. Roberto Frias,
4200-465 Porto, Portugal
[3] IT Peers, R. Eng. Frederico Ulrich 3210, 1 andar, s.101, 4470-605 Maia, Portugal

Abstract. Keeping high service levels of a fast-growing number of servers is crucial and challenging for IT operations teams. Online monitoring systems trigger many occurrences that experts find hard to keep up with. In addition, most of the triggered warnings do not correspond to real, critical problems, making it difficult for technicians to know which to focus on and address in a timely manner. Outlier and concept drift detection techniques can be applied to multiple streams of readings related to server monitoring metrics, but they also generate many False Positives. Ranking algorithms can already prioritize relevant results in information retrieval and recommender systems. However, these approaches are supervised, making them inapplicable in event detection on data streams. We propose a framework that combines event aggregations and uses a customized clustering algorithm to score and rank alarms in the context of IT operations. To the best of our knowledge, this is the first unsupervised, online, high-dimensional approach to rank IT ops events and contributes to advancing knowledge about associated key concepts and challenges of this problem.

Keywords: Big Data · Data Streams · Data Mining · IT Operations · Online Event Detection · Event Ranking

1 Introduction

In the context of the fast-growing digitization of businesses and institutions, high service levels of server computers are crucial to keeping business processes running. The crash or slowdown of server machines of, for example, a data centre, can

This work is partially financed by the ERDF - European Regional Development Fund through the Norte Portugal Regional Operational Programme - NORTE 2020 under the Portugal 2020 Partnership Agreement, within project OnlineAIOps, with reference NORTE-01-0247-FEDER-070104, and by National Funds through the Portuguese funding agency, FCT - Fundação para a Ciência e a Tecnologia, within project LA/P/0063/2020.

P. Quaresma et al. (Eds.): IDEAL 2023, LNCS 14404, pp. 345–355, 2023.
https://doi.org/10.1007/978-3-031-48232-8_32

affect the productivity of entire organizations, the availability of a vital public service, or even make them stop completely. To avoid these situations, organizations monitor their servers continuously, analyzing data streams concerning each machine's operational metrics. The metrics report on, among others, memory usage, disk access, number of running processes, etc. The idea is to detect anomalies in these metrics early enough to take action before a possible level of service degradation.

The detection task used to be performed by experts aware of the abnormal data values or patterns that could signal a problem. With the ever-growing number of machines in operation and the scarcity of human resources with the necessary expertise, automation is required to replace human intervention or decrease the load on the technical teams.

Outlier detection and concept drift detection techniques can be applied to data streams to detect abnormal values and patterns that could indicate a problem. However, many detected outliers or drifts in one metric may be misleading and not directly related to relevant problems but to noise or normal operational peaks. To mitigate this, previous approaches [4,10] introduce the notion of complex event detection that looks, for example, for combinations of outliers and/or concept drifts occurring simultaneously.

However, even when considering complex event detection instead of single metric events, results can still include false positives and/or sequences of several detections in a short period, which is a problem for the technical teams that need to decide, for example, which occurrences to address first. Our work addresses this problem using ranking techniques to order events by relevance.

Extracting knowledge from data streams introduces additional challenges compared to static data [12]. For example, streams keep coming indefinitely from the source, which means that the entire data set will not be available and is costly to store. The data arrival rate is high, which calls for processing each observation fast enough before the following one is received. And the data distribution is unknown and might change over time due to, for example, concept drift. Specific algorithms have been designed for this context to process observations individually as they arrive and learn incrementally [5].

Another challenge is the absence of labelled data for outlier or event detection, especially of abnormal events, as they are rare by definition. This is also the case in the context of Information Technology Operations (IT Ops), where the dedication of the same scarce experts whose time we are trying to spare would be needed to label data. The absence of labelling also applies to event ranking, which led us to focus our research on unsupervised techniques.

A final factor increasing the problem's difficulty is the observations' high dimensionality. Considering that each server can report over a hundred metrics and two types of occurrences in each of the metrics (ex: outlier and drift), the dimension of the feature space will be twice that number.

Our work proposes an unsupervised incremental algorithm to rank events in computer servers from multi-dimensional streaming data. To the best of our knowledge, our approach presents three new characteristics that make it highly suitable for fast, ever-changing streaming data. Events are processed one by one, the model is updated incrementally, and no particular data distribution is assumed. Additionally, instead of ranking events using a single metric, our algorithm processes combinations of simultaneous events in different metrics, covering the multi-dimensional context of server computers' operations.

Section 2 includes existing approaches to the problem, Sect. 3 describes our contributions, Sect. 4 characterizes the evaluation framework, results are presented and discussed in Sects. 5 and 6, respectively, and Sect. 7 contains the conclusions of the work.

2 Related Work

Several previous approaches addressed the problem of online event ranking in IT Ops. Jiang et al. [6] used manual rules based on comparisons with thresholds to classify alerts into different priority levels. Lin et al. [8] cluster tickets of alerts with semi-structured text and tickets of incidents with unstructured text. However, these linear methods cannot capture the complex relationship between the multiple dimensions of the events.

Other approaches focused on intrusion detection systems [2,9], which is a different context than ours.

Viswanathan et al. [13] use online methods to detect anomalies from data centres and rank them over concurrent windows. The proposed approaches use the probability of the Z-score and apply it to different anomaly detection techniques. Limitations of this technique include the assumption of a Gaussian distribution of the observations and a specific false positive rate. As stated above, the distribution is unknown in data streams and can change over time. The algorithm outputs rankings of metrics and servers in contrast with our method, which ranks multi-metric higher-dimensional events.

More recently, Zhao et al.'s AlertRank framework [14] automatically and adaptively identifies severe alerts occurring in large-scale online service systems. Specifically, AlertRank extracts a set of univariate and multivariate anomaly features for monitoring metrics, adopts the XGBoost ranking algorithm to identify the severe events out of all incoming alerts, and uses historical tickets' data to obtain labels for training and testing. Unfortunately, as we stated above, historical information about the events and their labels is often unavailable in these contexts, suggesting that unsupervised approaches are more adequate.

One interesting unsupervised ranking approach is TreeRank [3], which was applied to breast cancer data sets. The algorithm relies on the connection between anomaly ranking and supervised bipartite ranking, producing scoring rules described by oriented binary trees. This technique is tailored to large-scale and high-dimensional data. Although this approach did not adapt its model incrementally and was not applied to streams, it has the base requisites to adapt to an online scenario.

3 Proposed Methods

Figure 1 shows a schema of the framework designed for ranking events related to the operational metrics from computer servers. The original inputs are streams of metrics used for monitoring a server's operational status. The output is a rank of alarms according to the potential degree of impact on the service level. In the next subsections, we describe the proposed methods included in this framework.

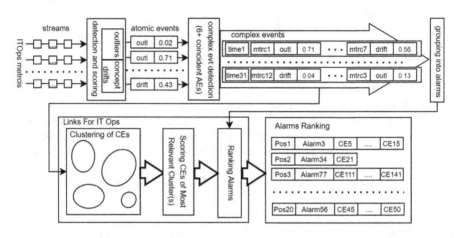

Fig. 1. Framework for ranking of alarms in a IT Ops' scenario

3.1 Atomic Events Detection

Outlier and concept drift detection techniques are applied to the data streams in the first processing phase. The output consisted of a list of atomic events. An atomic event (AE) is an event detected in one metric, which, in this case, can be an outlier or a concept drift. Additionally, a score is assigned to each atomic event specifically for the ranking component. The scores can be interpreted as a degree of "*outlierness*" or "*driftness*" of an outlier or concept drift detected, are computed as distances to the thresholds used by the detection techniques and normalized into a scale between zero and one. Our approach is agnostic regarding outlier and concept drift detection techniques as long as they are incremental and provide the above-mentioned scores.

3.2 Complex Events Detection

In a second phase, complex (composite) events (CEs) are formed by combining atomic events [1] coincident in time. CEs are used to focus the processing on more interesting higher-level events instead of going through every granular event triggered by (atomic) event detection systems. The goals for using CEs include making the processing lighter and reducing false positives. It is worth mentioning that, as this is a system where components are interconnected, their operation often depends on the well-functioning of the other components. So, it is likely that several metrics to be affected when there is a major problem in the system.

During the experiments, the team also observed that when a host had a problem, a series of consecutive complex events were triggered. These series are coherent with the notion that some problems remain affecting the metrics for some time, for example, until they are solved. In this case, we decided to aggregate these consecutive CEs into groups called alarms to avoid ranking several CEs that relate to the same problem.

3.3   Clustering for IT Ops

To build our incremental algorithm, we looked for existing batch unsupervised multi-dimensional ranking approaches that could be converted to an incremental version or for online approaches applied to other areas that could adapt to the IT Ops scenario. We identified two methods matching these criteria: Clémeçon at al.'s TreeRank [3] (see Sect. 2) and Mansfield et al.'s Links [11].

The Links algorithm was designed to perform online clustering in a high-dimensional Euclidean space. The algorithm is appropriate when it is necessary to cluster data efficiently as it streams in. Links has been successfully applied to embedding vectors generated from face images or voice recordings [11].

Although Links was not applied to the ranking problem, the clustering algorithm based on a score seemed interesting due to the possibility of using the score to produce a rank. Additionally, Links was already designed to be online, so we thought it would better fit the IT Ops context.

Links accepts unit vectors as inputs. In our case, the different possible atomic events correspond to the observations' space dimensions. Every complex event is represented as a vector in this space, using the scores of the atomic events as coordinates following each dimension. Once normalized, these vectors are fed to the algorithm for clustering.

The idea is to obtain a clear distinction between true positives and all the other events. Or, in other words, to have the true positives in a different cluster or clusters from the ones that include events that don't correspond to problems.

Links keeps a graph of subclusters. Each set of disconnected subclusters is a cluster. The cosine similarity to each subcluster's centroid is computed to assign a new vector to a cluster. The vector is added to the most similar subcluster, provided this similarity is above a subcluster similarity threshold.

The algorithm accepts three hyperparameters: Cluster Similarity Threshold (CST), Subcluster Similarity Threshold (SST) and Pair Similarity Threshold (PST). CST and SST regulate the granularity of clusters and subclusters. PST allows Links to adjust to anisotropy in the distribution [11].

To accomplish IT operations complex events' ranking, we developed additional components on top of the original approach.

Once the clustering is complete, a challenge is identifying the correct (true events) cluster to be ranked. We decided to select the cluster with the highest ratio of atomic events per complex event based on the experts' insight that, in server machines operations, the higher impact events affect a higher number of metrics on average. The algorithm does not store all the vectors for efficiency and scalability, so this ratio is updated incrementally for each cluster every time a new vector (CE) is processed. To calculate the AEs/CE ratio, we store and update each subcluster's total number of AEs and CEs.

The following step consists of calculating a ranking score for each complex event. We decided to develop four methods for this:

– the AESb score of an event corresponds to the summation of the atomic events *outlierness* and *driftness* (see Sect. 3.1) scores;
– SimCentr score is the summation of cosine similarities between the vector and the subclusters' centroids in the cluster to be scored. This score measures the centrality of each event in the cluster to be ranked. The closer a vector is to the other vectors, the higher its ranking will be (Fig. 2a).
– DstNorm is computed as the cosine dissimilarities between the vector and the subclusters' centroids of a cluster representing normality. It uses a measure of distance to normality. The more distant a vector is from normality, the higher its ranking will be (Fig. 2b);
– SCBatch score results from a batch version of SimCentr that stores the data set and is calculated as the summation of cosine similarities between the vector and the other vectors/CEs in the cluster to be ranked. We will use this score as a reference to evaluate our incremental scoring algorithms.

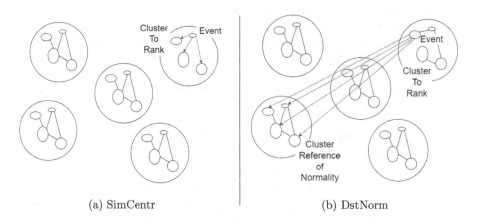

(a) SimCentr   (b) DstNorm

Fig. 2. Ranking score methods and distances used.

To identify the reference cluster for normality, we use the inverse of the method to identify the cluster to be ranked: we select the cluster with the lowest ratio of atomic events per complex event.

Finally, to address the triggering of sequences of consecutive complex events related to the same problem, we consider the aggregations of CEs in previously processed alarms and rank each (aggregated) alarm based on the top score of the constituent complex events.

As mentioned above, the incremental character of the approach is not compatible with storing the events on the data set. The algorithm keeps the top 15 positions on the ranking with their corresponding scores. The ranking information is stored, not only for the current cluster being ranked, but also for all other clusters, to cover the possibility of a future selection of the cluster to rank.

4   Evaluation Framework

We used data sets from two hosts. Readings were previously synchronized to a periodicity of 100 s, which means 864 daily readings per metric. For host A, we processed data from under six months (175 days) comprising 25 metrics and 3 780 000 total readings. We processed 71 days of 115 metrics for host B, totalling 7 054 560 samples.

The output of the atomic event detection component is a list of tuples, each including the time when the atomic event was detected, the metric where the event was detected, the event type (outlier or drift) and the *outlierness* or *driftness* score computed by the detection algorithm.

For the CEs detection component, after discussing the best balance between risks of false positives and false negatives in this context, the experts and we decided to consider combinations of six (five for host B) or more coincident atomic events to trigger (complex) events. The output of this component is a list of tuples, including the time when the CE has triggered and a list of 6 (5 for host B) or more atomic events.

Each CE is then converted into a vector to be fed to Links for ITOps. The AEs scores are the coefficients for each dimension. For host A, the dimensions were 50 (25 metrics times two types of atomic events). For host B we had 230 dimensions (115 × 2).

For alarm aggregation, the experts and we agreed to consider CEs as consecutive whenever the elapsed time in between was less than 10 min.

5   Results

The approach was applied to data from the hosts mentioned in Sect. 4. Table 1 shows the results of the complex event detection and clustering components. After trying several combinations of values, the best performance was achieved with Links hyper-parameters SST and PST tuned to 0.9 and 0.5, respectively.

For host A, 254 complex events (CEs) with six or more coincident atomic events were detected. IT Ops experts labelled 63 CEs as True Positives (TPs) and 191 as False Positives. The CEs were aggregated into 105 alarms, of which 17 were True Positives. 16 out of the 17 TP (94.11%) alarms were clustered together in the cluster to be ranked. The CST hyper-parameter value that delivered the best performance for host A was 0.7.

Table 1. Results of CE detection and clustering (SST = 0.9 and PST = 0.5).

Host	CEs		Alarms		CST	Ranked Cluster		
	Total	TPs	Total	TPs		Id	TPs	%TPs
A	254	63	105	17	0.7	0	16	**94,11%**
B	1086	593	387	17	0.6	1	9	**66.67%**

Host B data generated 1086 complex events (CEs) with five or more coincident atomic events. The experts labelled 593 True Positives and 493 False Positives. Total alarms were 387 with 17 True Positives. 9 out of the 17 TPs (66.67%) alarms were clustered together in the cluster to be ranked. For host B, the best performance was obtained with the CST hyper-parameter set to 0.6.

Table 2 presents the results of the rankings evaluations. The first three columns show the percentage of the ranking positions that contain True Positive (TP) alarms for the top 14, top 10 and top 5 places, respectively. The following three columns present the rate of positions that were filled with alarms that experts also ranked (labelled) in the same top k category. The three final columns consider the positioning of the alarms among the top k through the Normalized Discounted Cumulative Gain (NDCG) metric [7]. The experts considered alarms ranked below the top fourteen irrelevant and often very similar in importance, which made us limit the rankings to 14 positions.

Table 2. Rankings evaluation

Host	Method	% of TPs in Top			Precision			NDCG		
		14	10	5	@14	@10	@5	@14	@10	@5
A	DstNorm	42.86%	40.00%	20.00%	42.86%	40.00%	20.00%	0.25	0.22	0.04
A	SimCentr	71.43%	70.00%	80.00%	**78.57%**	**60.00%**	20.00%	0.65	0.44	**0.41**
A	SCBatch	**85.71%**	**90.00%**	80.00%	**78.57%**	50.00%	20.00%	**0.66**	0.43	**0.41**
A	AESb	57.14%	60.00%	80.00%	50.00%	40.00%	20.00%	0.51	**0.46**	0.32
B	DstNorm	42.86%	50.00%	**80.00%**	57.15%	50.00%	20.00%	**0.60**	**0.5**	0.24
B	SimCentr	42.86%	50.00%	40.00%	57.14%	50.00%	0.00%	0.51	0.39	0.00
B	SCBatch	42.86%	**60.00%**	60.00%	57.14%	**60.00%**	0.00%	0.57	**0.50**	0.00
B	AESb	42.86%	40.00%	60.00%	**71.43%**	50.00%	**40.00%**	0.59	0.49	**0.56**

For host A, 71.43% of the top 14 positions computed by the incremental SimCentr scoring algorithm contained TP alarms compared to 85.71% by the batch version and 57.14% when the summation of atomic events scores is utilized. The DstNorm variant performed below at 42.86%. Regarding the top 10 and top 5 positions, 70% and 80% consist of TP alarms when using the SimCentr approach, 90% and 80% for the batch approach, 60% and 80% when using the AESb variant and 40 and 20% when DstNorm was used.

Precision was 78.57%, 60% and 20% for the top 14, top 10 and top 5, respectively for the SimCentr approach, 78.75%, 50% and 20% using the batch version, 50%, 40% and 20% for the AESb algorithm and the DstNorm delivered 42.86%, 40% and 20%.

Incremental NDCG for SimCentr comes at 0.65, 0.44 and 0.41, compared to 0.66, 0.43 and 0.41 for the batch method, 0.51, 0.46 and 0.32 for the AESb baseline and 0.25, 0.22 and 0.04 from the DstNorm technique.

For host B, 42.86% of the top 14 positions computed by the DstNorm variant of the incremental algorithm contained TP alarms which matched the performances of all the alternative approaches. Regarding the top 10 and top 5 positions, 50% and 80%, respectively, are filled with TP alarms when using the DstNorm approach, 50% and 40% from SimCentr, 60% and 60% for the batch approach and 40% and 60% when using the AESb variant.

Precision was 57.14%, 50% and 20% for the top 14, top 10 and top 5, respectively for the DstNorm incremental approach, 57.14%, 50% and 0% for SimCentr, 57.14%, 60% and 0% using the batch version and 71.43%, 50% and 40% for the AESb algorithm.

Incremental NDCG comes at 0.60, 0.5 and 0.24 from DstNorm, compared to 0.51, 0.50 and 0 for SimCentr, 0.57, 0.50 and 0 for the batch version and 0.59, 0.49 and 0.56 for the AESb baseline.

6 Discussion

The SimCentr incremental approach produced the best results for host A and performed very close to the batch control version. Both the batch and incremental techniques beat the atomic-event-score-based approach in all metrics except NDCG@10. Conversely, DstNorm was the lowest performer, presenting weaker results in every metric.

Regarding host B, the scenario is different. The AESb and DstNorm approaches are the most competitive among the incremental approaches, even beating the batch variant in some metrics. Competition between DstNorm and the approach based on the atomic events scores is tight, with DSTNorm outperforming AESb in 4 metrics and falling behind in 3 others.

As expected, the batch version performs better generally than the SimCentr incremental algorithm at the cost of being computationally intensive and using undetermined, potentially infinite, memory space (depending on the number of CEs).

The encouraging performance of the centrality-based scoring function SimCentr in host A was not confirmed in host B. These results were discussed as being related to a much higher number of events triggered by this host and the subsequent higher complexity of the ranking task. The idea of DstNorm emerged from this discussion, as it probably presented a higher sensibility about what we were trying to measure.

The application of DstNorm to host B was successful as it clearly outperformed the SimCentr-based approaches and even AESb. However, the same didn't happen when applied back to host A.

We observed that this had to do with a requirement that DstNorm needs to perform well: a good reference (cluster) of normality to use for measurements. As host A had fewer events detected, the existing reference of normality didn't cover most of the dimensions in the context, leading to less accurate and less differentiating measurements. We could also observe the same effect when we forced a different (with a lower variety of vectors) normality reference cluster with the data from host B.

The need for a good normality reference can be interpreted as a limitation of the DstNorm approach. A probable way to mitigate this issue is to cluster more "normal" events, which seems feasible considering that normal data are highly available in event detection and abnormal data are the (very) rare exception.

The Links algorithm uses the cosine similarity criterium to address multidimensionality better. Because of this, the approaches SimCentr and DstNorm that use cosine similarity are expected to perform better when compared to more linear scoring functions like AESb. A better performance was observed with SimCentr in host A and partially with DstNorm in host B. We believe that once the constraints put by the normality reference cluster are minimized, a larger positive difference from the more linear scoring methods can be accomplished.

A limitation of this work is that we used data from just two machines. Application to other contexts is needed to test if the techniques are generalizable.

7   Conclusion

We implemented an unsupervised online event ranking framework for information technology operations that uses event detection techniques, aggregates atomic events into complex events and sequences of complex events into alarms. A clustering approach is used to calculate scores and produce a rank of alarms to prioritize these events for the IT personnel.

Unsupervised online event ranking for IT ops is a challenging task involving several components of different natures. This work contributes to clarifying what these components can consist of and how they can be combined.

Event aggregation and clustering can contribute to filtering false positives effectively. *outlierness* and *driftness* scores calculated by the detection algorithms help to rank and focus on the most urgent events to be addressed.

The approach built delivered the best results using incremental scoring methods based on measures of centrality and distance to normality. These methods are multidimensionally vector-based and consider the multi-dimensional relations in the IT ops context.

Interesting future research directions include building better references of normality to measure from, testing the generalisation of the techniques with new data sets, and building online versions of other unsupervised ranking algorithms.

References

1. Alaghbari, K.A., Saad, M.H.M., Hussain, A., Alam, M.R.: Complex event processing for physical and cyber security in datacentres - recent progress, challenges and recommendations. J. Cloud Comput. **11**, 65 (2022). https://doi.org/10.1186/S13677-022-00338-X

2. Alsubhi, K., Al-Shaer, E., Boutaba, R.: Alert prioritization in intrusion detection systems. In: NOMS 2008 - IEEE/IFIP Network Operations and Management Symposium: Pervasive Management for Ubiquitous Networks and Services, pp. 33–40 (2008). https://doi.org/10.1109/NOMS.2008.4575114

3. Clémençon, S., Baskiotis, N., Vayatis, N.: Anomaly ranking in a high dimensional space: the unsupervised TreeRank algorithm. In: Celebi, M.E., Aydin, K. (eds.) Unsupervised Learning Algorithms, pp. 33–54. Springer, Cham (2016). https://doi.org/10.1007/978-3-319-24211-8_2

4. Cugola, G., Margara, A.: Processing flows of information: from data stream to complex event processing. ACM Comput. Surv. **44**, 1–62 (2012). https://doi.org/10.1145/2187671.2187677

5. Gama, J.: Knowledge Discovery from Data Streams. CRC Press, Boca Raton (2010). https://doi.org/10.1201/EBK1439826119

6. Jiang, G., Chen, H., Yoshihira, K., Saxena, A.: Ranking the importance of alerts for problem determination in large computer systems. Cluster Comput. **14**, 213–227 (2011). https://doi.org/10.1007/S10586-010-0120-0

7. Järvelin, K., Kekäläinen, J.: Cumulated gain-based evaluation of IR techniques. ACM Trans. Inf. Syst. **20**, 422–446 (2002). https://doi.org/10.1145/582415.582418

8. Lin, D., Raghu, R., Ramamurthy, V., Yu, J., Radhakrishnan, R., Fernandez, J.: Unveiling clusters of events for alert and incident management in large-scale enterprise it. Proceedings of the ACM SIGKDD International Conference on Knowledge Discovery and Data Mining, pp. 1630–1639 (2014). https://doi.org/10.1145/2623330.2623360. https://dl.acm.org/doi/10.1145/2623330.2623360

9. Lin, Y., et al.: Collaborative alert ranking for anomaly detection. In: International Conference on Information and Knowledge Management, Proceedings, pp. 1987–1996 (2018). https://doi.org/10.1145/3269206.3272013. https://dl.acm.org/doi/10.1145/3269206.3272013

10. Luckham, D.: The power of events: an introduction to complex event processing in distributed enterprise systems. In: Bassiliades, N., Governatori, G., Paschke, A. (eds.) RuleML 2008. LNCS, vol. 5321, pp. 3–3. Springer, Heidelberg (2008). https://doi.org/10.1007/978-3-540-88808-6_2

11. Mansfield, P.A., Wang, Q., Downey, C., Wan, L., Moreno, I.L.: Links: a high-dimensional online clustering method (2018). https://arxiv.org/abs/1801.10123v1

12. Sadik, S., Gruenwald, L.: Research issues in outlier detection for data streams. SIGKDD Explor. Newsl. **15**, 33–40 (2014)

13. Viswanathan, K., Choudur, L., Talwar, V., Wang, C., Macdonald, G., Satterfield, W.: Ranking anomalies in data centers. In: Proceedings of the 2012 IEEE Network Operations and Management Symposium, NOMS 2012, pp. 79–87 (2012). https://doi.org/10.1109/NOMS.2012.6211885

14. Zhao, N., et al.: Automatically and adaptively identifying severe alerts for online service systems (2020)

A Subgraph Embedded GIN with Attention for Graph Classification

Hyung-Jun Moon[1] and Sung-Bae Cho[2]([✉])

[1] Department of Artificial Intelligence, Yonsei University, Seoul 03722, Korea
axtabio@yonsei.ac.kr
[2] Department of Computer Science, Yonsei University, Seoul 03722, Korea
sbcho@yonsei.ac.kr

Abstract. Graph neural networks (GNNs) have emerged as a powerful tool for analyzing graph data, where data are represented by nodes and edges. However, the conventional methods have limitations in analyzing graphs with diverse attributes and preserving crucial information during the graph embedding. As a result, there is a possibility of losing crucial information during the integration of individual nodes. To address this problem, we propose an attention-based readout with subgraphs for graph embedding that partitions the graph according to unique node attributes. This method ensures that important attributes are retained and prevents dilution of distinctive node features. The adjacency matrices and node feature matrices for the partitioned graphs go into a graph isomorphism network (GIN) to aggregate the features, where the attention mechanism merges the partitioned graphs to construct the whole graph embedding vector. Extensive experiments on six graph datasets demonstrate that the proposed method captures various local patterns and produces superior performance against the state-of-the-art methods for graph classification. Especially, on the challenging IMDB-MULTI dataset, our method achieves a significant performance gain of 27.87%p over the best method called MA-GCNN.

Keywords: Graph Neural Network · Attention-based Readout · Graph Partitioning · Graph Classification

1 Introduction

Graph classification is one of the fields where graphs are classified based on the attributes of their nodes and edges. With the recent advancements in deep learning, graph neural networks (GNNs) have been extensively studied to efficiently classify graphs [1]. GNNs update each node within its neighbors and then integrate all node information to create a single embedding vector for classification [2]. This GNN approach has shown impressive performance across a wide range of problems, making it highly effective for graph classification [1, 3, 4].

GNNs, despite their superior performance, suffer from the "smoothing phenomenon," causing nodes with low graph frequency to blend with neighboring nodes

© The Author(s), under exclusive license to Springer Nature Switzerland AG 2023
P. Quaresma et al. (Eds.): IDEAL 2023, LNCS 14404, pp. 356–367, 2023.
https://doi.org/10.1007/978-3-031-48232-8_33

during the node embedding process [5]. This diminishes GNNs' effectiveness in distinguishing unique node characteristics. The subsequent phase, which generates embedding vectors by combining node embeddings, often exacerbates the issue, especially for graph classification. To deal with this problem, recent studies aim to prevent smoothing by partitioning the graph into subgraphs. Yet, for large graphs like social networks, where node information is similar, this strategy may not be as effective, making the graph overly intricate and diminishing the methods' advantages [6–10].

To overcome these issues, we propose a method that involves partitioning the graph from multiple perspectives and assigns weights to the embedded subgraph embeddings when creating the graph embedding vector. By introducing these weights, we can prioritize the importance of each subgraph embedding based on its ability to capture the unique characteristics of the graph at hand. The proposed method starts with selecting node seeds based on the graph's internal attributes and then partitions the graph into multiple subgraphs. Each partitioned subgraph trained with node embedding updates through graph neural networks, ensuring that each graph reflects its characteristics. For classification, when creating a single graph embedding, we use attention to assign weights to subgraph embeddings that capture the distinctive attributes of each graph. This process ensures that the method allows for the effective extraction and preservation of local features within the graph.

To validate the proposed method, we have conducted experiments on six large-scale social graphs. The results from the experiments on all the datasets demonstrate the performance improvement, with a particularly noteworthy increase of 27.87%p compared to the previous state-of-the-art method of MA-GCNN on the IMDB-M dataset. Additionally, the ablation study analyzes the performance improvement of the proposed method by examining how the chosen seed values and attention-based readout affect the results.

2 Related Works

Recent studies have explored the use of graph neural networks for graph classification and recommendation systems in social networks [11]. Some approaches include graph convolutional neural networks, which learn structural graph features by aggregating neighbor nodes' features [12], and structural deep network embedding, which considers weights from Laplacian eigenmaps to extract additional local features with structural ones [13]. While these methods have advanced the field, they still have limitations in capturing detailed local information in the graph.

Early attempts to apply language models for graph classification include using the "deep walk" algorithm, which creates path samples of a fixed length from a certain point in the graph to extract local features [14]. However, this method is restricted in large-scale social networks and lacks consideration of the graph's structural properties.

Beyond model development, other recent methods focus on transforming input graphs to discover local or unobserved graph features. Examples include Y. Yang's work [15], which generates factor graphs through matrix factorization and utilizes independent GCNs to transfer features to the final output graph's nodes.

H. Peng's method [6] introduces the motif concept for selecting center nodes based on closeness-centrality, finding non-isomorphic subgraphs and combining them to preserve structural graph information. L. Cotta [16], P.A. Papp and Y. Rong [17] seek hidden local information by generating subgraphs using K nodes, removing K nodes, or removing edges from the input graph. J. Yang's recent work [7] focuses on capturing hierarchical graph representation using disentangled graph capsules.

The proposed method differs from these methods by considering three graph properties to extract new local features in social networks. While the MA-GCNN [6] model is similar to ours, it does not utilize the three properties in social networks and neglects the meaning of each subgraph. Our method observes the original graph from three different perspectives to obtain the local patterns in the graph. In the experiments, the proposed method is compared with the HGCN [7] and MA-GCCN models, both of which have achieved the state-of-the-art results in social network classification.

3   Proposed Method

In this section, the proposed method is described as a three-step structure: 1) Node selection process for subgraph partitioning, 2) subgraph embedding process, and 3) attention readout for subgraph embedding integration and classification. Figure 1 shows the overall structure of the proposed method.

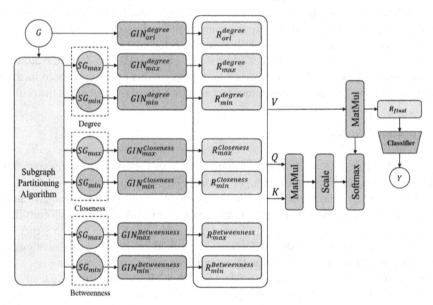

Fig. 1. The proposed method to learn graph representation with various local information for better classification.

3.1   Node Selection and Subgraph Partitioning Algorithm

As previously mentioned, to extract and preserve local information within the graph, we perform graph partitioning. For this purpose, we primarily utilize degree, closeness, and betweenness centrality, which are commonly employed in various graph analysis methods and graph clustering techniques [18–20]. These centrality measures serve as the basis for the graph partitioning process, facilitating the identification of pivotal nodes and aiding in the preservation of local characteristics within the graph. Selecting node seeds based on graph theory aims to model the structural attributes of the graph. Our ultimate objective is to preserve local information, which necessitates selecting specific reference nodes based on the graph's structural attributes [21]. To partition the graph based on these structural attributes, we adopt three criteria for node selection [22].

Algorithm 1 takes one of the three graph attributes and the adjacency matrix information as input variables [24]. Subsequently, the selected attribute, which holds scalar values for each node, allows us to choose two seeds based on their maximum and minimum values. Graph partitioning is performed using the following three attributes.

Degree-centrality in a graph represents its level of connectivity, indicating the number of edges it is linked to. Nodes with higher degrees are considered more connected and can serve as pivotal starting points for partitioning to preserve local characteristics within the graph.

$$Degree(v_i) = \Sigma_{j=1}^{g} x_{i,j}, i \neq j \tag{1}$$

The degree of a node v_i is the sum of all edges connecting it, where $i \neq j$. This degree indicates the node's connectivity level. Nodes with higher degrees have more connections, making them crucial reference points in sets with localized features. Such high-degree nodes are essential for graph partitioning, preserving local characteristics. In social networks, the nodes with top degree-centrality signify strong influence due to their many direct connections [23].

Closeness centrality in a graph represents the average distance between a node and all other nodes. It indirectly indicates a node's prominence and can be utilized for partitioning to limit information exchange and maintain node centrality within the graph.

$$Closeness(v_i) = \frac{N-1}{\Sigma_{j \neq i} l_{(i,j)}} \tag{2}$$

$\sum_{j \neq i} l(i, j)$ represents the sum of edges between node i and other nodes. High or low closeness centrality in graph neural networks indicates how often a node is referenced across network layers, suggesting its prominence in the graph. Using this for graph partitioning ensures centrality consideration and reduces information dilution. In social networks, nodes with top closeness-centrality significantly influence others [23].

Betweenness-centrality in a graph measures if a node lies on the shortest path between other nodes. The nodes with high betweenness centrality serve as communication bridges between subgroups, and their removal during partitioning creates distinct subgraphs with no communication between them.

$$Betweenness(v_i) = \frac{\Sigma_{s \neq v \neq t} l(s, v, t)}{l(s, t)} \tag{3}$$

where $l(s, v, t)$ is the shortest path from node s to t via node v, Eq. (3) calculates the ratio of paths passing through v to total paths. A node with high betweenness centrality in the graph acts as a bridge, linking smaller groups. Removing this node can isolate subgroups, making them distinct. In social networks, betweenness-centrality indicates nodes that bridge, rather than strongly influence, sparsely connected nodes.

Algorithm 1: Graph partitioning algorithm based on bubble framework.

Input: Graph $G = (V, E)$ and Seed Feature F (e.g., degree)
Output: Subgraph S = (S_1, S_2)
1: Initialize maxV = 0, minV=0 and S = []
2: **for** $v \epsilon V$ **do**
3:   Compute F(v)
4: maxV $\leftarrow \max_{v \epsilon V}(F(v))$
5: minV $\leftarrow \min_{v \epsilon V}(F(v))$
6: seed_node = [maxV, minV]
7: **while** True **do**
8:   **for** i=0: len(seed_node) **do**
9:     neighbor_node = G.neighbor(seed_node[i])
10:    S[i].append(neighbor)
11:    if G.check_all_node_tranversing(S) do
12:      break
13: **for** i=0: len(S) **do**
14:   S[i] = S.remove_node_duplication()
15: return S

By selecting seeds with distinct attribute values, we ensure mapping of nodes with dissimilar features. As nodes in the same subgraph commonly exhibit similar attributes, choosing the maximum and minimum values aids in capturing diverse characteristics. The process involves iteratively expanding from the two selected nodes using a neighbor-based extension until all nodes are traversed. Once all nodes are explored, the node expansion process terminates. Afterward, we proceed with the partitioning of the graph into two subgraphs, using the two selected nodes as the reference points. This method helps in constructing meaningful subgraphs with preserved local characteristics. Ultimately, the six constructed subgraphs are processed through the graph embedding stage, resulting in the formation of six individual vectors, each representing a graph.

3.2   Graph Embedding Network

The proposed network uses a graph isomorphism network (GIN) to receive adjacency matrices and node feature matrices to aggregate features of neighboring nodes and update features of each node. The detailed learning process is described as follows.

Through Eq. (4), the sum of the feature vectors of the node v and the neighboring node is obtained. Then, Eq. (5) updates each feature through MLP by adding the features of the neighbor nodes and the feature of the existing node v.

$$a_v^{k-1} = \Sigma_{\forall u \in N(v)} h_u^{k-1} \tag{4}$$

$$h_v^k = MLP^k(a_u^{k-1}, h_v^{k-1}) \tag{5}$$

For graph classification, the model produces an integrated vector representing the graph-level after the neural network layer aggregates data from each neighboring node. The graph neural network combines information from each node using a concatenation operation, which is then utilized for classification as shown in Eq. (6). The readout function combines all node information into a single value:

$$H_G = Concatenation(H_v^k, \forall v \in V | k = 0, 1, \ldots, K) \tag{6}$$

In the proposed method, each subgraph requires a separate model to learn unique local information. This ensures that local information for each criterion remains isolated and not mixed with other subgraphs. Additionally, each model takes the original graph (G) as input to extract various global information from its unique perspective. The proposed method allows for diverse local information extraction through individual models without weight sharing, while also considering global information from different perspectives. Thus, the extracted information from a subgraph model is defined as $h = (h_{Gp}, h_{sub1p}, h_{sub2p})$, and calculated accordingly. Equation (7) illustrates the process of updating nodes through subgraphs and the original graph using a graph neural network.

$$h_p = GNN_p(G, G_{sub1p}, GG_{sub2p}), \forall p \in Graph attributes \tag{7}$$

3.3 Attention-Based Readout for Classification

All local/global information from three individual subgraph models are expressed by $h = h_{Gp}, h_{sub1p}, h_{sub2p}$. Then, an integration network combines each subgraph's local and global features. The process of integrating local/global information (h_n) with attention weights (a_{ij}) is described in Eqs. (8), (9), and (10). Two subgraphs and the original graph feed into each model, and each graph converts into corresponding feature vectors. After that, we have seven feature vectors ($N = 7$): three global features (As their inputs are the same, only one is used.) and six local features before integration. If the size of feature vector is d, the dimension of extracted features would be $(7, d)$.

$$a_{ij} = softmax_j(h_{ij}) = \frac{\exp[score(h_i, h_j) \cdot W_v h_i]}{\sum_{n=1}^{N} exp[score(h_i, h_n) \cdot W_v h_i]} \tag{8}$$

$$score(h_i, h_j) = f(W_q h_i, W_k h_j) \tag{9}$$

$$c_i = \sum_{n=1}^{N} a_{in} \cdot h_n \tag{10}$$

At the end of the process, our model calculates an attention score to capture different influences or weights for aggregating the ten local features and three global features, which selectively activates a part of the network for different graphs.

Algorithm 2: Ensemble with self-attention

Input $h = \left\{ \left\{ h_{max}^p, h_{min}^p, h_{graph}^p \ \forall p \in three\ properties \right\} \right\})$

$\hat{C}$: $Classifier$, CE: $Cross\ Entropy$, lr: $learning\ rate$

1: Initialize W_V, W_Q, W_K, $a[len(h),\ len(h)]$, $c[len(h)]$

2: **for each** epoch **do**

3: **for** $i \epsilon h$ **do**

4: **for** $j \epsilon h$ **do**

5:   $score(R_i, R_j) = f(W_q R_i,\ W_k R_j)$

6:   $a_{ij} = \dfrac{exp[score(R_i, R_j) \cdot W_v R_i]}{\sum_{n=1}^{N} exp[score(R_i, R_n) \cdot W_v R_i]}$

7:   $c_i = \sum_{n=1}^{N} a_{in} \cdot h_n$

8: $c = \sum_{i=1}^{N} c_i$

9: $y_F = \hat{C}(c)$

10: $L_D = CE(y_F,\ y_{GT})$

11: $W_V, W_Q, W_K = (W_V - lr \nabla L_D),\ (W_Q - lr \nabla L_D),$

    $(W_K - lr \nabla L_D)$ // Update parameter

Algorithm 2 describes the process of using attention to merge the vectors of the subgraph and the original graph into a single integrated vector when they are output. The learning step is conducted by considering the degree of influence each graph information receives from other graphs through operations presented in line 5–7 of Algorithm 2. Classification is performed using the final vector c considering the degree of influence between each information, and W_V, W_Q, and W_K weights are updated to minimize the loss function.

4 Experimental Results

4.1 Datasets

We validate the effectiveness of the proposed method with six social network datasets: COLLAB, IMDB-Binary, IMDB-Multi, Dblp-ct1, Twitter-egos, and Reddit Binary. The specification for each dataset is as shown in Table 1.

Table 1. The detailed specification of the six graph datasets

Dataset	#Graph	#Avg. Node	#Avg. Edge	#Class
COLLAB [25]	5000	74.49	2457.50	3
IMDB-B [25]	1000	19.77	96.53	2
IMDB-M [25]	1500	13.00	65.94	3
Dblp_ct1 [26]	755	52.87	320.09	2
Twitch-egos [26]	127094	29.67	86.59	2
Reddit Binary [27]	200	429.63	497.75	2

4.2 Baseline Methods

The experiments are performed on Nvidia DGX station with 2,560 Nvidia tensor cores, Ubuntu desktop Linux OS, 4X Tesla V100 (64 GB), and 256 GB LRDIMM DDR4 memory. Ubuntu, Python 3.x, and TensorFlow 2.3 versions are used as software environments. We utilize graph isomorphism network (GIN-0), which achieves the SOTA performance for COLLAB, to demonstrate its usefulness through the proposed method rather than performance improvement through changes in model structure.

We implement GIN in Tensorflow environment with two graph convolution layers (filter size $= 32$, activation $=$ 'relu'), Adam optimizer with learning rate of 0.001, and cross-entropy loss function [28]. For graph deep learning methods, we compare with the following baselines: graph convolution network (GCN) [12], graph attention network (GAT) [29], graph isomorphism network (GIN-0) [28], GraphS sage network (Graph-SAGE) [30], FactorGCN [15], motif-based graph convolutional neural network (M-GCNN) [6], motif-based attentional graph convolutional neural network (MA-GCNN) [6], DropGNN [17], and hierarchical graph capsule network (HGCN) [7] to highlight the performance of the proposed method on both social and bioinformatics graph datasets. In the ablation study, we conduct experiments with different combination of seeds to examine the utility of each component in the proposed method. We perform additional experiments to demonstrate the effectiveness of the readout.

Table 2. Comparison of average classification accuracy and standard deviation on six social network datasets (bold: the highest and underline; the 2nd highest).

Model	COLLAB	IMDB-B	IMDB-M	Dblp-ct1	Twitch-egos	Reddit Binary
GCN [12]	79.36 ± 1.94	71.61 ± 2.11	50.61 ± 3.69	64.99 ± 2.93	66.94 ± 0.95	91.51 ± 1.23
GAT [29]	75.80 ± 1.60	70.50 ± 2.30	47.80 ± 3.10	63.53 ± 1.21	67.10 ± 0.72	92.60 ± 2.17
GIN-0 [28]	79.00 ± 1.70	74.00 ± 3.40	51.90 ± 3.80	61.67 ± 4.99	66.27 ± 1.66	90.40 ± 2.50
GraphSAGE [30]	80.26 ± 1.22	74.91 ± 1.96	51.66 ± 2.70	62.81 ± 1.88	68.41 ± 2.89	91.54 ± 1.99
FactGCN [15]	81.20 ± 1.40	75.30 ± 2.70	52.01 ± 4.28	61.17 ± 3.62	70.84 ± 4.08	90.33 ± 1.78
M-GCNN [6]	80.08 ± 2.12	75.10 ± 3.14	52.19 ± 2.66	62.34 ± 1.03	71.28 ± 1.42	88.06 ± 1.29
MA-GCNN[6]	83.13 ± 3.09	77.20 ± 2.96	53.77 ± 3.11	63.84 ± 2.31	73.70 ± 1.81	90.44 ± 2.18
DropGNN[17]	79.02 ± 3.08	75.70 ± 4.20	51.40 ± 2.80	61.15 ± 4.31	75.90 ± 2.13	80.08 ± 2.51
HGCN [7]	82.86 ± 1.81	77.20 ± 4.73	52.80 ± 2.45	65.50 ± 2.01	76.80 ± 1.15	93.15 ± 1.58
Ours	**86.35 ± 2.95** (+3.49%)	**90.43 ± 4.89** (+13.23%)	**81.64 ± 1.48** (+27.87%)	68.04 ± 1.96 (2.54%)	78.16 ± 0.66 (+1.36%)	93.25 ± 2.18 (0.10%)

4.3 Result Analysis

Overall Comparison on Social Network Datasets. The experiments are performed using 10-fold cross-validation on well-known benchmark datasets and with the latest graph neural network methods for classification. As these methods have been recognized to achieve the state-of-the-art performance, they serve as a metric to validate the significance of the proposed method.

Table 2 shows average accuracy and standard deviation of the performance. Our method, integrating the six local information with different attention weights, achieves considerable improvement in accuracy. As a result, the proposed method obtains enhanced accuracy by 3.49%p and 0.10%p compared to HGCN in COLLAB and Reddit Binary, respectively, and 13.23%p and 27.87%p compared to MA-GCCN in IMDB-Binary and IMDB-Multi, respectively. These two models are the current state-of-the-art methods for social network classification. Furthermore, for Dblp-ct1 and Twitter-egos datasets, the performance is improved by 2.54%p and 1.36%p, respectively, compared to HGCN, which is the SOTA for the datasets.

Ablation Studies on Three Graph Properties. Table 3 shows how performance changes based on the different starting points, or 'seeds', used for each dataset. The proposed method performs the best when using all the available information. This suggests that using a mix of seeds helps keep the detailed information in the graph, which in turn helps when creating graph summaries or embeddings. By making the most of all the graph's details, we get better results. Interestingly, when we use two specific types of starting points, the results are almost as good as using all three types together. This means these two starting points, on their own, capture different, yet complementary, details of the graph. Their combined effect really boosts the results in the proposed method for splitting up and integrating graphs.

Table 3. Comparison of average classification accuracy and standard deviation for A/B test on COLLAB, IMDB-B, IMDB-M, Dblp-ct1, Twitter-egos, and Reddit Binary. "De" represents Degree, "B" represents Betweenness, and "C" represents Closeness. The check mark signifies that the related vector was used for classification (bold: the highest; underline: the 2nd highest)

De	B	C	COLLAB	IMDB-B	IMDB-M	Dblp-ct1	Twitch-egos	Reddit Binary
✓			0.7850	0.7600	0.5540	0.6400	0.7039	<u>0.8000</u>
	✓		0.7750	0.7400	0.5237	0.6502	0.7035	<u>0.8000</u>
		✓	0.7650	0.7300	0.5015	0.6400	0.7637	0.7500
✓	✓		0.7866	0.7900	0.6021	0.6510	0.7264	0.8000
✓		✓	0.7843	0.7800	0.6018	<u>0.6525</u>	<u>0.7688</u>	0.7500
	✓	✓	0.7839	<u>0.8000</u>	<u>0.6169</u>	0.6518	0.7715	<u>0.8000</u>
✓	✓	✓	**0.7950**	**0.8300**	**0.6667**	**0.6533**	**0.7799**	**0.8500**

Ablation Studies on Attention-Based Readout. Table 4 displays the results of comparing the variations in graph features and how each graph integrates them. To see how our new attention-based method stacks up against other methods, we test them on various datasets. The results show that the proposed method performs better. Some other methods either overcomplicate or overly simplify the information when classifying, which can cause them to miss out on some local details. In contrast, our method pays close attention to each piece of local information on its own. This method leads to better

Table 4. Comparison of average classification accuracy and standard deviation for integration methods on six social network datasets

Method	COLLAB	IMDB-B	IMDB-M	Dblp-ct1	Twitch-ego	Reddit Binary
Concatenation	76.30 ± 1.53	70.07 ± 2.16	51.40 ± 2.96	55.54 ± 2.81	70.33 ± 0.61	83.71 ± 4.73
Ensemble with Average	77.46 ± 0.81	70.48 ± 5.26	48.33 ± 2.40	56.90 ± 0.92	71.37 ± 5.79	82.39 ± 1.13
Ensemble with Element-wise Attention	83.20 ± 1.65	87.15 ± 2.16	76.37 ± 3.64	65.15 ± 2.10	74.88 ± 4.00	87.70 ± 3.85
Ensemble with Channel-wise Attention (Ours)	**86.30 ± 1.89**	**90.15 ± 3.86**	**81.48 ± 2.44**	**67.54 ± 1.15**	**79.20 ± 0.80**	**90.99 ± 2.11**

results because it respects and retains the importance of each local detail, rather than just blending everything together.

5   Concluding Remarks

In this paper, we present a method that divides graphs using three key properties and uses separate graph convolution network for each property to capture the diverse local details in the graph. This method surpasses existing top-tier methods on six social networks by employing a self-attention mechanism that draws from three distinct measures. Using the attention distributed graph for every category, we can see which details are most effective in classifying the graph. Our findings emphasize the strength of a method that highlights various local patterns in social network data with fewer nodes. This is especially evident in social network classification that relies solely on the connections between nodes, without any specific node data.

In the future, we plan to enhance the method by reducing the time and space. We strongly feel that in real-world graph studies, the computation time should be as short as possible, so the primary focus stays on obtaining the most precise and meaningful results. With this perspective, we hope to develop methods that are not only efficient but also fit well with real-world uses. As we move forward in this field, our main goal remains: to provide practical and impactful graph-based solutions tailored to real-life situations.

Acknowledgements. This work was supported by IITP grant funded by the Korean government (MSIT) (No. 2020-0-01361, Artificial Intelligence Graduate School Program (Yonsei University)), an ETRI grant funded by the Korean government (23ZS1100, Core Technology Research for Self-Improving Integrated Artificial Intelligence System), and the Yonsei Fellow Program funded by Lee Youn Jae.

References

1. Wang, B., Gong, N.Z., Fu, H.: GANG: Detecting fraudulent users in online social networks via guilt-by-association on directed graphs. In: International Conference on Data Mining (ICDM), pp. 465–474. IEEE (2017)

2. Bianchi, F.M., Grattarola, D., Livi, L., Alippi, C.: Graph neural networks with convolutional arma filters. IEEE Trans. Pattern Anal. Mach. Intell. **44**, 3496–3507 (2021)

3. Jin, N., Young, C., Wang, W.: GAIA: graph classification using evolutionary computation. In: ACM SIGMOD International Conference on Management of Data, pp. 879–890 (2010)

4. Kim, J.-Y., Cho, S.-B.: A systematic analysis and guidelines of graph neural networks for practical applications. Expert Syst. Appl. **184**, 115466 (2021)

5. Park, K.-W., Cho, S.-B.: A residual graph convolutional network with spatio-temporal features for autism classification from fMRI brain images. Appl. Soft Comput. **142**, 110363 (2023)

6. Peng, H., Li, J., Gong, Q., Ning, Y., Wang, S., He, Li.: Motif-matching based subgraph-level attentional convolutional network for graph classification. Proc. AAAI Conf. Artif. Intell. **34**(04), 5387–5394 (2020)

7. Yang, J., Peilin Zhao, Y., Rong, C.Y., Li, C., Ma, H., Huang, J.: Hierarchical graph capsule network. Proc. AAAI Conf. Artif. Intell. **35**(12), 10603–10611 (2021). https://doi.org/10.1609/aaai.v35i12.17268

8. Park, K.-W., Bu, S.-J., Cho, S.-B.: Learning dynamic connectivity with residual-attention network for autism classification in 4D fMRI brain images. In: Yin, H., et al. (eds.) IDEAL 2021. LNCS, vol. 13113, pp. 387–396. Springer, Cham (2021). https://doi.org/10.1007/978-3-030-91608-4_38

9. Park, K.W., Cho, S.B.: A vision transformer enhanced with patch encoding for malware classification. In: Yin, H., Camacho, D., Tino, P. (eds.) Intelligent Data Engineering and Automated Learning – IDEAL 2022. IDEAL 2022. Lecture Notes in Computer Science, vol. 13756. Springer, Cham (2022). https://doi.org/10.1007/978-3-031-21753-1_29

10. Kim, K.-J., Cho, S.-B.: Personalized mining of web documents using link structures and fuzzy concept networks. Appl. Soft Comput. **7**, 398–410 (2007)

11. Wang, H., Zhang, F., Zhao, M., Li, W., Xie, X., Guo, M.: Multi-task feature learning for knowledge graph enhanced recommendation. In: World Wide Web Conference, pp. 2000–2010 (2019)

12. Kipf, T.N., Welling, M.: Semi-supervised classification with graph convolutional networks. arXiv preprint arXiv:1609.02907 (2016)

13. Wang, D., Cui, P., Zhu, W.: Structural deep network embedding. In: ACM SIGKDD International Conference on Knowledge Discovery and Data Mining, vol. 22, pp. 1225–1234 (2016)

14. Perozzi, B., Al-Rfou, R., Skiena, S.: Deepwalk: Online learning of social representations. In: ACM SIGKDD International Conference on Knowledge Discovery and Data Mining, pp. 701–710 (2014)

15. Yang, Y., Feng, Z., Song, M., Wang, X.: Factorizable graph convolutional networks. Adv. Neural. Inf. Process. Syst. **33**, 20286–20296 (2020)

16. Cotta, L., Morris, C., Ribeiro, B.: Reconstruction for powerful graph representations. Adv. Neural. Inf. Process. Syst. **34**, 1713–1726 (2021)

17. Papp, P.A., Martinkus, K., Faber, L., Wattenhofer, R.: DropGNN: random dropouts increase the expressiveness of graph neural networks. Adv. Neural. Inf. Process. Syst. **34**, 21997–22009 (2021)

18. Satuluri, V., Parthasarathy, S.: Symmetrizations for clustering directed graphs. Int. Conf. Extending Database Technol. **14**, 343–354 (2011)

19. Müller, E.: Graph clustering with graph neural networks. J. Mach. Learn. Res. **24**, 1–21 (2023)

20. Schaeffer, S.E.: Graph clustering. Comput. Sci. Rev. **1**, 27–64 (2007)

21. Azari, M., Iranmanesh, A.: Computing the eccentric-distance sum for graph operations. Discret. Appl. Math. **161**, 2827–2840 (2013)

22. Borgatti, S.P., Everett, M.G.: A graph-theoretic perspective on centrality. Soc. Netw. **28**, 466–484 (2006)

23. Zhang, J., Luo, Y.: Degree centrality, betweenness centrality, and closeness centrality in social network. In: International Conference on Modelling, Simulation and Applied Mathematics, vol. 2, pp. 300–303. Atlantis Press (2017)
24. Cho, S.-B., Shimohara, K.: Evolutionary learning of modular neural networks with genetic programming. Appl. Intell. **9**, 191–200 (1998)
25. Yanardag, P., Vishwanathan, S.: Deep graph kernels. In: ACM SIGKDD International Conference on Knowledge Discovery and Data Mining, vol. 22, pp. 1365–1374 (2015)
26. Rozemberczki, B., Kiss, O., Sarkar, R.: An api oriented open-source python framework for unsupervised learning on graphs. arXiv preprint arXiv:2003.04819 10 (2020)
27. Oettershagen, L., Kriege, N.M., Morris, C., Mutzel, P.: Temporal graph kernels for classifying dissemination processes. In: International Conference on Data Mining, pp. 496–504. SIAM (2020)
28. Xu, K., Hu, W., Leskovec, J., Jegelka, S.: How powerful are graph neural networks? arXiv preprint arXiv:1810.00826 (2018)
29. Veličković, P., Cucurull, G., Casanova, A., Romero, A., Lio, P., Bengio, Y.: Graph attention networks. arXiv preprint arXiv:1710.10903 (2017)
30. Hamilton, W., Ying, Z., Leskovec, J.: Inductive representation learning on large graphs. Adv. Neural. Inf. Process. Syst. **30**, 1025–1035 (2017)

A Machine Learning Approach to Predict Cyclists' Functional Threshold Power

Ronald Andrew Stockwell[1] and Andrea Corradini[2]([✉])

[1] Oakville, ON, Canada
andrew@thethreshold.co.za
[2] MCI, Universitätsstraße 15, 6020 Innsbruck, Austria
andrea.corradini@mci.edu

Abstract. The estimated Functional Threshold Power (eFTP) obtained from a cyclists best 20-min effort, expressed in watts, and multiplied by 95% is sometimes overstated, especially for non-elite level cyclists or cyclists just starting out. Since Functional Threshold Power (FTP) is used for training prescription, in terms of creating power zones to train in, it is of utmost importance that the correct FTP is used. If FTP is overstated, a cyclist will overtrain at a level that is unrealistic to achieve. This research project aims at validating eFTP using a large, big data dataset, as well as to build out a supervised machine learning model, in the form of a weighted logistic regression, that can be used to predict FTP (pFTP), to create predictor of FTP that's better than the currently accepted eFTP formula. Based on the data provided, our approach predicts FTP with far greater accuracy than eFTP, which turned out to overestimate cyclists' FTP. The model outcome for each athlete, pFTP, was evaluated versus eFTP, by comparing each of these outcomes to the actual FTP (aFTP) observed in the data. To do this, Mean Squared Error and Mean Absolute Error were computed with pFTP producing values of 314.87 and 10.38 respectively when compared to aFTP, and eFTP producing 6417.32 and 61.16. With more accurate FTP estimations, cyclists are more likely to train in the correct power zones. It also validates that a laboratory environment is not always required when trying to validate eFTP and that a large, big data dataset can act as a valid substitute.

Keywords: Functional Threshold Power · machine learning · weighted logistic regression · cyclist performance

1 Introduction

Functional Threshold Power (FTP) is defined as *"The highest power that a rider can maintain in a quasi-steady state without fatiguing for approximately one hour."* (Allen et al, 2019). Power is measured in wattage via a power meter training device that is attached to the bike. Essentially, FTP can be seen as the maximum average power output that a cyclist can maintain for a continuous 60-min effort.

Cyclists around the world rely heavily on FTP scores for training prescription and to measure improvement, however, the FTP test that is currently used, assumes that

P. Quaresma et al. (Eds.): IDEAL 2023, LNCS 14404, pp. 368–380, 2023.
https://doi.org/10.1007/978-3-031-48232-8_34

all cyclists are equal. This holds true with well-trained and elite cyclists but could not prove true for cyclists that are new to the sport of cycling or in between the spectrum of beginner to elite. The result of an FTP test is especially important, as it is used to prescribe the correct training zones for training prescription, as well as measure improvement in a cyclist's performance over time. Training zones are broken down into distinct levels, depending on what methodology is used. All zones are calculated as a percentage of FTP. For example, level 1 might be prescribed as <55% of FTP according to the Coggan's Classic Levels depicted in Table 1 (Allen et al, 2019).

Table 1. Power-Based Training Levels (Allen et al, 2019).

Level	Description	% Of FTP in Watt
1	Active Recovery	<55
2	Endurance	56–75
3	Tempo	76–90
4	Lactate Threshold	91–105
5	VO2 Max	106–120
6	Anaerobic Capacity	121–150
7	Neuromuscular Power	N/A

Having an accurate FTP, that is not overstated or understated, is critical for training prescription in all cyclists' training programs where a power meter is utilized. An FTP set too high would mean that a cyclist could potentially overtrain, whereas if the FTP is set too low it could mean that the cyclist is not training hard enough and may not reach their goals or their potential.

To calculate FTP, cyclists perform a 20-min maximum effort test and record the average power produced in watts. The reason a 20-min test is completed instead of a 60-min effort, as per the definition of FTP, is due to a 60-min test being far more taxing. The average wattage obtained for the 20-min test is then multiplied by 95%, and this new figure is assumed to be the cyclists FTP. We will refer to this calculated value as the estimated FTP (eFTP). For example: a cyclist performing a 20-min test and producing and average wattage of 300 W results in 285 W of eFTP.

Researchers have used a variety of methods to test and validate the eFTP methodology in a laboratory setting or in controlled field tests. However, many of the cyclists who participated in these experiments are either elite cyclists or club cyclists of above average strength and fitness. The sample size of cyclists used was also quite small.

The research into FTP in the past has always had a major limitation, namely, being able to include a large sample size of cyclists that reflects the greater cycling population. Valenzuela et al. conducted their study whereby 20 male cyclists where used, the study by MacInnes et al. including 8 male trained cyclists, and the study by Lillo-Beviá et al. with only 11 well-trained male cyclists. Including male and female cyclists of all levels would give a much more accurate view of where eFTP is falling short, as not all cyclists are at the same level. This is where a big data set and machine learning has the potential to help.

This research attempts at looking into whether a machine learning model can be built to predict a cyclist's actual FTP more accurately than the standard eFTP formula that is so widely used today, as well as to determine whether a model based on big data can successfully be used to do so, instead of tests in a lab or a controlled environment.

2   Research Background and Related Work

To understand why FTP is used today, it is necessary to look into the past to understand what was used previously, as well as how the current performance metric FTP, that Allen and Coggan pioneered (Allen et al, 2019), came to the fore. Anaerobic Threshold, Onset of Blood Lactate (OBLA), Maximal Lactate Steady State (MLSS) and Lactate Threshold (LT) are all measures that aim to predict the point at which lactate starts to accumulate in the blood, as this is an extremely reliable predictor of performance in all endurance athletes. Since the point at which LT is reached can be associated with a specific power output of a cyclist, it enables coaches and cyclists to define a training program based on physiological measures (Allen et al, 2019).

FTP testing became apparent due to its ease of execution and no laboratory environment being required, as all that is required is a bicycle power meter. Power meters allow cyclists to track their performance and quantitively evaluate strengths and weaknesses over time. They give rise to many benefits, such as better pacing strategies in races, and specific target power zones for interval training. However, the true benefit emanates from interpreting the data outputs of the power meter over time and ensuring that the correct training sessions are prescribed to maintain or improve fitness and strength (Allen et al, 2019). As power meters became more easily available and affordable, a shift took place from the LT testing protocol to that of FTP testing. This shift required that sufficient proof was provided that both methods produced comparable results. It is therefore important to understand how accurate FTP is relative to LT, as it has become a mainstream method for cyclists to measure improvement and assist in accurate training prescription, rather than using LT.

Valenzuela, P. L. et al. (Valenzuela et al, 2018) published a study that discussed whether FTP was a valid surrogate for LT. To investigate the difference or correlation between FTP and LT, they conducted a study with 20 male cyclists performing two tests: a Maximal Incremental Test to determine LT, and a 20-min time trial using a power meter, whereby the eFTP was calculated from the average wattage produced, respectively. Recruited participants needed to be cycling at least 4 h per week for at least 3 months prior to the actual test and also needed to have cycled regularly for more than 2 years. The researchers differentiated between recreational and trained cyclists, based on their fitness status at the time by using their Peak Power Output (PPO) to separate them. The study concluded that there was no difference between LT and FTP but cautioned that FTP and LT could not always be used interchangeably, due to the finding that recreational cyclists presented lower LT than their FTP. This finding is key, as it indicates that if a one size fits all formula is used to determined FTP, such as eFTP, it might not account for the difference in cyclists.

In another study by Denham et al. (Denham et al, 2020), they tried to understand whether different maximal power outputs could be used to predict FTP. The results of that study are hard to generalize as the participants were only elite cyclists. Open data sources, such as that used for the purposes of our research, will hopefully open the door for the inclusion of all types of cyclists, with the potential to help understand and create a better method for predicting FTP.

Some body of research into different machine learning methods that could be used for predicting FTP has been carried out. MacInnes et al. (MacInnes et al., 2019) investigated the relationship between 4-min and 20-min maximum efforts relative to FTP. Their goal was to test whether there was in fact a relationship between FTP and the two variables, in order to ascertain whether a shorter testing protocol could be used to determine FTP. Using a multiple linear regression model, this study found a strong correlation between the 60-min test conducted and the 4-min and 20-min tests, individually and collectively. When regressed individually, the maximum power output for 4-min and 20-min was able to explain 95% and 92% of the variation in the 60-min maximum power output, or FTP, recorded, respectively. When both of these variables were used in a linear regression, 98% of the variance was explainable. Interestingly, they noted that the 60-min maximum power output worked out to, on average, 75% of the cyclists' best 4-min maximum power output and 90% of the 20-min maximum power output, and not 95% as used in eFTP. They noted that undertaking a 60-min test could bring about considerable fatigue, which could potentially interrupt training. This study only recruited eight trained male cyclists between the ages of 18 and 40, who on average trained more than 10 h per week, hence the greater cycling population was not accurately represented. A big dataset would include thousands of cyclists and allow for more realistic results.

In another study (Lillo-Beviá et al, 2022) to validate whether FTP is valid surrogate for LT, only 11 well-trained male cyclists were used. Each of these cyclists performed three tests over a defined period. The first was a Maximal Lactate Steady State (MLSS) test, with the power at MLSS recorded. The next two tests were both 20-min maximal effort time trials using a power meter. The results were then used to determine how repeatable the 20-min test could be, as well as to determine the cyclists' eFTP. The study found that the 20-min time trial test was repeatable with a standard error of 3w and that 95% of a cyclists best 20-min time trial effort differed from MLSS intensity. It also found that a percentage closer to 91% would better predict MLSS.

Beside validating FTP versus LT, another key component to consider is the cyclist phenotype (Allen et al, 2019). Cyclist phenotypes are broken down into four distinct types: All-rounders, Time Trialists/Climbers/Steady-State Riders, Sprinters, or Pursuiter. The phenotype a cyclist falls into depends on the cyclist's weight and on four variables: their best effort 5-s, 1-min, 5-min, and FTP maximum power outputs. These phenotypes are crucial in understanding a cyclist's strengths and weaknesses, as well as help to define their training prescription.

3 Machine Learning Model to predict a cyclist's FTP

3.1 Data and Data Preprocessing

For the purposes of this research, there is no laboratory or field testing of cyclists. Instead, we used an open-source dataset comprising just over 1.26 million workouts that were submitted by thousands of cyclists of varying strength and fitness. This open source dataset can be found at Golden Cheetah (GoldenCheetah). Many cyclists fall short of being able to produce their eFTP for a sustained hour effort, particularly cyclists that are new to the sport of cycling, or do not come from a background of high-level training. This much larger sample size allows for cyclists of all abilities to be included in the data analysis but also required extensive cleaning with defined rules and assumptions being applied as to not skew any results of our study.

We summarized all data into one line per cyclist which included only their best produced power figures as well as some averages and summations metrics. Ultimately, for each cyclist we defined a feature vector containing the values shown in Table 2.

Table 2. Features determined from the power files.

Feature name	Description	Notes
id	unique identifier	
gender	cyclist's gender	
yob	cyclist's year of birth	
no_of_rides	total number of rides recorded from all power files	
avg_workout_time	average elapsed workout time	
avg_time_riding	average time spent riding	
avg_weight	average weight of the cyclist	
avg_power_all_rides	average power for all rides	
avg_non-zeropower_all_rides	average non-zero power for all rides	
max_power	maximum power recorded	
best_cp_setting	best FTP setting recorded	
avg_coggan_np	calculated average normalized power	
avg_coggan_variability_index	calculated variability index	
max_critical_power_<S>s	maximum S-second power recorded across all power files for the cyclist	for S = 1, 4, 10, 15, 20, and 30
max_critical_power_<M>m	maximum M-minute power recorded across all power files for the cyclist,	for M = 1, 2, 3, 5, 8, 10, 20, 30, and 60
avg_time_in_zone_<Z>	average time spent in zone Z as defined by Power-Based Training Levels table by (Allen et al, 2019)	for zone Z = L1, L2, L3, L4, L5, L6, and L7
max_peak_wpk_<S>s	maximum power to weight ratio calculated as max_critical_power_<S>s/avg_weight	for S = 1, 4, 10, 15, 20, and 30
max_peak_wpk_<M>m	maximum power to weight calculated as max_critical_power_<M>m/avg_weight	for M = 1, 2, 3, 5, 8, 10, 20, 30, and 60

Data in the power files might be corrupted with power figures that are not within the realm of human capabilities, we calculated what values are within human capabilities from the table for FTP defined in (Allen et al., 2019). Data with 5-s power above 25.18 watts/kg are excluded. Similarly, we excluded 1-min power data above 11.5 watts/kg, 5-min power data above 7.6 watts/kg, and 60-min power data above 6.6 watts/kg. All this data was removed from the analysis to make sure that no data is being skewed. To make sure all variables included in this summary had no outliers and followed sound logic, we also created histograms that could be visually interpreted. After visually inspecting the histograms, outliers were removed. Further, to complete the cleaning of the data, we included in our analysis only that data from cyclists who were between the ages of 16 and 100, had a recorded weight of between 50kg and 150kgs, and met a minimum number of 30 rides. This means that we set the lowest FTP to 49w. This value is calculated as 50 kg * 0.98 watts/kg where 0.98 is the lowest wattage per kilogram included in the FTP table defined in (Allen et al., 2019). The final cleaned and summarized dataset included 3487 unique cyclists.

3.2 Actual vs Estimated FTP

We started with an investigation of the accuracy of the estimated FTP (eFTP) relative to the actual best 60-min average maximum power outputs observed i.e., the actual FTP (aFTP), by calculating aFTP for each athlete using their data.

The values of the eFTP versus aFTP are visually represented on a scatter plot as seen in Fig. 1 (left). The plot shows that eFTP predicts aFTP very well, however, the data becomes sparse above 350 W. This is due to fewer cyclists having functional threshold power values this high. Interestingly, most of the data points on this visualization lie below the diagonal line, especially as wattage increases. This means that eFTP, for the most part, is outputting a higher value than is seen in the actual observed FTP (aFTP). This lines up with findings by MacInnes et al. (MacInnes et al., 2019) that found that 90% of the 20-min maximum power output is a more accurate predictor of aFTP than 95% of the 20-min maximum power output is. When reviewing the descriptive statistics in Table 3, it is evident that eFTP is overestimating aFTP in terms of the mean and median values. Dividing the mean value of aFTP (247.39w in Table 3) by the mean value for 20-min (285.22 W in Table 3) results in 0.8673 (i.e. 86.73%) average value estimator for aFTP. This result is significantly lower than the 0.95% (i.e. 95% rule) that is currently used to calculate eFTP.

3.3 Predicting FTP with Multiple Regression Models

To carry out our study we used the summarized dataset created and fed into a supervised machine learning model that predicts the numeric value of the FTP. Specifically, we used a Multiple Linear Regression Model that allows for the preprocessed data for each cyclist to be used as input variables, in order to test which ones are statistically significant in predicting their actual FTP. By allowing multiple variables as inputs to the model, it might help to differentiate between cyclists' phenotypes and therefore estimate their FTP differently with the use of different power duration variables that are statistically significant. For example, sprinters are known to have lower FTPs than time trialists.

Time trialists may have lower 5-s maximum power outputs, whereas sprinters might have higher 5-s power outputs. This is due to time trialists training to push a high wattage over a prolonged period, whereas sprinters train to push extremely high wattage over a much shorter period. If this variable, for example, is found to be statistically significant in predicting FTP and included in the model, it could help to differentiate between the different cyclists' predicted FTP (Allen et al, 2019).

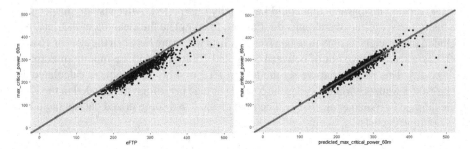

Fig. 1. (left) actual vs estimated FTP; (right) actual vs predicted FTP.

Model variable validation is an iterative process. Through each model build iteration, the variables were validated for statistical significance in terms of their corresponding p-value. The cleaned summarized data was split into a training dataset and a test dataset. The percentage split that was used, has 70% allocated to training and 30% allocated to testing. The first iteration model was built using all variables listed in Table 2. The result is depicted in Fig. 2. We also ran both Akaike's Information Criteria (AIC) and the Bayesian Information Criteria (BIC) functions in order to create measurable points to work against for future model iterations. The AIC and BIC are seen as measures of goodness of model fit, complexity and of how well a model can explain variability in the data. They can be compared between models with a lower AIC and BIC indicating a better fitting model. Our first model had values for AIC = 19937.58 and BIC = 20088.33. The Adjusted R-squared was recorded as 0.9151 and the Residual Standard Error as 14.39. These two metrics were compared against future iterations, with an increase in Adjusted R-squared and a decrease in Residual Standard Error indicating an improved model (Nwanganga and Chapple, 2020). Once the model was generated, we inspected all its variables. For generating the model in the next iteration, we selected only the variables with p-value < 0.05, due to their statistical significance. The variables included avg_weight, avg_power_all_rides, avg_coggan_np, max_critical_power_5s, max_critical_power_10s, max_critical_power_30m, perc_zone1, perc_zone3, max_critical_power_60m.

Model 2 was run resulting in AIC = 19892.34 and BIC = 19950.32, respectively. Both of these results were lower than those produced by Model 1, indicating an improvement. The Adjusted R-squared improved too, from 0.9151 to 0.9243, while the Residual Standard Error decreased from 14.39 to 13.76. A third model, Model 3, was then built with the new variables that emerged as statistically significant, according to their p-values from Model 2. The variables included avg_weight, avg_power_all_rides, max_critical_power_30m, perc_zone1, perc_zone3, max_critical_power_60m.

When compared to Model 2, Model 3 recorded an AIC and BIC as 20209.61 and 20250.19 respectively, indicating a deterioration in the model. The adjusted R-squared decreased to 0.9037, while the Residual Standard Error increased to 15.3. Thus, we decided to use Model 2 as the starting point for interpreting the residuals before making any more changes and iterating further with a fourth model.

Fig. 2. Summary results for Model 2 (left) and for the final model, Model 5 (right)

Not all the variables in Model 2 are statistically significant, with several having variables with p-values bigger than 0.05. However, we decided to keep them in for now as the variables with the higher p-values could be used to potentially differentiate cyclists from a phenotype perspective. We then evaluated Model 2 from a statistical perspective in terms of the residual diagnostics to determine whether:

- The residuals have a mean of zero.
- The residuals are distributed normally.
- There is heteroscedasticity in the model.
- Autocorrelation is present between the residuals.

Given Model 2, the first check was to ascertain whether they had a mean close to zero. The result computed at $1.209694e-15$, which is extremely close to zero, therefore satisfying this criterion. The residuals then needed to be tested for normal distribution to make sure all signals in the data have been captured and that the errors were random noise. To do this, we used a simple visual test by inspecting the histogram plotting the residuals (Nwanganga and Chapple, 2020). In the next test, we run a Breusch-Pagan statistical test to check that there was no heteroscedasticity in the model. As the test failed, it turned out that Model 2 was characterized by heteroscedasticity. To fix this issue, we used the weighted least squares regression methodology to produce a fourth model, Model 4, using the same variables used in Model 2. The weighted least squares regression methodology takes the model and assigns weights to the independent variables in the model that are trying to predict the dependent variable. Higher weights are then assigned to variables that have a lower variance, therefore decreasing the noise that the variables with high variance would cause (Strutz, 2016).

Model 4 showed a vast improvement in Adjusted R-squared and the Residual Standard Error, with a slight increase in AIC and BIC, coming out at 20187.44 and 20245.42 respectively, when compared with Model 2. Based on the outcome, we decided to run one more iteration, but to only include the variables that where significant (p-values < 0.05) in Model 4, and re-weighting these variables accordingly based on their variance, using the weighted least squares regression method. This produced the final model, Model 5, with the results summary depicted in Fig. 2 (right). Model 5 included the variables avg_power_all_rides, max_critical_power_5s, max_critical_power_30m, perc_zone3.

This final model, Model 5, was an improvement over Model 4 in both Adjusted R-squared and the Residual Standard Error. The weighted variables now accounted for the heteroscedasticity caused by the variables included in the model with high variance. The next step was to determine if there was multicollinearity between the independent variables. Linear regression models assume that variables used to predict the dependent variable are not correlated, if they are there is seen to be multicollinearity. The analysis of the Variance Inflection Factor (VIF) scores was used to rule out multicollinearity in our final model. The final model can be depicted as the following weighted linear regression equation:

$$
\begin{aligned}
max\_critical\_power\_60m = {} & 1.358652 + 0.276111 * (avg\_power\_all\_rides) - \\
& 0.002247 * (max\_critical\_power\_5s) \\
& 52.347384 * (perc\_zone3) + \\
& 0.789667 * (max\_critical\_power\_30m)
\end{aligned}
\tag{1}
$$

Equation (1) can then be tested and validated against the test data set aside, in order to establish whether the algorithm that produces pFTP, or the eFTP formula are more accurate predictors of actual FTP, which in this case, is the *max_critical_power_60m*. If the model is found to predict FTP better than eFTP, then it can be applied by many coaches to prescribe more suitable training zones to cyclists.

4 Evaluation of our Machine Learning Model

To ascertain whether our supervised machine learning model can be used to predict a cyclist's FTP more accurately than the standard eFTP formula, we evaluated the predicted FTP (pFTP). The test dataset that was held out, consisting of 1042 cyclists, was run through our final model weighted linear regression algorithm to produce the pFTP values. This dataset was used to evaluate the pFTP against the actual FTP (aFTP) and the formula-based FTP calculation of eFTP. To evaluate the outputs of the test dataset, aFTP versus pFTP was plotted in Fig. 1 (right) (i.e., actual vs predicted values). This gives a visual representation of accuracy that is otherwise difficult to see from tabular data.

Figure 1 (right) indicates that the values predicted (pFTP) by our model appear to be very accurate as all the dots are close to the diagonal line, unlike eFTP, where the dots were forming primarily below the diagonal line. The visualization also shows that over 350 W, the data is sparser once again, and that the model is not as accurate as it is below 350 W. Again, this is due to that fact that not many cyclists have functional threshold power that is so high. Therefore, with less observations in the training dataset, it would make sense that the model could be less accurate for cyclists with higher functional threshold power. If comparing this visualization to Fig. 1 (left), it is notable that the dots lie on both sides of the diagonal line, indicating that pFTP is not over predicting nor under predicting for all values. The values for pFTP can also be evaluated and compared from a descriptive statistics point of view when reviewing Table 3. The pFTP descriptive statistics appear to track aFTP much more accurately than eFTP does, with the mean, median, standard deviation and minimum values all being either in line or close. The only value that pFTP seems to be out of range on, is the maximum value. This could be due to few cyclists with extremely high functional threshold power being included in the analysis, therefore causing the model to struggle in predicting accurate FTP values for these cyclists. It is important to note that this analysis did not intentionally exclude cyclists with high FTP's, but that in a population, only a small subset of cyclists are able perform at a level superior enough to produce such high FTP values.

Table 3. Descriptive statistics for aFTP, pFTP, eFTP and 20 min best effort in watts

	N	Mean	StdDev	Median	Min	Max	Range
aFTP	1042	247,39	50,10	245,66	81,17	546,74	465,57
pFTP	1042	247,65	50,74	245,13	84,14	517,35	433,22
eFTP	1042	270,96	54,93	267,42	107,14	549,96	442,81
20-min	1042	285,22	57,82	281,49	112,78	578,90	466,12

In another evaluation (see Table 4), we calculated the Mean Square Error (MSE) between eFTP and the actual FTP as well as the Mean Absolute Error (MAE) between the predicted FTP (pFTP) and the actual FTP.

Table 4. Mean Squared Error (MSE) and Mean Absolute Error (MAE)

	Mean Squared Error	Mean Absolute Error
aFTP vs pFTP	314.87	10.38
aFTP vs eFTP	6417.32	61.16

The values obtained clearly show that the values pFTP produced by our model are a far better predictor of aFTP than eFTP for Mean Squared Error and Mean Absolute Error are far lower for pFTP. This indicates that, based on the data available, our supervised machine learning model can predict a cyclist's FTP better than the standard formula of eFTP.

As it can be seen from Eq. (1), the 20-min max power effort variable was found not to be predictive in the final model, but two other time-based power intervals of *max_critical_power_5s* and *max_critical_power_30m* were. This is a significant finding as it allows future researcher to dive deeper into these two variables to understand why they are more predictive than 20-min, which is so widely used. The fact that *perc_zone3* came through as statistically significant is also a significant discovery. This variable relates to Zone 3 found on Table 1. This requires further investigation, as this variable in our model's linear regression had a minus in front of it, meaning the higher the percentage of time spent in Zone 3, the lower the outcome for *max_critical_power_60m* (FTP). This could mean that cyclists who spend a higher percentage of their training time in this power zone, contribute negatively to their FTP.

5 Conclusion

In a sport where marginal gains are everything, having the correctly defined FTP is key to training prescription to obtain these gains. Cyclists training in the incorrect zones at the wrong time over long periods of time can be detrimental. Cyclists could either overtrain or undertrain.

With this research we determined that eFTP is not as accurate when compared to aFTP, and that a percentage of closer to 86% of a cyclists best 20-min effort, might be a more accurate predictor of aFTP. Next, we proposed a machine learning model, in the form of a weighted linear regression, that can predict aFTP more effectively than the current eFTP formula that is so widely used today. The reduction in the error that resulted from utilizing the model instead of using eFTP, can help coaches and cyclists to prescribe more effective and accurate training programs. It does, however, have its drawbacks for application in a real-world environment. For example, the aFTP that the model was trained off could or could not be a cyclist's best effort; this was the best available effort that was accessible in the dataset. This is where a laboratory environment may produce more accurate results.

There are a few weaknesses in the project that might be due to the dataset and only become apparent while cleaning and preprocessing the data. Nonetheless, these drawbacks are not insurmountable and can be mitigated. First, it is impossible to ascertain whether the dependent variable *max_critical_power_60m* (FTP) used to train the model against, is in fact the cyclist's best effort for 60-min. This holds true for all power variables in this analysis. There is no way of determining whether any of the variables are in fact the best efforts from any of the cyclists included. For some cyclists, the perfect set of numbers may be present, whereas with other cyclists, the numbers present may be nowhere near their best performance. This could cause the high variation in some of the significant variables found in our Model 2. Second, because the sample size included a wide range of cyclists, there were high variances that caused the heteroskedasticity in Model 2. This could be due to elite and recreational cyclists having been included in the data. For example, in a 10-s sprint, an elite cyclist may be able to maintain more than 1500 W, whereas a recreational cyclist may only be able to maintain 500 W or less. These large variances across the population suggest that building out future models that are based off a cyclist's current level, could yield more realistic results. This would mean running a model for recreational cyclists and a separate one for elite cyclists, therefore minimizing the variance and range of some of the power metrics. Another way to navigate this variance, would be to classify cyclists first, and then use their classification (elite cyclist or recreational cyclist) as a dummy variable. Third, most power meters have a claimed accuracy of 1–2%. In this analysis, there is no way to prove how accurate the power meters used actually were. It is also impossible to tell whether any methodology was followed before each ride to make sure the power meter had followed its brand's pre-ride calibration process. If a planned research study were to take place in future, whereby cyclists could submit their power files for use in a similar manner, it is advisable to introduce an onboarding process with guidelines and specifications for these cyclists. This onboarding would need to explain the steps cyclists need to take before each ride to ensure accurate data. Further, computing the algorithm of our model is not easy for a layman. The variables *max_critical_power_5s* and *max_critical_power_30m* are easy to obtain, however *perc_zone3* and *avg_power_all_rides* require summarization from all power files, or a subset for a defined period. This is something that cannot be completed without a tool that creates these data summaries for the athlete or coach. Lastly, there were not enough female cyclists included in the dataset. The final model might therefore be skewed to male cyclists. Cycling has been a male dominated sport for many years, with female riders earning significantly less, as well as their winnings not matching those of their male counterparts (Shaw et al., 2020). Had there been more data from female cyclists included, the gender variable might have come out as significant in predicting *max_critical_power_60m* (FTP).

Despite the above points, this research houses many strengths that could assist in further developing an even better model. Even though the power variables included in the analysis might not be the cyclists' best efforts, the methodology for creating our model still stands true and can therefore be replicated if this data was available from another source that had more controlled methodologies for data capturing. The fact that a shorter power time interval was found to be significant in the final model, namely *max_critical_power_5s*, shows that our model could potentially cater for the Sprinter phenotype. Sprinters have the highest value for 5 s in terms of power to weight (i.e. *max_critical_power_5s*), but also perform poorly compared to the other phenotypes when looking at FTP. This would not show in a 20-min effort (Allen et al, 2019). To calculate eFTP, all that is needed is the 20-min maximum power effort of an athlete. To use our model, the *max_critical_power_30m* maximum time interval is required. This is not a significant increase from 20-min, but still easier to compete versus a flat out 60-min test. As a final note, if our model was transferred into an online tool that already stored all of a cyclists' power files, it could be run every time a new power file was uploaded to calculate a dynamically changing modeled FTP from current and historical data.

References

Allen, H., Coggan, A., McGregor, S.: Training and Racing with a Power Meter, 3rd edn. VeloPress (2019)

Denham, J., Scott-Hamilton, J., Hagstrom, A.D., Gray, A.J.: Cycling power outputs predict functional threshold power and maximum oxygen uptake. J. Strength Cond. Res. **34**(12), 3489–3497 (2020)

GoldenCheeta: Golden Cheetah. https://github.com/GoldenCheetah/OpenData. Accessed on Sep 2023

Lillo-Beviá, J.R., Courel-Ibáñez, J., Cerezuela-Espejo, V., Morán-Navarro, R., Martínez-Cava, A., Pallarés, J.G.: Is the Functional Threshold Power a valid metric to estimate the maximal lactate steady state in cyclists? J. Strength Cond. Res. **36**(1), 167–173 (2022)

MacInnis, M.J., Thomas, A.C.Q., Phillips, S.M.: The reliability of 4-minute and 20-minute time trials and their relationships to functional threshold power in trained cyclists. Int. J. Sports Physiol. Perform. **14**(1), 38–45 (2019)

Combining Regular Expressions and Supervised Algorithms for Clinical Text Classification

Christopher A. Flores[(✉)] and Rodrigo Verschae

Institute of Engineering Sciences, Universidad de O'Higgins,
2841959 Rancagua, O'Higgins, Chile
{christopher.flores,rodrigo.verschae}@uoh.cl

Abstract. Clinical text classification allows assigning labels to content-based data using machine learning algorithms. However, unlike other study domains, clinical texts present complex linguistic diversity, including abbreviations, typos, and numerical patterns that are difficult to represent by the most-used classification algorithms. In this sense, sequences of character strings and symbols, known as Regular Expressions (RegExs), offer an alternative to represent complex patterns from the texts and could be used jointly with the most commonly used classification algorithms for accurate text classification. Thus, a classification algorithm can label test texts when RegExs produce no matches. This work proposes a method that combines automatically-generated RegExs and supervised algorithms for classifying clinical texts. RegExs are automatically generated using alignment algorithms in a supervised manner, filtering out those that do not meet a minimum confidence threshold and do not contain specific keywords for the classification problem. At prediction time, our method assigns the class of the most confident RegEx that matches a test text. When no RegExs matches a test text, a supervised algorithm assigns a class. Three clinical datasets with textual information on obesity and smoking habits were used to assess the performance of four classifiers based on Random Forest (RF), Support Vector Machine (SVM), Naive Bayes (NB), and Bidirectional Encoder Representations from Transformers (BERT). Classification results indicate that our method, on average, improved the classifiers' performance by up to 12% in all performance metrics. These results show the ability of our method to generate confident RegExs that capture representative patterns from the texts for use with supervised algorithms.

Keywords: Natural Language Processing · Regular Expressions · Text Classification

1 Introduction

In the biomedical area, the large amount of research articles and clinical texts makes it necessary to develop algorithms to classify such information automati-

Supported by ANID FONDECYT Postdoctorado 3220803 and ANID FONDEQUIP Mediano EQM170041.

cally. These algorithms must be highly accurate to correctly classify texts based on their content, which presents a challenge due to the inherent characteristics of natural language in clinical texts [15].

Some of the most widely-used algorithms for classifying clinical texts have been based on Random Forest (RF), Support Vector Machine (SVM), and Naive Bayes (NB) [12,19]. More recent algorithms based on Deep Neural Networks (DNNs), such as the pre-trained Bidirectional Encoder Representations from Transformers (BERT) model, provide a contextualized representation of the words, allowing fine-tuning these models to a specific classification task [7]. Although these classification algorithms have been widely used in the clinical domain, there is potential to improve their performance with other techniques closer to natural language, such as Regular Expressions (RegExs) [1,3,14,16]. In this context, RegExs are challenging since they do not always produce matches on unseen texts.

RegExs represent a set of symbols, including characters and metacharacters with special meanings (e.g., *,?,+), that allow defining patterns for searching substrings [18]. Since regular expressions are flexible enough to adapt to different study domains, including clinical, they have been used in text pre-processing, information extraction, and, to a lesser extent, in classification tasks [9,11,13,22]. However, combining RegExs and supervised classifiers have yet to be studied in depth. [4]. For instance, Bui and Zeng-Treitler propose a clinical text classifier termed Regular Expression Discovery (RED) that combines RegExs and a SVM [4]. Labeled RegExs are created by extracting common phrases from the texts in a supervised manner using the Smith-Waterman (SW) algorithm. At prediction time, if no RegEx matches a test text, SVM assigns a class. Classification results indicate that RED improved the SVM's performance by 3% in terms of Accuracy (ACC). In this regard, the authors suggest that an ensemble approach, i.e., combining RegExs and a classifier, could outperform the accuracy performance. On the other hand, Li et al. propose a hybrid classifier that combines RegExs and DNNs [13]. RegExs are constructed by identifying keywords from the clinical texts using a DNN algorithm termed Attention-based Bi-directional Long Short-Term Memory (ABLSTM) based on an attention mechanism and a Bidirectional Long Short-Term Memory (BiLSTM) neural network. Subsequently, keywords are combined using metacharacters and Boolean operators (e.g., AND, OR, NOT) to generate RegExs based on rules. At prediction time, RegExs are used to classify texts when the DNN predictions do not achieve minimum confidence. However, if no RegEx matches a test text, the non-confident predictions of the DNN model are considered. Classification results indicate that RegExs improved the ABLSTM's performance by up 5.38 % in terms of ACC. Other methods use RegExs as a pre-processing stage or for a feature extraction method to improve the classifiers' performance [8,17].

Given the above, we hypothesize that combining RegExs and a supervised learning algorithm could improve the classification of clinical texts. In this sense, the main contributions of this work are focused on a supervised feature space based on the automatic generation of RegExs and a classification method that combines RegExs and a supervised algorithm. We build a supervised feature space based on RegExs from Spanish clinical texts labeled for classification tasks. RegExs are generated for each class of the problem, using alignment algorithms to find common word sequences. In addition, it is possible to evaluate the confidence level of the RegExs according to the matches in the training set. At prediction time, if no RegEx matches a test text, a supervised learning algorithm assigns a class. Classification results indicate that our method, on average, improved the performance of the supervised classifiers by up to 12% in terms of ACC and F1-value (F1) metrics, especially in Datasets (DSs) containing more numeric attributes. Thus, we demonstrated the ability of our method to successfully combine the use of supervised algorithms when no confident RegEx matched a test text, which occurred in between 10% and 31% of the total cases.

The paper is organized as follows. Section 2 details the clinical texts and the proposed classification method. Section 3 presents the classification results. Section 4 reports conclusions and future work.

2 Materials and Methods

2.1 Clinical Texts and Pre-processing

Clinical texts in Spanish were used to assess the classification performance of the supervised algorithms. These DSs were provided anonymized by the Hospital Guillermo Grant Benavente (HGGB), Concepción, Chile, and then were labeled by Biomedical engineering students for three classification problems: OBESITY STATUS, OBESITY TYPES, and SMOKING STATUS (see Table 1 for further details) [9]. Additionally, the annotators provided domain-specific keywords for each DS based on the texts' content. In all cases, the agreement level among annotators (Kappa index) was almost perfect ($k > 0.81$) [21]. Finally, all texts were pre-processed by applying the following steps: lowercase conversion, white spaces normalization, and tokenization (i.e., separating words by whitespaces). For example, the tokens "patient", "suffers", "from", and "obesity" are extracted from the snippet text "Patient   suffers from OBESITY".

Table 1. General description of the datasets.

Dataset	Class distribution	Examples	Keywords
OBESITY STATUS	Negative (303), positive (858)	1,161	obes*, BMI, overweight, normal weight, weight
OBESITY TYPES	Moderate (185), severe (152), morbid (572)	909	obes*, BMI
SMOKING STATUS	Negative (505), positive (582)	1,087	smok*, tobac*, cigar*, pack*

* symbol indicates the root of a given word.

2.2   Classification Algorithm Based on Regular Expressions

We propose a classification method that combines a decision function based on automatically generated RegExs with a trained classifier as shown in Fig. 1. A feature space based on RegExs is generated in a supervised manner from training texts using alignment algorithms to capture representative patterns. Then, the most confident of these RegExs are used to predict labels in the test texts. A trained classifier predicts a label if no RegEx matches a test text.

2.3   Feature Space Based on Regular Expressions

RegExs are generated automatically from labeled training texts in a supervised manner. For each class of the problem, our method follows the next two steps. Firstly, each word of the pre-processed texts is grouped using hierarchical clustering and Levenshtein distance. Since Levenshtein distance calculates the minimum number of modifications to transform a string sequence into another one, it is possible to group similar words considering the number of deletions, insertions, and substitutions between two words [2]. Subsequently, all words belonging to a group are replaced by a representative pattern composed of the common letters of the cluster obtained by applying multiple alignments from a word base (e.g., the most frequent in the training set) and metacharacters (e.g., (?:\w)). Additionally, all numbers and punctuation are replaced by specific metacharacters to facilitate the extraction of common sequences (e.g., \d+, [∧a-z\d\s\+\−]). Secondly, each pair of text is aligned using the SW algorithm to extract sequences of similar tokens (words, numbers, punctuation). Subsequently, each sequence is converted to a RegEx by replacing whitespace with specific metacharacters (e.g., [\s]*). Finally, RegExs are filtered using the keywords from Table 1. Note that RegExs are generated from each text pair in the training set, so it is possible to assign labels to them.

2.4   Decision Function

During the training phase of the proposed classification method, the labeled RegExs are evaluated on the training set in terms of Precision (PR) score to obtain their confidence measure, filtering out those that do not satisfy this metric's minimum threshold (PR_{THR}). If a RegEx contains numbers (e.g., \d+), matches are saved to generate intervals based on the minimum and maximum values for matching purposes. Thus, n RegExs R, labels Y, and confidence values P, $(R, Y, P) = \{(r_1, y_1, p_1), \cdots, (r_n, y_n, p_n)\}$, are useful for text classification tasks. A classification algorithm f is also trained on the same training set. During the prediction stage (see Algorithm 1), our method applies in each test example all RegExs, assigning the class of the most confident one. An ϵ value, $0 < \epsilon < 1$, allows solving discrepancies when the same Precision values are obtained, considering the number of tokens of each RegEx in the decision function. We considered a ϵ value $= 10^{-4}$. Note that the function $\mathcal{L} : R \to Y$, maps a RegEx $r_i \in R$ to a class y_i, $y_i = \mathcal{L}(r_i)$, while the function $\Phi : R \to P$, maps a RegEx $r_i \in R$ to a

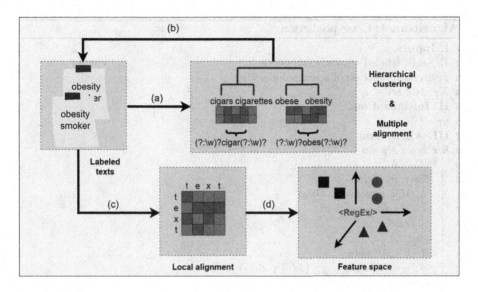

Fig. 1. Automatic generation of a feature space based on RegExs. (a) Input training texts. (b) clusters of similar words. (c) pre-processed texts. (d) sequence of common tokens.

precision value p_i, $p_i = \Phi(r_i)$. Finally, the trained classifier assigns a class if no RegEx matches a test example.

3   Experimental Results

To assess our method, we consider classifiers based on RF, SVM, NB, and BERT to use in combination with RegExs. In the case of classifiers based on RF, SVM, and NB, we considered a feature representation in terms of Term Frequency & Inverse Document Frequency (TfIdf) [20]. The hyperparameters of these classifiers were fine-tuned as indicated in Table 2. In the case of the classifier based on BERT, we considered the commonly used hyperparameters indicated in the state-of-the-art for text classification tasks [7,10].

On the other hand, to assess the performance of the classifiers, we performed a 5-fold cross-validation to average the ACC and F1 metrics [6]:

$$ACC = \frac{TP + TN}{TP + FP + TN + FN}, \tag{1}$$

$$F1 = \frac{2TP}{2TP + FP + FN}, \tag{2}$$

where TP and TN are True Positive and Negative values, and FP and FN are False Positive and Negative, respectively. In the case of the F1 metric, a weighted

---

Algorithm 1: Class prediction

---

1 **I. Input:**
2 R: set of labeled regular expressions
3 f: decision function of a trained classifier
4 X_T: test set
5 **II. Initialization:**
6 $y_T \leftarrow \emptyset$
7 **III. Algorithm:**
8 **for** x_t *in* X_T **do**
9      $R' \leftarrow \emptyset$
10   **for** r *in* R **do**
11   **if** r *matches* x_t **then**
12             $R' \leftarrow R' \cup r$
13   **end**
14   **end**
15   **if** $|R'| > 0$ **then**
16          $y \leftarrow \mathcal{L}(\mathrm{argmax}_{r \in R'}\{\Phi(r) + \epsilon|r|\})$
17   **end**
18   **else**
19          $y \leftarrow f(x_t)$
20   **end**
21      $y_T \leftarrow y_T \cup y$
22 **end**
23 **IV. Output:** y_T predicted labels on X_T

Table 2. Hyperparameters used to fine-tune classifiers.

Classifier	Hyperparameter	Values
SVM	*kernel*	RBF, Linear*
	C	10^0*, 10^1, 10^2, 10^3
RF	Criterion	Entropy*, Gini
	Estimators	10^1, 10^2, 5×10^2, 10^3*
NB	α	0, 0.25, 0.75*, 1
BERT	Epochs	4
	Batch size	8
	Dropout	0.2
	Optimizer	Adam
	Learning rate	2^{-5}

* symbol indicates the value used after fine-tuning in most cases.

average is calculated, considering the number of classes in the test set. Additionally, for each RegEx, we calculate a confidence metric based on the PR metric (see Algorithm 1) as follows:

$$PR = \frac{TP}{TP + FP},$$ (3)

where a TP means that a labeled RegEx matches a training text belonging to the same class. At the same time, an FP occurs when a labeled RegEx matches a text belonging to a different class. On the other hand, to evaluate classification error, the Zero-One-Loss (ZOL) metric was considered as follows [5]:

$$L(y_i, y_i') = \begin{cases} 1 \ y_i \neq y_i' \\ 0 \ \text{otherwise} \end{cases},$$ (4)

where y_i, and y_i' correspond to the $i - th$ label and prediction in the test set, respectively.

Table 3 shows the classification results at different precision thresholds (PR_{THR}) of the proposed classification method considering the cases when only

Table 3. RegExs classification results at different probability thresholds.

$PR_{THR}(\%) \geq$	OBESITY STATUS		OBESITY TYPES		SMOKING STATUS	
	ACC(%)	F1(%)	ACC(%)	F1(%)	ACC(%)	F1(%)
50	97.10	97.10	94.28	94.09	89.30	89.22
80	98.72	98.73	96.30	96.19	91.89	91.84
85	99.15	99.16	96.30	96.19	93.85	93.83
90	99.15	99.16	96.64	96.56	93.64	93.63
95	99.15	99.16	96.98	96.91	93.65	93.64
99	99.18	99.18	96.94	96.89	93.15	93.12

Bold value indicates the selected PR_{THR}.

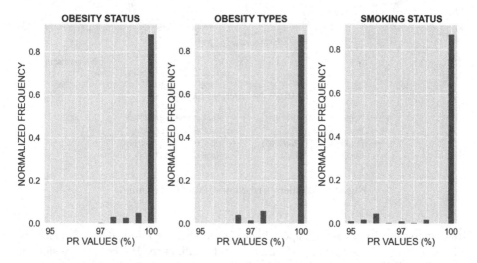

Fig. 2. Confidence values (PR) distribution of the RegExs during prediction.

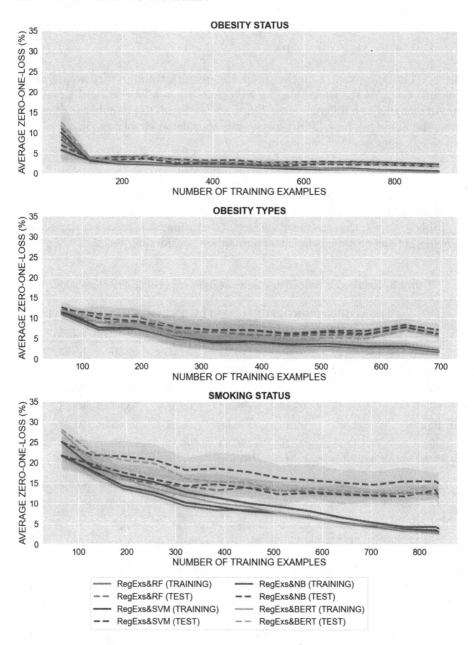

Fig. 3. Classifiers' error curves during training.

RegExs match a text test. It is possible to observe that a $PR_{THR} = 95\%$ achieves the best trade-off between performance and confidence. Hence, we consider a $PR_{THR} = 95\%$ as the minimum confidence threshold of the RegExs.

Figure 2 shows the confidence distribution of the RegExs during prediction in terms of PR. Noticeably, after filtering, RegExs have high precision values close to maximum.

Table 4 shows the performance of the base classifiers RF, SVM, NB, and BERT and our proposed method, i.e., combining a RegExs and a supervised algorithm (base classifier). Additionally, we consider a baseline classifier combining RegExs and a Random (RAND) labeling. It is possible to observe that BERT outperformed the rest of the classifiers in all classification tasks. On the other hand, It is noticeable that combining RegExs and a classifier outperformed the rest of the base classification algorithms ($\Delta > 0$). On average, i.e., considering all classifiers, our classification method improved by 3% to 12% the classifiers' performance in all performance metrics ($\overline{\Delta}$), especially in the classifier based on NB. Additionally, our classification method performed better than the baseline classifier in all cases.

Table 4. Average classification results.

Classifier	Type	OBESITY STATUS		OBESITY TYPES		SMOKING STATUS	
		ACC(%)	F1(%)	ACC(%)	F1(%)	ACC(%)	F1(%)
BERT	Base	96.47	96.50	90.76	90.43	87.12	87.11
	&RegExs	96.99	97.01	**94.50**	**94.36**	**88.68**	**88.67**
	Δ	0.52	0.51	3.74	3.93	1.56	1.56
RF	Base	96.38	96.38	82.95	81.37	84.73	84.74
	&RegExs	**97.24**	**97.25**	93.40	93.12	87.76	87.75
	Δ	0.86	0.87	10.45	11.75	3.03	3.01
SVM	Base	95.69	95.66	78.66	78.08	84.18	84.13
	&RegExs	97.24	97.24	92.41	92.19	88.59	88.57
	Δ	1.55	1.58	13.75	14.11	4.41	4.44
NB	Base	86.65	86.38	72.27	71.47	76.91	76.94
	&RegExs	96.90	96.93	92.85	92.61	85.28	85.28
	Δ	10.25	10.55	20.58	21.14	8.37	8.34
RAND	Baseline	95.78	95.82	90.42	90.31	80.40	80.40
$\overline{\Delta}$		3.29	3.38	12.13	12.73	4.34	4.34

Bold values indicate better performance in the corresponding dataset.
Δ indicates the difference between our proposed method (RegExs&classfier) and a respective base classifier.

Figure 3 shows the learning curves of the classifiers in terms of training and test error. At each iteration, we sampled a batch size of 64 training texts until all examples were selected. In all cases, it is possible to observe that the training error is lower than the test error, with no tendency to overfit, i.e., both curves decrease as the training size increases.

Finally, Fig. 4 shows the prediction error of our classification method in terms of ZOL metric. It is possible to observe that, in most cases, using RegExs&BERT outperformed the rest of the combinations. In this regard, RegExs&RAND performed lower than the other combinations. It is also necessary to point out that, on average, RegExs matched a test text in 69% of the cases in the SMOKING STATUS DS and over 90% in the case of the OBESITY and OBESITY TYPES DSs.

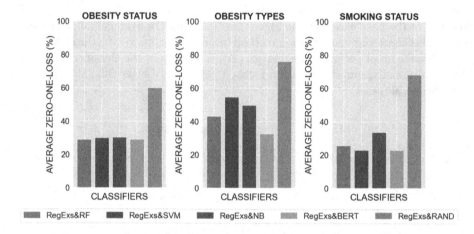

Fig. 4. Average classification error of the proposed method.

4   Conclusions and Future Work

This work proposed a clinical text classifier that combines RegExs and supervised algorithms. We aimed to improve the performance of classification algorithms by introducing confident RegExs (see Fig. 2) that capture representative patterns from the clinical texts.

Classification results indicate that our method improved the performance of the state-of-the-art classifiers such as RF, SVM, and NB, and more recent algorithms such as BERT in clinical text classification tasks (see Table 4). On the other hand, error curves (see Fig. 3) indicate that classifiers do not tend to overfit. This can be explained by the RegExs' ability to automatically capture representative patterns from the texts, including numerical ranges (e.g., BMI) and the linguistic diversity of words (e.g., typos).

RegExs matched, on average, over 69% of the texts on SMOKING STATUS DS and over 90% of the texts on OBESITY and OBESITY TYPES DSs during prediction. These results indicate that RegExs can better capture numerical attributes from the text. In the case of the SMOKING STATUS dataset, more examples are required to represent smoking-related terms (e.g., negations).

Finally, as future work, we plan to evaluate the automatically-generated RegExs on other study domains, including public datasets, or text mining tasks.

Acknowledgements. This work was partially supported by ANID FONDECYT Postdoctorado 3220803 and ANID FONDEQUIP Mediano EQM170041. The authors thank the Informatics Unit with the HGGB, Concepción, Chile, for providing datasets.

References

1. Abeyrathna, K.D., Granmo, O.C., Goodwin, M.: Extending the Tsetlin machine with integer-weighted clauses for increased interpretability. IEEE Access **9**, 8233–8248 (2021)
2. Arockiya Jerson, J., Preethi, N.: An analysis of Levenshtein distance using dynamic programming method. In: Gunjan, V.K., Zurada, J.M. (eds.) Proceedings of 3rd International Conference on Recent Trends in Machine Learning, IoT, Smart Cities and Applications. LNCS, vol. 540, pp. 525–532. Springer, Singapore (2023). https://doi.org/10.1007/978-981-19-6088-8_46
3. Balagopalan, A., Eyre, B., Robin, J., Rudzicz, F., Novikova, J.: Comparing pre-trained and feature-based models for prediction of Alzheimer's disease based on speech. Front. Aging Neurosci. **13**, 635945 (2021)
4. Bui, D.D.A., Zeng-Treitler, Q.: Learning regular expressions for clinical text classification. J. Am. Med. Inform. Assoc. **21**(5), 850–857 (2014)
5. Chen, S., Webb, G.I., Liu, L., Ma, X.: A novel selective Naïve Bayes algorithm. Knowl.-Based Syst. **192**, 105361 (2020)
6. Chicco, D., Jurman, G.: The advantages of the Matthews correlation coefficient (MCC) over f1 score and accuracy in binary classification evaluation. BMC Genomics **21**, 1–13 (2020)
7. Devlin, J., Chang, M., Lee, K., Toutanova, K.: BERT: pre-training of deep bidirectional transformers for language understanding. In: Burstein, J., Doran, C., Solorio, T. (eds.) Proceedings of the 2019 Conference of the North American Chapter of the Association for Computational Linguistics: Human Language Technologies, NAACL-HLT 2019, Minneapolis, MN, USA, 2–7, June 2019, vol. 1 (Long and Short Papers), pp. 4171–4186. Association for Computational Linguistics (2019). https://doi.org/10.18653/v1/n19-1423
8. Flores, C.A., Figueroa, R.L., Pezoa, J.E.: FREGEX: a feature extraction method for biomedical text classification using regular expressions. In: 2019 41st Annual International Conference of the IEEE Engineering in Medicine and Biology Society (EMBC), pp. 6085–6088. IEEE (2019)
9. Flores, C.A., Figueroa, R.L., Pezoa, J.E., Zeng-Treitler, Q.: CREGEX: a biomedical text classifier based on automatically generated regular expressions. IEEE Access **8**, 29270–29280 (2020). https://doi.org/10.1109/ACCESS.2020.2972205
10. Gal, Y., Ghahramani, Z.: Dropout as a Bayesian approximation: representing model uncertainty in deep learning. In: Balcan, M., Weinberger, K.Q. (eds.) Proceedings of the 33nd International Conference on Machine Learning, ICML 2016, New York City, NY, USA, 19–24, June 2016. JMLR Workshop and Conference Proceedings, vol. 48, pp. 1050–1059. JMLR.org (2016). http://proceedings.mlr.press/v48/gal16.html
11. Kashina, M., Lenivtceva, I., Kopanitsa, G.: Preprocessing of unstructured medical data: the impact of each preprocessing stage on classification. Procedia Comput. Sci. **178**, 284–290 (2020)

12. Kowsari, K., Jafari Meimandi, K., Heidarysafa, M., Mendu, S., Barnes, L., Brown, D.: Text classification algorithms: a survey. Information **10**(4), 150 (2019)
13. Li, X., Cui, M., Li, J., Bai, R., Lu, Z., Aickelin, U.: A hybrid medical text classification framework: integrating attentive rule construction and neural network. Neurocomputing **443**, 345–355 (2021)
14. Luo, X., et al.: Applying interpretable deep learning models to identify chronic cough patients using EHR data. Comput. Meth. Programs Biomed. **210**, 106395 (2021)
15. Rios, A., Kavuluru, R.: Convolutional neural networks for biomedical text classification: application in indexing biomedical articles. In: Proceedings of the 6th ACM Conference on Bioinformatics, Computational Biology and Health Informatics, pp. 258–267 (2015)
16. Saha, R., Granmo, O.C., Goodwin, M.: Using Tsetlin machine to discover interpretable rules in natural language processing applications. Expert. Syst. **40**(4), e12873 (2023)
17. Sharaff, A., Pathak, V., Paul, S.S.: Deep learning-based smishing message identification using regular expression feature generation. Expert Syst. **40**, e13153 (2022)
18. Subramanian, S., Thomas, T.: Regular expression based pattern extraction from a cell-specific gene expression data. Inform. Med. Unlocked **17**, 100269 (2019)
19. Vora, S., Yang, H.: A comprehensive study of eleven feature selection algorithms and their impact on text classification. In: 2017 Computing Conference, pp. 440–449. IEEE (2017)
20. Vranken, H., Alizadeh, H.: Detection of DGA-generated domain names with TF-IDF. Electronics **11**(3), 414 (2022)
21. Wennberg, S., Karlsen, L.A., Stalfors, J., Bratt, M., Bugten, V.: Providing quality data in health care-almost perfect inter-rater agreement in the Norwegian tonsil surgery register. BMC Med. Res. Methodol. **19**(1), 1–9 (2019)
22. Yao, L., Mao, C., Luo, Y.: Clinical text classification with rule-based features and knowledge-guided convolutional neural networks. BMC Med. Inform. Decis. Mak. **19**(3), 31–39 (2019)

Modeling the Ink Tuning Process Using Machine Learning

Catarina Costa[1]([envelope]) [ORCID] and Carlos Abreu Ferreira[1,2]([envelope]) [ORCID]

[1] Institute of Engineering, Polytechnic Institute of Porto - ISEP/IPP,
Porto, Portugal
{1180788,cgf}@isep.ipp.pt
[2] INESC TEC, Porto, Portugal

Abstract. Paint bases are the essence of the color palette, allowing for the creation of a wide range of tones by combining them in different proportions. In this paper, an Artificial Neural Network is developed incorporating a pre-trained Decoder to predict the proportion of each paint base in an ink mixture in order to achieve the desired color. Color coordinates in the CIELAB space and the final finish are considered as input parameters. The proposed model is compared with commonly used models such as Linear Regression, Random Forest and Artificial Neural Network. It is important to note that the Artificial Neural Network was implemented with the same architecture as the proposed model but without incorporating the pre-trained Decoder. Experimental results demonstrate that the Artificial Neural Network with a pre-trained Decoder consistently outperforms the other models in predicting the proportions of paint bases for color tuning. This model exhibits lower Mean Absolute Error and Root Mean Square Error values across multiple objectives, indicating its superior accuracy in capturing the complexities of color relationships.

Keywords: Ink Tuning · Machine Learning · Transfer Learning · Artificial Neural Network with a pre-trained Decoder

1 Introduction

Color is an abstract concept that, according to Ishikawa-Nagai et al. [8] consists of the attribute of visual perception produced by the light beam from the combination of chromatic and achromatic components. In fact, its definition, by the Munsell Ordering System, is based on the combination of three fundamental parameters, namely hue, luminosity and chroma [9], which admits presenting colors in three-dimensional space, shown in Fig. 1.

Several color spaces have been developed to express color in abstract mathematical models. Because of the uniformity of the color spectrum and the subtle variations in hue and intensity, CIELAB is progressively more widely used and accepted in all types of color representation systems. The Commission Internationale de l'Eclaiage (CIE) supports the color space on the three dimensions

P. Quaresma et al. (Eds.): IDEAL 2023, LNCS 14404, pp. 393–405, 2023.
https://doi.org/10.1007/978-3-031-48232-8_36

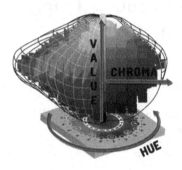

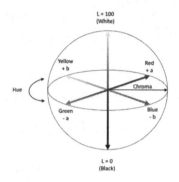

Fig. 1. Munsell Color Order System   **Fig. 2.** CIELAB color space diagram

of the Cartesian coordinate referential L, a, b, observable in Fig. 2, reflecting an approximately uniform system. In the CIELab color space diagram, the a axis indicates the red-green component of a color, while the yellow and blue components are represented on the b axis. The distance from the central L axis represents the saturation of the color [18].

According to Kenneth R. Fehrman & Cherie Fehrman [6], paint bases are essential for creating various shades by combining them in different proportions. Any change in the ratio will alter the resulting color. This paper is based on this premise and aims to determine the exact amount of bases to be added to the mixture to achieve a specific color. In the context of colorimetry, color measuring instruments can be used that provide the representation in CIELab color space of the desired color. Based on this codification, it is possible to determine the precise proportion of each paint base needed to be mixed to obtain the desired color.

Efforts to improve paint formulation prediction have led to the development of machine learning models to enable a greater understanding of the complex relationships between paint bases and desired colors. These models offer flexibility and can effectively capture complex relationships between paint bases and desired colors [11]. By leveraging the adaptability and nonlinearity of machine learning algorithms, these models have significant potential to improve the efficiency and accuracy of paint coloring processes.

The literature in this field is scarce, mainly focusing on metaheuristics [4] and artificial intelligence techniques, with a special focus on Artificial Neural Networks [3]. With an innovative approach, this study aims not only to contribute to the advancement of knowledge in the field of paint formula development, but also to automate paint formula development by considering color characteristics and desired finishes. A model known as Artificial Neural Network with a pre-trained Decoder was developed to accurately estimate the required proportions of each base. This model considers the color representation in the CIELab space, along with the desired finish, as input variables. By leveraging this model, precise predictions can be made regarding the necessary proportions of the 16 available bases for achieving the desired color.

In this paper, the sections are organized as follows: Sect. 2 offers a literature review on Multi-target Regression, including an overview of the problem and related work. Section 3 describes the methodology, including the four machine learning algorithms used and the training process. Section 4 discusses the prediction results. Finally, Sect. 5 presents the main conclusions.

2 Literature Review

Machine Learning techniques have stood out in several research areas, including colorimetry, due to its ability to detect complex patterns in the input data and convert them into knowledge to improve the prediction of future data [21]. With the adaptation process, the algorithm is able to progressively improve its architecture and performance.

The goal of predicting the amount of paint bases to mix to obtain the desired color fits into Multi-target Regression. Methods for multi-target regression enable the effective modeling of a dataset by capturing the implicit relationships between features and target variables, as well as the relationships among the target variables themselves [14]. Let $X = R^d$ be a d-dimensional input space of n instances, and let $Y = R^m$ be a m-dimensional output space. Multi-Output Learning aims to learn a function f, from the training set $D = (x_1, y_1), \ldots, (x_n, y_n)$, mapping each input to multiple outputs [24], as in Eq. 1 and Eq. 2.

$$f : \Omega x_1 \times \ldots \times \Omega x_m \rightarrow \Omega y_1 \times \ldots \times \Omega y_n \tag{1}$$

$$x = (x_1, \ldots, x_m) \rightarrow y = (y_1, \ldots, y_n) \tag{2}$$

where Ωx_i and Ωy_k refer to the space of predictor variables $x_i, \forall i \in \{1, ..., m\}$, and target variables $y_k, \forall k \in \{1, ..., n\}$.

Furthermore, Multi-target Regression methods typically utilize two approaches: Algorithm Adaptation and Problem Transformation. The first approach involves adapting a single model to simultaneously predict all the targets, taking into account the statistical dependencies among them. This allows for the exploration of correlations between the target variables and enhances the predictive capacity of the model. On the other hand, the second approach takes a different path, where independent models are developed for each target variable, and their results are concatenated. Each model is trained separately to predict a single target, treating the multi-target problem as multiple single-target regression problems. This approach enables more specific and personalized prediction for each target [2].

2.1 Related Work

Predicting ink base concentrations is a non-linear process of great importance in the painting, printing, and dyeing industries, since incorrect proportions can affect final product quality. This section explores relevant studies to understand effective approaches and techniques for accurate prediction in this field.

Some authors have proposed strategies based on metaheuristics. Chaouch et al. [4] evaluated the predicted concentrations and desired shades using the Ant Colony metaheuristic. The results revealed a satisfactory match between the predicted concentrations and the desired hues, with most color deviations below 0.7 units. These findings highlight the effectiveness of the algorithm, since successful color reproduction requires color deviations of less than 1 unit.

In the realm of artificial intelligence techniques, pioneering machine learning projects associated with colorimetry were carried out by Bishop, Usher et al. [3], who demonstrated the applicability of neural networks in the color industry. Some studies have shown good accuracy in color prediction, such as the analysis by Huang et al. [7], where the model was trained to minimize the residuals in predicting pigments for watercolor. Mei-Yun Chen et al. [5], on the other hand, developed an Intelligent Palette using a Deep Neural Network for pigment mixture recipe prediction. Both studies showed that about 85% of the data had color differences of less than 5 units in the test scores, which is below the threshold to be determined as a mismatch [5,7].

Westland [23] conducted a study comparing the performance of neural networks with the Kubelka-Munk prediction system, a mathematical model in the domain of colorimetry. The results revealed that machine learning outperformed the theoretical model. The neural networks obtained a color deviation of about 1 unit, while the K-M model showed a color deviation of about 1.5 units.

Direct comparison of the results of this study with related work is unfeasible due to the differences in the data sets used and the distinct problems addressed, since there are variations in the ink bases used, as well as the specific context of their application, whether as paint bases or watercolor pigments. In addition, the confidentiality of the paint formulation processes by the companies makes it difficult to provide data that is specific and relevant to the study area. Another significant difference lies in the nature of the comparison of the results. While previous studies focus on the color differences between the colors obtained and the reference, this study focuses on the analysis of the proportions provided in the paint formula.

3   Material and Methods

The paint tuning work presented in this paper follows the *Cross Industry Standard Process for Data Mining* (CRISP-DM) methodology [15]. This model comprises six stages: Business Understanding, Data Understanding, Data Preparation, Modeling, Evaluation and Deployment.

During the business understanding phase, specific needs of the industry are identified. Relevant data for the project is analyzed in the data understanding stage. Subsequently, the data is prepared for modeling, where different ML algorithms are explored. In the evaluation phase, the model that best fits the objectives is identified. Finally, the final model is implemented for easy access to the obtained results. In this study, the deployment phase was not carried out because main focus was on model development and evaluation. Deployment decisions will be made following optimization and validation by an expert.

3.1 Business and Data Understanding

This study utilized a dataset comprising 10,000 observations and 4 independent variables (L, a, b, and *Finish*) along with 16 dependent variables representing various ink base concentrations, including *Vivid Yellow, Light Orange, Ochre, Yellow Orange, Dark Orange, Carmine Red, Red, Bordeaux, Brown, Blue, Green, Light Green, White* and *Black* and the extenders *Matte* and *Shine*, which act as volumizing agents and do not influence the color. All variables are quantitative, except for the *Finish*, which is qualitative. It should be noted that each observation represents a unique color.

The analysis of the correlation matrix reveals significant relationships among the variables, with emphasis on the relationship between *Finish* and the type of extender used, either *Matte* (-99.6%) or *Shine* (99.7%). Furthermore, it is important to note that the *Mate* and *Shine* variables are mutually exclusive. It is observed that a higher value in the a coordinate is associated with a higher proportion of the *Light Orange* (43%) and *Red* (50%) bases, while it has a lower proportion of *Green* (-55%) and *Blue* (-39.5%). In turn, an increase in the b coordinate is related to a higher proportion of the bases *Light Orange* (45.0%), *Ochre* (35.6%), *Orange Yellow* (30.8%) and *Dark Orange* (29.3%), while presenting a lower proportion of *Blue* (-75.8%) and *Black* (-38.0%).

Regarding the L coordinate, it is possible to observe a positive correlation with the amount of *White* (81%) and a negative correlation with the amount of *Black* (-59.4%). In addition, it is important to highlight some inverse relationships between colors. For example, an inverse correlation is observed between the base *Light Orange* and the bases *Ochre* (-28.4%), *Blue* ($-29, 7\%$) and *Black* (-41.9%), as well as between the base *Ochre* and the base *Blue* (-42.0%), and between the base *Red* and the base *Green* (-37.2%).

As for the correlation between the input variables, the values reveal that the *Finish* has no significant correlation with the coordinates of the color space. Despite this, the positive correlation (45.5%) between the a and b variables, indicating a relationship between these coordinates in the color space. This can be justified by the existence of color segments with fewer observations.

3.2 Data Preparation

In this phase, the database underwent specific treatments to ensure data completeness and consistency. Firstly, noise values and missing values were removed, guaranteeing high-quality and reliable data. Additionally, the variable *Finish* has been converted into a binary format, representing two distinct categories and all observations were standardized.

According to the data explored in the 3.1 section, data argumentation was performed, duplicating the observations corresponding to the least represented color segments. Subsequently, the data was normalized using the Min-Max Scaler, to ensure that all feature values fall within the range of 0 to 1 [15].

3.3   Modeling

In order to predict the proportion of each paint base to be added to the mixture to achieve a specific color, an Artificial Neural Network with a pre-trained Decoder (ANND) was developed. As a term of comparison, some models commonly used in machine learning and appropriate to the problem were also developed, namely, Linear Regression (LR), Random Forest (RF) and Artificial Neural Network (ANN).

ANN With a Pre-trained Decoder (ANND):   The Autoencoder is a type of ANN that operates in two stages: encoding and decoding. During the encoding phase, the input data is transformed into a lower-dimensional representation, while in the decoding phase, the original data is reconstructed from the encoded representation. By minimizing the reconstruction error between the input and output data (tipically a lower dimension latent space), Autoencoders learn significant features within the data [1]. Previous investigations on Autoencoders explore different approaches associated with transfer learning [20]. Transfer learning is an approach that utilizes acquired knowledge from solving one problem and applies it to a new problem, thereby improving prediction accuracy [13]. In this study, a variant of Autoencoder inspired by the Transfer Learning technique, called ANND, was adopted. The Autoencoder was trained with the training set, acquiring complex knowledge about the relationship between the variables. In a unique approach, the initial input of the Autoencoder consisted of the 16 target variables $\{v_{1,1}, \ldots, v_{1,16}\}$, which were then compressed into 4 variables by the Encoder $\{v_{4,1}, \ldots, v_{4,4}\}$, trying to approximate the variables L, a, b and *Finish*. The Decoder reconstructed the original 16 variables. Figure 3 provides a visual representation of the ANND structure.

The Decoder was then extracted and used to construct a new network, similar to the Autoencoder previously used. The Decoder was frozen and integrated back into the data. Instead of applying the knowledge to a different problem as the Transfer Learning concept implies, the acquired knowledge was inserted into the data and retrained in the Autoencoder. An ANN will be trained using a pre-trained Decoder to map the input variables (L, a, b, and *Finish*) to the 16 target variables.

Linear Regression (LR):   LR assumes a linear relationship between variables and estimates coefficients through the least squares method. It provides information on variables' influence and direction on the target variable [25]. In order to predict all targets, 16 LR models were developed, one for each target, evidencing the adoption of a Problem Transformation approach.

Random Forest (RF):   RF is capable of combining several Decision Trees built to obtain a more accurate and stable prediction. In this method, each tree is exposed to a random subset to split a node, allowing the trees to capture different aspects and patterns of the data [10]. Employing a Problem Transformation approach, the methodology from the previous model is replicated in the development of 16 RF models, each focused on predicting a specific target variable.

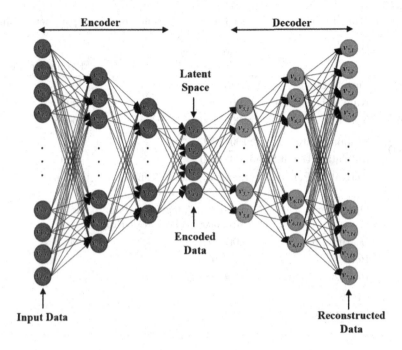

Fig. 3. Autoencoder Architecture

Artificial Neural Network (ANN): ANNs are computational networks inspired by the biological nervous system. They consist of interconnected neurons organized into layers, with information transfer between neurons through activation functions [16], ANNs establish empirical relationships between variables, extract knowledge from datasets, and refine predictive capabilities. This process is repeated until a final output value is obtained. In ANNs, learning involves iterative adjustments of weighted data signals [22]. The architecture of the ANN model is the same as the model that results from concatenating the pretrained Decoder with the original dataset. As in the ANND model, only one model was developed, simultaneously predicting 16 targets, evidencing the adoption of an Algorithm Adaptation approach.

For each model, the parameters that required tuning are given in Table 1.

To assess model performance, k-fold cross-validation (CV) exercises were conducted for all models, dividing the training data into $k-1$ subsets for training and one subset for testing. In this paper, k was set to 10, striking a balance between computational efficiency and accurate performance estimation [19].

3.4   Evaluation

The evaluation of regression models involves determining the error by comparing known values with estimated values. In addition to measuring computational times, the accuracy of the regression models was assessed using two common

Table 1. Tuned Parameters for each model and tuned optimum parameter value

Model	Parameters	Search Range	Optimal Parameters Setting
RF	Max Depth	5–30	30
	Num of Estimator	100–600	600
ANN and ANND	Epochs	1000–10000	10000
	Training Batch Size	100–1000	1000
	Validation Batch Size	100–1000	1000
	Activation Function	Sigmoid; Linear; Tahn; ReLU; SeLU; Softplus	Adam
	Optimizer	Adadelta; Adagrad; Adam; Adamax; Nadam; RMSProp	ReLU
	Learning Rate	0.001–0.5	0.4
	Beta 1	0.1–0.9	0.3
	Beta 2	0.1–0.9	0.2
	Epsilon	0.1–0.4	0.2
	Learning Rate Decay	0.0–0.1	0.0

statistical metrics on the test dataset, namely the mean absolute error (MAE) and the root mean squared error ($RMSE$) [12,17].

When evaluating regression models, MAE and RMSE provide valuable insights. MAE represents the average error, while RMSE measures the dispersion of prediction errors. Minimizing these metrics is crucial for optimizing results and improving prediction quality. By employing these statistical metrics, the best-fit regression model can be selected for more accurate results [17].

The equations of the statistical parameters are as shown in Eq. 3 and Eq. 4.

$$MAE = \frac{\sum_{i=0}^{n-1} |y_i - \hat{y}_i|}{n} \tag{3}$$

$$RMSE = \sqrt{\frac{\sum_{i=1}^{n} (y_i - \hat{y}_i)^2}{n}} \tag{4}$$

Where: n is the number of data points; $\hat{y}_i$ is the ith observed data; y_i is the ith modelled prediction; $\hat{y}$ is the n observed data.

4   Results and Discussion

In this section, it is reported the performance of the models and compare it in terms of different error metrics. Table 2 shows a comparison of MAE and RMSE on the test set in the development stage.

As can be seen from the results shown in Table 2, the models exhibit variations in performance depending on the specific target, with no single model

Table 2. Performance metrics of machine learning models

	LR		RF		ANN		ANND	
Target	MAE	RMSE	MAE	RMSE	MAE	RMSE	MAE	RMSE
Matte	0.19	1.00	0.03	0.61	0.04	0.70	0.00	0.03
Shine	0.21	1.88	0.02	0.52	0.05	1.16	0.09	0.13
Vivid Yellow	1.37	4.07	0.35	2.21	2.40	5.28	0.69	4.58
Light Orange	2.56	5.39	0.70	2.63	1.97	4.97	0.23	0.69
Ochre	6.77	10.31	1.88	4.19	5.24	7.85	0.95	1.52
Yellow Orange	1.54	4.17	0.29	1.89	0.93	3.08	0.64	4.41
Dark Orange	1.46	3.87	0.38	1.77	1.23	3.31	0.43	2.53
Carmine Red	1.81	4.50	0.43	1.78	2.14	4.69	0.61	3.71
Red	4.03	7.29	1.34	3.44	3.48	6.02	1.96	4.12
Bordeaux	2.01	5.33	0.50	2.38	3.81	6.43	0.76	4.40
Brown	1.12	3.31	0.35	1.60	1.06	3.09	0.59	3.44
Blue	4.78	7.94	0.81	2.77	2.73	4.75	0.46	1.30
Green	2.10	4.64	0.44	1.86	3.89	5.82	1.42	5.54
Light Green	0.17	1.25	0.08	1.08	0.12	1.25	0.08	1.26
White	7.20	9.42	2.87	5.78	5.78	8.37	1.11	2.15
Black	2.86	5.27	0.74	2.41	2.37	4.22	0.46	0.89
Model Performance	40.18	79.64	11.21	36.92	37.24	70.99	10.48	40.69

consistently outperforming in all of them. The ANND and RF were standout models for multiple targets, consistently exhibiting lower MAE and RMSE values compared to LR and ANN across the runs. The ANND demonstrated strong performance in *Matte, Ochre, Light Orange, Blue, White*, and *Black*. On the other hand, RF performed better in *Vivid Yellow, Yellow Orange, Dark Orange, Carmine Red, Red, Bordeaux, Brown* and *Green*. However, there was lack of agreement in performance measures for the *Shine* and *Light Green* targets. Some targets, such as *Ochre* and *White*, posed more challenges due to the intrinsic characteristics of the data and complex relationships involved.

In the last row of Table 2, the performance metrics MAE and RMSE are presented for all models, representing an overall average performance. Notably, the ANND and RF models show particularly noteworthy performance. The ANND and RF models demonstrate MAE and RMSE results that are at least 70% and 43% lower compared to the LR and ANN models. Importantly, there is a significant difference in results between ANND and ANN (that does not use Transfer Learning), highlighting the added value of incorporating the pre-trained Decoder in mapping the input, since both networks have the same architecture.

Therefore, a promising approach would be to develop an ensemble that leverages the best of all models. By combining the highlighted qualities of ANND and RF, a more robust and comprehensive system can be achieved. This ensemble strategy effectively addresses target variability and maximizes predictive accuracy.

The comparison between the machine learning models demonstrated the effectiveness of both the Problem Transformation approach and the Algorithm Adaptation approach in solving the prediction problem. However, it is not possible to conclude that one approach is universally better than the other.

As mentioned in Sect. 2.1, most previous studies used ANN to predict ink base concentrations. However, in this study, ANN do not adequately capture the complexities of the relationships between CIELab color parameters and ink base concentrations in this dataset.

Therefore, the main contribution of this study is the adoption of a diversified approach, using ANND and RF models that outperform ANN in predicting ink base concentrations accurately. These models effectively handle complex color data variations, enabling precise predictions in various scenarios and targets.

5 Conclusions

Paint bases have a fundamental role in creating a wide range of tones by combining them in different proportions. This color palette forms the foundation for paint formulation. However, finding the perfect combination of paint bases to achieve a specific color can be challenging.

In this context, an ANND was developed in order to predict the proportions of paint bases needed to achieve the desired color, considering color coordinates in the CIELAB space and the final finish. The proposed model is compared with some models commonly used in machine learning and appropriate to the problem were also developed, namely, LR, RF and ANN.

The experimental results demonstrate that RF and ANND consistently outperformed LR and ANN in predicting the proportions of paint bases, both for each target individually and as a whole, indicating their superior accuracy in capturing the complexities of color relationships. ANND and RF exhibited improvements of over 70% in MAE and 43% in RMSE compared to LR and ANN models. Moreover, based on the comparison between the machine learning models, both Problem Transformation and Algorithm Adaptation approaches were effective in addressing the problem, although no definitive overall superiority can be claimed between the two approaches.

With an innovative approach, this study contributes to advancing knowledge in paint formula development and automating the process of combining primary bases, considering color characteristics and desired finishes. This enables the creation of a predictive model capable of forecasting the optimal combination of bases and their quantities in multi-component mixtures, thereby enhancing the efficiency and quality of paint formulation processes.

The models' performance varies depending on the objective, with no single model outperforming all others consistently. ANND and RF excel in multiple objectives, outperforming LR and ANN. These models effectively handle color data complexity and variations, enabling accurate predictions across different scenarios and target colors.

Future research directions involve investigating the potential of ensembling ANND and RF models to enhance the reliability of predictions. Furthermore, exploring deep learning techniques, such as deep neural networks, to capture intricate color relationships and complexities could provide valuable insights. It is crucial to assess the applicability of these models in diverse industries. Additionally, developing a user-friendly interface for inputting desired color preferences holds promise for further research. The predicted values closely align with the observed values, but consulting an expert is recommended to assess the impact of these minor deviations on the final color outcome. These advancements aim to enhance automation and precision in paint formula development, providing efficient and tailored solutions for industry professionals.

Acknowledgements. This work is financed by National Funds through the Portuguese funding agency, Fundação para a Ciência e a Tecnologia (FCT), within project LA/P/0063/2020.

References

1. Abirami, S., Chitra, P.: Chapter fourteen - energy-efficient edge based real-time healthcare support system. In: Raj, P., Evangeline, P. (eds.) The Digital Twin Paradigm for Smarter Systems and Environments: The Industry Use Cases, Advances in Computers, vol. 117, pp. 339–368. Elsevier (2020). https://www.sciencedirect.com/science/article/pii/S0065245819300506
2. Barbon Junior, S., et al.: Multi-target prediction of wheat flour quality parameters with near infrared spectroscopy. Inf. Process. Agric. **7**(2), 342–354 (2020). https://www.sciencedirect.com/science/article/pii/S2214317318304554
3. Bishop, J., Bushnell, M., Usher, A., Westland, S.: Neural networks in the colour industry. In: Rzevski, G., Adey, R.A. (eds.) Applications of Artificial Intelligence in Engineering VI, pp. 423–434. Springer, Dordrecht (1991). https://doi.org/10.1007/978-94-011-3648-8_27
4. Chaouch, S., Moussa, A., Ben Marzoug, I., Ladhari, N.: Colour recipe prediction using ant colony algorithm: principle of resolution and analysis of performances. Color. Technol. **135**(5), 349–360 (2019)
5. Chen, M.Y., Huang, Y.B., Chang, S.P., Ouhyoung, M.: Prediction model for semi-transparent watercolor pigment mixtures using deep learning with a dataset of transmittance and reflectance. arXiv preprint arXiv:1904.00275 (2019)
6. Fehrman, C., Fehrman, K.: Color: The Secret Influence. 4th edn. Cognella, Incorporated, San Diego (2018). https://books.google.pt/books?id=vkBnuAEACAAJ
7. Huang, Y.B., Chen, M.Y., Ouhyoung, M.: Perceptual-based CNN model for watercolor mixing prediction. In: ACM SIGGRAPH 2018 Posters. SIGGRAPH 2018, Association for Computing Machinery, New York, NY, USA (2018). https://doi.org/10.1145/3230744.3230785
8. Ishikawa-Nagai, S., Yoshida, A., Sakai, M., Kristiansen, J., Da Silva, J.D.: Clinical evaluation of perceptibility of color differences between natural teeth and all-ceramic crowns. J. Dentist. **37**, e57–e63 (2009). https://www.sciencedirect.com/science/article/pii/S0300571209000906 . Journal of Color and Appearance in Dentistry
9. Korifi, R., Le Dréau, Y., Jean Francois, A., Valls, R., Dupuy, N.: CIEL * a * b * color space predictive models for colorimetry devices-analysis of perfume quality. Talanta **104**C, 58–66 (2013)

10. Kulkarni, P., Sreekanth, V., Upadhya, A.R., Gautam, H.C.: Which model to choose? performance comparison of statistical and machine learning models in predicting PM2.5 from high-resolution satellite aerosol optical depth. Atmos. Environ. **282**, 119164 (2022). https://www.sciencedirect.com/science/article/pii/S1352231022002291

11. Kusuma, K., et al.: The performance of machine learning models in predicting suicidal ideation, attempts, and deaths: a meta-analysis and systematic review. J. Psychiatr. Res. **155**, 579–588 (2022). https://www.sciencedirect.com/science/article/pii/S0022395622005416

12. Liu, X., Tang, H., Ding, Y., Yan, D.: Investigating the performance of machine learning models combined with different feature selection methods to estimate the energy consumption of buildings. Energy Build. **273**, 112408 (2022). https://www.sciencedirect.com/science/article/pii/S0378778822005795

13. Luo, S., Huang, X., Wang, Y., Luo, R., Zhou, Q.: Transfer learning based on improved stacked autoencoder for bearing fault diagnosis. Know.-Based Syst. **256**, 109846 (2022). https://www.sciencedirect.com/science/article/pii/S095070512200939X

14. Melki, G., Cano, A., Kecman, V., Ventura, S.: Multi-target support vector regression via correlation regressor chains. Inf. Sci. **415–416**, 53–69 (2017). https://www.sciencedirect.com/science/article/pii/S0020025517307946

15. Mirza, B., et al.: A clinical site workload prediction model with machine learning lifecycle. Healthc. Analytics **3**, 100159 (2023). https://www.sciencedirect.com/science/article/pii/S2772442523000266

16. Mohseni-Dargah, M., Falahati, Z., Dabirmanesh, B., Nasrollahi, P., Khajeh, K.: Chapter 12 - machine learning in surface plasmon resonance for environmental monitoring. In: Asadnia, M., Razmjou, A., Beheshti, A. (eds.) Artificial Intelligence and Data Science in Environmental Sensing, pp. 269–298. Cognitive Data Science in Sustainable Computing, Academic Press (2022). https://www.sciencedirect.com/science/article/pii/B9780323905084000125

17. Salehuddin, N.F., Binti Omar, M., Ibrahim, R., Bingi, K.: A neural network-based model for predicting Saybolt color of petroleum products. Sensors **22**, 2796 (2022)

18. Seymour, J.: Why does the cielab a* axis point toward magenta instead of red? Color Res. Appl. **45**(6), 1040–1054 (2020)

19. Song, T., Ding, L., Yang, L., Ran, J., Zhang, L.: Comparison of machine learning models for performance evaluation of wind-induced vibration piezoelectric energy harvester with fin-shaped attachments. Ocean Eng. **280**, 114630 (2023). https://www.sciencedirect.com/science/article/pii/S0029801823010144

20. Tan, C., Sun, F., Fang, B., Kong, T., Zhang, W.: Autoencoder-based transfer learning in brain-computer interface for rehabilitation robot. Int. J. Adv. Robot. Syst. **16**, 172988141984086 (2019)

21. Teixeira, B.M.F.: Análise e previsão de acidentes rodoviários usando data mining. Diploma thesis, Polytechnic Institute of Porto (2019). http://hdl.handle.net/10400.22/14860

22. Walczak, S., Cerpa, N.: Artificial neural networks. In: Meyers, R.A. (ed.) Encyclopedia of Physical Science and Technology. 3rd edn., pp. 631–645. Academic Press, New York (2003). https://www.sciencedirect.com/science/article/pii/B0122274105008371

23. Westland, S.: Artificial neural networks and colour recipe prediction. In: Proceedings of the International Conference and Exhibition: Colour Science, pp. 225–233 (1998)

24. Xu, D., Shi, Y., Tsang, I., Ong, Y., Gong, C., Shen, X.: Survey on multi-output learning. IEEE Trans. Neural Netw. Learn. Syst. **31**, 2409–2429 (11 2019)
25. Zhao, Y.: Chapter 5 - Regression. In: Zhao, Y. (ed.) R and Data Mining, pp. 41–50. Academic Press (2013). https://www.sciencedirect.com/science/article/pii/B9780123969637000052

Depth and Width Adaption of DNN for Data Stream Classification with Concept Drifts

XingZhi Zhou, Xiang Liu, and YiMin Wen[(✉)]

Guangxi Key Laboratory of Image and Graphic Intelligent Processing, Guilin University of Electronic Technology, Guilin 541004, China
ymwen@guet.edu.cn

Abstract. To handle data stream classification with concept drifts, recent studies have shown that a continuously evolving network structure can achieve better performance. However, firstly, they only change one hidden node at a time, which is not enough to alleviate underfitting or overfitting of network model, leading to model's inability to fit data well. Secondly, during the growth process of the network, they did not consider reducing hidden layers, which would affect the learning ability of deep neural network models. To overcome these shortcomings, an adaptive neural network structure (ANSN) is proposed to handle data stream classification with concept drifts. ANSN has a completely open structure, its network structure, depth, and width can evolve automatically in online mode. The experimental results on ten popular data stream datasets show that the proposed ANSN outperforms the comparison methods. The codes of the proposed algorithm is available on https://gitee.com/ymw12345/ansn.git.

Keywords: Deep Neural Networks · Supervised Learning · Concept Drift · Adaptive Method

1 Introduction

With the rapid development of the Internet and the Internet of Things, deep learning technology has made great progress in many fields through powerful nonlinear models, such as image classification [1], object detection [2], and machine translation [3]. However, deep learning still faces challenges in processing data streams with concept drifts. The main reasons are as follows: 1. DNN's fixed structure needs presetting of optimal model capacity, which may only fit one concept, and so the fixed capacity limits the model's performance for different concepts. 2. DNN's static structure prevents model from adapting to concept drifts.

Flexible structures with growing and pruning mechanisms are popular in DNN literature [4–6], like ADL [7], as they can change DNNs' structure as needed. DNN may also improve predictive performance and deal with changing data streams better than shallow network structures [8]. Pratama M [9] et al. highlighted DNN's adaptability in data stream research due to its fast drift response and appropriate model capacity.

© The Author(s), under exclusive license to Springer Nature Switzerland AG 2023
P. Quaresma et al. (Eds.): IDEAL 2023, LNCS 14404, pp. 406–417, 2023.
https://doi.org/10.1007/978-3-031-48232-8_37

Yuan [10] et al. reviewed how deep learning adapts to concept drift by updating model parameters [11–13] and structure [7, 9].

The drawback of the parameter update method is that it cannot adjust the model capacity to fit different data distributions better. The structure update method increases hidden layers by concept drift detection, which reduces model's convergence when the network depth becomes too deep. The challenge is how to reduce the depth adaptively without affecting the current classification performance. The network width is adjusted using a network saliency formula, which generally changes one node at a time to alleviate overfitting or underfitting when new data arrives, but in reality, the degree of alleviation is low. The challenge lies in how to adaptively change multiple nodes simultaneously based on the relationship between the current data and the current network.

Inspired by ADL [7], we propose a data stream classification algorithm with an adaptive neural network structure (ANSN). ANSN can adjust its depth and width according to the relationship between data and model. The main contributions are as follows: 1. A heuristic formula is proposed to change the number of hidden nodes based on the input data feature dimension, which can fit different data complexity and avoid overfitting or underfitting. 2. An adaptive method is proposed to merge hidden layers based on their association with the current data, which uses the ELM [14] algorithm to calculate the weights of merged hidden layer. This can ensure that the current model classification performance for the current data is unchanged when merging hidden layers, but reduce the model complexity.

2 Related Work

At present, there are few algorithms that apply deep neural networks to data streams with concept drift, so we classify the methods in this field into two categories based on whether the network model capacity is adaptively adjusted: dynamic structure-based methods and static structure-based methods.

In the static structure-based algorithms, Sahoo et al. proposed the HBP [11], which connects an output layer to each hidden layer in the multilayer perceptron and integrates their results. Kauschke et al. proposed NN_Patching [12], which trains a patch with the samples misclassified by the base classifier and an error detector to decide whether to use the base classifier or the patch. Guo et al. proposed SEOA [13], which uses adaptive neurons to replace the hidden layer neurons and selectively integrates the classifiers of each layer based on the data stream fluctuations. These algorithms use different hidden layer information, but the network model capacity is preset, which may prevent optimal results for different data distributions, lowering the model classification and generalization.

Algorithms based on dynamic structures can be divided into algorithms that only adjust the network width [15, 16], algorithms that only adjust the network depth [6], and algorithms that adjust both the network depth and width [7, 9]. The drawback of algorithms that only adjust the network width is that their parameters will gradually increase as new concepts arrive. The drawback of algorithms that only adjust the network depth is that the model depth will become deeper as concept drift occurs frequently, leading to a gradual decline in model performance.

Algorithms that adjust both the network depth and width: Ashfahani et al. proposed the autonomous deep learning (ADL) [7] algorithm, which is the first dynamic network structure used for data streams with concept drift. It increases the hidden layer by concept drift detection, so that the model can learn more features to adapt to concept drift, and changes the number of hidden nodes by the network saliency (NS) formula [17]. ADL changes the hidden nodes one at a time, but in fact, the data that comes later will continue to change, which will cause the model to not be able to mitigate the network's overfitting or underfitting state well. Liu et al. proposed DANINE [10] to avoid catastrophic forgetting [18]. It changes learning rates of different layers to preserve different concepts and uses adaptive memory to keep some past data. But it also changes one node at a time, like ADL. ADL and NADINE both have mechanisms for increasing hidden layers, but they do not consider reducing hidden layers. When the depth grows too much, it is difficult to train the network due to the gradient vanishing problem [19].

3 Network Structure and Learning Strategy of ANSN

This section mainly introduces the ANSN network structure and its learning strategy.

3.1 Network Structure

Figure 1 (b) shows the ANSN network structure. ANSN is a new DNN structure, in which every hidden layer is connected to a softmax layer which produces a local output and the global output Y is obtained from aggregation of each local output using a classification weight vector β that is updated as in ADL [7]. The symbol $\otimes$ means multiplication, and $\oplus$ means addition. The ANSN initial structure has an input layer, a single-node hidden layer, and an output layer. The input and output layers' dimensions depend on the number of data attribute and class categories. Figure 1 (a) shows the initial network for data with 4 dimensions and 3 categories.

3.2 Classification Mechanism

The classification process of the model is shown in eq. (1).

$$C = \arg\max_{k=1,\dots,m} Y_k; \quad Y = \sum_{l=1}^{L} \beta^{(l)} * y^{(l)} \tag{1}$$

where $m \in Z_+$, m represents the number of categories of data. $L \in Z_+$, L represents the number of hidden layers. $\beta \in R_+$, $\beta^{(l)}$ refers to the classification weight of the classifier that the l-th hidden layer connects to. $y^{(l)} \in R^m$, $y^{(l)}$ represents the output value of the classifier connected to the l-th hidden layer. $Y \in R^m$. Y represents the integrated output of all classifiers, and the final predicted label C is determined by the index of the highest value in Y_k, where $C \in Z_+$ and $Y_k \in R$.

$$h^{(l)} = sigmoid(W^{(l)} * h^{(l-1)} + b^{(l)}), h^{(0)} = X \tag{2}$$

$$y^{(l)} = softmax(\theta^{(l)} * h^{(l)} + z^{(l)}), \forall l = 1, ..., L \tag{3}$$

(2) represents the calculation method for the output result of the l-th hidden layer. $W^{(l)} \in Z_+^{d(l) \times d(l-1)}$, $b^{(l)} \in Z_+^{d(l)}$, $h^{(l)} \in Z_+^{d(l)}$, where d(l) represents the number of nodes in the l-th hidden layer, $W^{(l)}$ and $b^{(l)}$ represent the weights and biases of the l-th hidden layer respectively, $h^{(l)}$ is the output of the l-th hidden layer, and $h^{(0)}$ represents the output of the input layer, namely the current input data X. (3) represents the calculation method for the output result of the l-th classifier. $\theta^{(l)} \in R^{m \times d(l)}$, $z^{(l)} \in R^m$, where $\theta^{(l)}$ and $z^{(l)}$ represent the weights and biases of the l-th classifier.

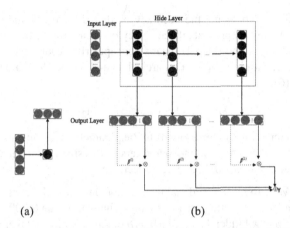

(a)   (b)

Fig. 1. The initial network structure (a) and the basic structure of the ANSN network (b).

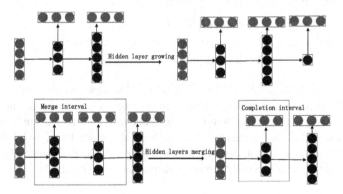

Fig. 2. The growing and merging process of the hidden layer.

3.3   Learning Strategy for Network Depth Adaption

The network depth is adaptively adjusted by hidden layer growing and merging, and the changing process of hidden layer is shown in Fig. 2.

Hidden Layer Growing. We employ the method in ADL to implement. If concept drift occurs, a hidden layer is added to the last layer of the current network and connected to a classifier. The network weights of the added hidden layer are randomly initialized using the Xavier function [20], and the weight of the classifier is set to 1. The purpose of adding hidden layers is to reduce the high bias from drift [21]. We change the detection method in ADL to improve detection level.

$$\hat{F} - \hat{G} > \varepsilon_D \tag{4}$$

$$\varepsilon_W < \hat{F} - \hat{G} < \varepsilon_D \tag{5}$$

In (4) and (5), $\hat{F}$ represents the accuracy predicted by the trained model on the data block D_{t-1}, while $\hat{G}$ represents the accuracy predicted by the trained model on the current data block D_t. Condition (4) means concept drift occurs, while condition (5) means concept drift warning. ε_W and ε_D are calculated by the Hoeffding error formula [7], as shown in (6).

$$\varepsilon_W, \varepsilon_D = \sqrt{(1/(4 * size)) * \log_2(1/\alpha)} \tag{6}$$

where the parameter α for ε_W corresponds to the parameter α_W, the parameter α for ε_D corresponds to the parameter α_D, and *size* refers to the size of the current data block. α_W and α_D are set as 0.0005 and 0.0001, respectively.

Hidden Layers Merging. To avoid making the model too deep by continuously adding hidden layers, we merge several classifiers with weights less than 0.0001 in consecutive hidden layers into a new hidden layer. The bias b of the new hidden layer is initialized to 0, and the weight W of the new hidden layer is computed by (7) to maintain the current model's performance on the current data block.

$$sigmoid(W * X + b) = Y \tag{7}$$

$$W = (\ln(Y/(I - Y)) - b) * X^{-1} \tag{8}$$

where (8) represents the solution value of (7), and X^{-1} represents the generalized inverse matrix of X. I is an all one matrix with the same dimension as Y. Assuming the range of hidden layers is $[i, j]$, that is, all hidden layers between the i-th and j-th hidden layers need to be merged. X is the data flowing into the i-th hidden layer, and Y is the data flowing out of the j-th hidden layer. Finally, the weight of the new hidden layer can be obtained through formula (8). If there are multiple ranges that meet the conditions, they all need to be merged.

3.4   Learning Strategy for Network Width Adaption

The ANSN model can also automatically generate and prune hidden nodes based on the network width controlled by the NS method [17]. The NS method estimated the network's generalization ability by measuring its bias and variance. We execute this on the l-th hidden layer connected to the classifier with maximum classification weight.

Hidden Nodes Growing. We take the ADL method to implement width adaptation, the main difference is the number of nodes that change each time. As follows, the mathematical expression of NS can be obtained from [7] as (9).

$$NS = Var(\tilde{y}^{(l)}) + Bias(\tilde{y}^{(l)})^2 = (E[\tilde{y}^{(l)^2}] - E[\tilde{y}^{(l)}]^2) + (E[\tilde{y}^{(l)}] - y)^2 \qquad (9)$$

where y represents the value of the true label and $\tilde{y}^{(l)}$ represents the predicted value of the l-th hidden layer for a sample, Var represents variance, and $Bias$ represents bias.

In order to calculate the values of $Var(\tilde{y}^{(l)})$ and $Bias(\tilde{y}^{(l)})^2$ in (9), we need to compute the values of $E[\tilde{y}^{(l)}]$ and $E[\tilde{y}^{(l)^2}]$. According to [7], we can obtain (10), (11), (12) and (13).

$$E[h^{(1)}] = sigmoid(W^{(1)} * \mu/(\sqrt{1 + \pi\sigma^2/8}) + b^{(1)}) \qquad (10)$$

$$E[h^{(l)}] = sigmoid(W^{(l)} * E[h^{(l-1)}]) + b^{(l)})\forall l = 2, ..., L \qquad (11)$$

$$E[\tilde{y}^{(l)}] = softmax(\theta^{(l)} * E[h^{(l)}]) + z^{(l)})\forall l = 1, ..., L \qquad (12)$$

$$E[\tilde{y}^{(l)^2}] = softmax(\theta^{(l)} * (E[h^{(l)}])^2) + z^{(l)})\forall l = 1, ..., L \qquad (13)$$

where μ and σ are the recursive mean and recursive standard deviation of data streams. And, the condition of the hidden nodes growing activated for the l-th hidden layer is obtained by (14).

$$\mu^t_{bias} + \sigma^t_{bias} \geq \mu^{min}_{bias} + k\sigma^{min}_{bias} \qquad (14)$$

Here, $k = 1.3\exp(-(Bias(\tilde{y}^{(l)}))^2) + 0.7$, while μ^t_{bias} and σ^t_{bias} represent the recursive average and standard deviation of $Bias^2$ up to t-th time instant, and μ^{min}_{bias} and σ^{min}_{bias} represent the minimum values of μ^t_{bias} and σ^t_{bias} up to t-th time instant but are reset whenever (14) is satisfied. Weights are randomly initialized using the Xavier function.

Hidden Nodes Pruning. The pruning mechanism for hidden nodes is similar to that of the growth mechanism, the pruning condition for triggering hidden nodes in the l-th hidden layer is given by (15) [7].

$$\mu^t_{var} + \sigma^t_{var} \geq \mu^{min}_{var} + 2\lambda\sigma^{min}_{var} \qquad (15)$$

Here, $\lambda = 1.3\exp(-(Var(\tilde{y}^{(l)}))^2) + 0.7$, while μ^t_{var} and σ^t_{var} represent the recursive average and standard deviation of Var up to t-th time instant, and μ^{min}_{var} and σ^{min}_{var} represent the minimum value of μ^t_{var} and σ^t_{var} up to t-th time instant but are reset whenever (15) is satisfied.

Adaptive Number of Hidden Nodes. We use adaptive methods to calculate how many hidden nodes to grow (*inode*) or prune (*rnode*), unlike ADL methods. The formula is (16).

$$inode = rnode = \lfloor \log_2(d) \rfloor \qquad (16)$$

Among them, d represents the size of the data dimension. The purpose of setting it by this way is to maintain a balance between low-dimensional and high-dimensional data. When the l-th layer triggers the need to increase nodes, it is necessary to add *inode* hidden nodes to that layer. When the deletion mechanism is triggered at the l-th layer, *rnode* nodes need to be deleted in this layer. These *rnode* hidden nodes have less impact on the output result than the other nodes in this layer. This effect is mainly computed by (16), which is to remove some hidden units with small contributions to alleviate overfitting. Figure 3 shows the change process for the nodes in a hidden layer.

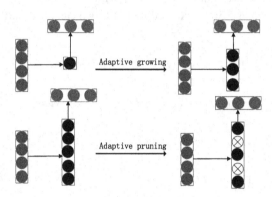

Fig. 3. The process of growing and pruning nodes from a hidden layer.

3.5   ANSN pseudocode

The pseudocode of ANSN is shown in Algorithm 1. In this pseudocode, lines 3 to 7 classify the data block and update the classification weights of the classifier, lines 11 to 14 are to perform the operation of adding a hidden layer, line 15 is to get the l-th hidden layer, which is connected to the classifier with the maximum classification weight, line 17 is to calculate the NS formula of the l-th hidden layer, lines 18 to 21 are to change the nodes of the l-th hidden layer, line 22 is to update the l-th hidden layer using data x_{ij}, line 23 is to train from the l-th hidden layer to the last hidden layer using data D_i, and line 24 is to perform the hidden layer merge operation.

Algorithm 1: ANSN Pseudocodes

Input: Hyperparameter $\alpha_W = 0.0005$, $\alpha_D = 0.0001$; Data stream: $D =$
$\{ D_1,D_2,...,D_N \}$; Data Block $D_i=\{x_{i1},x_{i2},...,x_{is}\}$; Number of classes: C; Number of
attributes: d ; Classification Result Buffer: $Y_1=\varnothing$, $Y_2=\varnothing$; Real label buffer: $\hat{Y}_1$, $\hat{Y}_2$;
Block size: $S=500$; Network Model: *net*;
Output: Classification results of D_i

1 *net*=initialize_network(d,C)
2 **for** D_i in D **do**
3   **for** x_{ij} in D_i **do**
4   y_{ij}=classification(*net*, x_{ij})
5        $Y_1= Y_1 \cup y_{ij}$
6   get the true labels $\hat{Y}_1$ of data block D_i
7   *net*.β=update_classifier_weights(*net*.β, Y_1, $\hat{Y}_1$)
8   **if** Y_2 **is** $\varnothing$ **then**
9        $Y_2= Y_1$, $Y_1=\varnothing$, $\hat{Y}_2 = \hat{Y}_1$
10   **if** Y_1 **not is** $\varnothing$ **then**
11   *drift*=concept_drift_detection$(Y_1,Y_2,\hat{Y}_1,\hat{Y}_2,\alpha_W,\alpha_D)$
12       $Y_2= Y_1$, $Y_1=\varnothing$, $\hat{Y}_2 = \hat{Y}_1$
13   **if** *drift* $==$ 'DRIFT' **then**
14   *net*=add_hidden_layer (*net*)
15   get the index l of the hidden layer with the maximum classification weight β
16   **for** x_{ij} in D_i **do**
17   I_P_node=NS_formula (*net*, l, $\hat{Y}_1$, x_{ij})
18   **if** I_P_node = 'INCREASE' **then**
19   *net*=add_hidden_nodes (*net*, l, d)
20   **else if** I_P_node ='PRUNE' **then**
21   *net*=prune_hidden_nodes (*net*, l, d)
22   *net*=update_network (*net*, x_{ij}, l)
23   *net*=update_network1 (*net*, D_i, l)
24   *net*=merge_hidden_layers (*net*, D_i)

---

4 Experiment

4.1 Dataset

Real datasets have *Weather*, *Kddcup*, and *Rfid*. Synthesized datasets have *Pmnist* [22], *SEA*, *Hyperplane*, *Rmnist*, *mnist_overturn(mnist_ot)*, and *mnist_translation(mnist_ts)*. The detailed information of each dataset is shown in Table 1. Where *F*, *C*, *I*, *T*, *Nod*, and CD represent features, Class, Instances, Types, Number of drift, and Concept Drifts respectively.

4.2 Baseline

ADL: In this model, the concept drift warning parameter α_W is set to 0.0005, the concept drift occurrence parameter α_D is set to 0.0001, the weight step size ξ is set to 0.001, and the learning rate δ is set to 0.02.

Table 1. Experimental datasets

Dataset	F	C	I	T	Nod	CD
Hyperplane	4	2	120k	Incremental	1	1–2
AbSea	3	2	120k	Abrupt,Reappear	5	1–2–3–1–2–3
RaSea	3	2	120k	Gradual	3	1–2–3
Pmnist	784	10	64k	Abrupt	3	1–2–3–4
Rmnist	784	10	67.5k	Abrupt,Reappear	8	1–2–3–1–2–3–1–2–3
Mnist_ot	784	10	67.5k	Abrupt,Reappear	8	1–2–3–1–2–3–1–2–3
Mnist_ts	784	10	67.5k	Abrupt,Reappear	8	1–2–3–1–2–3–1–2–3
Weather	8	2	18k	Unknown	-	-
Rfid	3	4	280k	Unknown	-	-
Kddcup	41	2	500k	Unknown	-	-

NADINE: In this model, the concept drift warning parameter α_W is set to 0.0005, the concept drift occurrence parameter α_D is set to 0.00001, and the learning rate δ is set to 0.02.

SEOA: In this model, the depth is set to 8, the number of hidden nodes is set to 100, and the learning rate is set to 0.02.

DNN: In this model, the depths are set to 3, and 4, respectively, with the hidden node number set to 256 and the learning rate set to 0.02.

Among them, the download addresses for ADL, NADINE, and SEOA codes can be found in the corresponding papers. All the above methods use the SGD method to adjust the parameters of the network.

4.3 Experimental Setup

We set the parameters $[\alpha_W, \alpha_D, \xi]$ to $[0.0005, 0.0001, 0.001]$. The values of these parameters are using the values of ADL parameters. The purpose of choosing ADL parameters is to demonstrate that the superiority of our algorithm does not come from the setting of parameters.

We set the parameters for the fixed-structure DNN. Regarding the size of hidden nodes, we experimented with different node sizes and found that 256 is better. In the ADL, NADNIE, and SEOA algorithms, we used the parameters in their code or paper for their parameter settings.

We used data blocks of size 500 and analyzed from the second one. For artificial data sets, we use 10 experiments with different random data. For real data sets, we use 10 experiments with different network weights. We compared the cumulative accuracy of different models. We used the Wilcoxon signed-rank test for accuracy, where "●" means this baseline is significantly different from our algorithm.

4.4 Results

As can be seen from Table 2, our algorithm is superior to our comparison algorithm on the datasets *Hyperplane*, *SEA*, *PMNIST*, *Weather*, and *RFID*. Our algorithm is lower than the recent deep learning algorithm for dealing with concept data streams (SEOA) on the datasets *kddcup, Rmnist,* and *mnist_ts,* but our algorithm is superior to the recent dynamic deep neural network structure algorithm (ADL, NADINE) on these datasets. Overall, our algorithm outperforms the most recent best algorithm on more than half of the datasets, and is superior to the deep learning algorithms with dynamic structures on all of the datasets. In experiments with different depths of DNN, which can reflect that our adaptive network structure algorithm can better adapt to the network structure and improve network performance when dealing with different datasets. Table 3 shows the experimental results of our ablation experiments, where ANSN1 means that we removed our own adaptive node number method in ANSN, and ANSN2 means that we removed the hidden layer merging method in ANSN. The experimental results show the superiority of our methods.

Table 2. Classification results on ten datasets

	ANSN	ADL	NADINE	SEOA	DNN3	DNN4
Hyperplane	**92.5 ± 0.8**	91.1 ± 1.3●	90.0 ± 1.7●	91.9 ± 1.0●	91.2 ± 0.7●	91.1 ± 0.7●
AbSEA	**97.6 ± 0.1**	95.6 ± 0.4●	96.0 ± 0.2●	96.9 ± 0.0●	95.8 ± 0.1●	95.9 ± 0.1●
PMNIST	**90.5 ± 0.2**	82.3 ± 0.6●	84.7 ± 0.4●	89.2 ± 0.1●	76.7 ± 0.1●	71.7 ± 0.3●
Mnsit_ot	**88.1 ± 0.1**	80.2 ± 0.9●	80.0 ± 0.7●	**88.1 ± 0.2**	74.0 ± 0.4●	68.8 ± 0.7●
Mnist_ts	86.1 ± 0.2	80.0 ± 1.1●	69.8 ± 3.2●	**89.3 ± 0.2**●	74.1 ± 0.5●	69.2 ± 1.1●
Rmnist	83.2 ± 0.3	78.4 ± 0.9●	77.4 ± 0.7●	**87.5 ± 0.2**●	72.3 ± 0.5●	67.7 ± 0.6●
Weather	**75.4 ± 0.6**	73.8 ± 0.7●	73.4 ± 0.4●	72.5 ± 0.2●	73.0 ± 0.0●	72.9 ± 0.0●
Rfid	**99.3 ± 0.0**	99.1 ± 0.0	99.1 ± 0.0	98.0 ± 0.3●	97.0 ± 0.2●	99.1 ± 0.0●
kddcup	99.8 ± 0.0	99.8 ± 0.0●	99.8 ± 0.0●	**99.9 ± 0.0**●	99.6 ± 0.0●	99.7 ± 0.0●
RaSEA	**97.5 ± 0.2**	95.7 ± 0.2●	95.8 ± 0.3●	96.7 ± 0.0●	95.7 ± 0.1●	95.7 ± 0.1●

Table 3. Ablation experiments by replacing different modules in ANSN

	ANSN	ANSN1	ANSN2
Hyperplane	92.55 ± 0.82	91.91 ± 1.02●	92.06 ± 1.10●
AbSEA	97.69 ± 0.11	97.65 ± 0.12	97.72 ± 0.11
PMNIST	90.5 ± 0.21	84.78 ± 0.19●	88.93 ± 0.23●
Mnsit_ot	88.12 ± 0.19	82.58 ± 0.20●	87.05 ± 0.15●
Mnist_ts	86.10 ± 0.20	78.34 ± 0.22●	84.93 ± 0.16●
Rmnist	83.25 ± 0.39	79.42 ± 0.34●	81.89 ± 0.32●
Weather	75.41 ± 0.67	73.6 ± 0.53●	74.9 ± 0.61●
Rfid	99.33 ± 0.02	99.34 ± 0.02	99.32 ± 0.02
kddcup	99.86 ± 0.01	99.84 ± 0.01●	99.86 ± 0.01
RaSEA	97.5 ± 0.21	97.48 ± 0.20	97.49 ± 0.21

5 Conclusion

This paper presents a data stream classification algorithm with adaptive neural network structure, which is built on a flexible working principle where the network structure can be automatically constructed by learning. The network width adaptation depends on the estimation of bias and variance in hidden layer and uses an adaptive method to change the number of hidden nodes to speed up model's ability to adapt to different concepts. The network depth adaptation uses concept detection to expand and an adaptive method to merge hidden layers, which can effectively control model's depth. The combination of these two methods can improve the generalization ability of deep learning applied to data stream classification with concept drifts. Inspired by these results, we plan to integrate these methods with semi-supervised methods in future work to further improve performance.

Acknowledgments. This work was partially supported by the National Natural Science Foundation of China (62366011), the Key R&D Program of Guangxi under Grant (AB21220023), and Guangxi Key Laboratory of Image and Graphic Intelligent Processing (GIIP2306).

References

1. Simonyan, K., Zisserman, A.: Very deep convolutional networks for large-scale image recognition. In: Proceedings of the 3rd International Conference on Learning Representations. LCLR, Washington DC (2015)
2. Redmon, J., Farhadi, A.: Yolov3: an incremental improvement. arXiv preprint arXiv: 1804.02767 (2018)
3. Vaswani, A. et al.: Attention is all you need. In: Proceedings of the 31st Conference on Neural Information Processing Systems, pp. 5998–6008. NIPS, San Diego (2017)
4. Pratama, M., Dimla, E., Tjahjowidodo, T., Pedrycz, W., Lughofer, E.: Online tool condition monitoring based on parsimonious ensemble+. IEEE Trans. Cybern. **50**(2), 64–677 (2020)
5. Pratama, M., Pedrycz, W., Lughofer, E.: Evolving ensemble fuzzy classifier. IEEE Trans. Fuzzy Syst. **26**(5), 2552–2567 (2018)
6. Pratama, M., Pedrycz, W., Webb, G.I.: An incremental construction of deep neuro fuzzy system for continual learning of nonstationary data streams. IEEE Trans. Fuzzy Syst. **28**(7), 1315–1328 (2020)
7. Ashfahani, A., Pratama, M.: Autonomous deep learning: continual learning approach for dynamic environments. In: Proceedings of the 2019 SIAM International Conference on Data Mining, pp. 666–674. SIAM, Canada (2019)
8. Ashfahani, A., Pratama, M., Lughofer, E., Cai, Q., Sheng, H.: An online RFID localization in the manufacturing shopfloor. In: Lughofer, E., Sayed-Mouchaweh, M. (eds.) Predictive Maintenance in Dynamic Systems, pp. 287–309. Springer, Cham (2019). https://doi.org/10. 1007/978-3-030-05645-2_10
9. Pratama, M., Za'in, C., Ashfahani, A., Ong, Y. S., Ding, W.: Automatic construction of multi-layer perceptron network from streaming examples. In: Proceedings of the 28th ACM International Conference on Information and Knowledge Management, pp. 1171–1180. ACM, New York (2019)
10. Yuan, L., Li, H., Xia, B., Gao, C., Liu, M., Yuan, W. You, X.: Recent advances in concept drift adaptation methods for deep learning. In: Proceedings of the 31st International Joint Conference on Artificial Intelligence, pp. 5654–5661. IJCAI, Freiburg (2022)

11. Sahoo, D., Pham, Q., Jing, L., Steven, C. H.: Online deep learning: learning deep neural networks on the fly. In: Proceedings of the Twenty-Seventh International Joint Conference on Artificial Intelligence, pp. 2660–2666. IJCAI, Freiburg (2018)

12. Kauschke, S., Lehmann, D. H., Fürnkranz, J.: Patching deep neural networks for nonstationary environments. In: Proceedings of the International Joint Conference on Neural Networks, pp. 1–8. IEEE, Piscataway (2019)

13. Guo, H., Zhang, S., Wang, W.: Selective ensemble-based online adaptive deep neural networks for streaming data with concept drift. Neural Netw. **142**(10), 437–456 (2021)

14. Ding, S., Zhao, H., Zhang, Y., Xu, X., Nie, R.: Extreme learning machine: algorithm, theory and applications. Artif. Intell. Rev. **44**(1), 103–115 (2015)

15. Rusu, A.A., Rabinowitz, N.C., Desjardins, G., Soyer, H., et al.: Progressive neural networks. arXiv preprint arXiv:1606.04671 (2016)

16. Yoon, J., Yang, E., Lee, J., Hwang, S.J.: Lifelong Learning with dynamically expandable networks. In: Proceedings of the International Conference on Learning Representations. LCLR, Washington DC (2018)

17. Andri, A., Mahardhika, P., Edwin, L., Ramasamy, S., Yew-Soon, O.: DEVDAN: deep evolving denoising autoencoder. Neurocomputing **390**(5), 297–314 (2020)

18. Jidong, H., Yujian, L.: A review of catastrophic forgetting in neural network models. J. Beijing Univ. Technol. **47**(5), 551–564 (2021)

19. Roodschild, M., Sardiñas, J.G., Will, A.: A new approach for the vanishing gradient problem on sigmoid activation. Prog. Artif. Intell. **9**(4), 351–360 (2020)

20. Glorot, X., Bengio, Y.: Understanding the difficulty of training deep feedforward neural networks. In: Proceedings of Machine Learning Research, pp. 249–256. Princeton, NJ (2010)

21. Montufar, G.F., Pascanu, R., Cho, K., Bengio, Y.: On the number of linear regions of deep neural networks. In: Proceedings of the 13th Conference on Neural Information Processing Systems, pp. 2924–2932. Curran Associates, New York (2014)

22. Kirkpatrick, J., Pascanu, R., Rabinowitz, N.: Overcoming catastrophic forgetting in neural networks. Proc. Natl. Acad. Sci. **114**(13), 3521–3526 (2017)

FETCH: A Memory-Efficient Replay Approach for Continual Learning in Image Classification

Markus Weißflog[1]([✉]) [iD], Peter Protzel[1] [iD], and Peer Neubert[2] [iD]

[1] Faculty of Electrical Engineering and Automation Technology,
Chemnitz University of Technology, Chemnitz, Germany
`markus.weissflog@etit.tu-chemnitz.de`
[2] Institute for Computational Visualistics, University of Koblenz, Koblenz, Germany

Abstract. Class-incremental continual learning is an important area of research, as static deep learning methods fail to adapt to changing tasks and data distributions. In previous works, promising results were achieved using replay and compressed replay techniques. In the field of regular replay, GDumb [23] achieved outstanding results but requires a large amount of memory. This problem can be addressed by compressed replay techniques. The goal of this work is to evaluate compressed replay in the pipeline of GDumb. We propose FETCH, a two-stage compression approach. First, the samples from the continual datastream are encoded by the early layers of a pre-trained neural network. Second, the samples are compressed before being stored in the episodic memory. Following GDumb, the remaining classification head is trained from scratch using only the decompressed samples from the reply memory. We evaluate FETCH in different scenarios and show that this approach can increase accuracy on CIFAR10 and CIFAR100. In our experiments, simple compression methods (e.g., quantization of tensors) outperform deep autoencoders. In the future, FETCH could serve as a baseline for benchmarking compressed replay learning in constrained memory scenarios.

Keywords: Continual Learning · Replay Learning · Compressed Replay

1 Introduction

Humans are capable of learning to solve new tasks ever throughout their lives. Learning to incorporate new information is crucial in areas such as robotics and machine learning, yet this endeavor remains challenging despite the use of advanced techniques [20]. If no special measures are taken, a learning system quickly forgets old knowledge as soon as it is presented with new information [5]. This challenge, known as Catastrophic Forgetting (CF), is so difficult to overcome that Continual Learning (CL) has emerged as a discipline of machine

© The Author(s), under exclusive license to Springer Nature Switzerland AG 2023
P. Quaresma et al. (Eds.): IDEAL 2023, LNCS 14404, pp. 418–430, 2023.
https://doi.org/10.1007/978-3-031-48232-8_38

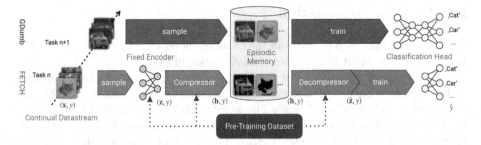

Fig. 1. Overview of the proposed method: GDumb [23] (red arrows) saves samples from the continual datastream in the episodic memory without preprocessing. A neural network is trained from scratch using only the samples from the memory. FETCH (blue arrows) fixes the early layers of the neural network after pre-training. The samples are saved efficiently using the compressor and decompressor. Only the classification head needs to be retrained.

learning. In recent years, many publications have approached the subject [5, 16,17,20]. Promising results were achieved by *replay techniques*, that keep an episodic memory of previously encountered samples to mitigate catastrophic forgetting [5]. Prabhu et al. [23] in particular presented *GDumb*, an approach that has attracted much attention in the community due to its simplicity and yet good performance.[1] Replay techniques can have high memory consumption, which has led to the development of *compressed replay methods* [8,9,27,28]. This work investigates whether the memory consumption of GDumb can be reduced using ideas from compressed replay and how the performance changes under constrained memory. Based upon this, we present FETCH (Fixed Encoder and Trainable Classification Head). A simplified schematic illustration can be found in Fig. 1. FETCH improves GDumb in several aspects:

- A pre-trained fixed encoder extracts general features from the images and enables knowledge transfer from a pre-training dataset. Additionally, the number of parameters in the trainable classification head is reduced.
- A compressor reduces the size of the samples in the episodic memory, thus reducing the overall memory footprint. This can either improve performance on a limited memory or reduce the memory footprint of the overall pipeline.
- In our experiments we assess various variations and components of FETCH and show improved performance over both GDumb and selected compressed replay techniques.

The paper is structured as follows: Sect. 2 introduces the problem of class-incremental continual learning. Section 3.1 presents general related work while Sect. 3.2 focusses on GDumb in particular. Section 4 details the design of FETCH. Section 5 summarizes implementation details. Section 6 presents our experiments. Section 7 concludes the paper. Code is available online[2].

---

[1] At the time of writing, the GDumb has received over 300 citations on Google Scholar.
[2] www.mytuc.org/pxwz.

2   Problem Formulation

Our proposed approach operates in the challenging online, class-incremental set-
ting. A learning agent is presented with a stream of tasks $\mathcal{T} = \{\mathcal{T}^1, \mathcal{T}^2, \ldots, \mathcal{T}^t,$
$\ldots, \mathcal{T}^T\}$ one task at a time. T is the total number of tasks and t is the current
task's identifier. Each task consists of multiple samples, i.e., images, $\mathbf{x}^t \in \mathcal{X}^t$ and
their corresponding labels $y^t \in \mathcal{Y}^t$. The agent's job is to find a model f that can
predict the label for each sample $f_t : \mathbf{x}^t_{\text{test}} \mapsto y^t_{\text{test}}$, where the samples $\mathbf{x}^t_{\text{test}}$ belong
to a never before seen test dataset $\mathcal{X}^t_{\text{test}}$. The labels of these samples consist of
the classes of the current task and all previous tasks $y^t_{\text{test}} \in \mathcal{Y}^t_{\text{test}} = \bigcup_{i=0}^{t} \mathcal{Y}^i$. To
achieve this, the agent is presented with a set of training examples belonging
to the current task $\mathcal{T}^t = (\mathcal{X}^t, \mathcal{Y}^t)$. The agent has the option of saving a pair of
image and label, or choosing to never see it again.

3   Related Work

3.1   Literature Review

Several publications offer an overview of CL. [5,17,20]. De Lange et al. [5] in
particular propose a widely adopted taxonomy, categorizing approaches and settings into parameter isolation, regularization, and replay methods. *Parameter
isolation methods* work by identifying important parts of the model for the different tasks. These parts can be, for example, gradually extended [1] or even
exchanged [12], as new tasks arise. *Regularization methods* introduce new terms
in the loss function to ensure that performance for previous tasks is not degraded
[2,4,7].

Replay methods work by storing exemplars in an episodic memory and inter-
weaving them into the stream of new data [24]. Sangermano et al. [26] use dataset
distillation in the replay memory. Chen et al. [3] use a database of unlabeled data
together with an episodic memory to improve learning performance. Gu et al.
[6] propose to better utilize samples from the continual datastream to mix with
the samples from the replay memory.

A special case of replay is *compressed replay*. As replay requires to store a
subset of exemplars in a memory, it is a sensible idea to compress these exemplars in order to reduce the overall memory footprint. Hayes et al. [8] and Wang
et al. [28] use different compression strategies to reduce the size of the images
in memory. Hayes et al. [9] extend these approaches by freezing the early layers of a neural network after initial pre-training and using the resulting feature
maps as exemplars for compressed replay. The decompressed samples are used
to train the remaining parts of the network. Wang et al. [27] extend this approach even further with an additional autoencoder. All methods operate in the
classical replay scenario where the continual datastream is mixed with the stored
exemplars.

3.2   Greedy Sampler and Dumb Learner

GDumb (Greedy Sampler and Dumb Learner) [23] was proposed as a simple baseline but still outperformed many previous methods. It uses an episodic memory with N free slots. During training, the memory slots are filled with samples from the datastream with a balancer ensuring equal class representation. When memory is full, new classes are added by removing exemplars from the largest class. After each task, GDumb retrains a backbone network from scratch using only the exemplars from the memory. Following [5], GDumb can be classified as online, class-incremental CL. GDumb's simple design comes with some drawbacks. Saving raw data isn't always feasible due to licensing and privacy concerns. Moreover, the images require significant storage space that might not be used efficiently. Therefore, we propose to combine the methodology of GDumb with the principles of compressed replay in order to exploit the advantages of both approaches.

4   Approach

Following GDumb [23], FETCH uses an episodic memory and retrains the classification head after each task. The blue arrows in Fig. 1 show an overview of the proposed pipeline. Whenever the input distribution changes, the data $\mathbf{x}$ is sampled from the continual datastream using GDumb's balanced greedy strategy. The data passes through the fixed encoder ($\mathbf{z}$) and the compressor ($\mathbf{h}$) before being stored in the episodic memory. During the inference phase, the classification head is trained from scratch using only the decompressed data ($\hat{\mathbf{z}}$) and corresponding labels y from the memory. Compression and decompression are not required for inference, so the data flows directly from the encoder to the classification head. The fixed encoder and some compressor-decompressor pairs must be pre-trained, so an additional pre-training dataset is used.

By comparing the red and blue arrows in Fig. 1, it becomes apparent that FETCH, unlike GDumb, leverages an additional fixed encoder, compressor, decompressor, and pre-training dataset. Both algorithms share the greedy sampler, memory, and retraining strategy for the classification head. Like GDumb, FETCH can be classified as online class-incremental CL. The following subsections describe the components in more detail.

4.1   Fixed Encoder and Trainable Classification Head

First, an encoding model, called *fixed encoder* in this work, converts the image to a latent representation $\mathbf{z}$. The classification model uses this representation $\mathbf{z}$ to predict the class $\hat{y}$ of the input data. If the encoding was successful, all the relevant information about the class is still present. As encoders, we utilize the early layers of a CNN, whose weights remain frozen, following prior works [9, 27]. To adhere to the paradigm of CL, we use different datasets for the pre-training of the encoder (called *pre-training dataset*) and the training and evaluation of FETCH.

This approach allows for transfer effects from the pre-training dataset and is computationally more efficient than GDumb, as data only passes through the encoder once instead of each epoch. Also, fewer parameters need to be updated as the encoder stays fixed. After the initial pre-training, the encoder's weights remain frozen. The fixed encoder is considered as the early layers of a CNN, so the *trainable classification head* can be regarded as the remaining layers. The classification head is trained using only the encoded data z from the memory.

For this work, different variations of the ResNet architecture [10] are used as fixed encoders and classification heads for multiple reasons: First, ResNets have a good performance in many classification tasks [10,17]. Second, implementations and pre-trained weights are available, and lastly, ResNets are used in many other publications, making comparisons fair [9,23,27]. If not stated otherwise, we used the layers up to conv4_x [10] as the fixed encoder for our experiments. The encoder pre-trained on ImageNet1k [25]. The weights were provided by PyTorch.[3] The remaining layers including conv4_x form the classification head.

4.2   Compressor and Decompressor

The encoded data z in memory may contain redundancies, so a second compression stage, called the *compressor*, is used to reduce their size on the actual hardware. It operates on the matrix/ tensor representation of the images or featuremaps. A *decompressor* restores the original encoded representation as close as possible. We have selected the approaches listed below for our experimental setup. Each method has its own hyperparameter k that controls the amount of compression.

- *Quantization.* Quantization describes the reduction of the tensor entries to a small number of k_{quant} discrete states, which can be represented with fewer bits than the actual values. For decompression, a lookup table is used. To get the discrete states, the pre-training dataset is analyzed. The range between the highest and lowest values in the whole pre-training dataset is split into k_{quant} equally sized intervals. For all experiments, TinyImagenet [14] was used as a pre-training dataset.
- *Thinning.* The basic idea of this method is to keep only the most important, i.e., the largest entries. The resulting tensor is sparse and can thus be stored more efficiently. Instead of storing all entries, only the non-zero entries are saved together with their corresponding index in the tensor. Decompression is done by setting the stored indices of the output tensor to their corresponding values. All other entries are assumed to be equal to zero. The parameter $k_{thin} \in [0, 1]$ describes the proportion of entries that are set to zero.
- *Autoencoding.* Convolutional autoencoders are a deep learning-based approach for dimensionality reduction [11,18] of images. The compressor and the decompressor are typically already part of their architecture. In this work, the compressor consists of two blocks of Conv2d layers with kernel size 3 and

---

[3] Online: https://pytorch.org/vision/stable/models.html.

Table 1. Computation of the total storage requirement for different compressors

Thinning	Quantization	Autoencoding
$s_\Sigma = s_{model}$	$s_\Sigma = s_{model}$	$s_\Sigma = s_{model}$
$+N \cdot n \cdot \lceil (1 - k_{thin}) \cdot s_{uint/float} \rceil$	$+k_{quant} \cdot s_{uint/float}$	$+s_{ae}$
$+N \cdot n \cdot \lceil (1 - k_{thin}) \cdot s_{addr} \rceil$	$+N \cdot \lceil \lceil \log_2 k_{quant} \rceil \cdot n \cdot \frac{1}{8} \text{ bytes} \rceil$	$+N \cdot k_{ae} \cdot n_h \cdot s_{float}$

padding 1, followed by ReLU activation and max-pooling with kernel size 2 and a stride of 2. The decompressor consists of two blocks of transposed two-dimensional convolution with kernel size 2, stride 2, and ReLU activation. For this architecture, the parameter k_{ae} describes the number of channels in the bottleneck. The autoencoder was pre-trained on the TinyImagenet [14] dataset.

4.3   Calculation of the Storage Consumption

The total amount of storage s_Σ that is consumed by the whole pipeline depends on several variables: First, s_Σ depends on the storage consumption of the model s_{model}, which is split into the fixed encoder and the classification head. We used ResNets for our experiments. As FETCH operates independently from the underlying model, s_{model} was set to zero in all evaluations. The resulting findings do not change since s_{model} is a constant value that gets added to all results. This is also in line with the literature [9, 22, 27].

Second, s_Σ depends on the used datatypes. In this work, we used single-precision floating point numbers, meaning $s_{float} = 4$ bytes. We used integers of size $s_{addr} = 2$ bytes as indices for matrices and arrays, as no matrix surpassed 2^{16} elements in our experiments. For raw images, we assumed RGB values with 8 bits per channel, therefore we used $s_{uint} = 1$ byte.

Third, s_Σ depends on the other components, one of which is the episodic memory with N slots. Additionally, the compressor's input influences s_Σ. If the full model is used as a classification head (e.g. the fixed encoder gets omitted), the input is equal to the raw images x with datatype uint8. When a fixed encoder is used, the compressor receives the encoded images z of type float32 as input. Let n refer to the number of elements in these tensors in the corresponding case and let $s_{uint/float}$ be the shorthand notation for the corresponding memory requirement.

Lastly, the total memory consumption depends on the compressor. The level of compression and, thus, the memory consumption of one exemplar can be adjusted using the compression parameter k. The resulting equation for the total memory requirement can be found in Table 1. For the autoencoder, n_h describes the number of elements in the spatial dimensions in the bottleneck and s_{ae} is the memory consumption of the autoencoder network, which varies between 4.6 KiB and 21.62 KiB, depending on the parameter k_{ae}.

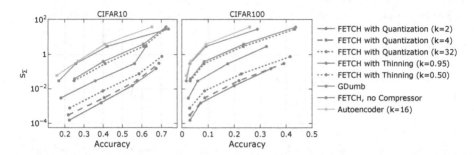

Fig. 2. Tradeoff between storage and performance. Lower is better.

5   Implementation Details

The implementation was based on GDumb's publicly available PyTorch-code[4] [23]. No hyperparameters were changed. The training was done using SGDR-optimization [15] with a batch size of 16 and learning rates in $[0.005 \ldots 0.05]$. Data regularization was done using normalization and cutmix [29] with $p = 0.5$ and $\alpha = 1$. Because only the data from the episodic memory is used for training, the backbone can be trained for multiple epochs without breaking the class-incremental paradigm. The performance is measured as the average accuracy on the test set after convergence and is always measured after the last task. We used two datasets and models for evaluation. For the CIFAR10 dataset [13], ResNet-18 [10] was used as the fixed encoder and the classification head. For CIFAR100, ResNet-34 was used as the fixed encoder and the classification head. Storage consumption was measured in mebibyte[5].

6   Experiments

6.1   Tradeoff Between Storage and Performance

This experiment aimed at finding a tradeoff between low memory consumption and high accuracy. The results can be seen in Fig. 2. The different methods were compared using a variable number of memory slots $N \in \{10; 100; 1000; 10000\}$.

Using a fixed encoder improves performance, as shown by the fact that most of the green curve is above the purple curves. This suggests that the transfer effects of the pre-trained fixed encoder are beneficial to the whole pipeline. GDumb demonstrates superior performance for high storage capacities on CIFAR10, which indicates that the impact of training data quantity on performance diminishes as the learner has more memory available. Nevertheless, model size remains a significant factor in determining performance. Restricting the number of adjustable parameters through pre-training can lead to poorer

---

[4] Online: https://github.com/drimpossible/GDumb.
[5] 1 MiB $= 2^{20}$ bytes $\approx 10^6$ bytes.

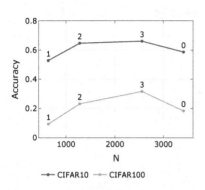

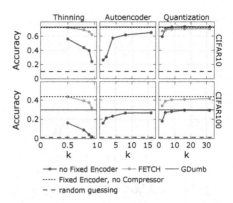

Fig. 3. Effect of pre-training early layers. The annotations correspond to the number of blocks in the encoder. 0 is the standard GDumb configuration, while 3 means that only the parameters of the last block were updated.

Fig. 4. Effect of compressing exemplars. Note that a higher compression parameter means *less* compression except for thinning, where a higher parameter means more compression. The vertical lines show the baselines without compression.

performance compared to full models. Additionally, the quality of the data can also affect performance, as encoded samples have fewer entries and therefore provide less information to the model. For the more complex CIFAR100 and ResNet-34, this is not the case. Here, a fixed encoder is always beneficial compared to GDumb.

The impact of compression depends on the setup. Positive effects can be seen in the blue curves, which are ordered by their compression parameter k_{quant}. Quantization increases performance by trading data quality for reduced storage size and thus increased storable exemplars N. The red curves show a different pattern. High compression parameters, such as $k_{thin} = 0.95$ show decreased storage consumption but the accuracy appears to approach an upper limit, as shown by the shape of the solid red curve in the left plot.

6.2 Ablation: Effect of the Fixed Encoder

This experiment aims to investigate the influence of the layer where the ResNet is split into encoder and classification head. For this reason, the compressor was omitted. Four different configurations were investigated: Using the whole model as a classifier (like GDumb), dividing after conv2_x, after con3_x, and after conv4_x [10]. For each configuration, the episodic memory was filled with as many samples as possible without exceeding a maximum storage of 10 MiB.

Figure 3 shows the results. Splitting ResNets at later layers appears beneficial, as shown by the fact that the performance is consistently higher. At later layers, the samples are more compressed and their representation also benefits from the encoder's pre-training, which is known to be beneficial [19,21]. Not splitting the

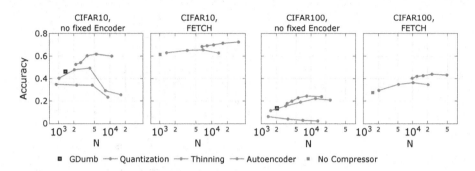

Fig. 5. Performance on a fixed memory budget

ResNet results in the highest number of memory slots, despite raw images **x** having more elements than encoded samples **z**. Raw images are smaller due to their representation using unsigned integers, taking up less space compared to floating point numbers used for encoded samples. The encoded samples have an advantage due to the encoder's pre-training, therefore they nonetheless improve the performance.

6.3   Ablation: Effect of Compression

The previous section examined the effect of an encoder in isolation, while this section examines the effect of a compressor in the same way. For the experiment, the number of memory slots was fixed at $N = 10000$, while the compression parameter k was varied.

The result can be seen in Fig. 4. The autoencoder could only be used for the experiments without the fixed encoder because the spacial dimensions of the encoded featuremaps (2×2) are too small to perform convolutions and pooling. It also becomes clear that the performance of the baseline cannot be reached using this architecture. Increased compression negatively impacts performance across all compressors, which is expected. Remarkably, the upper bound of the curves reflects the optimal accuracy achievable with this configuration. The quantization strategy approaches baseline performance, given a sufficiently high compression parameter. The best accuracy is reached between $k = 8$ and $k = 16$, which corresponds to compression of over 85 %. The thinning compression performs significantly better when a fixed encoder is used.

6.4   Performance on a Fixed Memory Budget

This experiment aims to show how well different configurations perform under a memory constraint. To replicate these conditions, the total memory consumption was fixed at 4 MiB for CIFAR10 and at 6 MiB for CIFAR100. The memory is filled with N samples up to the maximum available storage size.

The results can be seen in Fig. 5. Notably, almost all curves show a maximum, where the balance between the number and quantity of the samples in memory

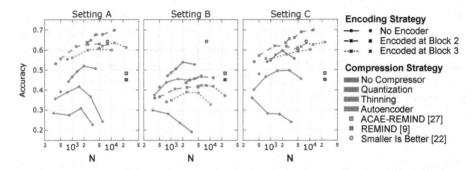

Fig. 6. Results for a varying compression parameter k and a fixed total memory of 1.536 MB, following [27]. The data for 'REMIND' and 'ACAE-REMIND' is sourced from [27]; the result for 'Smaller Is Better' was produced by the Code provided by [22]. Compared approaches are outlined.

is optimal. Exceptions include the quantization strategy in the first plot (where compression is always beneficial) and the thinning compressor in the second and fourth plots, where compression is always harmful. As previously discussed, the results show that encoding improves performance, as evidenced by the higher accuracy of the setups using FETCH.

6.5   Comparison with Other Approaches

We compare FETCH with the following state-of-the-art approaches: *REMIND* [9], freezes the early layers of a ResNet and performs product quantization on the resulting featuremaps, before storing them in memory. During continual learning, the data from the continual stream is mixed with samples from the memory. *ACAE-REMIND* [27] extends REMIND with an additional autoencoder to compress the samples even further. *'Smaller Is Better'*, the best-performing variant of an approach proposed in [22], involves resizing raw images to 8×8 pixels before storing them in the episodic memory. A network is retrained from scratch, whenever the input distribution changes. To the best of our knowledge, no method besides FETCH combines the benefits of freezing early layers of a convolutional neural network with compressed replay in the pipeline of GDumb.

We evaluate FETCH in three different settings: *Setting A*: the encoder is pre-trained on TinyImageNet. The classification head is initialized randomly. This approach was used in the other sections of this paper. *Setting B*: both the encoder and the classification head are pre-trained using the data from the first tasks (in the case of CIFAR10 the classes `Airplane` and `Automobile`). This setting is used by REMIND and ACAE-REMIND. *Setting C*: the encoder and classification head were pre-trained on TinyImageNet.

We vary the compression parameter and the layer up to which we keep the weights frozen. The memory is filled with N samples up to a maximum size of 1.536 MB. All compared methods use ResNet-18 and the CIFAR10 dataset. The results are shown in figure Fig. 6. Setting B shows that the simple quantization strategy performs similarly to REMIND and ACAE-REMIND, even outperforming both approaches for some configurations. Comparing settings A and B shows the positive influence of the diverse pre-training dataset. Setting A shows that pre-training enables FETCH to also outperform 'Smaller Is Better'. Our experiments in setting C show a positive effect of using in-distribution datasets for pre-training the classification head.

7   Conclusion

This work aimed at investigating the advantages of compressed replay in the context of GDumb. We evaluated the effect of different compression strategies as well as the effect of pre-training and freezing parts of the backbone. A combination of both techniques showed improved performance over both GDumb and selected compressed replay techniques in our experiments. These findings suggest that episodic memories with a large number of compressed exemplars and the transfer effects of pre-trained components benefit replay learning. However, FETCH has limitations, including the need to retrain the classification head whenever the data distribution changes. In future work, we wish to investigate the proposed two-step compression scheme in combination with other memory-based CL approaches outside computer vision. Although FETCH can be directly used in applications with limited memory, such as mobile robotics, this work intends to serve as a baseline for future research in the area of memory-constrained Continual Learning.

References

1. Ardywibowo, R., Huo, Z., Wang, Z., Mortazavi, B.J., Huang, S., Qian, X.: Vari-Grow: variational architecture growing for task-agnostic continual learning based on Bayesian novelty. In: Proceedings of the International Conference on Machine Learning (2022)
2. Bhat, P.S., Zonooz, B., Arani, E.: Consistency is the key to further mitigating catastrophic forgetting in continual learning. In: Proceedings of the Conference on Lifelong Learning Agents (CoLLAs) (2022)
3. Chen, T., Liu, S., Chang, S., Amini, L., Wang, Z.: Queried unlabeled data improves and robustifies class-incremental learning. Transactions on Machine Learning Research (TMLR) (2022)
4. Collins, J., Xu, K., Olshausen, B.A., Cheung, B.: Automatically inferring task context for continual learning. In: Conference on Cognitive Computational Neuroscience (CCN) (2019)
5. De Lange, M., et al.: A continual learning survey: defying forgetting in classification tasks. Trans. Pattern Anal. Mach. Intell. **44**(7), 3366–3385 (2022)
6. Gu, Y., Yang, X., Wei, K., Deng, C.: Not just selection, but exploration: online class-incremental continual learning via dual view consistency. In: Proceedings of the Conference on Computer Vision and Pattern Recognition (CVPR) (2022)
7. Guo, Y., Hu, W., Zhao, D., Liu, B.: Adaptive orthogonal projection for batch and online continual learning. In: Proceedings of the AAAI Conference on Artificial Intelligence (2022)

8. Hayes, T.L., Cahill, N.D., Kanan, C.: Memory efficient experience replay for streaming learning. In: International Conference on Robotics and Automation (ICRA) (2019)

9. Hayes, T.L., Kafle, K., Shrestha, R., Acharya, M., Kanan, C.: REMIND your neural network to prevent catastrophic forgetting. In: Vedaldi, A., Bischof, H., Brox, T., Frahm, J.-M. (eds.) ECCV 2020. LNCS, vol. 12353, pp. 466–483. Springer, Cham (2020). https://doi.org/10.1007/978-3-030-58598-3_28

10. He, K., Zhang, X., Ren, S., Sun, J.: Deep residual learning for image recognition. In: Proceedings of the Conference on Computer Vision and Pattern Recognition (CVPR) (2016)

11. Hinton, G.E., Salakhutdinov, R.R.: Reducing the dimensionality of data with neural networks. Science **313**(5786), 504–507 (2006)

12. Hossain, M.S., Saha, P., Chowdhury, T.F., Rahman, S., Rahman, F., Mohammed, N.: Rethinking task-incremental learning baselines. In: International Conference on Pattern Recognition (ICPR) (2022)

13. Krizhevsky, A., Hinton, G., et al.: Learning multiple layers of features from tiny images. University of Toronto, Technical report (2009)

14. Le, Y., Yang, X.: Tiny ImageNet visual recognition challenge. Technical report, Stanford University, CS231n (2015)

15. Loshchilov, I., Hutter, F.: SGDR: stochastic gradient descent with warm restarts. In: International Conference on Learning Representations (ICLR) (2017)

16. Mai, Z., Li, R., Jeong, J., Quispe, D., Kim, H., Sanner, S.: Online continual learning in image classification: an empirical survey. Neurocomputing **469**, 28–51 (2022)

17. Masana, M., Liu, X., Twardowski, B., Menta, M., Bagdanov, A.D., van de Weijer, J.: Class-incremental learning: survey and performance evaluation on image classification. Trans. Pattern Anal. Mach. Intell. **45**(05), 5513–5533 (2023)

18. Masci, J., Meier, U., Cireşan, D., Schmidhuber, J.: Stacked convolutional auto-encoders for hierarchical feature extraction. In: Honkela, T., Duch, W., Girolami, M., Kaski, S. (eds.) ICANN 2011. LNCS, vol. 6791, pp. 52–59. Springer, Heidelberg (2011). https://doi.org/10.1007/978-3-642-21735-7_7

19. Mehta, S.V., Patil, D., Chandar, S., Strubell, E.: An empirical investigation of the role of pre-training in lifelong learning. arXiv preprint: 2112.09153 (2021)

20. Parisi, G.I., Kemker, R., Part, J.L., Kanan, C., Wermter, S.: Continual lifelong learning with neural networks: a review. Neural Netw. **113**, 54–71 (2019)

21. Pawlak, S., Szatkowski, F., Bortkiewicz, M., Dubiński, J., Trzciński, T.: Progressive Latent replay for efficient Generative Rehearsal. In: Tanveer, M., Agarwal, S., Ozawa, S., Ekbal, A., Jatowt, A. (eds.) ICONIP 2022. Communications in Computer and Information Science, vol. 1791, pp. 457–467. Springer, Singapore (2023). https://doi.org/10.1007/978-981-99-1639-9_38

22. Pelosin, F., Torsello, A.: Smaller is better: an analysis of instance quantity/quality trade-off in rehearsal-based continual learning. In: International Joint Conference on Neural Networks (IJCNN) (2022)

23. Prabhu, A., Torr, P.H.S., Dokania, P.K.: GDumb: a simple approach that questions our progress in continual learning. In: Vedaldi, A., Bischof, H., Brox, T., Frahm, J.-M. (eds.) ECCV 2020. LNCS, vol. 12347, pp. 524–540. Springer, Cham (2020). https://doi.org/10.1007/978-3-030-58536-5_31

24. Robins, A.: Catastrophic forgetting, rehearsal and pseudorehearsal. Connect. Sci. **7**(2), 123–146 (1995)

25. Russakovsky, O., et al.: ImageNet large scale visual recognition challenge. Int. J. Comput. Vision **115**(3), 211–252 (2015). https://doi.org/10.1007/s11263-015-0816-y

26. Sangermano, M., Carta, A., Cossu, A., Bacciu, D.: Sample condensation in online continual learning. In: International Joint Conference on Neural Networks (IJCNN) (2022)
27. Wang, K., van de Weijer, J., Herranz, L.: ACAE-REMIND for online continual learning with compressed feature replay. Pattern Recogn. Lett. **150**, 122–129 (2021)
28. Wang, L., et al.: Memory replay with data compression for continual learning. In: International Conference on Learning Representations (ICLR) (2022)
29. Yun, S., Han, D., Oh, S.J., Chun, S., Choe, J., Yoo, Y.: CutMix: regularization strategy to train strong classifiers with localizable features. In: International Conference on Computer Vision (ICCV) (2019)

Enhanced SVM-SMOTE with Cluster Consistency for Imbalanced Data Classification

Tajul Miftahushudur[1,2]([✉]) [ID], Halil Mertkan Sahin[2] [ID], Bruce Grieve[2] [ID], and Hujun Yin[2] [ID]

[1] National Research and Innovation Agency (BRIN), Bandung, Indonesia
tajul.miftahushudur@gmail.com
[2] Department of Electrical and Electronic Engineering,
The University of Manchester, Manchester M13 9PL, UK

Abstract. The issue of imbalanced data in machine learning has gained significant attention in recent years. Imbalanced data, where one class has significantly fewer samples than others, can lead to poor performance for machine learning models, especially in detecting minority class samples. To address this problem, various resampling techniques have been proposed, including the popular SMOTE (Synthetic Minority Over-sampling TEchnique). However, SMOTE suffers from the overlapping problem and may misclassify samples near the separation boundaries. This paper presents a novel framework to optimise border-based-SMOTEs, including Borderline-SMOTE and SVM-SMOTE which were specifically developed to solve the problem of misclassifying border samples. The proposed method ensures that generated samples improve the decision boundaries and are free from overlapping issues. The proposed method is evaluated on synthetic and real-world datasets, and results demonstrate its effectiveness in enhancing the performance of machine learning models, particularly in classifying minority class samples.

Keywords: Imbalance dataset · Oversampling · Borderline · Suport Vector Machine · SMOTE

1 Introduction

Machine learning models require sufficient and balanced training data. However, real-world datasets often exhibit an inherent imbalance or it is challenging to obtain sufficient samples. Consequently, this issue has gained increased attention within the machine learning community. The presence of limited, rare or difficult-to-collect samples is a natural and unavoidable challenge in various real-world applications, such as plant health monitoring [17], medical diagnosis [16], fraud detection [4], oil spill detection [10], and manufacturing quality control [7].

Training a machine learning model with imbalanced datasets can negatively impact the model's performance, particularly in detecting minority class samples. The dominance of majority class samples during training can suppress the

© The Author(s), under exclusive license to Springer Nature Switzerland AG 2023
P. Quaresma et al. (Eds.): IDEAL 2023, LNCS 14404, pp. 431–441, 2023.
https://doi.org/10.1007/978-3-031-48232-8_39

model's ability to learn from and represent the minority class accurately [19]. As a result, the model may tend to classify samples into the majority class, making more minority samples misclassified than the majority samples. This issue can be crucial in applications like health monitoring or medical diagnosis, where the sensitivity of the machine learning model to detect minority class samples is more critical than identifying majority class samples.

Several approaches have been developed to solve the imbalance class problem, and they can be categorised into three groups: algorithm-level, data-level, and hybrid-level [9]. Among these three approaches, data-level is the most popular due to its simplicity and effectiveness. The data-level approaches attempt to reduce the imbalance degree through two sampling methods: undersampling and oversampling. Undersampling aims to decrease the number of majority samples by removing some of them. Unfortunately, such method has the potential of losing significant training samples. On the other hand, an oversampling method works by increasing the number of minority samples by generating synthetic data. SMOTE [2] is one of the most popular oversampling methods due to its ability to generate new instances and low chance of overfitting. However, the technique has the potential overlap problem that can lead to incorrect or bad decision boundaries. In addition, some machine learning algorithms such as Support Vector Machine (SVM) may have low sensitivity in detecting minority samples located near the majority class because these samples are more prone to misclassification than those located far from the majority class [13]. To solve this sensitivity issue, some modifications of SMOTE have been developed, such as Borderline-SMOTE [6] and SVM-SMOTE [13]. Unfortunately, these techniques are still vulnerable to the overlap problem, as they do not consider the existence of the majority class in generating new instances. In this paper we propose a new approach to optimise SVM-SMOTE. The approach aims to ensure that the generated samples have a positive impact on improving the decision boundary by making these samples free from the overlap problem and located as close as possible to the support vectors of the majority class.

The remainder of this paper is structured as follows: Sect. 2 provides a review of the existing literature on imbalanced data handling, highlighting the key features and challenges of the border-based oversampling techniques. Section 3 introduces the proposed method with details of the novel approach for improving SMOTE's performance. Section 4 presents extensive experimental evaluations and comparisons with existing solutions on various real-world and synthetic datasets. Finally, Sect. 5 concludes the paper and suggests potential directions for future research.

2   Border-Based SMOTE

There are many solutions to handle the imbalanced data problem, and the most popular technique is oversampling, which aims to add new instances on the minority class to balance the training set. The easiest oversampling technique is Random Oversampling (ROS), which tries to duplicate the minority samples. Unfortunately, ROS may potentially have an overfitting problem due to the excessive duplications of data. To resolve this problem, Chawla et al. [2] developed SMOTE, which works by randomly generating samples between two neighbourhood data samples. It becomes the most popular oversampling technique as it can intelligently generate artificial samples that are different from the original samples, hence reducing the overfitting problem while improving the classifier performance.

Despite the advantages of the SMOTE in dealing with imbalanced data problems, it also has several drawbacks, such as overlapping data problems and potentially amplifying noisy data [18], particularly when outliers are involved. The overlapping data problems may potentially occur when majority and minority classes are closely located and SMOTE may produce synthetic (minority) instances that fall into the majority class cluster. As a result of this problem, when the instances of different classes are intermingled, it can be difficult to define clear decision boundaries, making it difficult for the model to accurately separate instances of different classes.

To obtain optimal classification results, many algorithms, including SVMs, aim to learn data patterns to create a border that can optimally separate different classes. Samples located closer to the decision boundary are more likely to be misclassified compared to samples that are further away from the boundary [13]. To solve this problem and reduce the number of misclassified samples in the border area, several SMOTE modifications have been developed such as Borderline-SMOTE and SVM-SMOTE. These two SMOTE variants are designed to enhance the sensitivity of machine learning models towards minority samples in the border area of imbalanced datasets. Generally, both techniques operate by identifying minority samples located in the border area as reference points and generating synthetic samples around these reference samples. The key difference between Borderline-SMOTE and SVM-SMOTE lies in the approach used to select the reference samples. Borderline-SMOTE considers a sample to be on the border if its m-closest neighbours are dominated by the majority class. Conversely, SVM-SMOTE employs the SVM algorithm to find the support vector samples of the minority class and then uses these samples as references to generate new instances. In Borderline-SMOTE, noise data may be considered as reference points, as they may be surrounded by majority samples. In contrast, the support vector samples in SVM-SMOTE are free of noise because they are located closest to the border separating the majority and minority classes.

3 Materials and Methods

3.1 Datasets

The datasets used in this work consist of three different types: artificial, real-world, and hyperspectral. The artificial datasets consist of two 2 dimensional (2D) datasets with unique distributions, as visualised in Fig. 1. For simplicity, these datasets are named as "Circular Dataset" and "Moon Dataset", respectively. Both artificial datasets consist of two classes, totalling 500 samples, with

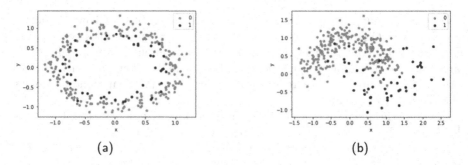

Fig. 1. Visualisation of Circular (a), and Moon (b) datasets.

400 samples belonging to the majority class and the remaining samples to the minority class.

The real-world datasets utilised in this experiment consist of several collections proposed in [5] and were obtained from the UCI repository [8]. The characteristics of these datasets are illustrated in Table 1.

Table 1. Summary of UCI datasets

Dataset	Imbalance Ratio	Number of samples	Number of features
Satimage	9.3 : 1	6435	36
Ecoli	8.6 : 1	336	7
US crime	12 : 1	1994	100
Oil	22 : 1	937	49
Letter img	26 : 1	20000	16
Libras mov	14 : 1	360	90
Webpage	13 : 1	34780	300
Mammography	34 : 1	2536	72

The hyspectral dataset is the Bonn-spec dataset [11,12]. This dataset contains spectral reflectance data extracted using 462 spectral bands distributed between 400 and 1050 nm from several sugar leaves under two conditions: healthy and unhealthy/diseased (infected by Cercospora, Rust, and Powdery Mildew). It contains 2516 spectral samples, where 860 are healthy and 1886 unhealthy.

3.2 Proposed Method

Herein, an extension of the SVM-SMOTE algorithm is proposed to address the weaknesses of SMOTE and optimise the original SVM-SMOTE algorithm. Specifically, this extension aims to solve the overlapping data problem and increase the influence of the generated data. The proposed method consists of four main steps, as illustrated in Fig. 2.

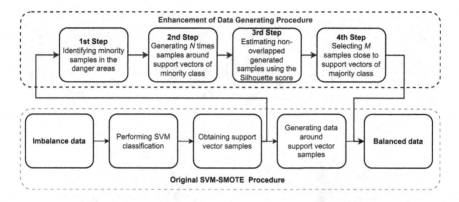

Fig. 2. Flowchart of proposed method.

The first step is to identify support vector samples from the minority class as reference points to generate synthetic data then determine their location around neighbouring samples. These locations are classified into two types of conditions: danger area and safe area. The danger area is where when the neighbouring samples of the reference points are dominated by the majority samples, whereas the safe area is dominated by intra-class samples (minority class). While generating data in the safe area is straightforward, this algorithm focuses on generating minority samples in the danger area, so to minimise mis-classification.

The second step involves the generation of artificial samples around the danger area. N times the required number of samples needed to balance the data are generated in the danger area using the original SVM-SMOTE algorithm. For instance, if M samples are needed to balance the data, then $M x N$ artificial samples will be produced during this step.

The third step aims to eliminate overlapping or mislabelled synthetic samples. These samples are estimated using the Silhouette score [15]. If the Silhouette score of an instance is negative or 0, it indicates that the instance is in overlap or mislabelled condition. Consequently, the instance is removed to avoid potential overlapping issues.

Lastly, support vector samples of the majority class or inter-class samples are located, and then M samples close to these support vector samples are identified.

In the experiments, shown in the next section, SVM was used as the main classifier to obtain the results and for fair comparisons. Also, the RBF kernel was selected due to its capability to provide a non-linear hyperplane and its low number of parameters.

4   Experimental Results and Discussion

It is important to note, as our study focuses on the imbalance classification problem, the measurement of accuracy metrics is not reliable for assessing classification performance due to its potential bias. For example, if the model correctly

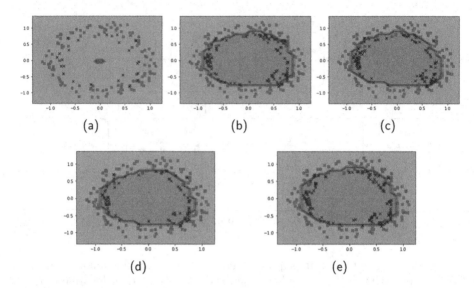

Fig. 3. Decision boundary illustration in Circular dataset when (a) without over-sampling, (b) SMOTE, (c) Borderline-SMOTE, (d) SVM-SMOTE, and (e) proposed method were used to train the SVM model.

classifies all majority samples but fails to classify all the minority samples, the accuracy metrics may yield a high result, which does not reflect the actual performance of the model, as it performs poorly on minority samples. Therefore, F measurement and Matthews Correlation Coefficient (MCC) scores were used as more appropriate measures for evaluating imbalance classification performance [1,3].

4.1 Artificial Datasets

The Circular dataset was divided with a proportion of 150 samples used as the training set, where 125 samples were of class 0 (majority class) and 25 samples were of class 1 (minority class). The remaining 350 samples were allocated to the testing set. For the Moon dataset, 90 samples were chosen as the training set, consisting of 75 samples of class 0 and 15 samples of class 1. The remaining 410 samples were used as the testing set.

Figure 3 and Fig. 4 illustrate how the decision boundary was formed using several over-sampling techniques: SMOTE, Borderline SMOTE, SVM-SMOTE, and the proposed work. For the circular dataset, it can be seen that the original dataset had a decision boundary that was heavily biased towards the majority class when SVM was used as the classifier. This bias was due to the dominance of majority class samples. Consequently, the classifier fails to classify any minority samples, as the decision boundary tends to classify samples as the majority class. This resulted in an MCC score of 0. As seen from the results in Fig. 3, the implementations of oversampling techniques effectively adjusted the deci-

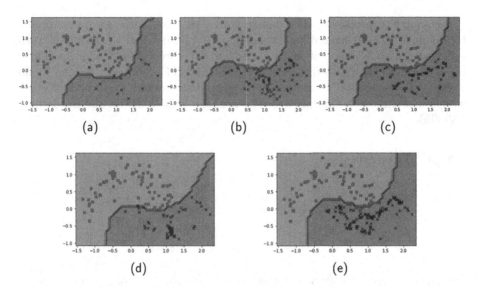

Fig. 4. Decision boundary illustration in Moon dataset when (a) without oversampling, (b) SMOTE, (c) Borderline-SMOTE, (d) SVM-SMOTE, and (e) proposed method were used to train the SVM model.

sion boundary to cover the minority samples and correctly classify them. Also, the MCC score significantly increased to 53.25% using SMOTE, 55.13% with Borderline-SMOTE, 55.68% with SVM-SMOTE, and the proposed work produced the highest MCC score of 56.04%.

For the Moon dataset. Without oversampling, SVM with the original imbalanced dataset produced an MCC score of 70.53%. However, over-sampling with SMOTE decreased slightly the MCC score to 69.14%. On the other hand, balancing the dataset with Borderline-SMOTE and SVM-SMOTE resulted in a small increase in the MCC score to 71.44% and 72.41%, respectively. The proposed work achieved the highest MCC score of 74.58%. To obtain more precise results, the experiments were repeated ten times and the results are provided in Table 2.

Table 2. Classification results on artificial datasets

Datasets	RAW		SMOTE		Borderline SMOTE		SVM SMOTE		Improved SVM SMOTE	
	MCC	F1	MCC	F1	MCC	F1	MCC	F1	MCC	F1
Circular	0	75.76	57.91	86.01	59.96	86.35	58.50	87.29	**60.91**	**89.26**
	±0	±0	±5.90	±1.60	±5.43	±2.12	±6.89	±2.21	**±4.61**	**±1.29**
Moon	79.07	94.15	80.57	94.07	80.98	94.24	81.28	94.46	**83.08**	**94.87**
	±4.05	±1.08	±3.71	±1.30	±0.68	±0.31	±4.08	±1.26	**±2.30**	**±0.83**

Overall, the proposed work demonstrated the most effective oversampling technique for the artificial datasets. For the Circular dataset, the proposed work achieved an MCC score of 60.91% and an F1 score of 89.25%, outperforming SVM-SMOTE, Borderline-SMOTE, and SMOTE. Similarly, for the Moon dataset, the highest MCC and F1 scores were observed with the proposed work, while the original SMOTE had a lower score than SVM-SMOTE and Borderline-SMOTE.

4.2   UCI Datasets

Table 3. Classification results on UCI datasets

Datasets	RAW		SMOTE		Borderline -SMOTE		SVM- SMOTE		Improved SVM-SMOTE	
	MCC	F1	MCC	F1	MCC	F1	MCC	F1	MCC	F1
Satimage	1.57	85.70	54.14	88.63	51.59	86.72	53.03	87.70	**56.62**	**91.66**
	±0.14	±0.07	±0.97	±0.53	±0.64	±0.34	±0.57	±0.27	**±0.93**	**±0.14**
Ecoli	40.88	89.36	59.43	90.27	59.97	90.62	59.37	90.67	**66.16**	**92.57**
	±6.48	±2.25	±3.56	±0.95	±4.05	±1.12	±3.70	±1.31	**±5.95**	**±2.22**
US crime	37.81	91.74	47.46	90.96	48.40	91.56	49.38	91.69	**50.61**	**92.82**
	±3.36	±0.31	±2.80	±0.55	±1.87	±0.34	±2.33	0.26	**±2.57**	**±0.23**
Oil	1.57	93.44	5.72	73.36	4.77	77.62	8.58	91.65	**10.01**	**93.04**
	±0.13	±0.05	±0.55	±0.26	±0.72	±0.23	±0.16	±0.35	**±0.25**	**±0.52**
Letter img.	81.99	98.72	85.62	98.86	86.41	98.95	83.65	98.66	**90.33**	**99.32**
	±0.92	±0.07	±0.96	±0.08	±0.59	±0.06	±1.04	±0.10	**±0.60**	**±0.04**
Libras mov.	64.28	95.49	82.78	97.80	83.07	97.81	83.07	97.81	**85.16**	**98.11**
	±6.09	±0.77	±3.60	±0.46	±3.99	±0.47	±3.99	±0.47	**±4.56**	**±0.56**
Webpage	64.70	98.09	57.12	97.50	57.25	97.56	59.87	97.66	**64.51**	**98.12**
	±0.92	±0.04	±1.74	±0.15	±1.96	0.17	±2.41	±0.21	**±1.38**	**±0.08**
Mammography	56.80	98.14	44.58	95.44	51.59	96.76	51.93	96.75	**64.71**	**98.42**
	± 1.14	±0.04	±2.01	±0.56	±3.21	±0.58	±4.18	±0.65	**±1.98**	**±0.11**

The *train test split* function from Scikit-learn [14] was utilised to randomly partition 30% of the total samples as the training set, while the remaining 70% as a testing set on each dataset in this study. The classification performances on the UCI datasets are presented in Table 3. Evaluation of all the datasets agreed that the proposed method provided the best classification performance in MCC and F1 score measures, followed by SVM-SMOTE, Borderline-SMOTE, and lastly the original SMOTE.

The proposed work showed a significant improvement in the Mammography dataset, achieving MCC scores up to 12%, 13% and 20% higher than SVM-SMOTE, Borderline-SMOTE, and SMOTE, respectively.

Interestingly, for the Mammography dataset, the original imbalanced data exhibited better performance than the implementation of SMOTE, Borderline-SMOTE, and SVM-SMOTE. One of the key reasons for these results is the significant presence of overlapping data generated by them. Nonetheless, the proposed method has effectively addressed this issue, hence resulting in a remarkable improvement in performance over the original imbalanced data, with nearly 10% higher accuracy.

4.3   Hyperspectral Dataset

Of the 2516 samples on the Bonn-spec dataset, 60% of the samples were used for training, consisting of 1131 unhealthy and 378 healthy samples, and the rest 40% for testing.

Classification results on the Bonn-spec dataset presented in Table 4. In general, the experimental results on the Bonn-spec dataset conform to the findings of the previous experiments, demonstrating that the proposed improvement of SVM-SMOTE produced better classification results than other oversampling techniques. The proposed method produced an MCC score of 44.33% and an F1 score of 76.77%, while the SVM-SMOTE only observed an MCC score of 41.04% and an F1 score of 65.76%, making it the second best.

Table 4. Classification results on Bonn-Spec dataset

Datasets	RAW		SMOTE		Borderline -SMOTE		SVM- SMOTE		Improved SVM-SMOTE	
	MCC	F1	MCC	F1	MCC	F1	MCC	F1	MCC	F1
Bonn-Spec	0.00 ±0.00	64.23 ±0.00	39.44 ±1.09	61.39 ±1.86	34.54 ±1.14	52.55 ±2.21	41.04 ±1.12	65.79 ±1.92	**44.33** ±1.84	**76.78** ±1.03

5   Conclusions and Future Work

In this work, a framework to enhance the performance of Border-based SMOTE was introduced. Specifically, the focus is on improving SVM-SMOTE to avoid overlapping data problems and reduce misclassified samples in the border area. The proposed framework consists of four steps. First, minority samples located in the border area are identified using the SVM algorithm. Next, a particular number of synthetic samples are generated using the SMOTE protocol. Then, any overlapping artificial samples are eliminated in based on their Silhouette scores. Finally, a specific number of samples that are closest to the support vector samples of the majority class are selected. The experimental results on various types of datasets, both artificial and real-world, show that the proposed work may enhance the performance of the machine learning model, especially in detecting minority class samples, compared to previous oversampling techniques that have existed. The proposed method may have two limitations. It can be sensitive to the parameter settings during locating the border samples and requires more time to process the steps in the enhancement of the data generating procedure. Therefore, future work will explore how to minimise its dependency and complexity.

Acknowledgment. Tajul Miftahushudur would like to acknowledge the Scholarship provided by the Indonesian Endowment Fund for Education (LPDP). Halil Mertkan Sahin would like to acknowledge the Scholarship provided by the Ministry of National Education of the Republic of Türkiye.

References

1. Baldi, P., Brunak, S., Chauvin, Y., Andersen, C.A.F., Nielsen, H.: Assessing the accuracy of prediction algorithms for classification: an overview. Bioinformatics **16**(5), 412–424 (2000). https://doi.org/10.1093/bioinformatics/16.5.412
2. Chawla, N.V., Bowyer, K.W., Hall, L.O., Kegelmeyer, W.P.: SMOTE: synthetic minority over-sampling technique. J. Artif. Intell. Res. **16**(1), 321–357 (2002). https://doi.org/10.1613/jair.953
3. Chicco, D., Jurman, G.: The advantages of the Matthews correlation coefficient (MCC) over F1 score and accuracy in binary classification evaluation. BMC Genomics (2020). https://doi.org/10.1186/s12864-019-6413-7
4. Dal Pozzolo, A., Boracchi, G., Caelen, O., Alippi, C., Bontempi, G.: Credit card fraud detection: a realistic modeling and a novel learning strategy. IEEE Trans. Neural Netw. Learn. Syst. **29**(8), 3784–3797 (2018). https://doi.org/10.1109/TNNLS.2017.2736643
5. Ding, Z.: Diversified ensemble classifiers for highly imbalanced data learning and its application in bioinformatics. Ph.D. thesis, USA (2011). aAI3486649
6. Han, H., Wang, W.-Y., Mao, B.-H.: Borderline-SMOTE: a new over-sampling method in imbalanced data sets learning. In: Huang, D.-S., Zhang, X.-P., Huang, G.-B. (eds.) ICIC 2005. LNCS, vol. 3644, pp. 878–887. Springer, Heidelberg (2005). https://doi.org/10.1007/11538059_91
7. He, Q., Pang, Y., Jiang, G., Xie, P.: A spatio-temporal multiscale neural network approach for wind turbine fault diagnosis with imbalanced SCADA data. IEEE Trans. Ind. Inf. **17**(10), 6875–6884 (2021). https://doi.org/10.1109/TII.2020.3041114
8. Kelly, M., Longjohn, R., Nottingham, K.: The UCI machine learning repository (2023). https://archive.ics.uci.edu
9. Krawczyk, B.: Learning from imbalanced data: open challenges and future directions. Prog. Artif. Intell. **5**(4), 221–232 (2016). https://doi.org/10.1007/s13748-016-0094-0
10. Ma, Y., Zeng, K., Zhao, C., Ding, X., He, M.: Feature selection and classification of oil spills in SAR image based on statistics and artificial neural network. In: 2014 IEEE Geoscience and Remote Sensing Symposium, pp. 569–571 (2014). https://doi.org/10.1109/IGARSS.2014.6946486
11. Mahlein, A.K., et al.: Development of spectral indices for detecting and identifying plant diseases. Remote Sens. Environ. **128**, 21–30 (2013). https://doi.org/10.1016/j.rse.2012.09.019

12. Mahlein, A.K., Steiner, U., Dehne, H.W., Oerke, E.C.: Spectral signatures of sugar beet leaves for the detection and differentiation of diseases. Precis. Agric. **11**(4), 413–431 (2010). https://doi.org/10.1007/s11119-010-9180-7

13. Nguyen, H.M., Cooper, E.W., Kamei, K.: Borderline over-sampling for imbalanced data classification. Int. J. Knowl. Eng. Soft Data Paradigm. **3**(1), 4–21 (2011). https://doi.org/10.1504/IJKESDP.2011.039875

14. Pedregosa, F., et al.: Scikit-learn: machine learning in Python. J. Mach. Learn. Res. **12**, 2825–2830 (2011)

15. Rousseeuw, P.J.: Silhouettes: A graphical aid to the interpretation and validation of cluster analysis. J. Comput. Appl. Math. **20**, 53–65 (1987). https://doi.org/10.1016/0377-0427(87)90125-7. https://www.sciencedirect.com/science/article/pii/0377042787901257

16. Roychowdhury, S., Koozekanani, D.D., Parhi, K.K.: DREAM: diabetic retinopathy analysis using machine learning. IEEE J. Biomed. Health Inform. **18**(5), 1717–1728 (2014). https://doi.org/10.1109/JBHI.2013.2294635

17. Sambasivam, G., Opiyo, G.D.: A predictive machine learning application in agriculture: Cassava disease detection and classification with imbalanced dataset using convolutional neural networks. Egypt. Inform. J. **22**(1), 27–34 (2021). https://doi.org/10.1016/j.eij.2020.02.007

18. Siriseriwan, W., Sinapiromsaran, K.S.: Adaptive neighbor synthetic minority oversampling technique under 1nn outcast handling. Songklanakarin J. Sci. Technol. **39**, 565–576 (2017). https://doi.org/10.14456/sjst-psu.2017.70

19. Zheng, M., Wang, F., Hu, X., Miao, Y., Cao, H., Tang, M.: A method for analyzing the performance impact of imbalanced binary data on machine learning models. Axioms 11(11), 607 (2022). https://doi.org/10.3390/axioms11110607. https://www.mdpi.com/2075-1680/11/11/607

Preliminary Study on Unexploded Ordnance Classification in Underwater Environment Based on the Raw Magnetometry Data

Marcin Blachnik[1]([⊠]) [iD], Piotr Ściegienka[1,2] [iD], and Daniel Dąbrowski[1,2]

[1] Department of Industrial Informatics, Silesian University of Technology, 44-100 Gliwice, Poland
{marcin.blachnik,piotr.sciegienka}@polsl.pl
[2] SR Robotics Sp. z o.o., Lwowska 38, 40-389 Katowice, Poland
daniel.dabrowski@srrobotics.pl

Abstract. Unexploded ordnance (UXO) dumped in water reservoirs pose a serious environmental and human safety hazard. Various ways of economically solving this problem are being sought. One of them is the use of machine learning methods for the automatic classification of dangerous objects based on the recorded signals. The paper presents the preliminary results on the use of machine learning methods applied to raw magnetometry data generated in a virtual environment based on the concept of a digital twin. This introduces a different approach to a standard approach, which is based on the inverse problem, where the signals are mapped to the magnetic dipole model. Conducted research points out that the highest performance can be obtained with neural networks, and a direct classification based on the raw signals allows to achieve accuracy of up to 93% when no remanent magnetization is present.

Keywords: UXO · magnetometry data · machine learning · classification · digital twin

1 Introduction

One of the key challenges currently related to the protection of the aquatic environment as well as the possibility of carrying out investments in coastal areas are unexploded ordnance (UXO). These UXO objects are mainly a remnant of past armed conflicts. Currently used methods of unexploded ordnance detection require costly seabed research and verification of found ferromagnets by qualified divers-sappers. Due to the fact that in the Baltic Sea alone (mainly in the Gotland and Bornholm Depths, but also in many other unidentified places) around 40 thousand tons of war remnants have been sunk, their efficient identification is crucial. The main sensor used for this task is the magnetometer, which detects magnetic anomalies caused by metal projectiles, but also by other

P. Quaresma et al. (Eds.): IDEAL 2023, LNCS 14404, pp. 442–447, 2023.
https://doi.org/10.1007/978-3-031-48232-8_40

magnetic objects. A typical measuring system includes 2 or more magnetometers - they are mounted on a special wing and towed behind a ship or their carriers are underwater vehicles. However, it is not a trivial task to distinguish between unexploded ordnance (UXO) and non-hazardous objects (non-UXO) with a registered magnetometric signal.

Methods using machine learning to deal with the above problem are gaining more and more popularity and effectiveness [1]. There are two approaches to that problem, one is based on the dipole model, which assumes that the signal of the UXO object is similar to the simple magnetic dipole, then the inverse transform of the recorded signals are mapped on the dipole and the parameters describing the dipole shape are extracted (the so called inverse problem) [3,5] and constitutes the training set. Another approach is based on the direct processing and classification of magnetic signals [4]. This approach however requires a large training set which is difficult to obtain. In this article we analyze preliminary results based on the second approach where the training data is obtained from the virtual model (a digital twin of UXO and nonUXO objects) described in [2]. Although, in this article, we use a simplified model which assumes no remanent magnetization. The evaluation of the dataset is performed using standard machine learning classifiers such as kNN, Random Forest, SVM, and MLP neural network [6] with a feature extraction procedure.

The article starts with a short description of the virtual environment and the dataset generation process, then the conducted experiments are described including the data preprocessing mechanism. The last section concludes the obtained results and draws future research directions.

2   The Virtual Environment and the Dataset

One of the key elements of building a machine learning system is appropriate training data. The training data $\mathbf{T}$ is a set of n pairs $(\mathbf{x}, y)$, such that $\mathbf{T} = [(\mathbf{x}_1, y_1), (\mathbf{x}_2, y_2), \ldots, (\mathbf{x}_n, y_n)]$, where $\mathbf{x} \in \Re^m$ and $y \in [c_1, c_2, \ldots, c_k]$, and k is the number of classes sample $\mathbf{x}$ can belong to. In machine learning a well-known rule is the more training data is delivered to the input of the model the more knowledge it gains. But for UXO identification one of the great challenges is data collection, which especially in the underwater environment can be very cost and labour inefficient. It must be conducted by a sapper diver who can work only for few hours per day due to the health consequences. In order to overcome this issue we created a virtual environment that allows us for easy data collection. The virtual environment and its properties were described in detail in [2]. It is based on the finite element method (FEM) where the influence of a 3D object on Earth's magnetic field is simulated. In the conducted experiments we simplified the model assuming no remanent magnetization and by defining UXO objects as:

- cylindrical / pipe-shaped objects with dimensions: diameter (corresponding to the caliber of the projectile) is in the range of 80 mm to 200 mm, and length between 350 mm and 750 mm.

and respectively by defining nonUXO objects using 4 numerical models:

- objects too small to be dangerous (marked as toSmall), these are usually objects resembling UXOs, but with a slightly smaller diameter compared to UXOs, i.e. 45 mm to 70 mm in diameter and 150 mm to 300 mm in length,
- UXO-shaped objects, however, too large (marked as toBig), i.e. pipe-type objects with the following parameters: diameter 250 mm to 400 mm and length 850 mm to 1200 mm
- objects whose shape can be approximated as a flat cylinder (e.g. mushroom anchor), with dimensions: diameter 110 mm to 400 mm and thickness 20 mm
- objects whose shape can be approximated as a flat cuboid (e.g. Danforth anchor, metal sheet), with dimensions in the range of length 200 mm to 900 mm, width 50 mm to 200 mm, and thickness 50 mm to 200 mm

For each type of object, 100 random shape parameters (e.g. diameter, length etc.) were generated and numerical simulations were performed. Next, in each simulation, the full 3D distribution of the magnetic field was recorded (a meshed cube). Then, for each cube, we simulated the flow of a formation of 9 virtual magnetometer sensors. In real application these magnetometers could be mounted on Autonomous Underwater Vehicles (AUV) or on a "wing" pulled behind the ship or some other underwater vehicle which allows the movement of the sensors. Due to the fact that each time a different set of signals may be recorded depending on the angle between north-south and the direction of the flow of the formation (east-west or north-south), 30 simulations of the flow over the test object were made, where the value of the angle was chosen randomly. The developed model also allowed for inaccuracy in the magnetometer positioning system, by sampling the signal with 20 cm accuracy. The goal of implementing this inaccuracy was to simulate a real underwater scenario where precise positioning is a real challenge. When creating the training set, it was also assumed that the height of the passage over the tested object is in the range of 1 m to 3 m, and the individual vehicles in the formation can be spaced from 0.5 m to a maximum of 2 m.

Finally, the data set consisted of a simulation of the formation of nine magnetic sensors: $Sens_1, Sens_2, \cdots, Sens_9$, flowing over the tested object at a certain depth (Fig. 1). For each of the sensors three components of the magnetic field vector were recorded, i.e. the X, Y, and Z components, and the values were recorded over a distance of 10 m with a resolution of 10 cm. In the calculations, it was assumed that the center of the tested object was in the middle of the section in which the data was recorded, i.e. 5 m before the tested object + 5 m after the tested object. As a result, a data matrix was created which consisted of $n = 5_{\text{types of obj.}} \cdot 100_{\text{simula per obj.}} \cdot 30_{\text{sens. passing simula.}} = 15,000$ samples each describing all of the recorded signals of all magnetometers

3   Experiments

In the experiments for the identification of UXO objects, we decided to use standard machine learning models including kNN, SVM, Random Forest, and

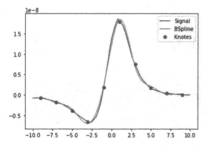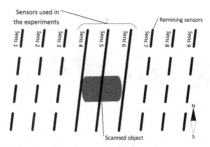

Fig. 1. Example of BSpline approx.   **Fig. 2.** UXO and scanning sensors

a fully connected MLP-typed network. A detailed description of the process is presented in the following subsection.

3.1 Dataset Processing and Evaluation Scheme

The evaluation process consisted of several steps. First, the dataset was filtered by removing unnecessary features. This includes meta features containing the name of the object, and redundant signals for simulated sensors included in the experiments. As introduced in [2] for each simulated object data for nine virtual sensors were recorded. These sensors form a formation of parallel moving magnetic sensors with a distance between each sensor of 0.5m. From all these collected data only 3 virtual sensors were selected for this experiment, these are sensors 4, 5, and 6. The remaining were removed from the experiment. The selection of only three virtual sensors derives from the fact that usually few, but more than one magnetometer sensor is used in real-life underwater scanning. The remaining, for example can be used to simulate scenarios when the object is scanned on the sides. These sensors included $Sens_5$ which passes directly over the scanned object and its adjacent sensors, which were 0.5m away on both sides, respectively $Sens_4$ and $Sens_6$ on the left and right side. Next, we calculated a module of the magnetic field for the X, Y, Z components for each sensor. After that, the feature space was very high. It consisted of $m = m_{len} \cdot m_{comp} \cdot m_{sens} + m_{meta}$ features, where $m_{len} = 101$ is the number of samples recorded for a single sensor, $m_{comp} = 4$ is the number of recorded components (X, Y, Z, and additionally M - the module of the X, Y, Z components), $m_{sens.} = 3$ is the number of magnetometer sensors considered in the experiments, and $m_{meta} = 2$ is the number of meta-features such as angle or height over the object. In total, it resulted in $m = 1,214$ feature space. In order to reduce the feature space we used BSplines to encode the signals by 10 knots and a polynomial of order 3. A signal and its approximation are shown in Fig. 2. This allowed for the reduction of the feature space by almost 1/3.

Next, a cross-validation procedure was executed. Within, first the data was normalized, here we normalized within each component of the signal of each sensor such that its maximum was equal to 1 and the minimum was equal to 0. On that data, we trained the prediction model. The test set was pro-

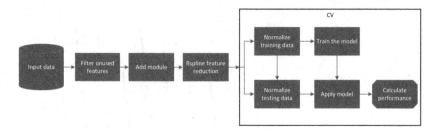

Fig. 3. The process of model evaluation

cessed accordingly, first, the test data was normalized based on the values obtained from the training stage, and then the prediction model was applied. In the experiments, we used the F1 measure which is more usable for unbalanced datasets (only 1/5 of the data belong to class UXO, and the remaining data belong to the nonUXO class). Model parameters were evaluated within the internal CV process where the grid search was used. For kNN we evaluated $k = [1, 3, 5, 7, 9, 15]$, for Random Forest $\#trees = [50, 100, 200]$ and depth $d = [5, 7, 9]$, for SVM $C = [0.01, 0.1, 1, 10, 100]$ and $\gamma = auto$, and for the MLP the following network structures (number of units in hidden layers) $S = [(10, 10), (100, 10), (1000, 10), (1000,)]$. The process is presented in Fig. 3. All of the experiments were implemented using Python and the scikit-learn package.

3.2 Results

The results obtained in the experiments - its average accuracy and standard deviation for each of the evaluated models are presented in Table 1 and visualized in Fig. 4.

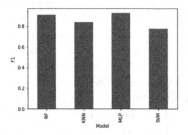

Fig. 4. Obtained performance of the evaluated classifiers

Table 1. Classification performance for different classifiers

Model	F1 Score	Std
RF	0.9114	0.0173
KNN	0.8403	0.0255
MLP	0.9280	0.0091
SVM	0.7736	0.0235

These results indicate that the MLP neural network obtained the highest performance of the evaluated models. Slightly worse results were obtained for the random forest classifier, then the kNN took third place and surprisingly the lowest performance was obtained for the SVM. An important benefit of the random forest classifier is the speed of the training and prediction process. In comparison the neural network was much slower, however, due to its best prediction performance, our work will go towards further development and design of a dedicated neural network structure involving various types of convolutional neural nets.

4 Conclusions

Within the conducted experiments we presented our preliminary research on UXO vs nonUXO classification using magnetometry data obtained using a concept of a digital twin embedded within a virtual environment. These experiments were simplified, for example, no remanent magnetization was assumed, but they were conducted in order to verify and acknowledge whether raw signals from the magnetometers can be used as an input to the machine learning models. Up to now, the most commonly used approach for UXO classification was based on adapting the recorded signals to the magnetic dipole model, and based on the dipole parameters the object was classified as UXO or nonUXO object. The results indicate that the use of raw signals is the right approach. Additionally, it is appropriate to conduct in-depth research on designing a dedicated neural network architecture for the UXO-specific task. Although, it must be noted that considering initial object magnetization may influence the classification performance because it changes the shape of the signals so that two identical objects can have significantly different magnetic images. The second aspect not covered within conducted research is the influence of the sensor position over the object. Here it was assumed that the magnetometer sensor located in the middle of the formation moves directly over the object.

References

1. Deschaine, L.M., et al.: Using machine learning to complement and extend the accuracy of UXO discrimination beyond the best reported results of the Jefferson proving ground technology demonstration. SIMULATION SERIES **34**, 46–52 (2002)
2. Blachnik, M., Przyłucki, R., Golak, S., Ściegienka, P., Wieczorek, T.: On the development of a digital twin for underwater UXO detection using magnetometer-based data in application for the training set generation for machine learning models. Sensors **23**, 6806 (2023). https://doi.org/10.3390/s23156806
3. Furey, J.S., Butler, D.K.: The physical dipole model and polarizability for magnetostatic object parameter estimation. J. Environ. Eng. Geophys. **16**, 49–60 (2011)
4. Bray, M.P., Link, C.A.: Learning machine identification of ferromagnetic UXO using magnetometry. IEEE J. Sel. Top. Appl. Earth Observations Remote Sens. 8(2), 835–844 (2015). https://doi.org/10.1109/JSTARS.2014.2362920
5. Wigh, M.D., Hansen, T.M., Døssing, A.: Inference of unexploded ordnance (UXO) by probabilistic inversion of magnetic data. Geophys. J. Int. **220**(1), 37–58 (2020)
6. Murphy, K.P.: Probabilistic Machine Learning: An Introduction. MIT Press, Cambridge (2022)

Efficient Model for Probabilistic Web Resources Under Uncertainty

Asma Omri[1(✉)], Djamal Benslimane[2], and Mohamed Nazih Omri[1]

[1] MARS Research Laboratory, University of Sousse, Sousse, Tunisia
omri.asmaaa@gmail.com, mohamednazih.omri@eniso.u-sousse.tn
[2] University Claude Bernard Lyon 1, LIRIS Research Laboratory,
Villeurbanne, France
djamal.benslimane@univ-lyon1.fr

Abstract. Individuals, organizations, and connected objects produce and publish vast data on the web. These data are then combined into mashups to produce new valuable data. However, combining data from different sources may lead to data uncertainty because these sources contain heterogeneous, contradictory, or incomplete information. In expert and intelligent systems, the word "uncertainty" is related to working with inaccurate data, imprecise and incomplete information, and unreliability of results that can lead to irrational decisions. An expert and intelligent system can consider all the uncertain information, and manage web services and web resources based on the best possible answer rather than on the quality of the exact answer to manage such problems. This paper proposes a new model to define, compute, and interpret uncertain web resources in the context of classical hypertext navigation and evaluation of data queries. The experimental study as well as the analysis of the results, which we carried out on different corpora, confirmed the usefulness of our approach. It provides the effect of the treatment of uncertainty on the execution time and the importance of calculating the confidence level of the response returned to the user. The analysis of the IT execution time shows that the time required for the composition of the services by our approach is negligible, as compared to that of the other approach studied. The impact of the variation in the number of nodes of type "mux" and type "ind" were also evaluated. Our algorithm checks all possibilities in polynomial time and can adapt to many possibilities of multiplexing values.

Keywords: Information Management · Web Resources · Web Application · Uncertainty · Probabilistic Approach

1 Introduction

Today's modern information systems enable individuals, connected objects, and organizations to produce and publish vast amounts of data on the Web through Web Application Programming Interface (APIs) and public endpoints [1]. The data are then collected from different sources and combined into mashups [2] to

© The Author(s), under exclusive license to Springer Nature Switzerland AG 2023
P. Quaresma et al. (Eds.): IDEAL 2023, LNCS 14404, pp. 448–457, 2023.
https://doi.org/10.1007/978-3-031-48232-8_41

generate added value. However, the integration process raises several problems because the data can be derived from heterogeneous, contradictory or incomplete sources [3].

The data extracted from the Web are full of uncertainties. In fact, they may contain contradictions resulting from inherently uncertain processes, such as data integration or retrieval. By Examining the sources of uncertainty - such as slight differences in the measurement of physical events or differences in information provided by different data sources describing the same entity - may help to understand its nature [6,7]. However, the main challenge of today's Web is to provide a solution to deal with this uncertainty. Instead of arbitrarily choosing a single version of information, we believe that the users should have a full range of possibilities to describe an entity. The current state of the Web enables the users to select a single representation for an entity [8]. In this paper, we present the notion of uncertain probabilistic Web resources, how they can be used to represent the data of the Web, and the complexity of the query evaluation within these resources. Actual data (for example, the World Wide Web) is often uncertain due to the inherent uncertainty of the data itself or the data collection, integration and management processes [9]. The main objective of this paper is to propose a theoretical framework to describe, manipulate and evaluate uncertain data on the Web and an algorithm that can help answer the queries and determine the degree of uncertainty of each answer [10]. We also present a model that helps define and interpret uncertain Web resources. We define an interpretation model and an algebra to compute uncertainty in the context of classical hypertext navigation and in the context of evaluating data queries.

The rest of this paper is organized as follows. In Sect. 2, we present the concept of a probabilistic Web resource and define its semantics. In Sect. 3, we present the programmatic representation of our probabilistic model. We presents the proposed probabilistic JSON parser principle. Section 4 presents and details the proposed approach for the evaluation of requests. In this section, we also describe the interpretation of a composition that has uncertain. In Sect. 5, we present the results of experimentation and evaluation of our approaches. The eighth and final section concludes this paper by invoking the perspectives of the different contributions presented in this paper.

2 Uncertain Web Resources

2.1 Definition

In the context of uncertain data, the semantics of uncertain Web resources is the way a resource is represented on the Web. In our approach, uncertain Web resources adopt the possible semantics of the Web based on a possible world theory [4]. An uncertain resource has several possible representations that can potentially and individually be interpreted as true. These possibilities can be interpreted as a set of possible worlds (PW1, ..., PWn) with a probability value (PWi). We call them possible Webs where the data are considered certain. As mentioned above, the REST resource principle will be adopted.

Based on the Web resource definition, we propose a specific representation that helps define uncertain resources. An uncertain resource is represented by the following notation:

$$R^P = \{URI_R, \langle A_j, V_j \rangle\}$$

With

- A_j : *Attribute   name*
- $V_i : V | IND(\langle V_1, P_1 \rangle, ..., \langle V_k, P_k \rangle) | MUX(\langle V_1, P_1 \rangle, ..., \langle V_k, P_k \rangle)$
- $i \in [1, n]$.
- $V_1, ..., V_k$: *Type   value   : String|Number|Objet|Array|False|null*

In this equation, the expression $\langle A_i, V_i \rangle$ represents the set of possible representations of RP. Since several representations of a resource cannot coexist at the same URI, these representations are mutually exclusive, and then we have $P_i \in]0, 1]$. In the next section, we will introduce a new model for probabilistic Web resources.

2.2   Model of an Uncertain Web Resource

We propose a probabilistic resource model that can effectively represent uncertain Web resources. This model is based on the principle of probabilistic XML, especially on the ProTDB approach [5] which is a probabilistic XML model proposed by Nieman and Jagadish in 2002. This approach differs from those of previous research in the development of probabilistic relational systems in that it builds a probabilistic XML database. This design is motivated by application needs involving data not readily available for relational representation.

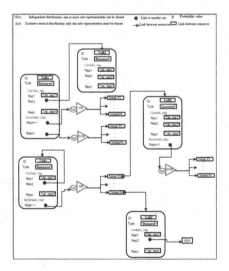

Fig. 1. General model of an uncertain resource.

Figure 1 shows our model for uncertain Web resources. We define the resource with a rounded rectangle in black each of which is identified by a unique URI, which is the identifier. The links are modeled through large dots and arrows in blue. These symbols in blue can indicate two different types of links: a link towards a new, certain or uncertain resource, either towards a "Mux" or "Ind" type operator. These two operators indicate the types of uncertainty. We will later explain the principle of these two nodes. As we have already explained, each resource can be defined by one or more representations. We can distinguish two types of representation: a certain representation represented by a simple green rectangle and an uncertain representation represented by a large blue dot. Finally, "P" represents the uncertainty value of a representation i.e. the probability value of having this value. The value of this probability P must belong to interval $]0, 1]$.

In our proposed approach to uncertain Web resources, we can distinguish two types of nodes. We should explain the principle of the two types of uncertainty mentioned above: "Mux" and "ind". We must also specify for each distributional node v, the probability distribution of choosing a subset of the children of v. We define two types of distribution nodes, each with a different way of describing this probability distribution. Figure 1 shows an example of an uncertain probabilistic Web resource that contains both types of nodes explained in the previous section.We note that the possible representations of Web resource scenario represented by Fig. 1 represent the different possible Webs in which the representation is certain. In each of these sites, the resources are used as conventional resources. It should be noteed that the interpretation of the probabilistic representation is completely independent of one resource to another. The associated model that we defined is the following: each resource is independent, but each URI identifies a unique resource that can have only one representation.

3   Programmatic Representation of Uncertain Resources

In order to provide a way to manage these uncertain resources, we have proposed a formalism to represent them physically. We propose a representation model where we provide all the possible representations of an uncertain resource. In this model, we give a probability to all these representations.

3.1   Parser JSON Probabilistic Model

As we have already presented, the probabilistic JSON model that we proposed considers 11 data types, namely a number, a probabilistic number, chain, probabilistic chain, Boolean, probabilistic Boolean, array, probabilistic array, object, probabilistic object and null. On the other hand, in the standard JSON model, only three types of data are taken into account, which are the scalar type (number, string, Boolean, null), the array type and the object type. Scalar types have just one value while arrays contain a list of arbitrary values. The objects constitute an unordered set of pairs (key, value), also called attributes different from those of XML, where the key is a string and the value can have an arbitrary type. Added types must also be modeled (Fig. 2).

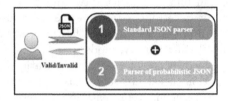

Fig. 2. Parser model JSON probabilistic.

In this part, we built a JSON parser using the parser library that we presented in the previous section. The parser is similar in nature to the tokenizer, except that it takes input tokens and pulls out the indices of the elements. As with the tokens, an element is marked by its position (starting index), its length and possibly its type of element. These numbers are stored in the same structure as the one used to store the tokens. The JSON grammar sets the correct format for JSON text. These are the rules we must follow to properly parse the JSON text. If the text does not follow the correct grammar, then, the parser has to throw an error message to indicate that the JSON text is impossible to parse.

Moreover, one must add the uncertain types that are the probabilistic table, probabilistic number, probabilistic Boolean, probabilistic String and probabilistic object. We also have to check the format of two *Mux* and *Ind* nodes and we have to calculate the sum of the probability values for the independent type of node and we have to check that this sum must be less than 1 (FIg. 3).

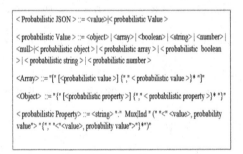

Fig. 3. Parser probabilistic JSON.

4   HTTP Request on Uncertain Resources

In order to manage uncertain Web resources, it is necessary for the client, who is querying an uncertain resource, to be able to understand this resource. We call an unsuspecting customer someone who can ask for and understand uncertain resources. Similarly, a client who does not know what an uncertain resource is should be able to work with this resource. In order to respect the principles of the Web and provide an opportunity to make the customer aware of the uncertainty, we rely on content negotiation to serve our uncertain resources.

Let R^P be an uncertain resource deployed on URI_{R^P}, we define the following representation expectations:

$$GET(URI_R) := Rep_R$$

$$GET^P(URI_{R^P}) := \langle A_i, V_i \rangle$$

With

- $V_i : V | IND(\langle V_1, P_1 \rangle, ..., \langle V_k, P_k \rangle) | MUX(\langle V_1, P_1 \rangle, ..., \langle V_k, P_k \rangle)$
- $i \in [1, n]$

Where

- $V_1, ..., V_k : type\ value\ String | Number | Objet | Array | False | null$

In our approach, GET^P does not define a new HTTP method since it is only a more complex notation of a standard GET with specific headers. The GET method acts as a standard GET with a specific HTTP header that we define as **Uncertainty-Processing: 1** and **Degree-Uncertainty: 0.2**. We choose to define a specific header to avoid any interference with the normalized use of the standard header. Indeed, the **"Uncertainty-Processing: 1"** header is used to inform the user that we can accept the uncertain Web resource processing and that the response we send to it is not certain. On the other hand, the heading **"Degree-Uncertainty: 0.2"** specifies the degree of uncertainty or accuracy of the sent response. It is a good practice to specify an ad hoc specific header to comply with HTTP standards.

4.1   Query Evaluation: Algorithm 1

In this section, we will present the simplest algorithm principle. It is the case where the query will return a whole JSON file. The principle of this method is presented by the following Fig. 4:

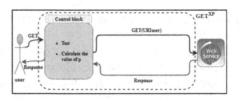

Fig. 4. Principle of the method GET 1.

This method is composed mainly of two layers: a user' layer and a server's layer. In the first layer, we find the user who may be either a certain or an uncertain customer. This client must send a GET request, which is normally in the form of an address of a representation of a requested resource. However, this request does not contain any information about the uncertainty or certainty of this representation. Then, at the end of this method, the sender receives the answer of his request accompanied by a header that contains the degree of uncertainty of the answer. For the second layer, we have the server side. First of all, the request must be sent to the control block which will simply send the request to the service concerned by the cleared address of the user's request. This service returns the representation of the requested resource in the form of a JSON file. This response must be sent to the control block that examines it, then calculates the probability value and then enrich the header of the answer to finally send the answer to the user. Algorithm 1 describes the evaluation of the queries through probabilistic Web resources and makes it possible to efficiently find all the possible representations presented in a single JSON file. In fact, this algorithm is divided into three parts. In the first part (line 1), the user sends his request to the server. Then, in the second part (line 2, lines 10–20) the control block is described. Besides, this part is divided into two main steps: the first step is represented by a block that receives the request sent by the user and sends it to the service to search for the requested resource representations. For the second step (line 10–20): the control block retrieves the response from the request, traverses it to clear the Mux and Ind nodes and to calculate the degree of probability of the response. Then Add the value of P to the response header and add information that indicates uncertainty in this resource and finally send the response to the user. And the third part presented by the lines (4–9) represents the function of the service responsible for the search for representation. First, it must browse all existing JSON documents to search for the requested resource, retrieve the requested representation, and send it to the control block.

Algorithm 1. Probabilistic GET 1

Require: URI
Ensure: $Prob_G ET_C as1$
1: Send Request (Req) to Control Service ($Cont\_Serv$).
2: Receive a GET type Req.
3: Send the Req to the corresponding service.
4: **for** All documents.JSON(Doc_josn) **do**
5:     **if** $Doci\_json = Doc\_requested(Doc\_Dem)$ **then**
6:         Recover the Doc_Dem, JSON (Res).
7:     **end if**
8: **end for**
9: Send Res to $Cont\_Serv$).
10: Proceed Res.
11: **if** We found Mux **then**
12:     $P = P \ XOR \ P\_Mux.$
13: **else if** We found Ind **then**
14:     $P = P * P\_Ind.$
15: **else if** End Res **then**
16:     Add the value of P to the header.
17:     Add an indication that he has an uncertainty in the header.
18:     Send the answer to the user.
19: **end if**
20: End.

---

5 Experimental Study and Analysis of the Results

All experiments are performed on a computer with a Windows 7 operating system and the Intel Core i5 processor and 6 GB of RAM. In order to analyze our approach and study its performance, we considered the execution time. We will present in this section some results and compare our uncertain approach with the certain semantic approach. The purpose of this experiment is to show that despite the great complexity of our uncertain invocation and web resources composition algorithm, queries can still be resolved in a reasonable amount of time. In this sense, we compare the performances of the certain composition algorithm which considers that wev ressources with certain semantics, with our proposition while increasing the number of web ressources, and varying the number of nodes of type ind per web resources. In this section we present two parts.

Experiment 1: Number of Node of Type Ind. In this part, we carry out a test to evaluate the impact of the variation of the number of the nodes of the type ind. We assume there is no parameter in the query path. As a result, the algorithm will recover all possibilities equal to n. For every n possibility in the array val, there is a generation of 2n possibilities to recover.

To improve the performance of our algorithm, we make some changes so that it only deals with one possibility and does not generate the other possibility.

The results of the Fig. 5 show that the execution time is proportional to the number of possibilities (1 possibility = 1000 ms) and that our probabilistic algorithm (GET) admits a reasonable execution time and can adapt to a large number of possibilities of nodes of type ind.

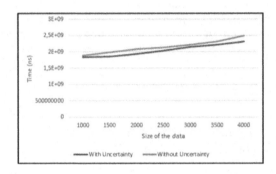

Fig. 5. Response time by varying number of nodes of ind type.

6   Conclusion

As a conclusion, we can say that it is necessary to provide a solution that helps Web users to manage data uncertainty while navigating hypertext. Therefore, the treatment of uncertainty will definitely improve the quality of the sent responses: we treat and trust the huge amount of information available on the Web. In this paper, we have discussed the need for a solution to manage data uncertainty when referencing and navigating resources on the Web. We also proposed a model that helps represent uncertain Web resources while considering uncertain resources as specific resources that could have multiple possible representations with probability. We defined these possible representations as mutually exclusive and rely on the possible world theory to interpret these representations as part of a set of possible Websites.

Moreover, we proposed a Parser to validate the probabilistic model that we proposed and help the developers to use this approach on the http browser. In addition to this, we rely on the uncertain Web resource model to propose an algebra for the interpretation and evaluation of data queries in uncertain resource compositions. In this context, we define data queries as Web resource paths, and propose algorithms to evaluate queries within uncertain resources. By respecting our probability model, we provided a computational algorithm which helps recover and calculate the uncertainty associated with the response to a data request.

References

1. Maleshkova, M., Pedrinaci, C., Domingue, J.: Investigating web APIs on the world wide web. In: 8th IEEE European Conference on Web Services (ECOWS 2010), 1–3 December 2010, Ayia Napa, Cyprus (2010). DBLP:conf/ecows/2010
2. Benslimane, D., Dustdar, S., Sheth, A.P.: services mashups: the new generation of web applications. IEEE Internet Comput. **12**, 13–15 (2008)
3. Halevy, A.Y., Rajaraman, A., Ordille, J.J.: Data integration: the teenage years. In: Proceedings of the 32nd International Conference on Very Large Data Bases, Seoul, Korea (2006)
4. Kharlamov, E., Nutt, W., Senellart, P.: Updating probabilistic XML. In: Proceedings of the 2010 EDBT/ICDT Workshops, Lausanne, Switzerland (2010)
5. Nierman, A., Jagadish, H.V.: ProTDB: probabilistic data in XML. In: VLDB 2002, Proceedings of 28th International Conference on Very Large Data Bases (2002)
6. Benslimane, D., Sheng, Q.Z., Barhamgi, M., Prade, H.: The uncertain web: concepts, challenges, and current solutions. TOIT **16**, 1–6 (2016)
7. Lemos, A.L., Daniel, F., Benatallah, B.: Web service composition: a survey of techniques and tools. ACM Comput. Surv. (CSUR) **48**, 1–41 (2016)
8. Malki, A., et al.: Data services with uncertain and correlated semantics. WWWJ **19**, 157–175 (2016)
9. Yu, Q., Liu, X., Bouguettaya, A., Medjahed, B.: Deploying and managing Web services: issues, solutions, and directions. VLDBJ **17**, 537–572 (2008)
10. Agrawal, P., Sarma, A.D., Ullman, J., Widom, J.: Fondements de l'intgration des donnes incertaines. Proc. VLDB Endow. **3**, 1080–1090 (2010)

Unlocking the Black Box: Towards Interactive Explainable Automated Machine Learning

Moncef Garouani and Mourad Bouneffa[✉]

Univ. Littoral Côte d'Opale, UR 4491, LISIC, Laboratoire d'Informatique Signal et Image de la Côte d'Opale, 62100 Calais, France
{moncef.garouani,mourad.bouneffa}@univ-littoral.fr

Abstract. Automated machine learning (AutoML) has transformed the process of selecting optimal machine learning (ML) models by autonomously searching for the most appropriate ones and fine-tuning associated hyperparameters. This eliminates the burdensome task of trial-and-error selection and parametrization of ML algorithms. Nonetheless, the lack of transparency and explainability poses a significant challenge when using AutoML, as it hampers user trust in the system's recommendations. Consequently, users often allocate more resources to the search process, resulting in reduced efficiency of the AutoML systems. To address this challenge, we propose an interactive and explainable AutoML framework that enables users to *understand* the reasoning behind the recommendations and *diagnose* any limitations of the suggested models using various explainable AI methods. Additionally, our framework provides the possibility of automated performance refinement. To operationalize the framework, we introduce AMLExplainer, an XAI system for interactive and interpretable AutoML that visualizes and performs all stages of the proposed pipeline(s) within the widely used Bootstrap Dash environment.

Keywords: Automated machine learning · Explainable AI · Transparency

1 Introduction

The rapid evolution of artificial intelligence (AI) technologies has resulted in the emergence of innovative use cases and futuristic applications that were previously unimaginable, thanks to the large amount of available data [1]. To alleviate the challenges of building machine learning models, automated ML methods have been introduced. Rather than manually searching for algorithms and fine-tuning hyperparameters, AutoML automatically explores various ML algorithms and optimizes hyperparameters within a pre-defined search space, which is essentially a set of feasible ML pipelines.

The interest in building complex AI models that can achieve unprecedented performance levels has been gradually replaced by a growing concern for

P. Quaresma et al. (Eds.): IDEAL 2023, LNCS 14404, pp. 458–469, 2023.
https://doi.org/10.1007/978-3-031-48232-8_42

alternative design factors leading to an improved usability of the resulting tools [2]. In many application areas, complex AI models become of limited practical utility because they focus solely on performance factors, leaving out important and even crucial aspects such as confidence, transparency, fairness, or accountability [3]. The absence of explanations for predicted performance factors makes AI models black boxes that only show input and output parameters but conceal inherent associations among them. In real-life applications such as industrial manufacturing processes, it is preferable to avoid such lack of transparency because these applications may involve critical decision choices. Therefore, the acceptance of and trust in an AutoML system depend heavily on the transparency of the recommendations [3].

Because of the lack of transparency in AutoML systems as Decision Support Systems (DSS), users tend to question the validity of automatic results, such as whether the AutoML ran long enough, whether it missed suitable models, whether it sufficiently explored [2,4] the search space, or whether the recommended configuration over- or underfit. Such queries may cause reluctance for users to apply the results of AutoML in more critical situations. Meanwhile, when AutoML provides unsatisfactory results, users are unable to reason and thus cannot improve the obtained results. They may only increase the computational budget, which can result in barriers to the effectiveness of AutoML. The initial goal of this work is to enhance the transparency, interpretability, and self-explanatory nature of the results produced by high-performing AutoML systems. By offering various visual summaries of the models and configurations they provide, AutoML decision support systems can become more dependable and practical. This could potentially improve the transparency and controllability of AutoML systems, leading to increased acceptance of these systems.

The rest of this paper is organized as follows: Sect. 2 provides an overview of the automated machine learning and the need for explainability in this domain. We discuss the current challenges and limitations associated with traditional AutoML approaches and highlight the importance of incorporating interactive and explainable elements into the process. Section 3 provides an overview of Explainable Artificial Intelligence (XAI) and its significance in addressing the black-box nature of AI systems. Section 4 focuses on Visual Analytics for AutoML, discussing the use of visual representations to enhance understanding and interpretation of AutoML processes. In subsections, we present a comprehensive framework for interactive explainable automated machine learning (AML-explainer). We describe the various components of our proposed framework, including data preprocessing, feature engineering, model selection, hyperparameter tuning, and model explanation. We explain how these components interact with each other to enable a user-friendly and transparent AutoML pipeline.

2 The Need for Transparency to Trust in AI and in AutoML

Black-box AI systems have been utilized in various fields [2]. Their involvement in critical domains, such as power consumption forecasting or supply chain management, often prioritizes overall system performance over important quality aspects like transparency and explainability. Even when these systems fail, such as the Quality Control System's inability to detect failures or the Equipment Failure Prevention system's limited capacity to identify the exact causes, they tend to generate false or inaccurate predictions. The consequences of such failures are underwhelming, especially in industrial critical applications where the lack of transparency in machine learning techniques can be disqualifying unless restricted [4]. A single incorrect decision can pose a significant risk to the entire production line, jeopardizing its safety (e.g., failure of a critical unit) and leading to substantial financial losses (e.g., non-compliant products). Therefore, relying on an incomprehensible black-box data-driven system is not the optimal choice. The lack of transparency is one of the primary concerns that question the adoption of AI models in the manufacturing industry.

Traditional predictive accuracy metrics such as precision and recall may not offer sufficient reliability in assessing the usefulness of a machine learning model [5]. To establish trust in an ML model and utilize it for real-world decision-making, it is crucial to understand the relationships the model has learned, its decision-making process, variations in decision logic across different feature spaces, potential biases in the data and model, and the collective impact of features on the model's outputs (see Fig. 1).

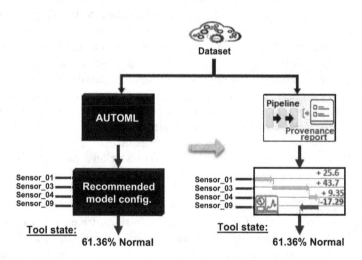

Fig. 1. From "Black-box" model recommendation and prediction to "White box" model with explanations.

Providing explanations for decisions and actions is a vital aspect of human interactions within the social sphere [4,6]. Just as explanations aid in building trust in human-to-human relationships, they should also be incorporated into human-to-machine interactions [7]. In this paper, we explore the feasibility and contributions of a process designed to make powerful AutoML decision support systems transparent, interpretable, and self-explanatory to foster the trust. This applies to situations where the AI system plays a supportive role (e.g., production planning) as well as scenarios where it provides guidance and decision-making (e.g., Quality Control, predictive maintenance, or autonomous driving). In the former cases, explanations offer additional information that helps the human involved gain a comprehensive understanding of the situation or problem at hand, facilitating decision-making [2]. Similar to an expert who provides a detailed report explaining their findings, a supportive AI system should provide detailed explanations instead of solely offering predictions or decisions.

3   Explainable Artificial Intelligence

Explainable AI refers to artificial intelligence technologies that can offer human-understandable explanations for their outputs or actions [8]. Users, by nature, often wonder about the underlying reasoning behind algorithms' decision-making processes. However, as AI algorithms and systems become more complex, they are often perceived as "black-boxes," lacking transparency and hindering comprehension of their reasoning. This lack of transparency negatively impacts users' trust in the systems. Model explainability can be categorized into two types: global explainability and local explainability. Global explainability enables users to understand the model based on its overall structure, while local explainability focuses on explaining a specific decision made by the model for a given input.

In recent years, there has been increasing attention towards developing methods for explaining, visualizing, and interpreting machine learning models within the XAI field [8–10]. Despite rapid advancements in XAI, there are still significant gaps that need to be addressed to generalize XAI approaches. Current major XAI methodologies are typically applicable to specific types of data and models, often requiring the pre-configuration of input parameters that are not easily implemented by non-experts (see Table 1).

Table 1. Properties of some XAI state of the art tools. *Level* is the interpretability coverage: local or global. *Dependency* specifies necessary inputs.

XAI method	Level		Dependency		Require
	Local	Global	Data	Model	preconfiguration
LIME [9]	●	○	●	○	●
ANCHORS [11]	●	○	●	○	●
SHAP [12]	●	●	○	○	●
Saliency [13]	●	○	●	●	●

One such method called LIME (Local Interpretable Model-Agnostic Explanations) is used to explain the importance of features by generating a linear surrogate model based on the output of a data sample [9]. ANCHORS [11], another technique, focuses on identifying influential input areas, referred to as anchors, to establish decision rules. Both LIME and ANCHORS are model-agnostic, relying on the sample inputs and outputs to explain the local decision boundaries created by the model.

Another type of XAI method, known as Saliency [12], constructs a visual representation highlighting the importance of features by masking aspects of each sample based on the model's perception of the input data. Unlike LIME and ANCHORS, Saliency is specific to artificial neural networks (models-specific). Some XAI methods offer low-abstraction capabilities, such as visualizing convolutional filters or illustrating data flow through computational graphs [14]. These methods are particularly beneficial for model developers seeking to enhance their models using low-abstraction XAI as a quality metric.

While these existing methods are specialized for specific use cases and cover various insights and application constraints, they only provide limited XAI methods without an interactive machine learning (IML) workflow [4]. Consequently, they cannot serve as XAI components in an AutoML DSS. An ideal system for explaining ML models should encompass a range of XAI levels and possess the flexibility to adapt to the output of AutoML, remaining agnostic to both the model and the data.

4   Visual Analytics for AutoML

Automated Machine Learning has emerged as a powerful approach to democratize the process of building machine learning models [1]. It aims to automate various stages of the machine learning pipeline, including data preprocessing, feature engineering, model selection, hyperparameter tuning, and model evaluation. While AutoML provides great convenience and efficiency, it often lacks transparency and interpretability, making it challenging for users to understand and trust the generated models. In recent years, there has been growing interest in developing explainable AutoML systems that can provide insights into the decision-making process of automated model building. Recent design recommendations prioritize intuitive interfaces, emphasizing clean and concise presentations, effective explanation features, and user-friendly interactions [15]. Visual Analytics (VA) can enhance the development and deployment of models in the IML workflow by providing customized visual interfaces that actively engage users, fostering deeper understanding and insight generation throughout the data analysis process.

4.1   Towards Interactive Explainable AutoML

Interactive Visual Analytics for AutoML leverages visualizations to enable users to explore and understand the AutoML pipeline. It combines the power of visual representations with ML algorithms to provide a comprehensive view of the automated model building process. By visualizing the data preprocessing steps, feature engineering techniques, model selection strategies, hyperparameters tuning, and evaluation metrics, users can gain a deeper understanding of the decision-making process and the impact of various choices on model performance. During our examination of existing systems for interactive machine learning and visual analytics, we have identified three stages of explanations based on their ability to cover the tracking and reporting of the AutoML's provenance. Additionally, we demonstrate the necessity of a comprehensive eXplainable Artificial Intelligence system that encompasses all stages and tasks to tackle the variations in interpreting black-box AutoML solutions.

The initial stage, referred to as the **understanding** stage, can be interpreted differently depending on the intended user group (see Fig. 2). For individuals with limited knowledge of machine learning models, an interactive VA system can serve as an "educational" tool to explain concepts. For instance, Harley, [16] visually presents changes in an image alongside the affected layers of an Artificial Neural Network. Similarly, Smilkov *et al.* [17] provide an interactive visual representation of an ANN. Various other works explore graphical representations of Deep Neural Networks. These examples highlight the importance of interactive exploration during the understanding phase. In contrast to novices, model users and developers require a deeper understanding of the inner workings of the model. Rauber *et al.* [18] focus on this aspect by visualizing the training process of ANNs, as well as the relationships between individual neurons and data. Based on insights from these studies, we conclude that providing customized explanations of AutoML at different levels of model abstraction is necessary to comprehend the recommendation and diagnosis processes of algorithms.

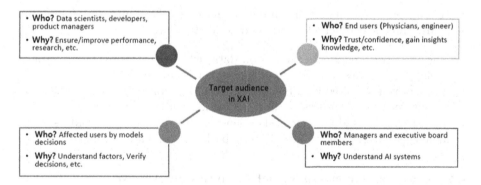

Fig. 2. Purposes of explainability in ML models sought by different audience profiles. Image partly inspired by the one presented in [19], used with permission from IBM.

Visual Analytics systems can address this gap by focusing on model **diagnosis** in an IML workflow to enable the detection of problems on different abstraction levels. Some systems facilitate model diagnosis by emphasizing the importance of features or examining the model's response to real or adversarial input examples [12]. Others focus on specific elements, such as neurons activation or hidden states of a cell [20], and action patterns of reinforcement learning algorithms to enable model-specific diagnosis.

An interactive explainable AutoML decision support system can go beyond the understanding and diagnosis phases, and target the **refinement** of ML models. We have identified works that are designed to diagnose and refine single ML models [21]. Others target multi-model visual comparison for refinement [22]. In addition to this distinction, various interactive refinement approaches are used in iterative cycles on medical images [23]. Such examples highlight the need for interactive and iterative refinement cycles in our self-explainable AutoML vision. Further, the ability to assist on the selection, configuration and refinement of adequate ML models is essential for assessing the quality of different models and selecting the most suitable for a given context.

In our system implementation, we attempt a transparent and interactive XAutoML pipeline that can cover different pathways through all of the addressed stages and tasks. Given a dataset, the AutoML support system automatically recommends the most suitable ML configurations and the proposed system explains the rationale traceability behind its recommendation. It is intended to support the analysis and inspection of all machine learning classification models without any data type or model dependency. The goal is manifold:

- To facilitate the inspection and evaluation of model performance by offering visual summaries and textual information.
- To offer a guided exploration of the reasoning behind the generated recommendations and provide possibilities for refining performances.

With our proposed approach, end users have the flexibility to explore the AutoML process at different levels, as described below:

- The Data-oriented level centered on data, which involves exploring various visualization levels to understand the properties of the data.
- The AutoML-oriented level focused on AutoML, which involves investigating the entire AutoML process from recommendation to refinement.
- The Model-oriented level focused on models provided by the AutoML such as analyzing model performance, decision path and conducting what-if analysis.

4.2   Key Components of Visual Analytics for AutoML

Data Exploration and Preprocessing Visualization. To enhance the transparency of AutoML, the proposed toolbox provides visualizations to help users explore and understand the input data. Users can visualize data distributions, identify outliers, and analyze missing values. Additionally, interactive visualizations can assist in preprocessing tasks such as feature scaling, imputation, and categorical variable encoding.

Features Importance Visualization. Features engineering plays a crucial role in building effective machine learning models. Visual Analytics for AutoML offers interactive visualizations to help users identify important features, and explore relationships between features. Users can leverage these visualizations to make informed decisions about feature selection and engineering techniques (see Fig. 3).

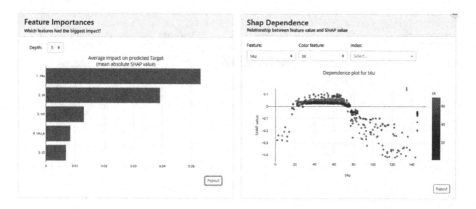

Fig. 3. Features importance Visualization.

Model Selection and Hyperparameters Tuning Visualization. Visual Analytics for AutoML provides visualizations to illustrate the performance of different models and hyperparameter configurations. Users can explore model accuracy, precision, recall, and other evaluation metrics to make informed decisions about the best model and hyperparameter settings. Interactive visualizations can assist in comparing models and analyzing trade-offs between performance metrics (see Fig. 4).

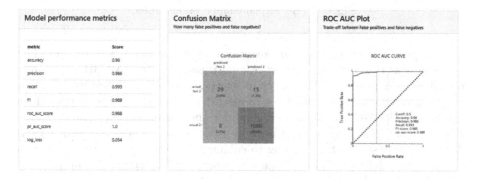

Fig. 4. Model Selection and Hyperparameters Tuning Visualization.

Model Explainability Visualization. To enhance the interpretability of AutoML models, we incorporated visualizations that explain the reasoning behind model predictions through contrastive explanation, what-if-analysis, decision paths, etc. Users can explore decision boundaries, and partial dependence plots to gain insights into how the model makes predictions. These visualizations can help users identify biases, understand the model's strengths and limitations, and build trust in the automated model building process (see Fig. 5).

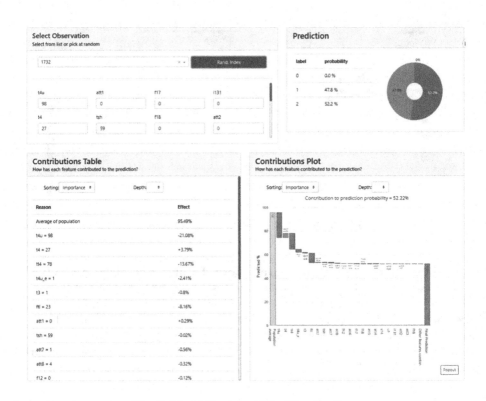

Fig. 5. Model Explainability Visualization.

The proposed toolbox is implemented as a client-server architecture. The server acts as the AutoML support system (referred to as AMLBID [4]), while the client-side interfaces allow users to interact with AutoML services through graphical elements. These interfaces connect data summaries with visualizations using multiple levels of visual summary for the recommended models. AML-Explainer also serves as a guide for end-users in cases where the results from AutoML are unsatisfactory, aiming to enhance predictive performance (refer to [24] for more details). This approach enhances the transparency, control, and reliability of the AutoML decision support system.

The following script outlines the steps involved in using the proposed toolbox to interpret any AutoML outcome. In line 6, we define the root directory for the dataset that needs to be loaded. On line 9, the AutoML system is introduced[1], which initializes the process aimed at identifying the pipeline with the highest score based on the desired performance criteria. Subsequently, the recommended pipeline is trained using the test set of the provided samples, as depicted on line 10. Once this code execution is completed, the interactive explanatory dashboard, line 13, is generated to enable the user to understand, diagnose and investigate the refinement possibilities of the recommend ML pipeline. The comprehensive documentation and a thorough list of features, along with an example that illustrates them are available in the Github repository: https://github.com/LeMGarouani/AMLBID.

```
 1 from AMLBID.recommender import AMLBID_Recommender
 2 from AMLBID.explainer import AMLBID_Explainer
 3 from AMLBID.loader import *
 4
 5 #Load dataset
 6 Data,X_train,Y_train,X_test,Y_test=load_data("Dataset.csv")
 7
 8 #Generate the optimal configurations
 9 model,config=AMLBID_Recommender.recommend(Data, metric="
     Accuracy")
10 model.fit(X_train, Y_train)
11
12 #Generate the interactive explanatory dash
13 Explainer = AMLBID_Explainer.explain(model, config,
14                          X_test, Y_test)
15 Explainer.dash()
```

Script 1.1. Illustrative code example of the proposed XAutoML toolbox.

5 Conclusion

There has been significant progress in democratizing the use of machine learning for individuals who are not data analysis experts by providing them with "off the shelf" available solutions. However, these solutions, while powerful, lack detailed instructions on recommended configurations and the inner workings of the models, making them less reliable and highly performant but opaque systems. In this paper, we presented a Visual Analytics toolbox for AutoML, an approach that combines interactive visualizations with automated ML techniques to enhance the explainability and interactivity of AutoML systems. Through visual representations, users can gain insights into the decision-making process of automated model building and make informed choices. The proposed toolbox opens up new possibilities for democratizing ML and improving the transparency and trustworthiness of AutoML. We investigated various levels of explanations, ranging

---

[1] In this example we made use of the AMLBID [4] AutoML tool, however, it could be replaced by any other AutoML system.

from individual decisions to the overall model's recommendations and predictions. The explanations provided by the predictive models and the what-if analysis were found to be effective in addressing real-world issues. Future research could explore advanced visual analytics techniques and user-centered design to enhance the usability and effectiveness of Visual Analytics for AutoML.

References

1. Garouani, M., Ahmad, A., Bouneffa, M., Hamlich, M.: Using meta-learning for automated algorithms selection and configuration: an experimental framework for industrial big data. J. Big Data **9**(1), 57 (2022). https://doi.org/10.1186/s40537-022-00612-4
2. Garouani, M., Ahmad, A., Bouneffa, M., Hamlich, M.: Towards big industrial data mining through explainable automated machine learning. Int. J. Adv. Manuf. Technol. **120**(1–2), 1169–1188 (2022). https://doi.org/10.1007/s00170-022-08761-9
3. Garouani, M.: Towards efficient and explainable automated machine learning pipelines design: application to industry 4.0 data, PhD thesis. Université du Littoral Cote d'Opale; Université Hassan II de Casablanca (2022)
4. Garouani, M., Bouneffa, M., Ahmad, A., Hamlich, M.: Version [2.0]- [amlbid: an auto-explained automated machine learning tool for big industrial data]. SoftwareX **23**, 101444 (2023). ISSN: 2352–7110. https://doi.org/10.1016/j.softx.2023.101444
5. Moradi, M., Samwald, M.: Post-hoc explanation of black-box classifiers using confident itemsets. Expert Syst. Appl. **165**, 113941 (2021). ISSN: 0957–4174. https://doi.org/10.1016/j.eswa.2020.113941
6. Heath, R.L., Bryant, J.: Human Communication Theory and Research: Concepts, Contexts, and Challenges, 2nd edn. Routledge, Mahwah (2000)
7. Samek, W., Müller, K.R.: Towards explainable artificial intelligence. In: Explainable AI: Interpreting, Explaining and Visualizing Deep Learning, Lecture Notes in Computer Science, pp. 5–22. https://doi.org/10.1007/978-3-030-28954-6_1
8. Gunning, D., Stefik, M., Choi, J., Miller, T., Stumpf, S., Yang, G.Z.: XAI—Explainable artificial intelligence. Sci. Rob. **4**(37), eaay7120 (2019). ISSN: 2470–9476. https://doi.org/10.1126/scirobotics.aay7120
9. Ribeiro, M.T., Singh, S., Guestrin, C.: Why should i trust you?: explaining the predictions of any classifier. In: Proceedings of the 22nd ACM SIGKDD International Conference on Knowledge Discovery and Data Mining, pp. 1135–1144. https://doi.org/10.1145/2939672.2939778
10. Castelvecchi, D.: Can we open the black box of AI? Nature News **538**(7623), 20 (2016). https://doi.org/10.1038/538020a
11. Ribeiro, M.T., Singh, S., Guestrin, C.: Anchors: high-precision model-agnostic explanations. In: Proceedings of the AAAI Conference on Artificial Intelligence (2018)
12. Lundberg, S.M., et al.: From local explanations to global understanding with explainable AI for trees. Nat. Mach. Intell. **2**(1), 56–67 (2020). ISSN: 2522–5839. https://doi.org/10.1038/s42256-019-0138-9
13. Simonyan, K., Vedaldi, A., Zisserman, A.: Deep Inside Convolutional Networks: Visualising Image Classification Models and Saliency Maps. arxiv.org/abs/1312.6034

14. Wongsuphasawat, K., Smilkov, D., Wexler, J.: Visualizing dataflow graphs of deep learning models in tensorflow. IEEE Trans. Vis. Comput. Graph. **24**(1), 1–12. ISSN: 1941–0506. https://doi.org/10.1109/TVCG.2017.2744878

15. Müller, J., et al.: A visual approach to explainable computerized clinical decision support. Comput. Graph. **91**, 1–11 (2020). ISSN: 0097–8493. https://doi.org/10.1016/j.cag.2020.06.004

16. Harley, A.W.: An interactive node-link visualization of convolutional neural networks. In: Bebis, G., et al. (eds.) ISVC 2015. LNCS, vol. 9474, pp. 867–877. Springer, Cham (2015). https://doi.org/10.1007/978-3-319-27857-5_77

17. Smilkov, D., Carter, S., Sculley, D., Viégas, F.B, Wattenberg, M.: Direct-manipulation visualization of deep networks. arXiv: 1708.03788 (2017)

18. Rauber, P.E., Fadel, S.G., Falcao, A.X., Telea, A.C.: Visualizing the hidden activity of artificial neural networks. IEEE Trans. Vis. Comput. Graph. **23**(1), 101–110. ISSN: 1941–0506. https://doi.org/10.1109/TVCG.2016.2598838

19. Francesca, R.: AI Ethics for Enterprise AI (2019)

20. Ming, Y., et al.: Understanding hidden memories of recurrent neural networks. In: 2017 IEEE Conference on Visual Analytics Science and Technology (VAST) (2017). https://doi.org/10.1109/VAST.2017.8585721

21. Pezzotti, N., Höllt, T., Van Gemert, J.: DeepEyes: progressive visual analytics for designing deep neural networks. IEEE Trans. Vis. Comput. Graph. **24**(1), 98–108. ISSN: 1941–0506. https://doi.org/10.1109/TVCG.2017.2744358

22. Murugesan, S., Malik, S., Du, F., Koh, E., Lai, T.M.: DeepCompare: visual and interactive comparison of deep learning model performance. IEEE Comput. Graph. Appl. **39**(5), 47–59. ISSN: 1558–1756. https://doi.org/10.1109/MCG.2019.2919033

23. Cai, C.J., et al.: Human-centered tools for coping with imperfect algorithms during medical decision-making. arxiv.org/abs/1902.02960 (2019)

24. Garouani, M., Ahmad, A., Bouneffa, M.: Explaining meta-features importance in meta-learning through shapley values. In: Proceedings of the 25th International Conference on Enterprise Information Systems - Volume 1: ICEIS, pp. 591–598 (2023). https://doi.org/10.5220/0011986600003467

Machine Learning for Time Series Forecasting Using State Space Models

Jose M. Sanchez-Bornot[1] and Roberto C. Sotero[2(✉)]

[1] School of Computing, Engineering and Intelligent Systems, Ulster University,
Derry/Londonderry, UK
jm.sanchez-bornot@ulster.ac.uk
[2] Department of Radiology, and Hotchkiss Brain Institute, University of Calgary,
Calgary, AB, Canada
roberto.soterodiaz@ucalgary.ca

Abstract. State-space models (SSMs) are becoming mainstream for time series analysis because their flexibility and increased explainability, as they model observations separately from unobserved dynamics. Critically, using SSMs based on multivariate autoregressive equations enhances understanding of system evolution and its dynamical interactions. However, some challenges remain unsolved, such as estimation in large-scale scenarios, with very noisy data, and model selection. Here, we explore a state-space alternating least squares (SSALS) algorithm for time series forecasting, demonstrating its application with simulated and real data, and how to solve model selection in noisy scenarios with a novel cross-validation technique. Altogether, testing this methodology with time series forecasting is ideal to demonstrate its strengths and weaknesses, and appreciate its advantages compared to current methods.

Keywords: machine learning · time series analysis · state space models

1 Introduction

Time series forecasting is a challenging and practical research area to test many developed methods and explore novel scientific ideas. It also shares many features with other research areas, which makes it helpful to develop methods that can be applied to other disciplines. Its complexities and attractiveness come from the necessity to estimate unknown generative models with linear or nonlinear characteristics that may ignore many unidentified, or subjective, but critical variables, as can be observed in many applications for economic and ecological time series forecasting. In the absence of a ground truth, the use of state-space models (**SSMs**) can bring significant modeling diversity and simplicity at the same time. Therefore, SSMs are becoming popular for time series analysis, including applications such as the modeling of ecological time series [1], functional

© The Author(s), under exclusive license to Springer Nature Switzerland AG 2023
P. Quaresma et al. (Eds.): IDEAL 2023, LNCS 14404, pp. 470–482, 2023.
https://doi.org/10.1007/978-3-031-48232-8_43

connectivity estimation (e.g., Granger causality [2]) and neuroinformatics [3–7]. However, one critical open issue for SSMs estimation is model selection, which is typically carried out with cross-validation techniques [8].

Despite its controversial use in time series analysis, cross-validation has shown its usefulness for model selection. Some limitations may arise from assuming that observations are uncorrelated, which can be in contradiction with observed temporal correlations. Moreover, a typical application will use data from the beginning, middle, or end of the segment (randomly), as training data to make prediction on the remaining test samples. This contradicts the causality principle. Consequently, many different cross-validation variants have been developed to accommodate these criticisms [9]. However, the classical application of k-fold cross-validation method has been shown to be more successful than alternatives [9].

Bergmeir et al. [9] demonstrated that using cross-validation can be justified with the use of multivariate autoregressive (**MVAR**) modeling as long as fitting the MVAR model renders the residual noise uncorrelated. This is achievable by risking overfitting with the use of a large-order MVAR model. However, overfitting can be ameliorated by using regularization on the autoregressive coefficients [9]. In contrast, Auger-Methe et al. [1] suggested to avoid the model overparameterization as model identifiability is challenging, as may be the case for large-scale time series analysis. Clearly, Bergmeir and colleagues were more concerned about time series forecasting than identifiability [9,10]. Because the SSMs application for time series forecasting is scarce [1], we mainly provide empirical demonstrations for the use of cross-validation in MVAR-based SSMs, extending from Bergmeir's ideas [9].

The paper is organized as follows. In Sect. 2, we present SSMs based on MVAR equations to solve time series forecasting problems, while related methodologies are briefly reviewed in Sect. 3. Then, a novel data-driven framework to solve SSMs based on an extension of K-fold cross-validation is introduced in Sects. 4 and 5. In Sect. 6, we present for first time a state-space alternating least squares (SSALS) algorithm for time series forecasting. Succinctly, we have extended the methodology from a recently developed data-driven regularization framework in our group [6,7]. We illustrate the usefulness of this methodology for time series forecasting with simulated and real data analysis in Sect. 7. Section 8 concludes our study and provides final remarks.

2   State Space Models

We present MVAR-based SSMs as follows for applications in this study:

$$\mathbf{y}_t = \mathbf{B}\mathbf{x}_t + \mathbf{c} + \mathbf{w}_t; \text{for } t = 1, 2, \ldots, T, \tag{1}$$

$$\mathbf{x}_t = \sum_{p=1}^{P} \mathbf{A}_p \mathbf{x}_{t-p} + \mathbf{v}_t; \text{for } t = P + 1, \ldots, T, \tag{2}$$

where $\mathbf{y}_t \in R^{M \times 1}$ and $\mathbf{x}_t \in R^{N \times 1}$ represent the observations and processes (latent variables) dynamics, for $t = 1, 2, \ldots, T$ time instants, and $\mathbf{c} \in R^{M \times 1}$

is the intercept. As shown in Eq.(2), process dynamics are modelled using MVAR equations as convenient to capture process-to-process interaction and for forecasting applications [9]. The MVAR's coefficients $\mathbf{A}_p \in R^{N \times N}$ represent the influences from process past's dynamics over present time, as indicated for lagged interaction $p = 1, \ldots, P$. In general, $\mathbf{B} \in R^{M \times N}$ is a mixing matrix; therefore, observations can be a mixed representation of latent dynamics. The model also includes noise terms $\mathbf{v}_t \in R^{N \times 1} \sim N(0, \sigma_s^2 \mathbf{I}_{N \times N})$ and $\mathbf{w}_t \in R^{M \times 1} \sim N(0, \sigma_o^2 \mathbf{I}_{M \times M})$ to represent uncontrolled random perturbations in a real-life scenario. For simplicity, noise terms are assumed as identically and independently (i.i.d.) Gaussian distributed processes with variances σ_o^2 and σ_s^2.

In the present case of time series forecasting, $\mathbf{B} = I_{N \times N}$ (identity matrix), which makes the application of these models more straightforward in this area. Otherwise, in neuroimaging analysis $\mathbf{B}$ corresponds to a lead field matrix which is calculated for an underdetermined algebraic problem, i.e., with more variables (brain sources) than observations (magneto/electroencephalographic sensor data) [3,4]. In contrast, in dynamic factor analysis, it is assumed that $\mathbf{B} \in R^{M \times N}, M \geq N$ has a lower triangular matrix structure because this problem also requires estimating $\mathbf{B}$ [13], together with the typical state space model parameters $\mathbf{x}_t$ and $\mathbf{A}_p$. The lower triangular matrix structure guarantees more "unique" solutions. As shown below, the unicity is not guaranteed because of local-minima solutions, mainly in large-scale and noisy scenarios; however, the use of a regularization approach greatly guarantees the estimation of robust solutions.

3   Time Series Forecasting Methodology

We consider below several statistics in time series forecasting, following closely the notation from a recent study [9]. For a particular model, define the multivariate generative formula as:

$$\mathbf{y}_t = \mathbf{g}(\{\mathbf{x}_{t-P}, \ldots, \mathbf{x}_{t-1}\}, \boldsymbol{\Theta}) + \mathbf{w}_t,$$

where $\mathbf{x}_t$, $\mathbf{y}_t$, the innovations term $\mathbf{w}_t$, and the model parameters $\boldsymbol{\Theta}$, are defined as above, and $\mathbf{g}(\cdot)$ represents the true model that we aim to replicate, i.e., by cloning its statistical properties with the surrogate model $\mathbf{g}_s(\cdot)$. Here, the model relies on the knowledge of the immediate preceding past, i.e., $\{\mathbf{x}_t\}_{t-P}^{t-1} = \{\mathbf{x}_{t-P}, ..., \mathbf{x}_{t-1}\}$, in order to accurately predict the future. For estimation purposes, $\mathbf{g}_s$ can have a more flexible representation than the original model despite of representing accurately the same linear or nonlinear interactions. For example, when using the regularization approach it is convenient to write the model in a general form $\mathbf{g}_s(\{\mathbf{x}_t\}_{t-P}^{t-1}, \boldsymbol{\Theta}, \lambda)$, which also relies on the estimation of a set of hyperparameters, which are contained in the vector λ; or more succinctly, as $\mathbf{g}_\lambda(\{\mathbf{x}_t\}_{t-P}^{t-1}, \boldsymbol{\Theta})$, where different hyperparameter subsets (e.g., sparse priors [5,6]) determine different models. With this last notation, we assume that the model can be unequivocally determined by the selected hyperparameters, which may not be always true, e.g., when local minima issue persists despite of regularization.

Now, suppose that $\tilde{\mathbf{y}}_1, \ldots, \tilde{\mathbf{y}}_T$ are measurements obtained from the oracle model $\mathbf{g}(\cdot)$. In time series forecasting analysis, these samples are split into a training data $\{\tilde{\mathbf{y}}_1, \ldots, \tilde{\mathbf{y}}_{T_{tr}}\} \in R^{M \times T_{tr}}$, used to estimate $\mathbf{g}_s$, and a validation data $\{\tilde{\mathbf{y}}_{T-H+1}, \ldots, \tilde{\mathbf{y}}_T\} \in R^{M \times H}$, with an horizon forecast $H > 0$ for $T_{tr} < T - H$, for model evaluation [9]. The validation data is commonly used for out-of-sample (**OOS**) forecast evaluation as it is independent from the data used to train the model, and thereby it must be less prone to produce biased evaluation due to the data double dipping that normally occurs in the application of cross-validation or model selection techniques. Specifically, for MVAR models of order $P, T_{tr} < T - H - P + 1$. Typically, in cross-validation applications, double dipping can be avoided by using nested cross-validation technique [11]. However, in time series forecasting, using OOS validation accomplishes the same goal.

With this convention, the performance of estimated models $\hat{\mathbf{g}}_s$ can be quantified using the estimated predicted error $\widehat{PE}_{OOS}$, which is evaluated for the OOS data. For example, for models that statistically depend only on the P previous time instants, such as autoregressive models, it can be calculated, together with the true predicted error PE_{OOS} (i.e., calculated using the ground-truth model and data), as follows:

$$\widehat{PE}_{OOS} = \frac{1}{NH} \sum_{t=T-H+1}^{T} \|\tilde{\mathbf{y}}_t - \hat{\mathbf{g}}_\lambda(\{\hat{\mathbf{x}}_t\}_{t-P}^{t-1}, \hat{\mathbf{\Theta}})\|_2^2,$$

$$PE_{OOS} = \frac{1}{NH} \sum_{t=T-H+1}^{T} \|\tilde{\mathbf{y}}_t - \mathbf{g}(\{\mathbf{x}_t\}_{t-P}^{t-1}, \tilde{\mathbf{\Theta}})\|_2^2.$$

In general, when the ground-truth model is unknown, estimated models can be evaluated using the symmetric Mean Absolute Percentage Error (*sMAPE*) [12]:

$$sMAPE(i) = \frac{1}{H} \sum_{t=T-H+1}^{T} \frac{2|\tilde{y}_t^{(i)} - \hat{g}_\lambda^{(i)}(\{\hat{\mathbf{x}}_t\}_{t-P}^{t-1}, \hat{\mathbf{\Theta}})|}{|\tilde{y}_t^{(i)}| + |\hat{g}_\lambda^{(i)}(\{\hat{\mathbf{x}}_t\}_{t-P}^{t-1}, \hat{\mathbf{\Theta}})|},$$

which must be evaluated individually for each time series $i = 1, \ldots, M$ for the OOS data. For instance, we have $\hat{\mathbf{x}}_t = \tilde{\mathbf{y}}_t$ in the case of MVAR models, where $\mathbf{\Theta}$ and the hyperparameter values must be estimated and evaluated using the training data.

As an alternative, despite the increased cost, $\mathbf{\Theta}$ can be re-estimated incrementally by considering also the OOS data until the current forecasting point, i.e., using $\{\tilde{\mathbf{y}}_1, \ldots, \tilde{\mathbf{y}}_{t-1}\}$ in order to predict $\mathbf{y}_t$, for $t = T - h + 1, \ldots, T$, which should produce more accurate forecasting. In this case, only the hyperparameters are fixed to the values that were assessed during the training step.

Additionally, in MVAR-based SSMs, the state dynamics are also unknown but critical to predicting future observations. Therefore, the internal processes must be estimated incrementally with the update of model parameters, i.e., $\{\mathbf{x}_1, \ldots, \mathbf{x}_{t-1}\}$ and $\mathbf{\Theta}$ must be re-estimated using the observation samples

$\{\tilde{\mathbf{y}}_1, \ldots, \tilde{\mathbf{y}}_{t-1}\}$ and the known hyperparameter values (assessed during training), and then the estimates $\{\hat{\mathbf{x}}_{t-P}, \ldots, \hat{\mathbf{x}}_{t-1}\}$ and $\hat{\boldsymbol{\Theta}}$ are used to predict $\mathbf{x}_t$ with the MVAR Eq. (2), prior to the final prediction $\hat{\mathbf{y}}_t = \mathbf{B}\hat{\mathbf{x}}_t$, for $t = T - h + 1, \ldots, T$. Otherwise, without any new samples, when requested to predict the future blindly, we can use the estimate $\hat{\boldsymbol{\Theta}}$, obtained during the training step, and propagate the processes' future dynamics using only the estimated MVAR model as generator.

Another useful statistic used here to evaluate the performance of estimated forecasting models is the Mean Absolute Predictive Accuracy Error ($MAPAE$), which relies on Monte-Carlo (**MC**) simulations, defined as [9]:

$$MAPAE = \frac{1}{N_{MC}} \sum_{k=1}^{N_{MC}} |\widehat{PE}_{OOS}(k) - PE_{OOS}(k)|,$$

where N_{MC} is the number of MC simulations. However, to compare methods using multiple time series datasets, we prefer to use their modified "scale free" version based on the relative (R) predicted error correction, as follows (percentage unit):

$$MRAPAE = \frac{100}{N_{MC}} \sum_{k=1}^{N_{MC}} |\frac{\widehat{PE}_{OOS}(k) - PE_{OOS}(k)}{PE_{OOS}(k)}|.$$

4   Regularization Framework for State Space Models

To derive a regularization framework for the estimation of state space models, it is helpful to represent the observation and state equations using conditional probability distribution functions. For example, assuming that $\mathbf{y}_t \mid \mathbf{B}, \mathbf{c}, \mathbf{x}_t \sim N(\mathbf{B}\mathbf{x}_t + \mathbf{c}, \sigma_o^2 \mathbf{I}_M)$, $\mathbf{x}_{t>P} \mid \mathbf{A}, \mathbf{x} \sim N(\sum_{p=1}^{P} \mathbf{A}_p \mathbf{x}_{t-p}, \sigma_s^2 \mathbf{I}_N)$, and $\mathbf{x}_{t \leq P} \sim N(\mathbf{0}_N, \sigma_s^2 \mathbf{I}_N)$, lead to the state space model in Eqs.(1,2). Furthermore, an a priori distribution can be incorporated into the model to add stability for estimators, e.g., $vec(\mathbf{A}) \sim N(\mathbf{0}_{N^2P}, \sigma_A^2 \mathbf{I}_{N^2P})$. Other a priori information have been adopted in the literature, such as sparse priors [5,6]; however, for simplicity we use only the ones mentioned above.

For these assumptions, the maximum a posteriori (**MAP**) estimate can be obtained by solving the optimization problem:

$$F = \frac{1}{2}\left(\frac{1}{T}\left(\sum_{t=1}^{T} ||\mathbf{y}_t - \mathbf{B}\mathbf{x}_t - \mathbf{c}||_2^2 + \lambda \sum_{t=P+1}^{T} ||\mathbf{x}_t - \sum_{p=1}^{P} \mathbf{A}_p \mathbf{x}_{t-p}||_2^2 + \right.\right.$$

$$\lambda \sum_{t=1}^{P} ||\mathbf{x}_t||_2^2) + \lambda_2 ||\mathbf{A}||_F^2) \tag{3}$$

$$\hat{\mathbf{x}}, \hat{\mathbf{A}}, \hat{\mathbf{c}} = \arg \min_{\mathbf{x}, \mathbf{A}, \mathbf{c}} F(\mathbf{y}, \mathbf{x}, \mathbf{A}, \mathbf{c}, \lambda, \lambda_2). \tag{4}$$

In general, we must assess the values of hyperparameters λ and λ_2 before estimating MVAR-based SSMs parameters, which can be done in practice using cross-validation approaches [1,6,7,9–11], as demonstrated next.

5   Model Selection with K-Fold Cross-Validation Based on Imputed Data

Cross-validation techniques are frequently used to assess the regularization parameters in regression and classification problems [8]. Despite its apparently inappropriateness for time series analysis, Bergmeir and colleagues [9,10] have shown that K-fold cross-validation can produce better results than other model selection approaches. However, one challenging obstacle is how to appropriately implement cross-validation with SSMs. In MVAR models, simply after transforming its equations into the linear model $\mathbf{y} = \mathbf{X}\beta + \epsilon$, the elements of the response variable $\mathbf{y}$ and the corresponding rows of the design matrix $\mathbf{X}$ can be randomly assigned to the different K folds during model estimation. However, in SSMs the latent variables are unknown for the hold-out data and therefore it is not straightforward how to use them to predict the response.

As shown next, a recent development that combined cross-validation with data imputation can be used for MVAR-based SSMs selection [6]. This approach is applied here for first time to time series forecasting analysis. The method is named cross-validation based on imputed data (K-fold CVI) after observing in SSMs estimation that the latent variables $\mathbf{x}_t$ must be estimated first before calculating the cross-validation prediction error for the observation samples that were hold out in each k-fold iteration (e.g., $\{\tilde{\mathbf{y}}_t^{(k)}\}$ for $k = 1, \ldots, K$). This can be done by using a loss function to modify the aforementioned optimization problem [6,7], as follows:

$$F_k = \frac{1}{2}\left(\frac{1}{T}\left(\sum_{t=1}^{T} L_t^{(k)}||\mathbf{y}_t - \mathbf{B}\mathbf{x}_t - \mathbf{c}||_2^2 + \lambda \sum_{t=P+1}^{T} ||\mathbf{x}_t - \sum_{p=1}^{P} \mathbf{A}_p\mathbf{x}_{t-p}||_2^2 + \right.\right.$$

$$\lambda \sum_{t=1}^{P} ||\mathbf{x}_t||_2^2) + \lambda_2||\mathbf{A}||_F^2) \tag{5}$$

$$\hat{\mathbf{x}}^{(k)}, \hat{\mathbf{A}}^{(k)}, \hat{\mathbf{c}}^{(k)} = \arg\min_{\mathbf{x},\mathbf{A},\mathbf{c}} F(\mathbf{y}, \mathbf{x}, \mathbf{A}, \mathbf{c}, \lambda, \lambda_2), \tag{6}$$

where $L_t^{(k)} = \begin{cases} 0 & \text{if } \tilde{\mathbf{y}}_t \in S_k, \\ 1 & \text{otherwise.} \end{cases}$ is the loss function introduced to remove the contribution of the hold-out data samples $\tilde{\mathbf{y}}_t \in S_k$ for each k-fold run. Note that the loss function is applied only to the observation term in Eq.(5).

After solving this modified optimization problem, the K-fold CVI prediction error can be calculated as follows (recall that $\mathbf{B} = \mathbf{I}_N$ in forecasting problems):

$$\widehat{PE}_{CV} = \frac{1}{(T-P)MK} \sum_{k=1}^{K} \sum_{\tilde{\mathbf{y}}_t \in S_k} ||\tilde{\mathbf{y}}_t^{(k)} - \mathbf{B}\hat{\mathbf{x}}_t^{(k)} - \hat{\mathbf{c}}^{(k)}||_2^2.$$

Figure 1 shows an example of the data partition for $K = 5$ fold CVI. Particularly, for SSMs relying on MVAR equations of order P, we remove the first P samples from each k-fold subset, as logically they cannot be predicted by the

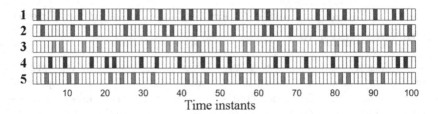

Fig. 1. Example of time-sequence partition for 5-fold CVI with $T = 100$ time instants. Time samples selected for the K folds are highlighted with different colors.

model, and then randomly distribute the remaining time instants into the folds while avoiding sending two or more consecutive samples to the same fold, in order to avoid large prediction errors due to the temporal correlations in time series.

6   State Space Alternating Least Squares (SSALS) Algorithm

Recently, an alternating least squares (ALS) algorithm was developed to estimate MVAR-based SSMs in noisy and large-scale data analysis [6]. For the first time, we use this method here for time series forecasting, where application is more straightforward as $\mathbf{B} = \mathbf{I}_N$. In summary, the SSALS algorithm allows for the iterative alternated estimation of SSM parameters, e.g., $\{\mathbf{x}_t\}$, for $t = 1, \ldots, T$, and $\{\mathbf{A}_p\}$, for $p = 1, \ldots, P$.

For the first step, at i-th iteration, conditioning on the previous solution for the autoregressive coefficients (i.e., $\{\hat{\mathbf{A}}_p^{(i)}\}$), we obtain the closed-form solution for $\{\mathbf{x}_t\}$ as follows:

$$\hat{\mathbf{X}}_V^{(i+1)} = (\mathbf{D}_L^T \mathbf{D}_L \otimes \mathbf{B}^T \mathbf{B} + \lambda(\mathbf{I}_{TN} - \mathbf{W})^T$$
$$(\mathbf{I}_{TN} - \mathbf{W}))^{-1}(\mathbf{D}_L^T \mathbf{D}_L \otimes \mathbf{I}_N)vec(\mathbf{B}^T \mathbf{Y}), \qquad (7)$$

where $\mathbf{Y} = \{\tilde{\mathbf{y}}_T, \tilde{\mathbf{y}}_{T-1}, \ldots, \tilde{\mathbf{y}}_1\} \in R^{M \times T}$ (time-reversal order) is the matrix containing the measured data, $\hat{\mathbf{X}}_V = vec(\hat{\mathbf{X}})$ is the vectorized representation ($\hat{\mathbf{X}} \in R^{N \times T}$), $\otimes$ is the Kronecker's product, $\mathbf{D}_L$ contains the values of the loss function $L_t^{(k)}$ in its diagonal, and $\mathbf{W}$ is represented as follows [6]:

$$
\mathbf{W} =
\begin{bmatrix}
\mathbf{0}_N\ \mathbf{A}_1\ \cdots\ \mathbf{A}_P\ \mathbf{0}_N & \ddots \\
\mathbf{0}_N\ \mathbf{0}_N\ \mathbf{A}_1\ \cdots\ \mathbf{A}_P\ \mathbf{0}_N & \ddots \\
\ddots\ \mathbf{0}_N\ \mathbf{0}_N\ \mathbf{A}_1\ \cdots\ \mathbf{A}_P \\
\ddots & \cdots & \ddots \\
\mathbf{0}_N\ \mathbf{0}_N\ \mathbf{A}_1\ \cdots\ \mathbf{A}_P \\
\cdots\ \mathbf{0}_N\ \mathbf{0}_N\ \mathbf{0}_N\ \mathbf{0}_N \\
\cdots\ \cdots\ \mathbf{0}_N\ \mathbf{0}_N\ \mathbf{0}_N \\
\cdots\ \cdots\ \mathbf{0}_N\ \mathbf{0}_N\ \mathbf{0}_N
\end{bmatrix},
$$

which is a diagonal-block matrix containing the autoregressive matrices in the upper-diagonal blocks, for each $p = 1, \ldots, P$. In Eq.(7), for simplicity, the intercept coefficient is ignored as we can subtract it from measurements before the estimation step. Notice that these calculations are made in time-reversal order to be consistent with the structure of $\mathbf{W}$.

For the second step, given the numerical values $\hat{\mathbf{X}}_V^{(i+1)}$, then

$$
\hat{\mathbf{A}}^{(i+1)} = \mathbf{X}_{P+1:T} \mathbf{Z}^T (\mathbf{Z}\mathbf{Z}^T)^{-1}, \tag{8}
$$

is the classical least-squares solution for MVAR equations, used here for simplicity, where $\mathbf{A} = \{\mathbf{A}_1, \ldots, \mathbf{A}_P\} \in R^{N \times NP}$ and $\mathbf{Z} = \{\mathbf{X}_{P:T-1}; \ldots; \mathbf{X}_{1:T-P}\} \in R^{NP \times (T-P)}$, where $\mathbf{X}_{P+1:T} = \{\mathbf{x}_{P+1}, \ldots, \mathbf{x}_T\} \in R^{N \times (T-P)}$. Note that, similar as with Matlab notation, "," and ";" are used to denote horizontal and vertical concatenation operations. In numerical implementations, however, the Yule-Walker equations or a more efficient algorithm should be used to solve the MVAR model [2]. Note that, for the present application, the more computational demanding operation is given by Eq.(7).

Lastly, for each iteration, the intercept is updated as

$$
\hat{\mathbf{c}}^{(i+1)} = \frac{1}{\sum_{t=1}^{T}(L_t^{(k)} L_t^{(k)})} \sum_{t=1}^{T} L_t^{(k)} (\tilde{\mathbf{y}}_t - \mathbf{B}\hat{\mathbf{x}}_t^{(i+1)}). \tag{9}
$$

7 Validation with Simulated and Real Data

Here, we created a small-scale bivariate simulation to initially test the SSMs framework with synthetic data in noisy scenarios. The simulation parameter values are $T = 200$, $M = 5$, $N = 2$, and $P = 1$, and the time series are generated using Eqs.(1,2) by setting $\sigma_o = 1.0$ or $\sigma_o = 3.0$, for two different noisy scenarios, where $\sigma_s = 1$ for each case. We fixed the ground-truth autoregressive matrix as $\mathbf{A}_1 = \begin{bmatrix} -0.5 & 0 \\ 0.7 & -0.5 \end{bmatrix}$. We also tested different methods for a bivariate time series data of predator-prey population dynamics. This last example is discussed in great detail in Matlab documentation.

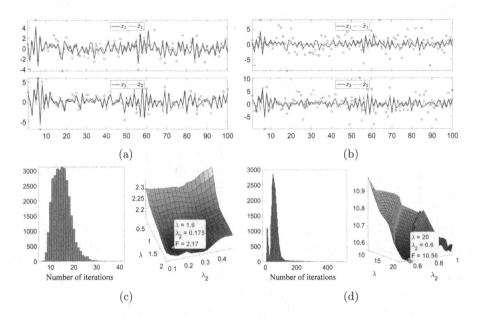

(a)   (b)

(c)   (d)

Fig. 2. Time series estimation with SSMs-based data-driven framework. (**a, b**) Estimated bivariate time series $\hat{\mathbf{x}}_t$ for the $\sigma_o = 1.0$ and $\sigma_o = 3.0$ simulated scenarios, respectively. The blue continuous curves highlight the ground-truth while the red discontinuous curves highlight the estimated time series. The open-circle markers denote the noisy observations $\tilde{\mathbf{y}}_t$. (**c, d**) Corresponding outcome for the cases shown in (a, b) respectively, highlighting the number of iterations for SSALS convergence, and the hyperparameter selection with the help of calculated cross-validation error surfaces.

7.1   Small Scale Simulation with Noisy Data

Figure 2 shows the results for this initial analysis. As can be seen by visual inspection, despite the high observation noise used in the simulations, in both noise conditions we were able to estimate the latent variables and parameters with high accuracy (2a, 2b). Although, the problem was more difficult to solve for the highest noise scenario ($\sigma_o = 3.0$) as the algorithm needed a much higher number of iterations (2d).

In this example, we did not optimize much for the selection of the hyperparameter values grid, as we try to use general grids that do not vary much from one noisy scenario to another and check how it may affect the solution quality in the worst cases. Here, for $\sigma_o = 1.0$, we used the grid $\lambda = 0.3 : 0.1 : 2.0$ and $\lambda_2 = 0.1 : 0.025 : 0.5$; whereas for $\sigma_o = 3.0$, we used $\lambda = 10 : 0.5 : 20$ and $\lambda_2 = 0.5 : 0.025 : 1.0$ (Matlab's notation). Note that due to the aforementioned probabilistic argument used to introduce the MVAR-based SSMs regularization framework, in the case that model parameters are known, then $\hat{\lambda} \approx \frac{\sigma_o^2}{\sigma_s^2}$. Thus, λ is expected to increase with the noise in the data. Also, we had to increase the search range for λ_2 as more noise should demand more regularization. Importantly, the presented regularization framework automatically reflects the noise

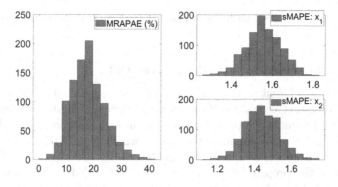

Fig. 3. Time series forecasting statistics calculated for the simulated bivariate time series example.

increment, which can be observed in the cross-validation error surfaces (Figs. 2c, 2d).

Figure 3 shows the results of the calculation of time series forecasting statistics using Monte-Carlo simulations with $N_{MC} = 1000$ replications. This was done only for the highest noise scenario ($\sigma_o = 3.0$). Notice that the stepwise prediction error is mostly below 30%, as reflected by the MRAPAE statistic, despite the high noise conditions. Similarly, the sMAPE statistic shows reasonably low values for the error on predicted time series by separate.

Interestingly, with these simulations, we noticed that the SSALS algorithm can converge to different local minima. In general, it is guaranteed that ALS algorithms produce iterative solutions correspondingly to strictly monotonic decreasing evaluations of the objective function, which converges to a fixed point [14]. However, as we noticed for the noisier scenarios, using separately the all-zeros and ground-true autoregressive coefficient values as initialization for the estimation of $\hat{\mathbf{X}}_V^{(1)}$ in the first iteration (Eq.(7)), the final solutions may not converge to the same estimator. Besides, we checked if these final solutions were local minima or saddle points by introducing small random perturbations in the neighbourhood of the obtained estimators and using it as initialization, but successive runs still did not move from the achieved solutions, except when the added perturbations were significantly high. This problem was not present when the observation noise was relatively small.

This uncertainty, caused by the solution being trapped in different local minima due to different initializations, was controlled by fixing always the same all-zeros initialization for the whole cross-validation procedure, but even so this issue appears for different cross-validation partitions despite of using the same hyperparameter values. Therefore, as a more robust strategy, we repeat the cross-validation partition for $R = 10$ replications. That is, in fact we run an $R \times K$ fold CVI procedure. Finally, instead of averaging the $R \times K$ calculated cross-validation prediction error curves, for each combination of the hyperparameter values, we calculated the median, as it is more robust to overfitting or gross errors due to the corresponding solutions being trapped in a suboptimal local solution.

7.2 Application to Prey-Predator Real Data

Here, we tested different methods for bivariate time series data analyses of prey-predator population dynamics. This example discusses models for both linear (**ARMA**) and nonlinear interactions (Gaussian processes and a grey-box models). It also includes a nonlinear AR model, which is an ARMA extension for bilinear interactions (Matlab's version 2022a).

As part of the preprocessing, the data was detrended and the model order was estimated with Matlab function "arxstruc", which is only necessary for the application of ARMA model. The estimated autoregressive order is $P = 8$. Subsequently, for all the models, the population time series was split between a training (first 120 samples) and validation (last 81 samples) subsets, and for all the models it was required to forecast the next 100 samples using only the testing data.

Figure 4 shows the results for the four mentioned methods (see more detail in Matlab documentation). Additionally, for estimating the MVAR-based SSMs demonstrated here we used a much higher order, $P = 50$, as we can control for overfitting during the model selection. The outcome shows that the proposed approach (black-colored forecast curves) can provide an accurate forecasting estimate for the future dynamical changes of these populations, despite the apparent nonlinearities and large MVAR model order in the analysis.

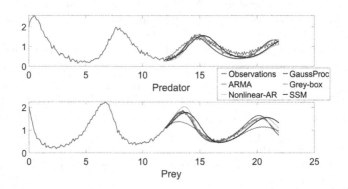

Fig. 4. Forecasting of dynamical changes in predator-prey population dynamics using five models: (i) ARMA (red curve), (ii) Nonlinear AR (yellow), (iii) Gaussian processes model, (iv) nonlinear Grey-box model (green) and (v) SSMs. (Color figure online)

8 Conclusion

We have applied for first time a MVAR-based SSMs regularization framework for analysis of time series forecasting. Mainly, the combination of a recently developed SSALS algorithm with a cross-validation approach [6,7] allowed the application of SSMs methodology to solve this problem, as demonstrated with simulated and real data. Although it was not demonstrated here, this framework has been extensively validated with large-scale neuroimaging data from synthetic and real magneto/electroencephalogram (MEG/EEG) datasets, which included comparison against state-of-the-art approaches [6,7]. Therefore, we believe that the proposed methodology can have a significant impact in time series forecasting and other similar applications, particularly for large-scale data analysis.

Acknowledgments. The authors are grateful for access to the Tier 2 High-Performance Computing resources provided by the Northern Ireland High Performance Computing (NI-HPC) facility funded by the UK Engineering and Physical Sciences Research Council (EPSRC), Grant No. EP/T022175/1. RCS was supported by RGPIN-2022-03042 from Natural Sciences and Engineering Research Council of Canada.

References

1. AugerMéthé, M., et al.: A guide to statespace modeling of ecological time series. Ecol. Monogr. **91**(4), e01470 (2021)
2. Barnett, L., Seth, A.K.: The MVGC multivariate Granger causality toolbox: A new approach to Granger-causal inference. J. Neurosci. Methods **223**, 50–68 (2014)
3. Puthanmadam, S.N., Tronarp, F., Särkkä, S., Parkkonen, L.: Joint estimation of neural sources and their functional connections from MEG data. Preprint, pp. 1–24, (2020). https://doi.org/10.1101/2020.10.04.325563
4. Van de Steen, F., Faes, L., Karahan, E., Songsiri, J., Valdes-Sosa, P., Marinazzo, D.: Critical comments on EEG sensor space dynamical connectivity analysis. Brain Topogr. **32**(4), 643–654 (2019). https://doi.org/10.1007/s10548-016-0538-7
5. Manomaisaowapak, P., Nartkulpat, A., Songsiri, J.: Granger causality inference in EEG source connectivity analysis: a state-space approach. IEEE Trans. Neural Netw. Learn. Syst. **33**(7), 1–11 (2021)
6. Sanchez-Bornot, J.M., Sotero, R.C., Kelso S., Özgür Ş., Coyle, D.: Solving large-scale MEG/EEG source localization and functional connectivity problems simultaneously using state-space models. Submitted to Neuroimage (under major revision). Preprint: https://arxiv.org/abs/2208.12854v1 (2022)
7. Sanchez-Bornot, J.M., Sotero, R.C., Coyle, D.: Dynamic source localization and functional connectivity estimation with state-space models: preliminary feasibility analysis. In: ICASSP 2023 IEEE International Conference on Acoustics, Speech and Signal Processing (ICASSP 2023). https://sigport.org/documents/dynamic-source-localization-and-functional-connectivity-estimation-state-space-models
8. Hastie, T., Friedman, J., Tibshirani, R.: The Elements of Statistical Learning. In: Springer Series in Statistics. Springer, New York (2001). https://doi.org/10.1007/978-0-387-21606-5
9. Bergmeir, C., Hyndman, R.J., Koo, B.: A note on the validity of cross-validation for evaluating autoregressive time series prediction. Comput. Stat. Data Anal. **120**, 70–83 (2018)
10. Bergmeir, C., Benítez, J.M.: On the use of cross-validation for time series predictor evaluation. Inf. Sci. **191**, 192–213 (2012)
11. Varoquaux, G., et al.: Assessing and tuning brain decoders: cross-validation, caveats, and guidelines. Neuroimage **145**, 166–179 (2017)

Causal Graph Discovery for Explainable Insights on Marine Biotoxin Shellfish Contamination

Diogo Ribeiro[1,2,3], Filipe Ferraz[1,2,3], Marta B. Lopes[4,5] iD,
Susana Rodrigues[6] iD, Pedro Reis Costa[6] iD, Susana Vinga[3] iD,
and Alexandra M. Carvalho[1,2,7(✉)] iD

[1] Instituto Superior Técnico (IST), 1000-049 Lisbon, Portugal
alexandra.carvalho@tecnico.ulisboa.pt
[2] Instituto de Telecomunicações (IT), 1000-049 Lisbon, Portugal
[3] INESC-ID, Instituto Superior Técnico, Universidade de Lisboa,
1000-029 Lisbon, Portugal
[4] Center for Mathematics and Applications (NOVA Math),
Department of Mathematics, NOVA SST, 2829-516 Caparica, Portugal
[5] UNIDEMI, Department of Mechanical and Industrial Engineering,
NOVA SST, 2829-516 Caparica, Portugal
[6] IPMA - Instituto Portuguës do Mar e da Atmosfera, 1495-165 Algés, Portugal
[7] Lisbon Unit for Learning and Intelligent Systems, Lisbon, Portugal

Abstract. Harmful algal blooms are natural phenomena that cause shellfish contamination due to the rapid accumulation of marine biotoxins. To prevent public health risks, the Portuguese Institute of the Ocean and the Atmosphere (IPMA) regularly monitors toxic phytoplankton in shellfish production areas and temporarily closes shellfish production when biotoxins concentration exceeds safety limits. However, this reactive response does not allow shellfish producers to anticipate toxic events and reduce economic losses. Causality techniques applied to multivariate time series data can identify the variables that most influence marine biotoxin contamination and, based on these causal relationships, can help forecast shellfish contamination, providing a proactive approach to mitigate economic losses. This study used causality discovery algorithms to analyze biotoxin concentration in mussels *Mytilus galloprovincialis* and environmental data from IPMA and Copernicus Marine Environment Monitoring Service. We concluded that the toxins that cause diarrhetic and paralytic shellfish poisoning had more predictors than the toxins that cause amnesic poisoning. Moreover, maximum atmospheric temperature, DSP toxins-producing phytoplankton and wind intensity showed causal relationships with toxicity in mussels with shorter lags, while *chlorophyll a* (*chl-a*), mean sea surface temperature and rainfall showed causal associations over longer periods. Causal relationships were also found between toxins in nearby production areas, indicating a spread of biotoxins contamination. This study proposes a novel approach to infer the relationships between environmental variables to enhance decision-making and public health safety regarding shellfish consumption in Portugal.

Keywords: Causal Discovery · Mussels Contamination · Biotoxins

© The Author(s), under exclusive license to Springer Nature Switzerland AG 2023
P. Quaresma et al. (Eds.): IDEAL 2023, LNCS 14404, pp. 483–494, 2023.
https://doi.org/10.1007/978-3-031-48232-8_44

1   Introduction

In recent years, the production and harvest of shellfish have increased from the north to the south coast of Portugal, with a significant impact on the national economy [1]. Since shellfish are filter-feeding organisms, they depend on the natural primary production available in the water column, such as microalgae (also known as phytoplankton). However, some microalgae species produce biotoxins that can accumulate and contaminate shellfish. Harmful Algal Blooms (HABs) are a natural phenomenon that occurs when environmental conditions (such as atmospheric, oceanographic and biological) are favourable, resulting in the rapid and uncontrolled growth of toxic phytoplankton [2,3]. Through these events, shellfish may accumulate high concentrations of biotoxins, making them unsafe for human consumption.

Contaminated shellfish pose a risk to human health and, depending on the type of poisoning, can lead to significant health problems. The most common types in Portugal are the Paralytic Shellfish Poisoning (PSP), the Amnesic Shellfish Poisoning (ASP), and the Diarrhetic Shellfish Poisoning (DSP). As the name implies, it can cause paralysis, amnesia and diarrhoea [4]. To protect public health, the European Parliament has implemented legal limits on contaminated bivalve molluscs placed on the market for human consumption. These limits are set at 160 µg AO equiv. kg^{-1} for DSP, 20 mg DA equiv. kg^{-1} for ASP, and 800 µg STX equiv. kg^{-1} for PSP [5].

In Portugal, the Portuguese Institute for the Ocean and Atmosphere (IPMA) carries out the official control of shellfish production areas. When biotoxin concentrations exceed safety limits, the harvest and trade of shellfish are prohibited, resulting in temporary closures of shellfish production areas. However, this reactive response causes economic losses, particularly affecting local farmers. Therefore, it is essential to develop proactive strategies that mitigate the damage caused by the proliferation of harmful phytoplankton and provide advance warnings to farmers.

Identifying the variables that contribute to shellfish contamination and understanding when production areas can be re-opened is crucial. Causal discovery techniques can play a pivotal role by inferring causal relationships between variables in the data. Understanding causal relationships brings us closer to comprehending and characterizing dynamic systems, allowing us to potentially foresee the effects of environmental system changes before they happen. The growing interest in causality algorithms to explore life sciences [6–8] has led to the development of several state-of-the-art time series causality discovery algorithms, including Granger causality [9], Peter-Clark Momentary Conditional Independence (PCMCI) [10], Vector Autoregressive Linear Non-Gaussian Acyclic Model (VAR-LiNGAM) [11], and DYNOTEARS [12].

Unravelling the underlying causal structure can result in ideal features for prediction and warning systems [13,14]. Indeed, most of the studies make use of forecasting models to predict shellfish toxicity levels and HAB events. Cruz et al. (2021) conducted a comprehensive literature review of the latest forecasting methods to predict HABs and shellfish contamination [15]. For the Portuguese coast scenario, Cruz et al. (2022) [16], focused on predicting DSP toxins concentrations in mussels one to four weeks in advance with multiple forecasting models. The most accurate response was achieved using a bivariate long short-term memory (LSTM) network predicting biotoxin contamination 1 week in advance based on toxic phytoplankton and biotoxin concentration time-series data.

Several environmental variables have been pointed as potential predictors of biotoxins shellfish contamination, e.g. atmospheric temperature, salinity, sea surface temperature (SST), rainfall, phytoplankton or wind direction [16–18].

Shellfish contamination has been studied on the Portuguese mainland coast, focusing on its seasonal contamination patterns and how they may be related to the seasonality of the environmental variables. Hence, it was verified that the toxicity in shellfish is higher between April/May and September/October [19]. Patrício et al. (2022) [20] determine that correlations between the same species in different areas are more impactful than correlations of different species inside the same production area since the accumulation of biotoxins in bivalves varies among species. The authors also reported that some bivalve species accumulate toxins faster than others, which is in accordance with the concept of indicator species. This refers to the shellfish species that has the highest rate of toxin accumulation for a given production area because accumulates biotoxins at the fastest rates, namely the mussel species *Mytilus galloprovincialis* [21,22].

This study aims to infer causal relationships from observational multivariate time series. Specifically, it employs causality discovery algorithms to identify the environmental variables with the most significant influence on current mussels *M. galloprovincialis* contamination based on their past values, something that has not yet been done. This way, a novel perspective for inferring the relationships that govern biotoxin accumulation is proposed, which can aid in decision-making and improve seafood safety regarding mussels consumption in Portugal.

2   Background

Granger Causality. One of the oldest and most well-known concepts to infer causal relationships from time series data is the Granger causality [9]. This statistical concept verifies whether a time series can be used to improve the prediction about another time series, respecting temporal precedence. Consider two univariate time series X and Y, X Granger causes Y if X contains unique and statistically significant information on the past observations of Y that is not available in Y. This can be formalized using a vector autoregressive (VAR) model [23].

The Granger Causality test is available in the *Python* package called *statsmodels*, using the function *grangercausalitytests* [24]. It is important to note that this test is bivariate. Therefore, when applied in a multivariate context, it may result in some spurious correlations.

PCMCI. The PCMCI algorithm [10] is a constraint-based approach that exploits the conditional independencies embedded in the data to build the underlying causal graph that follows several assumptions: causal Markov condition, causal sufficiency, causal stationarity and faithfulness. This causal discovery algorithm is able to detect time-lagged linear and nonlinear causal relations in time-series data and can be framed into two main phases: the Peter-Clark (PC) phase and the Momentary Conditional Independence (MCI) phase. In the first stage, the PC algorithm is applied to uncover dependencies between each variable and estimate a set of parents for every variable. The resulting skeleton is constructed based on conditional independence tests and the unnecessary edges, such as indirect links or independent links, are removed to avoid conditioning on irrelevant variables. At this stage, the goal is to identify direct causal relationships and remove spurious correlations. In the second phase, an MCI test is conducted to determine causal relationships between the previously selected time-shifted parents and each pair of variables by testing for independence, while taking into account autocorrelation. Basically, the direction of causal links is established and the flow of information is identified. The PCMCI is available in the *Python* package called *Tigramite* using the function *run_pcmi* [10].

VAR-LiNGAM. The VAR-LiNGAM [11,25] is an algorithm that belongs to the noise-based approach, where causal models are described by a set of equations. Each equation describes one variable of the causal structure plus some additional noise.

However, given two variables, it is not easy to distinguish cause from effect. There must be a way to capture the asymmetry between them. The VAR-LiNGAM is a temporal extension assuming linear, non-Gaussian, and acyclic models (LiNGAM), to uncover the underlying asymmetries, i.e., causal relations. Without these additional assumptions, it would not be possible to identify the direction of the causal relationship. The main idea of this algorithm is to estimate the least-squares of the Structural AutoRegressive Model (SVAR) to determine the causal order based on the independence between the residuals and predictors. This step is carried out iteratively by first identifying the predictor that is the most independent from the residuals of its target variables. Then, the effects of the previously identified predictors are removed. The VAR-LiNGAM is available in the *Python* package called *lingam* using the function *VARLiNGAM* [25].

DYNOTEARS. The DYNOTEARS algorithm [12] is a score-based approach that aims to learn an SVAR model in time-series data. This model can be considered a class of Dynamic Bayesian Networks (DBNs), which are unable to capture temporal dynamics. DYNOTEARS differs from other existing algorithms by simultaneously learning both contemporaneous (intra-slice) and lagged (inter-slice) dependencies between variables, instead of applying these steps successively. The adjacency matrices for the inter-slice and intra-slice dependencies are learned by minimizing a loss function based on the Frobenius norm of the residuals of a linear model. However, an acyclicity constraint is still necessary to model contemporaneous links. The DYNOTEARS is available in the *Python* package called *CausalNex* using the function *from_pandas_dynamic* [12].

3   Data Preprocessing

Data Description. Time-series data were collected from 2015 to 2020, provided by IPMA and Copernicus Marine Environment Monitoring Service (CMEMS). The variables selected for this analysis were chosen based on the literature, i.e. whether they were identified as potential predictors of shellfish contamination by DSP, ASP, PSP or toxic phytoplankton (which are in-situ measurements). IPMA provided the referenced variables, as well as meteorological elements from various meteorological stations, such as minimum, mean and maximum atmospheric air temperature, wind speed, wind direction (encoded in cardinal directions), mean wind direction (encoded in degrees) and rainfall. Additionally, the CMEMS provided measures obtained by remote sensing (satellite), such as SST and *chl-a*.

Data Selection. The objective of this project is to assess potential drivers of contamination in mussels *M. galloprovincialis*. This species is by far the most frequently analysed in Portugal and is used as indicator species in many production areas. When toxin levels in mussels are above the EU regulatory limits, other shellfish species of the production area are monitored in order to confirm if the biotoxins levels are below the regulatory limits and harvest is allowable.

Out of the 40 shellfish production areas available for analysis, 8 were selected based on the number of mussel samples and their geographical location: RIAV1, RIAV4, ETJ1, L5b, LAL, L7c1, LAG and POR2.

After the data acquisition process, it was necessary to determine which variables were worth exploring, regarding the amount of missing data present. In this way, to ensure consistency in testing the same predictors across different production areas, only variables with less than one-third of missing data in up to three production areas we considered. If a variable had more than one-third of missing data in four or more production areas, the variable was discarded. Table 1 provides details on the selected variables in each area.

Table 1. Variables selected to each production area.

Variable	Description	Unit	Type
DSP toxins	DSP toxins concentration	μg AO equiv. kg^{-1}	Continuous
ASP toxins	ASP toxins concentration	mg DA equiv. kg^{-1}	Continuous
PSP toxins	PSP toxins concentration	μg STX equiv. kg^{-1}	Continuous
DSP phyto	DSP toxins-producing phytoplankton	cell L^{-1}	Continuous
ASP phyto	ASP toxins-producing phytoplankton	cell L^{-1}	Continuous
PSP phyto	PSP toxins-producing phytoplankton	cell L^{-1}	Continuous
SST	Mean sea surface temperature	K	Continuous
Chl-a	Mean of *chlorophyll-a* concentration	mg m^{-3}	Continuous
Max temperature	Maximum air temperature	°C	Continuous
Wind intensity	Mean wind speed	m s^{-1}	Continuous
Wind direction	Mode wind direction (N,S,E...)	-	Discrete
Rainfall	Accumulated precipitation	mm	Continuous

Datasets Integration. To aggregate the data from IPMA and CMEMS, we considered the sampling times. The meteorological and satellite data were measured approximately every day, while the in-situ data were approximately weekly. However, there were weeks without any measurements and weeks with multiple measurements. Thus, the data were resampled over 312 weeks, capturing the worst situation of toxins and toxic phytoplankton each week and the weekly mean for the other variables.

Regarding the imputation of missing values, the monthly mode was used for wind direction, which is the only discrete variable. For the continuous variables, an algorithm was employed to select the best imputation method that minimizes the root mean squared error (RMSE) between the ground truth (longest period without missing data) and the same period inserted with missing data. The *interpolate* function from the *pandas* library in *Python* was used with the following options: *linear*, *quadratic*, *cubic*, *spline*(order = 2), *spline*(order = 3), *locf*, *nocb*, *nn*, and *knn*. For instance, the production area L7c1 obtain as imputation method: *linear*, *locf*, *nn*, *linear*, *spline3*, *linear*, *linear*, *spline2*, *spline3*, *spline3* and *knn* for the following variables, respectively: DSP toxins, ASP toxins, PSP toxins, DSP toxins-producing phytoplankton, ASP toxins-producing phytoplankton, PSP toxins-producing phytoplankton, SST, *chl-a*, maximum temperature, wind intensity and rainfall.

We conducted a stationary analysis, as non-stationary data can lead to spurious correlations. For analysing stationarity, a combination of the augmented Dickey-Fuller (ADF) test [26] test and the Kwiatkowski-Phillips-Schmidt-Shin (KPSS) test [27] was performed. This system can produce one of four results: (i) the time series is stationary; (ii) the time series is trend stationary: detrend by applying a moving average; (iii) the time series has a stochastic trend: difference one time; (iv) the time series is not stationary: difference n times (other transformations may be necessary). In practice, only the first three were found, and when necessary, we performed detrending (ii) or differencing (iii) to make the time series stationary.

Finally, normalization techniques were performed using the *minmaxScaler* so that variables with different scales could contribute equally to the model.

4   Results and Discussion

Model Development. Since we lack prior knowledge to develop models tailored to the characteristics of the real system, we cannot determine the best algorithms for our case study. However, we can group the results obtained from various models and evaluate the most commonly occurring causal relationships. To this end, five methods were employed: (i) Granger causality; (ii) PCMCI with partial correlation (ParCorr) as independence test; (iii) PCMCI with Gaussian processes and distance correlation (GPDC) as independence test; (iv) VAR-LiNGAM; and (v) DYNOTEARS. Additionally, to achieve a more precise analysis with less uncertainty, we only considered oriented edges to infer the direct relationship between two variables. Undirected and bidirected edges were discarded.

We consider a cause to be a predictor of shellfish contamination if multiple models identify the same relationships among the data. For example, if two out of the five models detected that the maximum temperature causes the concentration of DSP toxins with a 1-week lag (Max temp$(t-1)\rightarrow$DSP toxins(t)), we can conclude that changes or variations in the maximum temperature provide useful information for explaining changes in the DSP toxins 1-week ahead.

Causal Association Between Toxins Contamination and Environmental Variables. Herein, we discuss the cause-effect relationships between the toxicity in mussels, either by DSP, ASP or PSP toxins, and the biological (toxins-producing phytoplankton), oceanographic (SST and *chl-a*) and meteorological (maximum temperature, wind intensity, wind direction and rainfall) variables.

Figure 1 represents the causal associations detected in several mussels production areas by at least two models with temporal lags of up to four weeks. It is worth noting that thicker edges represent causal connections that were detected in several production zones. Thus, the links discovered in more areas were the following: (i) DSP toxins autocorrelation at a 1-week lag was revealed in all production areas; (ii) PSP autocorrelation at a 1-week lag was found in 7 production areas; (iii) ASP autocorrelation at a 1-week lag was found in 5 production areas; (iv) PSP autocorrelation at a 2-week lag was found in the half production areas.

Regarding the number of causal associations for each biotoxin, the DSP toxins have the fewest links (9), however, with more different predictors: *chl-a*, maximum temperature, DSP toxins-producing phytoplankton, wind intensity, and wind direction (the first three of which are only unique to DSP toxins). ASP toxins have the highest number of links (11), with the following predictors: wind intensity, wind direction, SST and ASP toxins-producing phytoplankton (only the last one is unique to ASP toxins). PSP toxins have 10 links, with the predictors rainfall and PSP toxins-producing phytoplankton being unique to this toxin, and sharing SST with ASP toxins. The fact that DSP toxins show the fewest causal links in terms of quantity can be explained by the lack of causal relationships with DSP toxins-producing phytoplankton. Note that, ASP and PSP toxin-producing phytoplankton influence ASP and PSP toxins with lags of 1, 2, 3 and 4 weeks, while DSP toxin-producing phytoplankton only show causal relationships of 1 week. However, it is not surprising. According to Table 2, DSP toxins are the biotoxins that most often exceed safety limits, which may explain the difficulty in finding a causal link with phytoplankton. Additionally, the quantification of toxic phytoplankton may not be representative of the entire water column where mussels inhabit, other DSP-toxins producing algae may be present, and more plausible is the differential toxins accumulation dynamics of shellfish that may not mirror algae abundance, as have been observed in controlled experiments in the laboratory [28,29].

This approach alone does not allow us to isolate which predictors contribute most to mussel contamination above safe limits, but it does allow us to infer in a more general way which are the most important variables in the toxicity ecosystem and which temporal lag impacts mussel toxicity. Thus, we can conclude that the maximum temperature, DSP toxins-producing phytoplankton and

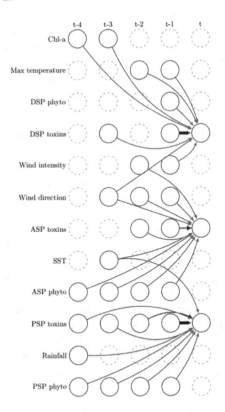

Fig. 1. Causal relationships detected with a temporal lag up to four weeks. Grey edges represent that they were detected in a maximum of 3 production areas. Black edges means they were detected in 4 or 5 areas. Thick black edges denote that they were detected in 7 or 8 production areas. (Color figure online)

Table 2. Number of mussel samples exceeding the legal limit in each production area between 2015 and 2020.

Toxin	RIAV1	RIAV4	ETJ1	L5b	LAL	LAG	POR2	L7c1
PSP	11	4	8	11	7	0	1	1
ASP	1	1	0	0	0	0	0	1
DSP	187	92	50	125	27	43	45	45

wind intensity are predictors of toxin concentration for shorter-term dependencies (1-week and 2-week lags), while *chl-a*, SST and rainfall are predictors of toxin concentration for longer-term dependencies (3-week and 4-week lags).

Causal Association Between Toxins Contamination and Toxins-Producing Phytoplankton Among Mussels Production Areas.

Herein, we discuss cause-and-effect relationships between biotoxin contamination and toxins-producing phytoplankton in geographically close locations,

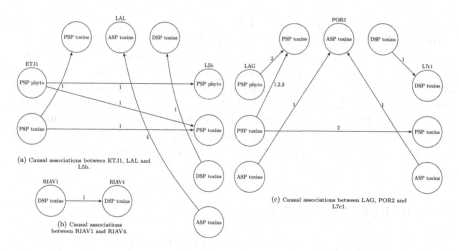

Fig. 2. Causal relationships among biotoxin contamination and toxins-producing phytoplankton in geographically production areas.

namely (a) LAG, POR2 and L7c1; (b) ETJ1, L5b and LAL; (c) RIAV1 and RIAV4. Figure 2 illustrates the significant causal connections detected between biotoxins contamination and toxins-producing phytoplankton in the mentioned areas. For this analysis, as only the toxin concentration in mussels and toxic phytoplankton cell counts from different production areas are used, only causal links that had been detected by at least 3 algorithms were considered, in order to decrease the spurious correlations.

For the mussels production areas of ETJ1, LAL and L5b (more centrally located in Portugal), it was determined that PSP toxins originating from ETJ1 contribute to PSP contamination in production areas LAL and L5b with a time-lag of 1 week, and that DSP and ASP toxins spread from L5b to LAL with a 1-week and 4-week lags, respectively. For production areas RIAV1 and RIAV4 (further north in Portugal), DSP toxins originating from RIAV1 and spread to RIAV4 with a 1-week lag. Finally, for the most southern production areas in Portugal LAG, POR2 and L7c1, ASP toxins originating in LAG and L7c1 have been found to contribute to the contamination present in POR2 with 1-week lag. Furthermore, PSP toxins originating from the LAG production area spread to POR2 with a 1-week, 2-week and 3-week lags and to L7c1 with a time-lag of 2 weeks. Lastly, DSP toxins from the POR2 production area were found to contribute to the contamination in L7c1. These results are in accordance with previous field observations on regional synchronism, spatial distribution and spread of *Dinophysis* (DSP toxins-producing algae) and *Gymnodinium catenatum* (PSP toxins-producing algae) blooms over shellfish production areas along the Portuguese coast [30, 31].

The interpretation of the spread of mussel contamination between production areas aims to improve warning systems based on the specific characteristics of each production area. For example, if PSP toxins in ETJ1 contribute to PSP contamination in production areas LAL and L5b, whenever PSP toxin events occur in ETJ1, local producers in LAL and L5b should be warned in advance. It should be noted that this analysis is more impactful in production areas that regularly exceed safety limits. As causality methods applied to time series do not allow to isolate only the predictors of contamination above the safety limits, any cause-effect below the limits are also detected, namely the causal associations seen in LAG, POR2 and L7c1 for ASP and PSP toxins. Thus, by analysing production areas where toxicity is regularly high, such as DSP in RIAV1 and RIAV4, we are able to get closer to the primary goal of contamination above safety limits.

5   Conclusion and Future Work

The temporary and unpredictable closure of shellfish production areas due to high biotoxin concentrations continues to pose a challenge. Identifying the contamination phenomenon and the impact of collected climate variables can aid in constructing more precise forecasting systems. These systems can function as warning mechanisms to detect mussel toxicity.

This work proposed using five causality models to investigate the causal connections between mussel toxicity and climate variables, including Granger causality, PCMCI with ParCorr and GPDC, VAR-LiNGAM and DYNOTEARS.

We initiated the study by examining the potential connections between marine biotoxins and environmental variables. Among the toxins, ASP toxins had the fewest predictors and were the least likely to surpass safety thresholds. Notably, DSP and PSP toxins exhibited a robust 1-week lag autocorrelation, which could enhance the 1-week ahead forecasting system.

In general, the variables can be differentiated based on their predictive role. Maximum temperature, DSP toxins-producing phytoplankton and wind intensity are useful for predicting toxin contamination over shorter periods. Meanwhile, *chl-a*, SST and rainfall exhibit causal relationships over longer periods.

Lastly, we examined the spread of toxicity among production zones and identified several significant pairs. Detecting causal relationships between production areas can serve as a warning for potential zone closures. For example, the RIAV1 production area appears to influence the concentration of DSP toxins in the RIAV4 area with a 1-week lag. Therefore, if we identify high toxicity cases in RIAV1, local producers in RIAV4 can be warned in advance.

Acknowledgements. This work was supported by national funds through FundaÇão para a Ciência e a Tecnologia (FCT) through projects UIDB/00297/2020 and UIDP/00297/2020 (NOVA Math), UIDB/00667/2020 and UIDP/00667/2020 (UNIDEMI), UIDB/50008/2020 (IT), UIDB/50021/2020 (INESC-ID), and also the project MATISSE (DSAIPA/DS/0026/2019), and CEECINST/00042/2021, PTDC/CCI-BIO/4180/2020, and PTDC/CTM-REF/2679/2020. This project has received funding from the European Union's Horizon 2020 research and innovation programme under grant agreement No 951970 (OLISSIPO project).

References

1. Mateus, M., et al.: Early warning systems for shellfish safety: the pivotal role of computational science. In: Rodrigues, J.M.F., et al. (eds.) ICCS 2019. LNCS, vol. 11539, pp. 361–375. Springer, Cham (2019). https://doi.org/10.1007/978-3-030-22747-0_28

2. Lee, T., Fong, F., Ho, K.-C., Lee, F.: The mechanism of diarrhetic shellfish poisoning toxin production in prorocentrum spp.: physiological and molecular perspectives. Toxins 8, 272 (2016)

3. Dale, B., Edwards, M., Reid, P.: Climate Change and Harmful Algal Blooms (2006)

4. Grattan, L.M., Holobaugh, S., Jr. Morris, J.G.: Harmful algal blooms and public health. Harmful Algae 57(B), 2–8 (2016)

5. European Parliament, Council of the European Union. Commission Regulation (EC) No 853/2004 of the European Parliament and of the Council of 29 April 2004 Laying down specific hygiene rules for food of animal origin. Off. J. Eur. Union 2004, L226, 22–82 (2004). https://www.ipma.pt/pt/bivalves/docs/index.jsp

6. Runge, J., Bathiany, S., Bollt, E., et al.: Inferring causation from time series in Earth system sciences. Nat. Commun. 10, 2553 (2019)

7. Kretschmer, M., Coumou, D., Donges, J., Runge, J.: Using causal effect networks to analyze different arctic drivers of midlatitude winter circulation. J. Clim. 29(11), 4069–4081 (2016)

8. McGowan, J.A., et al.: Predicting coastal algal blooms in southern California. Ecology 98, 1419–1433 (2017)

9. Granger, C.W.J.: Investigating causal relations by econometric models and cross-spectral methods. Econometrica 37, 424–438 (1969)

10. Runge, J., Nowack, P., Kretschmer, M., Flaxman, S., Sejdinovic, D.: Detecting and quantifying causal associations in large nonlinear time series datasets. Sci Adv. 5(11), eaau4996 (2019). https://github.com/jakobrunge/tigramite

11. Hyvärinen, A., Zhang, K., Shimizu, S., Hoyer, P.: Estimation of a structural vector autoregression model using non-Gaussianity. J. Mach. Learn. Res. 11, 1709–1731 (2010)

12. Pamfil, R., et al.: DYNOTEARS: structure learning from time-series data (2020). https://github.com/quantumblacklabs/causalnex

13. Davidson, K., et al.: HABreports: online early warning of harmful algal and biotoxin risk for the scottish shellfish and finfish aquaculture industries. Front. Mar. Sci. 8, 631732 (2021)

14. Silva, A., et al.: A HAB warning system for shellfish harvesting in Portugal. Harmful Algae 53, 33–39 (2016)

15. Cruz, R.C., Reis, C., Vinga, S., Krippahl, L., Lopes, M.B.: A review of recent machine learning advances for forecasting harmful algal blooms and shellfish contamination. J. Mar. Sci. Eng. 9, 283 (2021)

16. Cruz, R., Reis, C., Krippahl, L., Lopes, M.: Forecasting biotoxin contamination in mussels across production areas of the Portuguese coast with artificial neural networks. Knowl Based Syst. 257, 109895 (2022)

17. Mudadu, A.G., et al.: Influence of seasonality on the presence of okadaic acid associated with Dinophysis species: a four-year study in Sardinia (Italy). Ital. J. Food Saf. 10(1), 8947 (2021)

18. Vale, P.: Two simple models for accounting mussel contamination with diarrhoetic shellfish poisoning toxins at Aveiro lagoon: control by rainfall and atmospheric forcing. Estuar. Coast. Shelf 98, 94–100 (2012)

19. Braga, A.C., Rodrigues, S.M., Lourenço, H.M., Costa, P.R., Pedro, S.: Bivalve shellfish safety in Portugal: variability of faecal levels, metal contaminants and marine biotoxins during the last decade (2011–2020). Toxins **15**, 91 (2023)

20. Patrício, A., Lopes, M.B., Costa, P.R., Costa, R.S., Henriques, R., Vinga, S.: Time-lagged correlation analysis of shellfish toxicity reveals predictive links to adjacent areas, species, and environmental conditions. Toxins **14**, 679 (2022)

21. Vale, P., Gomes, S.S., Botelho, M.J., Rodrigues, S.M.: Monitorização de PSP na costa portuguesa através de espécies-indicadoras. In: Avances y tendencias en Fito-plancton Tóxico y Biotoxinas, Gilabert, J. (Ed.), U. P. de Cartagena (2008)

22. Vale, P., Botelho, M.J., Rodrigues, S.M., Gomes, S.S., Sampayo, M.A.D.M.: Two decades of marine biotoxin monitoring in bivalves from Portugal (1986–2006): a review of exposure assessment. Harmful Algae **7**(1), 11–25 (2008)

23. Assaad, C.K., Devijver, E., Gaussier, E.: Survey and evaluation of causal discovery methods for time series. J. Artif. Int. Res. **73** (2022)

24. Seabold, S., Perktold, J.: Statsmodels: econometric and statistical modeling with python. In: Proceedings of the 9th Python in Science Conference (2010). https://github.com/statsmodels/statsmodels

25. Ikeuchi, T., Ide M., Zeng Y., Maeda T.N., Shimizu S.: Python package for causal discovery based on LiNGAM. J. Mach. Learn. Res. **24**, 14 (2023). https://github.com/cdt15/lingam

26. Dickey, D.A., Fuller, W.A.: Distribution of the estimators for autoregressive time series with a unit root. J. Am. Stat. Assoc. **74**, 427–431 (1979)

27. Kwiatkowski, D., Phillips, P.C.B., Schmidt, P., Shin, Y.: Testing the null hypothesis of stationarity against the alternative of a unit root. J. Econ. **54**, 159–178 (1992)

28. Reguera, B., et al.: Dinophysis toxins: causative organisms, distribution and fate in shellfish. Mar. Drugs **12**, 394–461 (2014)

29. Braga, A.C., et al.: Invasive clams (ruditapes philippinarum) are better equipped to deal with harmful algal blooms toxins than native species (R. Decussatus): evidence of species-specific toxicokinetics and DNA vulnerability. Sci. Total Environ. **767**, 144887 (2021)

30. Moita, M.T., Oliveira, P.B., Mendes, J.C., Palma, A.S.: Distribution of chlorophyll a and gymnodinium catenatum associated with coastal upwelling plumes off central Portugal. Acta Oecologica **24**, S125–S132 (2003)

31. Moita, M.T., Pazos, Y., Rocha, C., Nolasco, R., Oliveira, P.B.: Toward predicting dinophysis blooms off NW Iberia: a decade of events. Harmful Algae **53**, 17–32 (2016)

Special Session on Federated Learning and (Pre) Aggregation in Machine Learning

Adaptative Fuzzy Measure for Edge Detection

C. Marco-Detchart[1]([✉]) [iD], G. Lucca[2] [iD], G. Dimuro[2] [iD], J. A. Rincon[1,3] [iD], and V. Julian[1,3] [iD]

[1] Valencian Research Institute for Artificial Intelligence (VRAIN),
Universitat Politècnica de València (UPV),
Camino de Vera s/n, 46022 Valencia, Spain
{cedmarde,vjulian}@upv.es, jrincon@dsic.upv.es
[2] Centro de Ciências Computacionais, Universidade Federal do Rio Grande,
Av. Itália km 08, Campus Carreiros, Rio Grande 96201-900, Brazil
{giancarlo.lucca,gracalizdimuro}@furg.br
[3] Valencian Graduate School and Research Network of Artificial Intelligence
(VALGRAI), Universitat Politècnica de València (UPV),
Camí de Vera s/n, 46022 Valencia, Spain

Abstract. In this work, an analysis of the influence of the fuzzy measure for Choquet integral and its generalizations is presented. The work has been done in the context of feature fusion for edge detection with gray-scale images. The particular case of adaptive fuzzy measure is considered, testing a variety of approaches. We have tested our proposal using the power measure adapting the exponent depending on the local information of each particular image. For comparison purposes and to test the performance of our proposal, we compare our approach to the results obtained with the Canny edge detector.

Keywords: Edge detection · Feature extraction · Choquet integral · Fuzzy measure · Aggregation

1 Introduction

Boundary detection has been a subject of extensive research in computer vision, resulting in diverse literature featuring various inspirations and computational structures. In early works, edges were primarily regarded as fundamental components of low-level features, with the understanding that they played a crucial role in the overall operations. This concept was supported by neurophysiological experiments conducted by Hubel and Wiesel [10,11], demonstrating that contour detection in the human visual system relies on early receptive fields. These findings led to significant advancements in edge detection, such as the neuro-inspired Laplacian operators proposed by Marr and Hildreth [15,17]. In fact, edges play a crucial role in Marr's *Primal Sketch* [16].

Edge detection methods often rely on gradient-based approaches, which involve measuring the intensity variation along specific directions in the pixel's

P. Quaresma et al. (Eds.): IDEAL 2023, LNCS 14404, pp. 497–505, 2023.
https://doi.org/10.1007/978-3-031-48232-8_45

neighbourhood. These gradients are typically computed by convolving the original image with a filter. Several well-known edge detectors utilize this approach, including Sobel [21], Prewitt [20], and Canny [7].

The increasing complexity of boundary detection methods has created a need for process normalization that allows for a systematic understanding of boundary detection as a series of configurable and interpretable steps. It also facilitates qualitative and quantitative comparisons among different methods beyond their end results. Moreover, process normalization enables testing specific techniques in individual phases of the process, aiming to improve and optimize existing methods.

One notable proposal for process normalization in edge detection was presented by Bezdek et al. [5] The framework proposed, known as the Bezdek Breakdown Structure (BBS), delineates the edge detection process into four distinct phases: conditioning, feature extraction, blending, and scaling. The conditioning phase enhances the image quality and removes unnecessary information, feature extraction computes the variations between pixels, blending fuses the extracted features to identify edges, and scaling thins the edges to obtain a binary edge image.

This work focuses on the concept of boundary feature fusion in the blending phase of the BBS. Specifically, we analyze the behaviour of an adaptive fuzzy measure in the fusion process using the Choquet integral and some of its generalizations. These functions were chosen as the fusion method due to their ability to effectively capture the relationship between neighbouring pixels.

The structure of this paper is as follows: Sect. 2 provides the necessary preliminaries for our proposed approach, Sect. 3 describes the experimental study, including the dataset, performance metrics, along with the analysis of the obtained results, and finally, Sect. 4 summarizes the conclusions drawn from our research and suggests future work.

2   Preliminaries

This section recovers some important concepts needed to understand the approach presented in this work. Concretely we recall definitions for fuzzy measure, Choquet integral and some of its generalisations, as well as the concept of aggregation function.

Definition 1. *[19] A function* $\mathfrak{m} : 2^N \rightarrow [0, 1]$ *is a* fuzzy measure *if, for all* $X, Y \subseteq N = \{1, \ldots, n\}$*, the following properties hold: (i) if* $X \subseteq Y$*, then* $\mathfrak{m}(X) \leq \mathfrak{m}(Y)$*; (ii)* $\mathfrak{m}(\emptyset) = 0$ *and* $\mathfrak{m}(N) = 1$*.*

Definition 2. *[8] The discrete Choquet integral, related with the fuzzy measure* $\mathfrak{m}$*, is the function* $\mathfrak{C}_\mathfrak{m} : [0, 1]^n \rightarrow [0, 1]$*, defined, for all* $x \in [0, 1]^n$*, by*

$$\mathfrak{C}_\mathfrak{m}(x) = \sum_{i=1}^{n} \left(x_{(i)} - x_{(i-1)} \right) \cdot \mathfrak{m}\left(A_{(i)} \right), \tag{1}$$

where $(x_{(1)}, \ldots, x_{(n)})$ *is an increasing permutation of* x*,* $x_{(0)} = 0$ *and* $A_{(i)} = \{(i), \ldots, (n)\}$ *is the subset of indices of* $n - i + 1$ *largest components of* x*.*

Definition 3. *[4] A mapping* $M : [0,1]^n \to [0,1]$ *is an aggregation function if it is monotone non-decreasing in each of its components and satisfies the boundary conditions,* $M(\mathbf{0}) = M(0,0,\dots,0) = 0$ *and* $M(\mathbf{1}) = M(1,1,\dots,1) = 1$.

Definition 4. *An aggregation function* $T : [0,1]^n \to [0,1]$ *is said to be a t-norm if, for all* $x,y,z \in [0,1]$, *the following conditions hold: Commutativity:* $T(x,y) = T(y,x)$; *Associativity:* $T(x,T(y,z)) = T(T(x,y),z)$; *Boundary conditions:* $T(1,x) = T(x,1) = x$.

Definition 5. *[13] Let* $\mathfrak{m} : 2^N \to [0,1]$ *be a fuzzy measure and* $T : [0,1]^2 \to [0,1]$ *be a t-norm. Taking as basis the Choquet integral, we define the function* $\mathfrak{C}_\mathfrak{m}^T : [0,1]^n \to [0,n]$, *for all* $\mathbf{x} \in [0,1]^n$, *by:*

$$\mathfrak{C}_\mathfrak{m}^T(\mathbf{x}) = \sum_{i=1}^{n} T\left(x_{(i)} - x_{(i-1)}, \mathfrak{m}\left(A_{(i)}\right)\right), \tag{2}$$

where $\left(x_{(1)},\dots,x_{(n)}\right)$ *is an increasing permutation on the input* $\mathbf{x}$, *that is,* $x_{(1)} \leq \dots \leq x_{(n)}$, *with the convention that* $x_{(0)} = 0$, *and* $A_{(i)} = \{(i),\dots,(n)\}$ *is the subset of indices of the* $n - i + 1$ *largest components of* $\mathbf{x}$.

Definition 6. *[14] Let* $F : [0,1]^2 \to [0,1]$ *be a bivariate fusion function and* $\mathfrak{m} : 2^N \to [0,1]$ *be a fuzzy measure. The Choquet-like integral based on* F *with respect to* $\mathfrak{m}$, *called* C_F-*integral, is the function* $\mathfrak{C}_\mathfrak{m}^F : [0,1]^n \to [0,1]$, *defined, for all* $\mathbf{x} \in [0,1]^n$, *by*

$$\mathfrak{C}_\mathfrak{m}^F(\mathbf{x}) = \min\left\{1, \sum_{i=1}^{n} F\left(x_{(i)} - x_{(i-1)}, \mathfrak{m}\left(A_{(i)}\right)\right)\right\}, \tag{3}$$

where $\left(x_{(1)},\dots,x_{(n)}\right)$ *is an increasing permutation on the input* $\mathbf{x}$, *that is,* $x_{(1)} \leq \dots \leq x_{(n)}$, *with the convention that* $x_{(0)} = 0$, *and* $A_{(i)} = \{(i),\dots,(n)\}$ *is the subset of indices of the* $n - i + 1$ *largest components of* $\mathbf{x}$.

One of the most simple and widely used fuzzy measures is the power measure, which is adopted in this work:

$$\mathfrak{m}_q(X) = \left(\frac{|X|}{n}\right)^q, \text{ with } q > 0, \tag{4}$$

where $|X|$ is the number of elements to be aggregated, n the total number of elements and $q > 0$. We have selected this measure as it is the one achieving the highest accuracy in classification problems [13,14]. In this work, we consider fixed values for q and an adaptative approach depending on the values found at each position.

3 Experimental Setup

In this section, we briefly explain the dataset used for the experiments along with the evaluation metrics selected to measure the performance of our proposal. We also expose the result obtained with the different configurations tested.

Table 1. Families of C_T and C_F-integrals used in this work, as generalizations of the Choquet integral.

Choquet-like integral	Base Function
$\mathfrak{C}_m^{C_F}$	$C_F(x,y) = xy + x^2y(1-x)(1-y)$ (copula [1])
$\mathfrak{C}_m^{O_B}$	$O_B(x,y) = \min\{x\sqrt{y}, y\sqrt{x}\}$ (overlap function [3,6])
$\mathfrak{C}_m^{F_{BPC}}$	$F_{BPC}(x,y) = xy^2$ (aggregation function)
$\mathfrak{C}_m^{\text{Hamacher}}$	$T_{HP}(x,y) = \begin{cases} 0 & \text{if } x = y = 0 \\ \frac{xy}{x+y-xy} & \text{otherwise} \end{cases}$ (t-norm)

3.1 Dataset and Evaluation of the Proposal

We analyze our proposal using the Berkeley Segmentation Dataset (BSDS500) [2], composed of three different partitions, *train*, *test* and *val*. For this experiment, we only use the *test* partition composed of 200 images. Each one of the images comes with 4 to 9 ground-truth images defined by experts (Fig. 1).

In this context, images are regarded as functions $D : R \times C \mapsto L$, where $R = 1, ..., r$ and $C = 1, ..., c$ represent the sets of rows and columns, respectively. L defines the type of image, which refers to the number of tones present in the image. For binary images, $L = 0, 1$, indicating two possible tones: 0 and 1. On the other hand, for gray-scale image pixels, $L = 0, ..., 255$, representing a range of tones from 0 to 255. In the case of colour images, L becomes the Cartesian product of tonal palettes for each colour stimulus. For example, $L = 0, ..., 255^3$ for RGB images. For the purpose of this study, gray-scale images are the focus of consideration.

The performance evaluation of an edge detector involves treating it as a binary classification task, where the ground-truth image indicates the presence or absence of an edge at each pixel. In our experiments, the obtained binary edges are compared with the ground truth to assess the detector's accuracy using the evaluation procedure suggested by Estrada and Jepson [9]. This method makes individual decisions on the validity of each boundary pixel, using information from the surrounding region in both detected edges and ground truth images. Given that potential disparities might exist between the detected presence of edges and the actual edges, it becomes necessary to account for a certain tolerance level, essential to categorize some of the detected edges as valid positives, even when a minor discrepancy exists. In this study, we established a spatial tolerance equivalent to 2.5% of the length of the image's diagonal.

To detect whether our solution is correct, a confusion matrix is built following Martin *et al.* approach [18], where True Positive (TP), True Negative (TN), False Positive (FP) and False Negative (FN) are extracted. Then, to quantify the results the following well-known Precision (PREC), Recall (REC) and F_α measures are considered: $\text{PREC} = \frac{\text{TP}}{\text{TP}+\text{FP}}$, $\text{REC} = \frac{\text{TP}}{\text{TP}+\text{FN}}$ and $F_\alpha = \frac{\text{PREC}\cdot\text{REC}}{\alpha\cdot\text{PREC}+(1-\alpha)\cdot\text{REC}}$. We opt for the most frequently utilized descriptor, namely the $F_{0.5}$ measure, as documented in the literature [12,18].

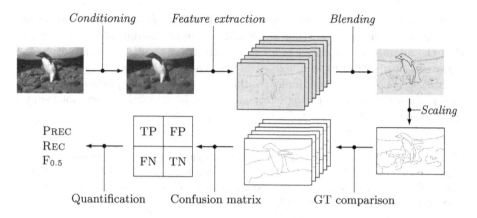

Fig. 1. Schematic representation of the BBS followed in this article, along with the performance evaluation process.

3.2 Framework and Results

Edge detection is commonly approached by convolving a kernel with the image to extract cues indicating the presence of edges. In this work, image features are computed using a sliding window technique (of size 3×3) applied to each pixel, where we extract the intensity difference between the central pixel and its eight neighbours. These features are subsequently combined to generate the final edge image. Since the obtained features typically comprise multiple values (in our case, eight features per pixel), it is necessary to represent them using a single value.

To achieve this, we employ a fusion process based on the Choquet integral and some of its extensions that have shown good results in classification problems (Table 1). For the q value of the fuzzy power measure, we use both fixed and adaptative values to study its behaviour. The adaptative q value consists in using a different q for each window. The q value is selected by using different approaches. Specifically, we test the maximum (max), the minimum (min) and the mean (mean). We test our proposal with two variations of the conditioning phase, using two different σ values for the Gaussian filter, $\sigma = 1$ (S_1) and $\sigma = 2$ (S_2).

As we can observe in Table 2 when using the S_1 smoothing, with less blurring, the best performer in terms of $F_{0.5}$ is the Choquet generalization with F_{BPC} combined with the adaptative fuzzy measure where q is obtained with the maximum value at each window. With this combination, we are even over the Canny edge detector. If we look at the remaining approaches, all of them are near Canny except two of them that obtain the same result ($C_{\max}$ and $\mathfrak{C}_{\max}^{C_F}$), both using the adaptative q obtained with the maximum. In an overall way, the use of the adaptative approach for obtaining q increases the performance of edge detection. Remark that the only generalization that does not benefit from the adaptative fuzzy measure is the one using Hamacher, which obtains the best results with $q = 1$ for all the positions of the image. If we analyze the results when a higher level of blurring is applied in the smoothing phase we observe that the adaptative approach tends to obtain worse results than one using the fuzzy measure with a fixed q, except in the case of the Choquet generalization with F_{BPC}. This behaviour indicates that the use of an adaptative fuzzy measure is sensitive to the blurring effect and could be beneficial when not using a regularization step (Fig. 2).

Table 2. Comparison of the different configurations for the power measure with the Choquet integral and its generalizations, along with the Canny method approach in terms of PREC, REC, and $F_{0.5}$. The experiments have been done applying Gaussian smoothing prior to the process with $\sigma = 1$ and $\sigma = 2$.

Method	Smoothing S_1			Smoothing S_2		
	PREC	REC	$F_{0.5}$	PREC	REC	$F_{0.5}$
C_1	.704	.756	.709	.730	.744	.719
C_{max}	.671	.818	.719	.745	.699	.705
C_{min}	.707	.738	.702	.760	.646	.681
C_{mean}	.693	.777	.714	.751	.669	.691
$\mathfrak{C}^{C_F}_{p^{0.8}}$	.704	.754	.708	.734	.739	.718
$\mathfrak{C}^{C_F}_{max}$	.673	.815	.719	.744	.698	.704
$\mathfrak{C}^{C_F}_{min}$	.707	.739	.702	.759	.647	.681
$\mathfrak{C}^{C_F}_{mean}$	.694	.774	.714	.750	.669	.691
$\mathfrak{C}^{O_B}_{p^1}$	.704	.760	.711	.738	.728	.714
$\mathfrak{C}^{O_B}_{max}$	.690	.783	.716	.751	.671	.692
$\mathfrak{C}^{O_B}_{min}$	.708	.735	.701	.759	.646	.681
$\mathfrak{C}^{O_B}_{mean}$	.702	.756	.708	.756	.655	.685
$\mathfrak{C}^{F_{BPC}}_{p^{0.4}}$	.703	.757	.709	.731	.742	.718
$\mathfrak{C}^{F_{BPC}}_{max}$	.655	.850	**.722**	.730	.744	**.721**
$\mathfrak{C}^{F_{BPC}}_{min}$	.706	.740	.702	.759	.651	.683
$\mathfrak{C}^{F_{BPC}}_{mean}$	.672	.817	.720	.743	.694	.702
$\mathfrak{C}^{Hamacher}_{p^1}$	.693	.776	.714	.750	.682	.698
$\mathfrak{C}^{Hamacher}_{max}$	.704	.748	.706	.760	.646	.681
$\mathfrak{C}^{Hamacher}_{min}$	.709	.734	.700	.762	.642	.680
$\mathfrak{C}^{Hamacher}_{mean}$	.707	.740	.703	.761	.644	.680
Canny	.742	.730	.719	.747	.704	.707

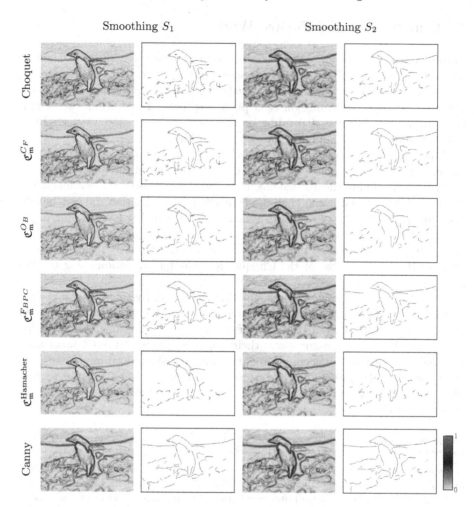

Fig. 2. Image features and final edges extracted from original image 106005 of the BSDS test set with the different approaches tested that obtained the best results compared to the classic Canny edge detector.

In terms of visual results, we can see that most of the methods obtain very similar features and edges. The most similar method to the result obtained by Canny is when using $\mathfrak{C}_{max}^{F_{BPC}}$, being the only one that detects the edge between the sky and the ground. In addition, the majority of the Choquet-based approaches detect the eye of the penguin while the best performer with smoothing S_1 does not delineate it, increasing its similarity to the ground-truth images.

4   Conclusions and Future Work

In this work, we have analyzed the role of the fuzzy measure within the Choquet integral and its generalizations in the context of edge detection. We have tested the fuzzy power measure with a fixed q value and an adaptative approach where the value changes depending on each pixel neighbour. In addition, to see how the adaptative q behaves depending on the regularization, we have tested two levels of smoothing by applying Gaussian smoothing.

The experiment conducted has yielded interesting results and is promising in enhancing edge detection performance. Overall, the use of the adaptative approach for determining the q value has improved the edge detection performance. However, it should be noted that the Hamacher generalization did not benefit from the adaptative fuzzy measure. When applying a higher level of blurring in the smoothing phase, the adaptative approach generally yielded poorer results than using a fixed q, which indicates that the adaptative fuzzy measure is sensitive to the blurring, except the Choquet integral generalization using overlap functions.

As future research lines, further exploring the adaptative fuzzy measure approach and investigating its performance under different blurring conditions would be interesting, especially in the case of using overlap functions. Additionally, exploring other fuzzy measures and their combinations with the Choquet integral and its generalizations could provide valuable insights into improving edge detection accuracy. Furthermore, considering alternative techniques for feature representation and fusion processes may also lead to advancements in the field of edge detection.

Acknowledgements. This work was partially supported with grant PID2021-123673OB-C31 funded by MCIN/AEI/10.13039/501100011033 and by "ERDF A way of making Europe", Consellería d'Innovació, Universitats, Ciencia i Societat Digital from Comunitat Valenciana (APOSTD/2021/227) through the European Social Fund (Investing In Your Future) and grant from the Research Services of Universitat Politècnica de València (PAID-PD-22).

References

1. Alsina, C., Frank, M.J., Schweizer, B.: Associative Functions: Triangular Norms and Copulas. World Scientific Publishing Company, Singapore (2006)
2. Arbelaez, P., Maire, M., Fowlkes, C., Malik, J.: Contour detection and hierarchical image segmentation. IEEE Trans. Pattern Anal. Mach. Intell. **33**(5), 898–916 (2011)
3. Bedregal, B.C., Dimuro, G.P., Bustince, H., Barrenechea, E.: New results on overlap and grouping functions. Inf. Sci. **249**, 148–170 (2013)
4. Beliakov, G., Bustince Sola, H., Calvo, T.: A Practical Guide to Averaging Functions, Studies in Fuzziness and Soft Computing, vol. 329. Springer, Cham (2016)
5. Bezdek, J., Chandrasekhar, R., Attikouzel, Y.: A geometric approach to edge detection. IEEE Trans. Fuzzy Syst. **6**(1), 52–75 (1998)

6. Bustince, H., Fernandez, J., Mesiar, R., Montero, J., Orduna, R.: Overlap functions. Nonlinear Anal. Theory Methods Appl. **72**(3–4), 1488–1499 (2010)
7. Canny, J.F.: A computational approach to edge detection. IEEE Trans. Pattern Anal. Mach. Intell. **8**(6), 679–698 (1986)
8. Choquet, G.: Theory of capacities. Ann. l'Institut Fourier **5**, 131–295 (1953–1954)
9. Estrada, F.J., Jepson, A.D.: Benchmarking image segmentation algorithms. International J. Comput. Vis. **85**(2), 167–181 (2009)
10. Hubel, D.H., Wiesel, T.N.: Integrative action in the cat's lateral geniculate body. J. Physiol. **155**(2), 385–398 (1961)
11. Hubel, D.H., Wiesel, T.N.: Receptive fields, binocular interaction and functional architecture in the cat's visual cortex. J. Physiol. **160**(1), 106–154 (1962)
12. Lopez-Molina, C., De Baets, B., Bustince, H.: Quantitative error measures for edge detection. Pattern Recogn. **46**(4), 1125–1139 (2013)
13. Lucca, G., et al.: Preaggregation functions: construction and an application. IEEE Trans. Fuzzy Syst. **24**(2), 260–272 (2016)
14. Lucca, G., Sanz, J.A., Dimuro, G.P., Bedregal, B., Bustince, H., Mesiar, R.: CF-integrals: A new family of pre-aggregation functions with application to fuzzy rule-based classification systems. Inf. Sci. **435**, 94–110 (2018)
15. Marr, D.: Vision. MIT Press, Cambridge (1982)
16. Marr, D.: Early processing of visual information. Philos. Trans. Roy. Soc. London. B Biol. Sci. **275**(942), 483–519 (1976)
17. Marr, D., Hildreth, E.: Theory of edge detection. Proc. Roy. Soc. London. Ser. B. Biol. Sci. **207**(1167), 187–217 (1980)
18. Martin, D.R., Fowlkes, C.C., Malik, J.: Learning to detect natural image boundaries using local brightness, color, and texture cues. IEEE Trans. Pattern Anal. Mach. Intell. **26**(5), 530–549 (2004)
19. Murofushi, T., Sugeno, M., Machida, M.: Non-monotonic fuzzy measures and the Choquet integral. Fuzzy Sets Syst. **64**(1), 73–86 (1994)
20. Prewitt, J.: Object enhancement and extraction (1970)
21. Sobel, I., Feldman, G.: A 3x3 isotropic gradient operator for image processing. Hart, P.E., Duda, R.O. Pattern Classif. Scene Anal., 271–272 (1973)

Special Session on Intelligent Techniques for Real-World Applications of Renewable Energy and Green Transport

Prediction and Uncertainty Estimation in Power Curves of Wind Turbines Using ε-SVR

Miguel Ángel García-Vaca[1](✉), Jesús Enrique Sierra-García[2], and Matilde Santos[3]

[1] Computer Science Faculty, University Complutense of Madrid, 28040 Madrid, Spain
magvaca@ucm.es
[2] Department of Digitalization, University of Burgos, 09006 Burgos, Spain
jesierra@ubu.es
[3] Institute of Knowledge Technology, University Complutense of Madrid, 28040 Madrid, Spain
msantos@ucm.es

Abstract. One of the most important challenges in the field of wind turbines is the modeling of the power curve, since it serves as an adequate indicator of their performance and state of health. This curve relates the electrical power generated by the turbine to the available wind speed. Due to its highly complex nature, one of the approaches to address this issue is through machine learning techniques. In this paper we use epsilon support vector regression (ε-SVR) to predict this power curve. Equally important to the model is the uncertainty associated with this prediction. To estimate the uncertainty, probabilistic analysis of the model residuals is applied. This model has been compared with the Gaussian process regression (GPR) model, widely used in various scientific fields. The results show that the ε-SVR model with uncertainty estimation is able to faithfully characterize the shape of the power curve and the corresponding prediction uncertainty. Furthermore, this model improves the results obtained with GPR in terms of some evaluation metrics, while achieving a better adjustment of the uncertainty and requiring a lower computational cost.

Keywords: Wind turbine · Power curve · Support vector regression · Uncertainty estimation · Gaussian process regression

1 Introduction

Due to the importance and prominence that climate change is acquiring, during the last decades significant efforts have been made to progressively move from different traditional energy sources to more sustainable and environmentally friendly alternatives, as well as to provide them with technological improvements [1]. Among the most used renewable energy sources is wind energy. Wind turbines (WT) take advantage of the power of the wind by rotating the blades to generate electricity. These wind turbines can be installed both on floating or fixed marine platforms and on land installations [2].

To make electricity generation profitable with these infrastructures, it is necessary to provide them with algorithms that allow evaluating the health of the system and, therefore, maximizing its performance, thus reducing the costs associated with its operation and maintenance (O&M) [3, 4]. The most used method to evaluate the operating

© The Author(s), under exclusive license to Springer Nature Switzerland AG 2023
P. Quaresma et al. (Eds.): IDEAL 2023, LNCS 14404, pp. 509–517, 2023.
https://doi.org/10.1007/978-3-031-48232-8_46

conditions of a wind turbine is to model its power curve as accurately as possible [5]. The power curve relates the electrical power generated to the wind speed available at a given time and is specific to each particular turbine.

One of the sources of inaccuracy in wind energy prediction lies in the wind turbine power curve model used. Precise modelling of the power curve in a wind farm is of great interest for the electrical network and to adapt production to demand and for its impact on the cost of the service. On the other hand, an accurate forecast improves the price if electricity is traded through the energy exchange [6]. Additionally, wind farm operators can operate more efficiently with these models when scheduling wind turbine maintenance.

A first approach to modeling the power curve is to use physical parameters of the system [7]. However, the equation that relates these parameters does not take into account other important variables that affect the shape of the curve, such as turbulence intensity, wind direction, etc. Furthermore, modeling this curve is not a simple task due to its non-linear and complex nature, and depends largely on the environmental conditions of the location and the methods used to acquire the variables involved [8]. An alternative method is to use the power curve provided by the manufacturer based on wind tunnel measurements. However, due to the real conditions of the turbines, this curve differs significantly from the real power curve. Therefore, a more precise approach would be to use data-driven algorithms [9].

This article is focused on the application of an epsilon-support vector regression (ε-SVR) model using real-world data extracted from the integrated SCADA system of an onshore WT. Using this methodology the shape of the power curve is obtained, which will allow us to predict the electrical power value for each wind speed. To estimate the uncertainty associated with this prediction, the model residuals will be fitted to a Laplace distribution. The proposed technique is compared with one of the most commonly used methods: Gaussian process regression (GPR) [10].

The structure of this work is as follows. Section 1 gives a summary of the problem. Section 2 presents a brief state-of-the-art. Section 3 describes the methodology applied to obtain the model of the power curve. Section 4 presents the data used and the pre-processing. In Sect. 5 results are discussed. The paper ends with the conclusions.

2   State-of-Art

Using different mathematical approaches, numerous techniques have been applied to tackle the problem of power curve characterization. In [4], a review of the latest data-driven modeling techniques is presented, including neural networks, support vector machines, probabilistic models, and decision trees. Among all of them, the ones that focus on these data-driven algorithms are highlighted in the following paragraph.

In [11], for an onshore turbine, Gaussian process regression is proposed to estimate both the reference power curve and variables such as pitch angle and rotor speed as function of wind. In [12], it is demonstrated how empirical copulas can model the probability density function of the power curve. This is applied to two fault-free onshore wind turbines located in Scotland. Similarly, in [13], a self-organizing neural network called generalized mapping regressor, a multilayer perceptron, and a general regression neural network are applied to estimate the power curve. These models are applied to a wind farm during a year. On the other hand, in [14], a multiclass support vector machine is applied to model a wind turbine with simulated data in various turbine control-related failure scenarios.

As observed, various techniques have been applied to power curve modeling, mainly using machine learning (ML) techniques. However, it is desirable not only to obtain a model but also to estimate the associated uncertainty, which is the focus of this work.

3 Methodology

3.1 Epsilon Support Vector Regression (ε-SVR)

Within ML, one of the most widely used is Support Vector Machines (SVM) [15]. SVM relies on the use of different types of kernels, selecting one or another depending on the characteristics of the problem at hand. Its strength lies in producing accurate predictions without incurring significant computational costs. While SVM is extensively used for classification tasks, we will focus on its regression version, using symmetrical epsilon insensitive loss function based on [16, 17].

Let's consider a training dataset $\{(x_1, y_1), \ldots, (x_n, y_n)\} \subset X \times Y$ where X represents the input space, $X = \mathbb{R}^d$, and $Y = \mathbb{R}$ represents the output data. In our case $d = 1$. The goal in ε-SVR is to find the function $f(x, w)$ that has a maximum deviation ε from the targets y_i for all the training data, while remaining as flat as possible. That requires minimizing Eq. (1).

$$\frac{1}{2}||w||^2 + C \sum_{l=1}^{l} (\xi_i - \xi_i^*) \tag{1}$$

Here, $||w||$ represents the norm, C is a penalty constant, and ξ_i, ξ_i^* are slack variables, which correspond to dealing with the ε-insensitive loss function $|\xi_\epsilon|$, Eq. (3). w denote the coefficients in the function f, thus $||w||$ is used to optimize the flatness.

$$|\xi_\epsilon| := \begin{cases} 0 \ if \ \text{if} \ |\xi| \le \varepsilon \\ |\xi| - \varepsilon \ otherwise \end{cases} \tag{2}$$

Various kernel functions can be used to train the model. In our case, we use the radial basis function (Gaussian) kernel, defined by (3).

$$K(x_i, x_j) = e^{-\gamma ||x_i - x_j||^2} \tag{3}$$

where γ is the scale parameter.

3.2 Uncertainty Estimation in ε-SVR

Once the ε-SVR model is trained and the power curve function $\overline{f}(x)$ is obtained, we need to determine a confidence interval I, within which we can infer that a certain $y \in I$ with a certain probability. Namely, given a specific point x and the ε-SVR model, we want to determine the probability $P(y|x, \overline{f}(x))$. To achieve this, we use a simple approximation that allows us to provide an estimation of these prediction intervals [18].

Let $\zeta_i = y_i - \overline{f}_j(x_i)$ represents the out-of-sample residuals in predicting the point (x_i, y_i) for each j-fold cross-validation. It is verified that these residuals, ζ_i fit a Laplace distribution with mean zero which coincides with the probability, $P(y|x, \overline{f}(x))$, (4).

$$P(z) = \frac{1}{2\sigma} e^{-\frac{|z|}{\sigma}} \tag{4}$$

where σ is the scale parameter. Assuming that ζ_i are independent from each other, this scale parameter, σ, is calculated following Eq. (5), where l is the number of points:

$$\sigma^2 = \frac{\sum_{i=1}^{l} |\zeta_i|}{l} \tag{5}$$

3.3 Gaussian Process Regression (GPR)

Gaussian Process Regression (GPR) models are non-parametric probabilistic models based on kernels [10]. A Gaussian Process (GP) is a collection of random variables, such that any finite combination of them follows a joint Gaussian distribution. In other words, if $f(x)$ is a GP, then taking n observations from the process, $x_1, x_2, \ldots x_n$, the joint distribution of the random variables $f(x_1), f(x_2), \ldots f(x_n)$ will be Gaussian.

A GP is defined by its mean $m(x)$ and covariance $k(x, x')$ functions. Therefore, if $f(x)$ is a GP then $E(f(x)) = m(x)$ and $Cov[f(x), f(x')] = k(x, x')$. Based on GPs, the output of GPR models is given by (6) where $f(x)$ is a GP with zero mean and covariance function $k(x, x')$, $h(x)$ is a set of basic functions that transforms the input space, and β is a vector of coefficients.

$$y(x) = h(x)^T \beta + f(x) \tag{6}$$

As a probabilistic model, the output y_i at time i can be modeled by expression (7).

$$P(y_i|f(x_i), x_i) \sim N(y_i|h(x_i)^T \beta + f(x_i), \sigma^2) \tag{7}$$

Different basis functions and kernel types can be used. In our case, linear basis functions along with the squared exponential kernel given by (8) were used.

$$k(x, x') = \sigma^2 \exp\left(-\frac{(x - x')^2}{2l^2}\right) \tag{8}$$

4 Dataset and Preprocessing

We use a dataset from an onshore wind farm located at Penmanshiel, United Kingdom [19]. This wind farm consists of Senvion MM82 wind turbines pitch regulated, and their main characteristics are: Rated power (kW) = 2050; Rated wind speed (m/s) = 14.5; Cut-out wind speed (m/s) = 5, and Cut-in wind speed (m/s) = 2.5.

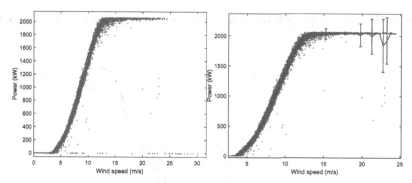

Fig. 1. Raw dataset (left); after removal of data points with zero or negative power (right)

Raw data are obtained every 10 min through a SCADA system. The following variables have been used, which have been averaged: timestamp, generated power (kW), wind speed corrected with air density (m/s) and ambient temperature (°C). For the case study, data corresponding to January and February 2017 from WT01 have been used, with a total of 8,352 data points. Figure 1 shows the raw data used (left).

Raw data are first cleaned to remove anomalous values. This data cleaning process is divided into two parts.

1. Removal of data points with zero or negative generated power, as well as data points with indeterminate values.
2. Removal of outliers. These outliers are characterized by being isolated and not exhibiting any clear pattern. Since average data are acquired every 10 min, during that time interval the turbine may have transitioned between operational states (on or off or vice versa), causing an erroneous average power value. To remove these points the power curve is split into intervals of 0.5 m/s, and their mean and standard deviation are calculated according to International Electrotechnical Commission, IEC 61400-12-1:2017 [20]. Any data point that deviates ±2.5σ from its corresponding interval is considered an outlier and removed from the dataset.

Figure 1 (right) shows the dataset after part 1 of the preprocessing, where the red line represents the mean values of the intervals, and the error bars represent ±2.5σ around the mean of each interval. The final clean data are show in Fig. 2 (left). These data points constitute the dataset on which the two considered models will be trained. A total of 7479 data points are obtained.

5 Model and Results

In Fig. 2 (right), the results of the ε-SVR model are shown. To obtain these data ε has been set to 8.93. Overlaid on the dataset, the red line represents the obtained regression curve, and the black dots represent the confidence interval above and below the regression curve corresponding to a 95% confidence level. It is possible to see how the regression curve closely fits the data in the entire range of the power curve. Regarding the confidence interval, at the initial and final parts of the curve this interval overestimates the data variance. However, as we will see next, this phenomenon is also present in the GPR model.

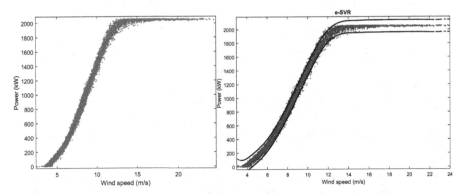

Fig. 2. Cleaned data (left); ε-SVR model obtained (right)

In order to establish a comparison, Fig. 3 presents the GPR model obtained (red line, regression curve; black dots, 95% confidence interval). The regression curve is also closely aligned with the data. Again, the confidence interval is overestimated at the initial and final parts of the curve.

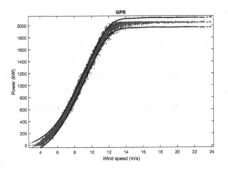

Fig. 3. GPR model obtained

Figure 4 shows the predictions: left (ε-SVR) and right (GPR), where the black line represents the points where the real value coincides with the predicted value. The results are visually quite similar for the two models.

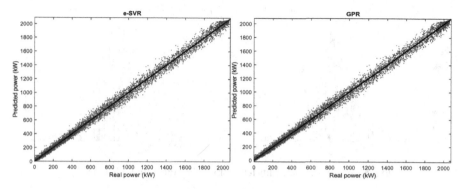

Fig. 4. Comparison ε-SVR vs GPR

In order to quantify the differences and determine which model performs better, four evaluation metrics are calculated, specifically RMSE, S-MAPE$^{0-200}$, MAE and MSE. These results are shown in Table 1 where the best results are bolded.

Table 1. Evaluation metrics of both models

	RMSE	S-MAPE	MAE	MSE
ε-SVR	41.44	*7.96*	*29.62*	*1716*
GPR	*41.43*	8.28	29.79	1716

Based on these results, it can be seen that GPR produces a slightly better RMSE value compared to ε-SVR. However, this difference is practically negligible. Regarding S-MAPE and MAE, ε-SVR gives better values. Thus, we can conclude that the ε-SVR model outperforms GPR.

Regarding the estimation of uncertainty, the ε-SVR model requires to calculate the distribution of out-of-sample residuals and fit them to a Laplace distribution. In our case, we obtained a value of $\sigma = 29.62$. Figure 5, left, shows the histogram of the distribution of residuals for ε-SVR, and the red line represents the obtained Laplace distribution. It is noted that this distribution matches our residuals quite well.

For the GPR model, the uncertainty is obtained directly from the model following a Gaussian distribution. In our case, we obtained a parameter $\sigma = 40.42$. Figure 5, right, presents the histogram with the distribution of residuals, where the red line represents the Gaussian distribution derived from the model. It is shown how the Gaussian distribution is not able to accurately fit the residuals obtained from the GPR model. We can conclude that the Laplace distribution provides a much more faithful fit to the residuals, indicating that the uncertainty associated with that regression curve is more precise and faithful to the distribution of the real power curve.

Another aspect to consider is the computational cost required for calculating both models. Both models were developed using Matlab software running on the same

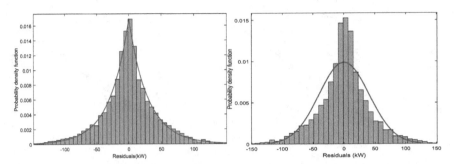

Fig. 5. Residuals and its fits **left)** for ε-SVR **right)** for GPR model

machine (online version of Matlab was used). Results in terms of time (seconds) are 425 and 1465 for ε-SVR and GPR, respectively, and 468 and 674 kilobytes for memory.

The computational cost, both in time and memory space, is significantly reduced for the proposed ε-SVR model in comparison to the GPR model. This difference is particularly evident in terms of time, which is one order of magnitude lower.

6   Conclusions and Future Works

In conclusion, it is shown how is possible to apply the epsilon insensitive Support Vector Regression (ε-SVR) model for the modeling of wind turbine power curves, which is able to capture the nonlinear and complex nature of this curve, as evidenced by the quantitative and qualitative results obtained. Additionally, the estimation of the uncertainty for this model provides a fairly accurate fit to the real data. Besides, this is achieved without a significant high computational cost. Finally, in comparison to the widely used GPR model, the proposed model outperforms it in terms of the error metric used, computational cost, and uncertainty estimation.

As future work, it would be interesting to compare this ε-SVR model with uncertainty estimation to other different models. Additionally, the inclusion of additional variables such as rotor speed, blade pitch, or turbulence intensity could further improve the model performance [21]. Furthermore, this model could be applied to evaluate different types of turbine failures and the probability of them being true faults.

Acknowledgments. This work has been partially supported by the Spanish Ministry of Science and Innovation under project MCI/AEI/FEDER number PID2021-123543OB-C21.

References

1. Strategic Energy Technology Information System (SETIS). https://setis.ec.europa.eu/ind ex_en. Last accessed 01 Jul 2023
2. Zhou, B., Zhang, Z., Li, G., Yang, D., Santos, M.: Review of key technologies for offshore floating wind power generation. Energies **16**(2), 710 (2023)

3. Sierra-García, J.E., Santos, M.: Redes neuronales y aprendizaje por refuerzo en el control de turbinas eólicas. Revista Iberoamericana de Automática e Informática industrial **18**(4), 327–335 (2021)
4. Pandit, R., Astolfi, D., Hong, J., Infield, D., Santos, M.: SCADA data for wind turbine data-driven condition/performance monitoring: a review on state-of-art, challenges and future trends. Wind Eng. **47**(2), 422–441 (2023)
5. Uluyol, O., Parthasarathy, G., Foslien, W., Kim, K.: Power curve analytic for wind turbine performance monitoring and prognostics. In: Annual Conference of the PHM Society, vol. 3, no. 1 (2011)
6. Barthelmie, R.J., Murray, F., Pryor, S.C.: The economic benefit of short-term forecasting for wind energy in the UK electricity market. Energy Policy **36**(5), 1687–1696 (2008)
7. Gray, C.S., Watson, S.J.: Physics of failure approach to wind turbine condition based maintenance. Wind Energy **13**(5), 395–405 (2010). https://doi.org/10.1002/we.360
8. Pandit, R., Infield, D., Santos, M.: Accounting for environmental conditions in data-driven wind turbine power models. IEEE Trans. Sustain. Energy **14**(1), 168–177 (2022)
9. Long, H., Wang, L., Zhang, Z., Song, Z., Xu, J.: Data-driven wind turbine power generation performance monitoring. IEEE Trans. Ind. Electron. **62**(10), 6627–6635 (2015)
10. Rasmussen, C.E., Williams, C.K.I.: Gaussian Processes for Machine Learning. The MIT Press. ISBN 0–262–18253-X (2006)
11. Pandit, R., Infield, D.: Gaussian process operational curves for wind turbine condition monitoring. Energies **11**(7), 1631 (2018)
12. Gill, S., Stephen, B., Galloway, S.: Wind turbine condition assessment through power curve copula modeling. IEEE Trans. Sustain. Energy **3**(1), 94–101 (2012)
13. Marvuglia, A., Messineo, A.: Monitoring of wind farms' power curves using machine learning techniques. Appl. Energy **98**, 574–583 (2012)
14. Vidal, Y., Pozo, F., Tutivén, C.: Wind turbine multi-fault detection and classification based on SCADA data. Energies **11**(11), 3018 (2018)
15. Vapnik, V.N.: The Nature of Statistical Learning Theory. Springer-Verlag, New York (1995)
16. Smola, A.J., Schölkopf, B.: A tutorial on support vector regression. Stat. Comput. **14**(3), 199–222 (2004)
17. Chang, C.C., Lin, C.J.: LIBSVM: a library for support vector machines. ACM Trans. Intell. Syst. Technol. **2**(3), 1–27 (2011)
18. Lin, C.-J., Weng, R.C.: Simple probabilistic predictions for support vector regression. Technical report, Department of Computer Science, National Taiwan University (2004)
19. Penmanshiel Wind Farm Data. https://doi.org/10.5281/zenodo.5946808. Last accessed 01 Jul 2023
20. International Electrotechnical Commission.: Wind energy generation systems—Part 12–1: Power performance measurements of electricity producing wind turbines. International Electrotechnical Commission (IEC), IEC Central Office, vol. 3, pp. 2017–03 (2017)
21. Ramos-Teodoro, J., Rodríguez, F.: Distributed energy production, control and management: a review of terminology and common approaches. Revista Iberoamericana de Automática e Informática industrial **19**(3), 233–253 (2022)

Glide Ratio Optimization for Wind Turbine Airfoils Based on Genetic Algorithms

Jinane Radi[1($\boxtimes$)], Abdelouahed Djebli[1], Jesús Enrique Sierra-Garcia[2($\boxtimes$)], and Matilde Santos[3]

[1] Energetic Laboratory, Department of Physics, Faculty of Sciences, University of Abdelmalek Saadi Tetouan, Tétouan, Morocco
jinane.radi@etu.uae.ac.ma
[2] Automation and Systems Engineering, University of Burgos, 09006 Burgos, Spain
jesierra@ubu.es
[3] Institute of Knowledge Technology, University Complutense of Madrid, 28040 Madrid, Spain
msantos@ucm.es

Abstract. The main objective in the design of wind turbine blades is the use of suitable airfoils to increase the aerodynamic performance and decrease the cost of energy. The objective of this research is to employ numerical optimization techniques to develop airfoils for wind turbines. To achieve this objective, a mathematical model is formulated by combining genetic algorithms with the XFOIL flow solver. Three different airfoils with different characteristics are created using the genetic algorithm. Throughout the optimization procedure, XFOIL software is used to determine the lift and drag coefficients. This study confirms the feasibility and effectiveness of the innovative design approach. Furthermore, it provides a valuable design concept that can be effectively applied to the airfoils of medium-thickness wind turbines.

Keywords: Airfoil · Wind Turbine · Optimization · Glide ratio · Genetic algorithm

1 Introduction

Wind energy is increasingly being used as a renewable energy source in many countries. There are all kinds of different designs, sizes and implementations of wind turbines, in particular blades, depending on the location and wind conditions, as well as the power capacity and other structural properties of the wind turbines. The efficiency and performance of wind turbines depend to a large extent on the design of their airfoils, so this is a crucial aspect to be considered. The amount of energy that can be produced by the wind can be significantly affected by the shape and properties of the airfoil of the blades. Consequently, the design of optimal airfoils is a relevant area of research in the field of wind energy [1–3].

The continued advancement of wind turbines depends to a large extent on the development and design of airfoils specifically adapted to the blade application. NACA,

© The Author(s), under exclusive license to Springer Nature Switzerland AG 2023
P. Quaresma et al. (Eds.): IDEAL 2023, LNCS 14404, pp. 518–526, 2023.
https://doi.org/10.1007/978-3-031-48232-8_47

NREL, RISO and DU airfoil types are still frequently used in wind turbines. However, there is a growing trend towards the development of new airfoil families designed specifically for wind turbine applications [4]. Most airfoils for wind turbines have traditionally been developed using conventional inverse procedures, in which the desired airfoil surface flow characteristics are prescribed under specific operating conditions, and the airfoil shape is determined to achieve these desired conditions with the blade surface [5]. Therefore, it is desirable to develop a model that offers sufficient flexibility to describe numerous types of airfoils [6]. This paper proposes to employ numerical optimization techniques to develop airfoils for wind turbines.

Genetic algorithms were used to explore and identify the optimal airfoil shape. A genetic algorithm works by generating a population of potential solutions to a given problem and then choosing the most suitable individuals from that population to develop the next generation of solutions. The process is repeated until a satisfactory solution is found. Several investigations have used genetic algorithms in the optimization of wind turbine airfoils. For example, Grasso [4] presents a comprehensive discussion of the numerical design optimization technique for wind turbine airfoils, whose optimization process is based on the combination of genetic algorithms and RFOIL software.

Quan Wang presents a novel integrated design approach for mid-thickness airfoil, aerodynamic performance and structural stiffness characteristics [7]. Rui Yin provides in his paper an optimal design strategy for a small wind turbine blade through multi-objective airfoil optimization using genetic algorithms and CFD investigations on steady-state and non-steady-state aerodynamic performance [8]. Bizzarrini et al. designed a wind turbine specific airfoil for the blade tip region using genetic algorithms [9]. Also, Ram et al. designed and optimized the USP07-45XX family of airfoils for a 20 kW wind turbine using genetic algorithms [10]. In [11] a blade airfoil for a horizontal axis sea turbine is designed. Several NACA airfoils are studied as a function of lift and drag coefficients. Similarly, in [12] the sizing of a sea turbine is also optimized but in this case with economic criteria.

The main objective of this work is to design a wind turbine airfoil using numerical optimization techniques. The airfoil design method used is based on the work of Ziemkiewicz [6], and the airfoil performance is determined using the XFOIL code. The airfoil design presented in this work provides a more direct and accurate approach compared to the traditional inverse design method [13, 14].

With the proposed design procedure, three airfoils named AQ-10, AQ-13 and AQ-24 have been obtained, each with a specific thickness to chord ratio of 0.10, 0.13 and 0.24, respectively.

The structure of the rest of the paper is as follows. In Sect. 2, the numerical optimization problem for the airfoil of the wind turbine blades is presented. The results of the shapes obtained by genetic algorithms are shown in Sect. 3. The paper ends with conclusions and future work.

2   Numerical Optimization Procedure

An optimization algorithm is a procedure used to discover the optimal solution to a given problem. The goal of an optimization algorithm is to minimize or maximize an objective function, which mathematically describes what is to be optimized. There is a wide range of optimization algorithms, each with its own strengths, weaknesses and areas of application.

Some common optimization algorithms include: local model-based optimization, differential evolution (DE), particle swarm optimization (PSO), multi-objective optimization, etc. The method chosen to optimize the wind turbine airfoil in this research is genetic algorithms. This technique is based on the natural evolution of species and use techniques such as mutation, crossover and selection to evolve a population of candidate solutions towards an optimal solution [14–16]. They have been successfully used in different engineering problems [17–19].

The blade profiles used in wind turbines have their origin in those used in aviation. They are typically grouped into standardized families whose name identifies certain characteristic parameters of the airfoil.

Several approaches can be employed in airfoil design. The inverse design process, for example, is a widely used strategy. The selection of airfoil parameterization using design variables is one of the most crucial elements in numerical optimization, for which multiple mathematical formulations have been suggested [4]. In the present work, the equations used are based on the work of Ziemkiewicz [6], described parametrically by the expressions (1) and (2). These expressions formalize the variation of the coordinates (X, Y) of each point of the blade profile as a function of the angle θ.

$$X(\theta) = 0.5 + 0.5 \frac{|\cos(\theta)|^B}{\cos(\theta)} \tag{1}$$

$$Y(\theta) = \frac{T}{2} \frac{|\sin(\theta)|^B}{\sin(\theta)} \left(1 - X^P\right) + C sin\left(X^E \pi\right) + R sin(X(\theta) 2\pi) \tag{2}$$

where B, T, P, E, C, R are the characteristics that define the aerodynamic profile of the blade. Each change in a parameter affects the profile of the blade differently.

These parameters are explained below:

B – Base shape coefficient: This parameter mainly affects the leading edge.
T – Thickness as a fraction of the chord: This term refers to a dimensionless value that represents the thickness of the airfoil at a specific point along its chord length. This parameter is important in airfoil design as it influences its aerodynamic performance and characteristics such as lift and drag forces. Airfoils with different thickness to chord ratios are used in specific applications to achieve the desired performance requirements, such as high lift or low drag.
P – exponent of conicity. For a higher value, the profile narrows more sharply near the trailing edge.
C – curvature, this parameter indicates the ratio between maximum curvature and the length. Curvature plays a crucial role in airfoil design, as it affects aerodynamic characteristics and performance.
R – mirroring parameter. Positive value generates a reflected trailing edge, while negative value emulates flaps.
E – curvature exponent. It refers to the position along the chord length at which the maximum curvature or deviation from a straight line occurs on the upper surface of the airfoil. It represents the distance from the leading (front) edge of the airfoil to the point of

maximum camber. The optimum position of the maximum camber is determined by the specific requirements of the airfoil application, including operating conditions, desired lift/drag ratio and overall efficiency.

The optimization of turbine performance is carried out through a series of sequential steps (Fig. 1). In each iteration, the GA generates a set of values (B, T, P, C, R, E), then the airfoil is created using the Eqs. (1) and (2); this airfoil is evaluated with XFOIL to obtain the lift and drag coefficients; finally the drag and lift values are used to compute the fitness function.

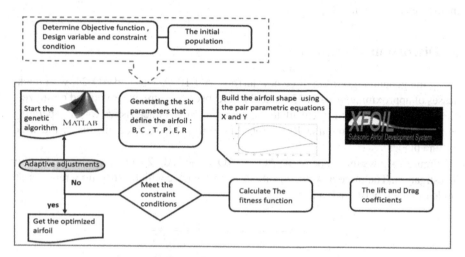

Fig. 1. Optimization process

This way, the genetic algorithm works by encoding the airfoil parameters (B, T, P, C, R, E) into a genetic representation, which is then subjected to evolutionary operations such as selection, crossover and mutation. This allows the algorithm to explore and exploit the design space, gradually improving the airfoil performance over successive generations.

An important element in the whole airfoil design procedure is the criteria of high lift and low drag. For this reason, in this work the inverse of the glide ratio (GLR) is used as cost function (3). In this way, the GLR is maximized, seeking to obtain high lift (C_L) and low drag (C_D). A high ratio between lift and drag coefficients can increase energy capture and reduce its cost. As the C_L and C_D varies with the angle of attack and XFOIL provides a set of values, the GLR is computed for all of them, and the fitness function considers the maximum one.

$$f = \min(\frac{1}{GLR}) = \min(\frac{C_D}{C_L}) \tag{3}$$

Through the iterative application of these genetic operators, the algorithm converges to an optimal airfoil design, resulting in improved efficiency and increased power output of the wind turbine. This optimization procedure has been summarized in the diagram shown in Fig. 1. In general, the design of wind turbines using genetic algorithms aims to achieve optimal energy extraction and higher overall turbine efficiency. As a final step, lift and drag are evaluated using the XFOIL analysis tool which is a computer program designed for the design and analysis of airfoil shapes, primarily in the context of aerodynamics and aircraft design. It was developed by Mark Drela at the Massachusetts Institute of Technology (MIT) and has gained extensive utilization in both academic and industrial applications [20].

3   Discussion of Results

Using the genetic algorithm, three new airfoils, AQ-10, AQ-13 and AQ-24, with thicknesses of approximately 10%, 13% and 24%, respectively, are generated and optimized. XFOIL software is used to calculate the aerodynamic characteristics for the operating conditions (Reynolds number Re = 1e6, Ma = 0) of these three airfoils of the wind turbine blades.

Figure 2 shows the shape of the AQ-10, AQ-13 and AQ-24 airfoils to be used in this investigation. The three new airfoils are designed to have a high lift-to-drag ratio for an angle of attack between 0 and 15°.

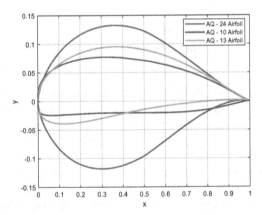

Fig. 2. AQ-10, AQ-13 and AQ-24 airfoils

Figures 3 and 4 graphically represent the lift and drag coefficients of the three airfoils. The lift coefficient of airfoil AQ-24 increases to 1.647 with increasing angle of attack. The second airfoil, AQ-13, reaches a maximum lift coefficient of 1.404 with an angle of attack of 15. The lift coefficient of the third airfoil rises to 1.586.

Similar characteristics are observed in Fig. 4 for the drag coefficient, we can clearly see that it increases to 0.05161, 0.0397 and 0.03 respectively for airfoils AQ-24, AQ-13 and AQ-10.

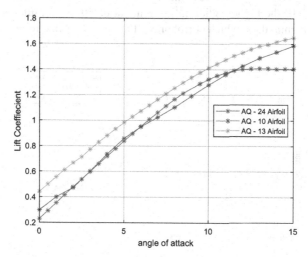

Fig. 3. Lift coefficient of AQ-10, AQ-13 and AQ-24 airfoils.

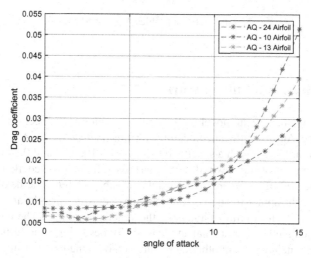

Fig. 4. Aerodynamic drag coefficient of AQ-10, AQ-13 and AQ-24 airfoils

Figure 5 shows the GLR of the three airfoils at different angles of attack. The maximum lift/drag ratio for the new airfoils is up to 107.4 for the AQ-24 airfoil, this maximum value appears at the angle of attack of 8. And for the AQ-13 the maximum value of the glide ratio is obtained at 133.9 corresponding to an angle of attack of 3.5. Regarding the third airfoil AQ-10 the highest glide ratio occurs at 86.59. The results obtained agree with the experimental data, which corroborates their accuracy and reliability.

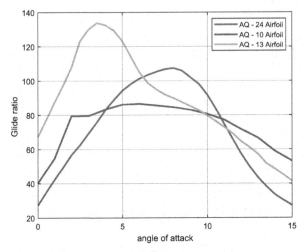

Fig. 5. Glide ratio of profiles AQ-13 and AQ-24

4   Conclusions and Future Works

In this work, a mathematical model with only six parameters has been used to describe the airfoil shape. Applying this method, three airfoils named AQ-10, AQ-13 and AQ-24 have been designed, each with a specific thickness of 0.10, 0.13 and 0.24 respectively.

The performance of the newly designed airfoils (lift and drag coefficients) is evaluated using the XFOIL software at a Reynolds number of 1e6. The flexibility of this approach allows the approximation of a wide range of existing airfoils, and its simplicity makes it well suited for optimization using the genetic algorithm. The results indicate that the proposed method is practical and feasible for designing wind turbine airfoils.

Future work includes using other equations to model the blade profiles and testing other optimization techniques such as particle swarm algorithms.

Acknowledgments. This work has been partially supported by the Spanish Ministry of Science and Innovation under project MCI/AEI/FEDER number PID2021-123543OBC21.

References

1. Bashir, M.B.A.: Principle parameters and environmental impacts that affect the performance of wind turbine: an overview. Arabian J. Sci. Eng. **47**, 7891–7909 (2021). https://doi.org/10.1007/s13369-021-06357-1
2. Radi, J., et al.: Design And Simulation of a Small Horizontal Wind Turbine Using MAT-LAB and XFOIL. In: WWME 2022 IV. Jardunaldia-Berrikuntza eta irakaskuntza energia berriztagarrien aurrerapenetan. Argitalpen Zerbitzua (2023)
3. Radi, J., Djebli, A.: Optimal design of an horizontal axis wind turbine using blade element momentum theory. E3S Web Conf. **336**, 00008 (2022). https://doi.org/10.1051/e3sconf/202233600008
4. Grasso, F.: Usage of numerical optimization in wind turbine airfoil design. J. Aircraft **48**(1), 248–255 (2011)
5. Chen, J., Wang, Q., Zhang, S., Eecen, P., Grasso, F.: A new direct design method of wind turbine airfoils and wind tunnel experiment. Appl. Math. Model. **40**(3), 2002–2014 (2016)
6. Ziemkiewicz, D.: Simple analytic equation for airfoil shape description. arXiv preprint arXiv:1701.00817 (2016)
7. Wang, Q., Huang, P., Gan, D., Wang, J.: Integrated design of aerodynamic performance and structural characteristics for medium thickness wind turbine airfoil. Appl. Sci. **9**(23), 5243 (2019)
8. Yin, R., Xie, J.-B., Yao, J.: Optimal design and aerodynamic performance prediction of a horizontal axis small-scale wind turbine. Math. Probl. Eng. **2022**, 1–19 (2022). https://doi.org/10.1155/2022/3947164
9. Bizzarrini, N., Grasso, F., Coiro, D.P.: Genetic algorithms in wind turbine airfoil design. EWEA, EWEC2011, Bruxelles, Belgium, 14 (2011)
10. Ram, K.R., Lal, S.P., Ahmed, M.R.: Design and optimization of airfoils and a 20 kW wind turbine using multi-objective genetic algorithm and HARP_Opt code. Renew. Energy **144**, 56–67 (2019)
11. Olivares, I., Santos, M., Tomás Rodríguez, M.: Análisis para el diseño de las palas de una turbina marina. In: XXXIX Jornadas de Automática, pp. 430–435. Área de Ingeniería de Sistemas y Automática, Universidad de Extremadura (2018)
12. Lillo, D., Santos, M., Esteban, S., López, R., Guiijarro, M.: Modelización, simulación y evaluación técnico-económica de una turbina de mar. In: XL Jornadas de Automática, pp. 24–31. Universidade da Coruña, Servizo de Publicacións (2019)
13. Wang, X., Wang, L., Xia, H.: An integrated method for designing airfoils shapes. Math. Problems Eng. **2015**, 1–12 (2015). https://doi.org/10.1155/2015/838674
14. Hajek, J.: Parameterization of airfoils and its application in aerodynamic optimization. WDS **7**, 233–240 (2007)
15. Galletly, J.: An overview of genetic algorithms. Kybernetes **21**(6), 26–30 (1992)
16. Zheng, Y., et al.: Optimization problems and algorithms. Biogeography-Based Optimization: Algorithms and Applications, 1–25 (2019)
17. Abajo, M.R., Enrique Sierra-García, J., Santos, M.: Evolutive tuning optimization of a PID controller for autonomous path-following robot. In: González, H.S., López, I.P., Bringas, P.G., Quintián, H., Corchado, E. (eds.) SOCO 2021. AISC, vol. 1401, pp. 451–460. Springer, Cham (2022). https://doi.org/10.1007/978-3-030-87869-6_43
18. Bayona, E., Sierra-García, J.E., Santos, M.: Optimization of trajectory generation for automatic guided vehicles by genetic algorithms. In: Bringas, P.G., et al. (eds.) 17th International Conference on Soft Computing Models in Industrial and Environmental Applications (SOCO 2022): Salamanca, Spain, 5–7 Sep 2022, Proceedings, pp. 484–492. Springer Nature Switzerland, Cham (2023). https://doi.org/10.1007/978-3-031-18050-7_47

19. Serrano, C., Sierra-Garcia, J.E., Santos, M.: Hybrid optimized fuzzy pitch controller of a floating wind turbine with fatigue analysis. J. Marine Sci. Eng. **10**(11), 1769 (2022)
20. Drela, M.: XFOIL: An analysis and design system for low Reynolds number airfoils. In: Low Reynolds Number Aerodynamics: Proceedings of the Conference Notre Dame, Indiana, USA, 5–7 June 1989, pp. 1-12. Springer Berlin Heidelberg, Berlin, Heidelberg (1989)

Special Session on Data Selection in Machine Learning

Detecting Image Forgery Using Support Vector Machine and Texture Features

Garrett Greiner[1] and Eva Tuba[1,2](✉) (iD)

[1] Trinity University, San Antonio, TX, USA
`ggreiner@trinity.edu`
[2] Singidunum University, Belgrade, Serbia
`etuba@ieee.org`

Abstract. In the past decades, image manipulation software has become more and more accessible. In addition, the software has improved greatly over the years so it has become harder to detect image forgery with the naked eye as a result. Therefore, this paper proposes a method for analyzing digital images and determining whether or not images have been forged, where two types of image forgery were considered: copy move and splicing. There are three crucial parts for successful forgery detection: quality of data, feature selection, and classification method. Images come in various different sizes, leading to an issue with extracting the same number of features from each image. Thus, we propose two forgery detection models: a size dependent, and a size independent model. We detect forgery using several features such as local binary pattern (LBP), kurtosis, and skewness of pixel values. These features are fed into a support vector machine (SVM) that was further tuned for the considered problem. The best model has an accuracy of 97.90% which is comparable or better in performance when compared to other models from the literature.

Keywords: Image Forgery · Splicing · Support Vector Machine · SVM · Feature Extraction · Texture Features · Local Binary Pattern · Kurtosis · Skewness

1 Introduction

In this day and age, we come into contact with digital images through various forms of media in our everyday lives. Whether we see them in the news, social media, or articles we read, we interact with and produce images more than ever. In addition, different software are more distributed and accessible than ever before. Such software allows users to edit digital images in various ways, so it is not uncommon that images are manipulated, either by mistake or with intention, resulting in information being misconstrued. Anyone can do simple image manipulation given the right software and some time to learn how to use it. Although it is not exactly clear what implications this widespread ability to manipulate images has, one important implication is that it could be used

© The Author(s), under exclusive license to Springer Nature Switzerland AG 2023
P. Quaresma et al. (Eds.): IDEAL 2023, LNCS 14404, pp. 529–537, 2023.
https://doi.org/10.1007/978-3-031-48232-8_48

maliciously to spread misinformation. Therefore, it is important that we develop methods that enable us to detect when an image is forged. This allows us to determine the credibility of a given image. There are many different ways that images can be manipulated, and there are a wide variety of classifications for image forgeries. The main three categories of digital image forgery are copy-move, image splicing and image resampling.

Copy-move forgery occurs when a portion of an image is transposed into a different location within the same image [3,4]. Image splicing forgery occurs when a region of an image is transposed onto another image [3]. Image resampling forgery occurs when images undergo some type of geometric transformation such as scaling, stretching, rotation, flipping, etc. These transformations make specific correlations that can be used for recognition of the mentioned forgery.

In this paper, we focus on detecting copy-move and image splicing forgery, as these are two common types of forgery. Example of images that undergone these forgeries are presented in Fig. 1.

Fig. 1. (**a, c, e**): Authentic images. (**b, d, f**): Forged Images

Considering the importance of distinguishing genuine images from forged images, it is an active research topic and many different methods for this problem have been proposed. In general, it can be said that there are two main types of forgery detection: active and passive.

An active detection approach uses information which is inserted into image prior to distribution. For example, using information such as digital signatures or digital watermarks [7]. These methods are only applicable for the images that have these information embedded, but are useless for majority of images used in everyday life. On the other hand, a passive approach employs statistical methods to detect alterations in the image. For example, methods that extract features such as texture features, or statistical measures, would classify as a passive approach [7]. The passive detection methods are applicable for any given image, so they are more widely used, however, they are also more complex.

Al-Azawi et al. in [2] proposed a texture based detection method. The proposed method included preprocessing, feature extraction, and classification. The features that were used were: fractal entropy (FrEp), local binary patterns, skewness, and kurtosis. In [13], a method for forgery detection was proposed based on texture features of the object shadows. More recent methods for forgery detection include deep learning [8]. Some methods completely rely on deep learning, while others use deep learning for feature extraction [1].

In this paper, we propose a passive approach for distinguishing authentic images from forged images. The proposed method is based on textural (LBP) and statistical (skewness, kurtosis) features. A support vector machine (SVM) is used to classify the images as forged or authentic. Additionally, we considered the issue of data used for training. Most classification methods expect the data to be of the same size, however, images can vary in their resolution. For image forgery detection, a common approach is to extract features from small regions in the image, allowing for the model to gain local information about the image. In some cases, attempting to use such feature extraction methods on images of different sizes will result in a different number of extracted features. One solution is to resize all images to the same size, but the downside is that valuable information would be lost in the process. In this paper, we analyzed the effect of modifying the feature extraction process by resizing and flipping images. The proposed method was tested on the CASIA 2.0 dataset [5]. This dataset is composed of a total of 12614 images. There are a total of 7491 authentic images, and 5123 forged images. There are both examples of copy-move and spliced images among the forged images.

The rest of the paper is organized as follows. Section 2 describes the proposed method, including the features used and the classifier. Simulation results and analysis of the obtained results are presented in Sect. 3. Section 4 concludes the paper and proposes future research.

2 The Proposed Method for Digital Image Forgery Detection

In this paper, we proposed two frameworks for digital image forgery detection, one which is image size dependent, and the other is image size independent. In the used dataset, CASIA 2.0 dataset, most images have the same resolution, so we proposed a model that can predict images of this common resolution. In general, any dataset that contains images of uniform size can use this framework. However, since this is not always the case, we also wanted to create a model that could be used on all images in a dataset, and any given image after.

The proposed method consists of several steps including preprocessing, feature extraction and classification. These steps are detailed in the following sections. Flow chart of the proposed method is presented in Fig. 2.

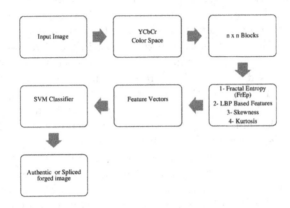

Fig. 2. Flow chart of the proposed framework for digital image forgery detection

2.1   Preprocessing

The preprocessing method employed does not depend on the size of the image, and it is common for both approaches. It is a step that prepares the image for feature extraction. During the preprocessing step, the images are converted from the RGB color space (or any other color space) into the YCbCr color space, since it has been found that this color space is more suitable for forgery detection [6]. Then, the Cb and Cr color channels are extracted, as forgery is more detectable in these channels than in the intensity channel. This is due to the fact that the Cb and Cr color channels contain the information for the chrominance color. The information held in these channels does not have any visible affects on the image, so changes to these channels are more difficult to see, which is why the images are manipulated/forged there. Further, it has been shown that analysis of these channels can reveal manipulated parts of images [2].

It is important to note that this step does not result in the loss of any information. The preprocessing step modifies the image with the intent to enable the model to extract more meaningful features for a specific purpose. In the case of image forgery detection, it is important to not further modify the image and potentially lose valuable data for detecting altered parts of the image [13].

2.2   Feature Extraction

Feature extraction is one of the most important steps in the case of classification. In this paper, commonly used features for forgery detection were used [2,4,13].

The main analysis concerns the behavior of the classification method depending on different input data.

Size Dependent Model. In the CASIA 2.0 dataset, it was noticed that most of the images have a dimension of 384×256, so we decided to make a method that focuses on predicting images of these sizes. The proposed method tries to simplify the process of detection from existing methods. We tried to use less features while still maintaining comparable classification accuracy.

For feature extraction, the image is split into 3×3 non-overlapping blocks. The following features are extracted from each block: local binary pattern (LBP), kurtosis, and skewness for both of the Cb and Cr channels.

Local binary patterns (LBP) is a widely used method for describing the texture of an image [11,12]. A binary number is obtained by using a central pixel for thresholding neighboring pixels, i.e. smaller values than the central pixel become 0 and larger or equal values become 1. These binary values can then be transformed into a decimal number.

Kurtosis and skewness are both statistical measures. The first step for finding these measurments is to create histograms for Cb and Cr channels. Once we have a normalized histogram, statistical measures can be applied. In statistics, kurtosis is a measure used to describe the peakness or flatness of a probability distribution. Skewness is a statistical quantity of the asymmetric distribution of a variable [2], i.e. it measures symmetry. For negative skewness values, we can conclude that the histogram is negatively skewed. In the case of digital image histogram, that means that darker shades are less concentrated as compared to the light shades. Both kurtosis and skewness are descriptive statistics, and they can help describe the distribution of a histogram, but not necessarily the data. Since there is a strong correlation between pixels in the image, different distributions in different blocks can potentially detect altered parts.

Size Independent Model. Although one approach is to create a model that is size dependent, we also want to make a model that works for all images. For these reasons, we additionally propose a size-independent model. The same set of features were extracted, but instead of extracting features from 3×3 non-overlapping blocks, they were extracted from the entire image. The resulting feature vector includes six features (LBP, skewness, kurtosis for Cb and Cr channel).

2.3 Classification

Once the features have been extracted, we can then use a classification model to determine the authenticity of an image. In this study, we wish to simply determine whether a given image is either forged or authentic. This is a binary classification task, so models such as support vector machine (SVM) are well suited for this task [10]. In addition, other studies have used support vector machines in their implementations, and obtained better results as compared to other popular classification methods [2].

3   Simulation Results

Various tests were conducted for the size dependent and size independent model. We tested both models on balanced and unbalanced portions of the dataset.

3.1   Data

We trained and tested our proposed frameworks with the CASIA 2.0 dataset [5]. This dataset is composed of a total of 12614 images. There are a total of 7491 authentic images, and 5123 forged images. There are both examples of copy-move and spliced images in the forged data. This dataset has been extensively used in studies covering image forgery, and therefore, can be seen as a credible benchmark to evaluate our model against. The CASIA 2.0 dataset has been described as a *sophisticated and realistic dataset for tamper detection* [9].

As explained previously, a majority of these images have the same width and length, with dimensions of 384×256. For our size dependent model, we considered images of this size. There were a total of 4529 authentic images that had the dimensions 384×256, while there were only a total of 1164 forged images that had these dimensions. In the dataset, there are some images that have dimensions of 256×384. In order to maximize the number of images that have the dimensions of 384×256, we flipped the images of size 256×384. By flipping these images, we do not lose any of the information contained in them, as we simply transpose the pixel matrix. We were able to flip a total of 1898 authentic images, and 215 forged images. This gave us a total of 6427 authentic images, and 1397 forged images with dimensions of 384×256. It can be noticed that data in set created this way is very unbalanced. To test the importance of balanced data, we applied our framework to both unbalanced and balanced portions of the dataset. In order to create a balanced dataset, we randomly choose 1397 authentic images, so that the number of authentic and forged images was the same.

The main metric used to determine the performance of our model was accuracy. In our case, accuracy is simply defined as the number of correctly classified images divided by the total number of images that we attempted to classify:

$$acc = \frac{\#correct}{\#total} \cdot 100 \tag{1}$$

For the size dependent model, tests were run using images of the size 384×256 with the data from both before, and after images were flipped. For the size independent model, we used the whole CASIA 2.0 dataset. We run all tests for all available data, but also for balanced data where we randomly choose images from the class with more training examples to match the number of images from the other class. The results are summarized in Table 1.

Table 1. Classification accuracy with different input data

Size dependent	Unbalanced	Balanced
Original (before flipping)	96.1	**97.7**
After flips	93.5	94.0
Resizing	93.1	94.2
Size independent	79.0	55.9

We can compare our results to the method previously introduced by Al-Azawi et al. In their paper, they demonstrated that their model performed better than a few other relevant methods that used the CASIA 2.0 dataset for training [2]. By comparing our model to this previous model, we can tell how well our model performs relative to other models that have the same purpose.

3.2 Size Dependent Model

Before Image Flips. Before we flipped any images, we trained the model using only images of the same size from the dataset. In total, there are 4529 authentic, and 1164 forged images of size 384×256 in the dataset. As it can be seen, the resulting training data is rather unbalanced. After training the SVM with these data, the resulting accuracy was 96.1%. To test the importance of balanced data, we randomly chose 1500 of the authentic images, and used only them for training. In that case, the model was trained on 1500 authentic images, and 1164 forged images. Further, 80% of the data were used for training, and 20% for testing. This resulted in an accuracy of 97.7% which is comparable to [2] whose accuracy was 98% [2]. In our model, we used one less feature, and we have achieved similar success.

After Image Flips. Once we flipped the images that had the size 256×384, we trained the model on all of the images of the size 384×256. The best result that was achieved was when the model was trained on 1379 authentic images, and 1379 forged images. All images were of the size 384×256. Similarly, we used 80% of the data for training, and 20% for testing. This resulted in an accuracy of 94%. When we trained the model with various other portions of data, the accuracy ranged from approximately 90%-93%.

Size Dependent Model with Image Resizing. The last simulation that was done included using all images from the dataset with the size dependent model, i.e. with the model that assumes that all images are of the same size. In order to do so, we simply resized all of the images to the most common size in the dataset, 384×256. By resizing images, some information will be changed/lost. If the image is scaled up, some approximation was performed based on the neighboring pixels. If it was a forged part, the number of forged distributions is increased, so the classifier can now consider this as an authentic patch. On the other hand, if the image was downsized, certain resolution reduction occurs, and potentially, forged blocks are lost. The accuracy obtained on the dataset created this way was 93.1%, which is less than the accuracy of the size dependent model without image resizing and with balanced data, 97.7%. This loss of the accuracy can be explained by the mentioned information loss. Further investigation by including other measurements like specificity and precision will be done.

Looking at the results, the highest accuracy was obtained by training the model on the data that did not include any flipped images. Training the model on 1500 authentic images and 1166 forged images resulted in an accuracy of 97.7%. This accuracy was comparable to the accuracy of the previous implementation. However, our model uses one less feature. Based on these two simulations, it can be concluded that having balanced dataset improved the ability of the classifier to correctly find a connection between authentic and forged images.

3.3   Size Independent Model

For this model, we consider images of any size. The best result that was obtained was when the model was trained on 7437 authentic images, and 2064 forged images. We used 80% of the data for training, and 20% for testing. This resulted in an accuracy of 79%. If we used the equal number of the authentic and forged images, the accuracy of the model went down to 55.9%. Both of these models are very close to random classification, or better to say, to accuracy if we assign one class to all instances. Attempting to apply size independent method does not result in accurate models. Instead, resizing is still a better option, as it allows us to get local data, even if it results in slightly damaged data.

4   Conclusion

In this paper, we proposed a method that determines the authenticity of an image by using the SVM as classifier, and texture and statistical features as input. The RGB images were converted into the YCbCr color space, and then the Cb and Cr color channels were used. Next, three features were extracted: Local Binary Patterns (LBP), Kurtosis, and Skewness. Finally, the SVM was used to classify the image. Two models were created: one model that is size dependent, and one that is size independent.

The best result that was achieved for the size dependent model was 97.7%, which was comparable to that of the previous implementation. Additionally, the size dependent model was trained on the entire dataset when all images were resized to the same size. The accuracy of that model was 93.1%. This decrease in accuracy as compared to the accuracy of the same model when the original size images were used can be explained by lost of information during the image resizing step. The best result that was achieved for the size independent model was only 79%. Our size dependent model was able to perform at a similar level to the previous implementation while utilizing one less feature. We were able to reduce the complexity of previous methods while maintaining similar accuracy. Future work will include a more detailed analysis of the results, including different metrics. Also, slight alterations in the feature extraction process will be considered.

References

1. Agarwal, R., Verma, O.P.: An efficient copy move forgery detection using deep learning feature extraction and matching algorithm. Multimed. Tools Appl. **79**(11–12), 7355–7376 (2020)
2. Al-Azawi, R.J., Al-Saidi, N., Jalab, H.A., Ibrahim, R., Baleanu, D., et al.: Image splicing detection based on texture features with fractal entropy. Comput. Mater. Continua **69**, 3903–3915 (2021)
3. Ali, S.S., Ganapathi, I.I., Vu, N.S., Ali, S.D., Saxena, N., Werghi, N.: Image forgery detection using deep learning by recompressing images. Electronics **11**(3), 403 (2022)
4. Brajic, M., Tuba, E., Jovanovic, R.: Ovelapping block-based algorithm for copy-move forgery detection in digital images. Int. J. Comput. **1** (2016)
5. Dong, J., Wang, W., Tan, T.: Casia image tampering detection evaluation database. In: 2013 IEEE China Summit and International Conference on Signal and Information Processing, pp. 422–426. IEEE (2013)
6. Kasban, H., Nassar, S.: An efficient approach for forgery detection in digital images using Hilbert-Huang transform. Appl. Soft Comput. **97**, 106728 (2020)
7. Kaur, G., Singh, N., Kumar, M.: Image forgery techniques: a review. Artif. Intell. Rev. **56**(2), 1577–1625 (2023)
8. Mehrjardi, F.Z., Latif, A.M., Zarchi, M.S., Sheikhpour, R.: A survey on deep learning-based image forgery detection. Pattern Recogn., 109778 (2023)
9. Shinde, A., Patil, G., Kumar, S.: Document image forgery and detection methods using image processing techniques-a review **8**(4), 1077–1101 (2022)
10. Tuba, E., Capor Hrosik, R., Alihodzic, A., Jovanovic, R., Tuba, M.: Support vector machine optimized by fireworks algorithm for handwritten digit recognition. In: Simian, D., Stoica, L.F. (eds.) MDIS 2019. CCIS, vol. 1126, pp. 187–199. Springer, Cham (2020). https://doi.org/10.1007/978-3-030-39237-6_13
11. Tuba, E., Strumberger, I., Bacanin, N., Tuba, M.: Analysis of local binary pattern for emphysema classification in lung CT image. In: 11th International Conference on Electronics, Computers and Artificial Intelligence (ECAI), pp. 1–5. IEEE (2019)
12. Tuba, E., Tomic, S., Beko, M., Zivkovic, D., Tuba, M.: Bleeding detection in wireless capsule endoscopy images using texture and color features. In: 26th Telecommunications Forum (TELFOR), pp. 1–4. IEEE (2018)
13. Tuba, I., Tuba, E., Beko, M.: Digital image forgery detection based on shadow texture features. In: 24th Telecommunications Forum (TELFOR), pp. 1–4 (2016)

Instance Selection Techniques for Large Volumes of Data

Marco Antonio Peña Cubillos[✉] and Antonio Javier Tallón Ballesteros[✉]

Department of Electronic, Computer Systems and Automation Engineering,
International University of Andalusia, Huelva, Spain
promarcope@gmail.com, antonio.tallon@diesia.uhu.es

Abstract. Instance selection (IS) serves as a vital preprocessing step, particularly in addressing the complexities associated with high-dimensional problems. Its primary goal is the reduction of data instances, a process that involves eliminating irrelevant and superfluous data while maintaining a high level of classification accuracy. IS, as a strategic filtering mechanism, addresses these challenges by retaining essential instances and discarding hindering elements. This refinement process optimizes classification algorithms, enabling them to excel in handling extensive datasets. In this research, IS offers a promising avenue to strengthen the effectiveness of classification in various real-world applications.

Keywords: Data mining · Instance selection · Data preparation · Knowledge discovery in databases

1 Introduction

The advancement of technologies and the various existing classification algorithms have made it necessary for researchers to improve the performance of these algorithms to improve applications in today's world. It is very common to need to use these methods on large datasets, therefore, algorithms employed with a high degree of complexity may be problematic. The complexity of the model training and its application are usually the two problems that arise. However, enhancing classifiers performance is a challenging mission. There are several ways to improve classifiers accuracy such as preprocess dataset, enhancing algorithms performance, and post-process the classifiers results.

Data sampling is one of data preprocessing techniques. These data preprocessing techniques have been previously covered in other works, such as "Data Preprocessing in Data Mining" [3], the authors present a perspective on data preprocessing where data reduction plays a significant role. As a central concept in this data preprocessing approach, various data reduction techniques are introduced. The author highlights specific cases and suggests that when dealing with redundant or conflicting data, one solution is instance selection.

This research boosts the classification algorithms by enhancing selection of training sample as a preprocess step. The remaining of this paper is organised as follows. Section 2 reviews different concepts about data preprocessing,

P. Quaresma et al. (Eds.): IDEAL 2023, LNCS 14404, pp. 538–546, 2023.
https://doi.org/10.1007/978-3-031-48232-8_49

the previous methedology before teh application and presents the motivation of this work. Section 3 shows a review of the literature and the applied process. Section 4 describes the proposed approach. Section 5 depicts the experimental results. Lastly, Sect. 6 draws some conclusions.

2 Data Preprocessing

As mentioned in the ML literature, the input data is the cornerstone of ML. Algorithms need accurate and well-structured datasets to be able to perform their training. In reality and unfortunately, these datasets are often not well structured, various external factors often affect the consistency of the data in the real world, e.g. the presence of noise, superfluous data, the immense amount of attributes and instances in the datasets [13]. For this reason data preprocessing tasks occupy an important role over time in ML Workflows. They are usually grouped into two groups of commonly used techniques such as data preparation and data reduction.

Formally, data preparation is the process of collecting, combining, structuring, and organizing data, but data preparation is more than organizing data. Data preparation comprises the process of cleaning and transforming the raw data prior to processing and respective analysis. It is a fundamental step before processing, which usually involves reformatting the data, making corrections to the data, and combining data sets to enrich them (Fig. 1).

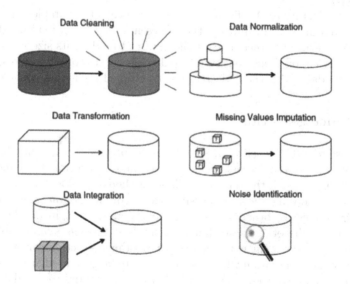

Fig. 1. Data preparation. Reproduction from the book Data Mining Preprocessing in Data Mining (García, Luengo, y Herrera, 2015)

The goal of data reduction is to reduce the difficulty and improve the quality of the resulting data sets by reducing their size. Thus data reduction identifies and in turn leads to discarding information that is irrelevant or repeating. The objective of reducing the data is not to lose the extracted information, but to increase the efficiency of ML when the data sets are quite large [14]. This is considered to be the most critical point when performing data reduction. The following are several pertinent techniques, with some of them grouped together:

- Discretization of data. It helps to partition the continuous attributes and reduce the size of the dataset, thus preparing it for its respective analysis. The data discretization process can also be considered part of the data preparation stage. The decision to include it as a data reduction task is explained in [3]. The discretization stage maps the data from a range of numerical values into a reduced subset of categorical values.
- Feature selection. It is a technique of reducing the dimensionality of the data set by eliminating irrelevant or redundant attributes [9]. The objective of attribute selection [3] is to obtain a representative subset of attributes that has a smaller number of attributes, such that the resulting probability distribution of the output attributes of the data is as close as possible to the original distribution obtained using all attributes. This facilitates the understanding of the extracted patterns and increases the speed of learning.
- Feature extraction. It is a process by which an initial set of data is reduced to a subset to improve processing. These methods select or merge subsets, create new artificial attributes, include modifications to attributes or remove attributes.
- Instance generation. Artificial examples or surrogate examples are created or adjusted in order to improve the representation of decision boundaries in supervised learning, since it helps the dataset to reduce its size [5].
- Instance selection. This consists of finding the most representative subset of the initial dataset, without decreasing the original prediction capacity.

3   Instance Selection

The purpose of instance selection algorithms, as previously explained in the literature, is to reduce the complexity of the learning algorithms by reducing the number of instances in the datasets. These methods are oriented to the intelligent way of choosing the best possible subset of the original data by using rules or heuristics. In other words, if we train one algorithm on the original set of data and the other algorithm with the best possible subset, both algorithms should have similar performance [10]. Instance selection can be taken as a case of instance generation where the instances to be generated are constrained from the original instances. These methods take an essential role in data reduction processes. Attribute selection or discretization processes reduce complexity, instance selection methods reduce the size of the data set [4]. The intention of these algorithms is to be able to extract the subset with the highest significance of instances by leaving aside those instances that do not provide relevant or valuable information. Reducing the data set reveals two main advantages: a) Reduce the amount of space it occupies on the system [15]. b) Decrease processing time for learning tasks

The set of selected instances can be used to train any type of classifier but, in the past some instance selection algorithms have been created for the k nearest neighbor classifier [1], or kNN for short. In such a way for the instance selection process the term prototype selection [3] is also used. In this research the term instance selection is used to refer to the process that involves the selection of a subset of instances from the original data set, regardless of the algorithm that is subsequently used.

When analyzing datasets or large volumes of data in the real business world, the need to use an instance selection algorithm becomes more noticeable. Nowadays the size of datasets is larger and as technology advances these datasets become larger and larger. On the other hand, data sets in reality often contain noisy instances, missing values, outliers and anomalies. There have been numerous attempts to train a classifier on datasets of millions of instances, as this is often a complex or even intractable task for the classifier. Therefore proper selection of a subset of instances is in such a way a good option to reduce the sample size [3], and improve the subsequent sample processing.

4   The Proposed Methodology

This paper presents an approach in the instance selection algorithms aim to reduce the size of datasets and select the most representative subset. When performing instance selection, it is important to consider that these are multi-objective problems. On one hand, reduction needs to be taken into account, and on the other, accuracy. However, according to the literature [7], these two objectives often go in opposite directions. This section details the methodology used in the instance selection context.

When estimating the accuracy of a predictive model using a classifier or regressor, it is essential for evaluating instance selection algorithms. Model accuracy must be estimated correctly, and a method with low variance and bias should be selected [6]. According to the literature, the reason there are several methods is that not all methods are suitable depending on the conditions of the study [12]. Considering the literature with [3], before applying our instance selection technique, it is necessary to train our dataset.

The flow of our proposal comprises the following steps (a) first, the dataset is divided into training and test sets, (b) second, the training set is passed through a classification algorithm, (c) third, the initial training sets are applied two instance selection methods on two set sample sets, (d) now, the new training data is ready to be tested by classification algorithms. It is important to note that the proposal is feasible for classification problems.

The research has made a substantial contribution to the field of instance selection algorithms. In particular, we have focused on improving the effectiveness and efficiency of these algorithms, addressing the fundamental challenge of optimising datasets for machine learning applications.

One noteworthy achievement is the development of a novel improved pre-processing that integrates instance selection techniques. This innovation allows users to achieve an optimal balance between dataset reduction and preservation of model accuracy.

5   Experimental Results

In this section we describe the datasets, instance selection methods and classification algorithms used to perform the experiments assisted by WEKA (free software), RapidMiner, Keel and Python tools. To obtain the datasets, we rely on the database repositories of the University of California at Irvine, UCI ML (usual repository for experiments in ML), in the Keel repository. Test results are also reported.

Table 1 describes the datasets selected for the analysis of different instance selection algorithms, subsequently the dataset resulting in the instance selection algorithm will be previously analyzed with a classification algorithm. These datasets have large volumes of data, were previously selected for this thesis in order to show the performance of the instance selection algorithms.

Table 1. Characterization of the data sets adopted for the analysis of instance selection and classification models.

Dataset	Instances	Features	Classes
Shuttle	58000	9	7
Volcanoes-b6	10130	3	5
Connect-4	67557	42	3
Volcanoes-d3	9285	3	5
Volcanoes-b4	10190	3	5
PishingWebsites	11055	30	2

The previously selected datasets possess a large number of data since the central theme of this study is to test the instance selection algorithms on large datasets. The results of each algorithm vary when selecting smaller datasets, being significantly worse than when used for large datasets.

As instance selection methods, the following ones have been applied:

- Kennard-Stone: This was originally designed for calibration and test data sets from an original data set [8]. Here we use this method to reduce the size of data sets to improve the performance of a classification model. The method selects a subset of samples that provides uniform coverage over the dataset and includes samples at the boundary of the dataset. In such a way it performs the division of the original data set into the calibration and validation subset, such that each of them contains the samples that can manage to capture the maximum variability of the original set.

- Random sampling method (Random sampling method RS) [11]: This is a method that chooses the elements that will constitute the subsets of the sample in a random way. It is also known as probability sampling, it is a method that allows randomizing the selection of the sample. In this thesis we used the simple random sampling method (Simple random sampling) [2] which is a subset of a statistical population in which each member of the subset has the same probability of being selected. A simple random sample is one of the sample selection techniques that is intended to be an unbiased representation of a group.

We describe the classification algorithms used for the purpose of comparison; these will be used at the end of the instance selection process following the literature according to [3]. These algorithms, namely, Simple Logistic (SL), SMO, Random Tree (RT) and Random Forest (RF) were implemented in WEKA, without hyperparameter setting for the classification tasks.

The experimental part starts with the original datasets mentioned in Table 1. Taking into account the literature cited above, the experiments used a k=4 cross validation. The first stage will result from the application of the Kennard-stone instance selection and Random sampling techniques, which we will describe in two experimental phases. The first experimental phase consists of reducing the trained data set by applying the mentioned techniques in 5000 samples to subsequently build the classification model which were tested with the classifiers mentioned in the previous section, so that the performance of these techniques and the results presented will be evaluated with three significant metrics such as AUC, Accuracy and Cohen's kappa.

The second phase of experimentation of the first stage will give the result of the application of the Kennard-stone and Random sampling instance selection techniques using a value of 2500 samples to subsequently build the classification model and evaluate its performance through the results collected taking into account the metrics and parameters previously mentioned in the first phase.

Figure 2 shows the test results.

Firstly, we have incorporated the Simple Logistic (SL) algorithm, which assumes a linear relationship between the natural logarithm of the odds ratio and the measurement variable. This methodological choice enables us to accurately model certain underlying phenomena in our data, revealing relationships that might go unnoticed under other approaches.

Furthermore, we have employed the Sequential Minimal Optimization (SMO) algorithm, based on support vector machines, to address classification problems. This minimal optimization approach has proven particularly effective in class separation in our data, enhancing the accuracy of our predictions.

Our study also benefited from the use of the Random Tree (RT) algorithm, which, unlike traditional decision trees, considers random attributes at each node. This allowed us to explore non-linear and complex relationships in our data, leading to intriguing discoveries.

Finally, the implementation of the Random Forest (RF) algorithm, consisting of an ensemble of decision trees, enriched our research by providing a more

Dataset	Metrics	Classifiers				Classifiers KS Phase 1				Classifiers RS Phase 1				Classifiers KS Phase 2				Classifiers RS Phase 2			
		SL	SMO	RT	RF	SL	SMO	RT	RF	SL	SMO	RT	RF	SL	SMO	RT	RF	SL	SMO	RT	RF
Shuttle	AUC	0.991	0.953	0.999	1.000	0.990	0.825	1.000	1.000	0.991	0.922	0.999	1.000	0.986	0.824	0.998	1.000	0.992	0.900	0.995	1.000
	Accuracy	96.10	96.84	99.95	99.97	93.14	92.19	99.96	99.96	96.17	95.60	99.86	99.84	92.95	92.20	99.85	99.95	95.91	94.74	99.67	99.75
	Cohen's kappa	0.886	0.910	0.998	0.999	0.787	0.747	0.999	0.999	0.890	0.8713	0.996	0.995	0.776	0.747	0.9959	0.998	0.881	0.843	0.999	0.993
Volcanoes-b6	AUC	0.917	0.500	0.684	0.890	0.915	0.500	0.672	0.869	0.915	0.500	0.663	0.862	0.914	0.500	0.675	0.847	0.914	0.500	0.674	0.861
	Accuracy	96.56	96.17	94.39	96.44	96.68	96.17	93.88	96.44	96.68	96.17	94.19	96.32	96.52	96.17	94.55	96.21	96.60	96.17	94.23	96.24
	Cohen's kappa	0.294	0.000	0.252	0.350	0.334	0.000	0.222	0.350	0.348	0.000	0.229	0.319	0.252	0.000	0.262	0.355	0.313	0.000	0.227	0.304
Connect-4	AUC	0.857	0.723	0.727	0.934	0.779	0.666	0.621	0.825	0.846	0.846	0.712	0.864	0.768	0.580	0.598	0.837	0.835	0.703	0.626	0.837
	Accuracy	75.86	76.07	70.24	81.92	71.61	73.12	59.69	71.95	75.15	75.12	63.51	73.44	68.70	67.58	59.00	73.38	74.52	74.70	62.18	73.38
	Cohen's kappa	0.445	0.451	0.402	0.583	0.271	0.329	0.201	0.291	0.426	0.427	0.266	0.398	0.168	0.085	0.164	0.327	0.407	0.415	0.223	0.327
Volcanoes-d3	AUC	0.500	0.503	0.585	0.789	0.500	0.503	0.607	0.787	0.500	0.503	0.568	0.773	0.500	0.503	0.611	0.777	0.500	0.503	0.593	0.777
	Accuracy	94.40	94.40	89.75	94.27	94.40	94.40	90.48	94.27	94.40	94.40	90.65	94.09	94.40	94.40	88.50	94.27	94.40	94.40	91.21	94.27
	Cohen's kappa	0.000	0.000	0.114	0.126	0.000	0.000	0.157	0.141	0.000	0.000	0.106	0.115	0.000	0.000	0.119	0.063	0.000	0.000	0.140	0.063
Volcanoes-b4	AUC	0.846	0.500	0.678	0.783	0.840	0.500	0.696	0.809	0.846	0.500	0.659	0.759	0.840	0.500	0.688	0.778	0.849	0.500	0.661	0.793
	Accuracy	96.47	96.19	94.15	96.15	96.35	96.19	93.75	96.27	96.35	96.19	94.27	96.03	96.35	96.19	94.27	96.19	96.35	96.19	93.72	95.95
	Cohen's kappa	0.297	0.000	0.259	0.292	0.279	0.000	0.249	0.354	0.257	0.000	0.227	0.300	0.279	0.000	0.274	0.318	0.257	0.000	0.212	0.285
PhishingWebsites	AUC	0.986	0.935	0.967	0.995	0.986	0.940	0.967	0.996	0.986	0.934	0.954	0.993	0.985	0.937	0.922	0.993	0.985	0.936	0.913	0.990
	Accuracy	93.77	93.70	95.51	97.14	93.66	94.03	95.69	96.92	93.59	93.63	94.06	96.70	93.66	93.59	90.81	95.15	93.37	93.81	90.84	95.98
	Cohen's kappa	0.873	0.872	0.909	0.941	0.871	0.879	0.913	0.937	0.870	0.870	0.879	0.933	0.872	0.870	0.815	0.902	0.865	0.874	0.814	0.918

Fig. 2. Test results

comprehensive and robust view of the data. This tree ensemble technique contributed to reducing bias and improving the generalization of our models.

Collectively, our work highlights the effective application of these classification algorithms, which has enabled significant results and a deeper understanding of the phenomena under study. These contributions not only complement prior findings but also pave the way for future research in this ever-evolving field.

If we look at volcanoes-b4, we see that we have an AUC of 0.846 in the original situation (starting point). With RS 1, we get the same result, which indicates that we maintain the same performance, while with RS 2 the performance is improved and with the remaining method it is not maintained.

Continuing within the same row, we observe that for Random Trees (RT) and Random Forest (RF), the optimal choice is KS 1.

SMO maintains its performance in all methods, so to determine in this situation who has better performance, we will opt for the option that retains fewer instances and has shorter execution time.

6   Conclusions

As a result of this empirical study involving the application of instance selection techniques on datasets of various types and sizes, we have observed that the exponential growth of data generation necessitates the discovery and utilization of techniques that can significantly reduce the volume of data while maintaining the capacity for effective classification when constructing models.

Two distinct instance selection techniques were employed in this study, namely Kennard-Stone, Random Sampling, and Random Sampling, to construct classification models. The aim was to empirically determine the most appropriate classification methods.

The obtained results serve as a foundation for future research, enabling a concentrated exploration of a single technique for data reduction and manipulation. In subsequent research endeavors, the selection of techniques should be deliberate and purposeful.

The researcher who wants to continue contributing to this field of research must have a clear guide and objective, It should be borne in mind that these techniques vary according to the conditions to which they are subjected. For this reason, these techniques were selected in order to demonstrate and provide new research in fields that are not so similar to each other. The techniques were therefore selected in order to demonstrate and provide new research in fields not so similar and with different uses not previously applied. By conducting an empirical study allows future researchers to initiate and focus their research on instance reduction techniques. reduction techniques in an easier way and with different fields used to widen the spectrum of use and making it possible to continue to spectrum of use and further expand the line of research.

References

1. Cover, T., Hart, P.: Nearest neighbor pattern classification. IEEE Trans. Inf. Theory **13**(1), 21–27 (1967)
2. Etikan, I., Bala, K.: Sampling and sampling methods. Biometrics Biostatistics Int. J. **5**(6), 00149 (2017)
3. García, S., Luengo, J., Herrera, F.: Data Preprocessing in Data Mining, vol. 72. Springer, Cham (2015)
4. García, S., Luengo, J., Herrera, F.: Tutorial on practical tips of the most influential data preprocessing algorithms in data mining. Knowl.-Based Syst. **98**, 1–29 (2016)
5. Garcia, S., Luengo, J., Sáez, A., Lopez, V., Herrera, F.: A survey of discretization techniques: taxonomy and empirical analysis in supervised learning. IEEE Trans. Knowl. Data Eng. **25**(4), 734–750 (2012)
6. Kohavi, R., et al.: A study of cross-validation and bootstrap for accuracy estimation and model selection. In: Ijcai, vol. 14, pp. 1137–1145. Montreal, Canada (1995)
7. Leyva, E., González, A., Pérez, R.: Three new instance selection methods based on local sets: A comparative study with several approaches from a bi-objective perspective. Pattern Recogn. **48**(4), 1523–1537 (2015)
8. Li, T., Fong, S., Wu, Y., Tallón-Ballesteros, A.J.: Kennard-stone balance algorithm for time-series big data stream mining. In: 2020 International Conference on Data Mining Workshops (ICDMW), pp. 851–858. IEEE (2020)
9. Li, Y., Li, T., Liu, H.: Recent advances in feature selection and its applications. Knowl. Inf. Syst. **53**(3), 551–577 (2017). https://doi.org/10.1007/s10115-017-1059-8
10. Nanni, L., Lumini, A.: Prototype reduction techniques: a comparison among different approaches. Expert Syst. Appl. **38**(9), 11820–11828 (2011)
11. Rendon, E., Alejo, R., Castorena, C., Isidro-Ortega, F.J., Granda-Gutierrez, E.E.: Data sampling methods to deal with the big data multi-class imbalance problem. Appl. Sci. **10**(4), 1276 (2020)
12. Schaffer, C.: A conservation law for generalization performance. In: Machine Learning Proceedings 1994, pp. 259–265. Elsevier (1994)
13. Triguero, I., Sáez, J.A., Luengo, J., García, S., Herrera, F.: On the characterization of noise filters for self-training semi-supervised in nearest neighbor classification. Neurocomputing **132**, 30–41 (2014)
14. Yıldırım, A.A., Özdoğan, C., Watson, D.: Parallel data reduction techniques for big datasets. In: Big Data: Concepts, Methodologies, Tools, and Applications, pp. 734–756. IGI Global (2016)
15. Zhang, S., Zhang, C., Yang, Q.: Data preparation for data mining. Appl. Artif. Intell. **17**(5–6), 375–381 (2003)

Author Index

P. Quaresma et al. (Eds.): IDEAL 2023, LNCS 14404, pp. 547–549, 2023.
https://doi.org/10.1007/978-3-031-48232-8

Printed in the United States
by Baker & Taylor Publisher Services